Handbook of Experimental Pharmacology

Volume 153

Springer
Berlin
Heidelberg
New York
Hong Kong
London
Milan
Paris
Tokyo

Stereochemical Aspects of Drug Action and Disposition

Contributors

S.K. Branch, C.M. Brett, P.-A. Carrupt,
C. Dressler, M. Eichelbaum, G. Geisslinger
K.M. Giacomini, D.J.W. Grant, A. Gross,
C.-H. Gu, P.J. Hayball, K. Hostettmann,
A.P. Ijzerman, G. Klebe, C. Leisen,
D. Mauleón, J.M. Mayer, R.J. Ott, M. Reist, S. Rudaz,
R.R. Shah, A. Somogyi, W. Soudijn,
H. Spahn-Langguth, I. Tegeder,
C. Terreaux, B. Testa, C. Valenzuela, N.P.E. Vermeulen,
J.-L. Veuthey, P. Vogel, I. van Wijngaarden,
K. Williams

Editors
Michel Eichelbaum, Bernard Testa and Andrew Somogyi

Springer

Prof. Dr. Michel Eichelbaum
Dr. Margarete-Fischer-Bosch
Institut für Klinische Pharmakologie
Auerbachstr. 112
70376 Stuttgart
Germany
e-mail: michel.eichelbaum@ikp-stuttgart.de

Prof. Dr. Bernard Testa
University of Lausanne
School of Pharmacy
1015 Lausanne
Switzerland
e-mail: Bernard.Testa@ict. unil.ch

Prof. Dr. Andrew Somogyi
University of Adelaide
Dept. of Clinical
and Experimental Pharmacology
5005 Adelaide
Australia
e-mail: andrew. somogyi@adelaide.edu.au

With 96 Figures and 32 Tables

ISBN 3-540-41593-9 Springer-Verlag Berlin Heidelberg New York

Cataloging-in-Publication Data applied for

Bibliographic information published by Die Deutsche Bibliothek Die Deutsche Bibliothek lists this publication in the Deutsche Nationalbibliografie; detailed bibliographic data is available in the Internet at <http://dnb.ddb.de>.

Springer-Verlag Berlin Heidelberg New York
a member of Bertelsmann Springer Science + Business Media GmbH

http://www.springer.de

Typesetting: SNP Best-set Typesetter Ltd., Hong Kong
Cover-Design: design & production GmbH, Heidelberg
Printed on acid-free paper 27/3020/kk 5 4 3 2 1 0

Preface

State-of-the-Art Chirality

Stereochemistry in general and chirality in particular have long been recognized as major structural factors influencing pharmacological activity and pharmacokinetic behavior. For more than a century, relevant information in these fields has been accumulating at an accelerating pace, leading to rationalizations, concepts and theories of increasing breadth and depth.

Frequently, fundamental advances in stereochemical aspects of molecular pharmacology, drug disposition and pharmacochemistry have been translated into corresponding progress in clinical pharmacology and pharmacotherapy. There have been exceptions, however, since some extrapolations from the biochemical and in vitro situations to the in vivo human situation have proven premature. This notion resulted in the now appeased, but far from closed, debate regarding racemic versus enantiopure drugs, which saw some proponents state that "in many cases, only one isomer contributes to the therapeutic action while the other, the 'isomeric ballast', only contributes to the side effects and toxicity" (Ariëns 1986, 1989, 1992). Other authors, in contrast, have cautioned against hasty generalizations and advocated a more pragmatic, case-by-case and evidence-based view (Caldwell 1995; De Camp 1989; Szelenyi et al. 1998; Testa 1991; Testa and Trager 1990; Testa et al. 1993).

A survey conducted a few years ago (Millership and Fitzpatrick 1993) found that the vast majority of natural and semisynthetic drugs are chiral and marketed as a single isomer. In contrast, only half of synthetic drugs are chiral, and among these more than half are marketed as the racemate. To avoid the complications of developing chiral drug candidates, some medicinal chemists give systematic preference to achiral compounds. This attitude is in sharp contrast with a global trend, clearly detectable in the pharmaceutical industry and its suppliers, to face head-on the synthetic, analytical, pharmacokinetic and pharmacological problems specific to enantiopure drugs, in a sustained effort towards innovation and originality; an approach which is built on current knowledge, but clearly requires a greater understanding of the complex stereochemical interplay between chemical/drug and body macromolecule that impacts at the pharmaco-dynamic, -toxicologic and -kinetic level (Stinson 1995, 1998, 1999, 2000).

The results of such an approach have been inconsistent. For example, there are numerous instances of recognized therapeutic advantages to single-isomeric drugs (HANDLEY 1999), not to mention commercial advantages when a patent granted to an enantiopure drug can replace an expiring patent to the racemate (AGRANAT and CANER 1999; STINSON 1997). These recognized or expected benefits have led to a number of "chiral switches", whereby chiral drugs marketed as racemates are redeveloped and marketed as single enantiomers (i.e., as the eutomer). Successful examples include (*R*)-albuterol, (*S*)-omeprazole and (*S*)-ketamine.

In contrast, the pharmaceutical industry is confronted with some chiral switches that have ended in failure and market recall. (*R*)-Fluoxetine and (*S*)-fenfluramine are respective examples of a pre- and post-marketing failure caused by unwanted and obviously unexpected side effects. These failures are all the more frustrating since the fundamental mechanisms underpinning such side effects remain poorly understood. In these cases, the pharmacodynamic and/or pharmacokinetic interactions between distomer and eutomer in the racemate may have prevented a side effect caused by the eutomer administered alone; however, the mechanism(s) remain to be established.

Clearly, the above paradigm could lead the community of drug researchers to the conclusion that the problem of racemic-versus-enantiopure drugs knows only exceptions and no rule. In a reaction against such a pessimistic and even defeatist attitude, we have accepted Springer's invitation to edit a monograph on the stereochemical aspects of drug action and disposition.

To map both our vast knowledge and deep ignorance on the subject, we have been fortunate to receive the enthusiastic support of a number of colleagues whose combined expertise covers this multidisciplinary field. Quite naturally, the book spans the subject from the molecular to the clinical level. The first section on chemical aspects contains chapters on chemical synthesis, analysis, natural products, chiral stability (racemization) and physical properties. The second section is on experimental pharmacology, with chapters on drug-receptor interactions, chiral recognition, ion channels and molecular toxicology. The third section focuses on drug disposition, with chapters on absorption, distribution, protein binding, metabolism and elimination. The final section is dedicated to regulatory and clinical aspects.

With such a broad yet structured monograph, we hope to convince our readers in the pharmaceutical and biotechnology industries and academia that there are reasons for both optimism and caution in investigating stereoisomeric drugs. When used as probes or as medicines, stereoisomeric drugs offer invaluable pharmacological insights or innovative therapeutic strategies. But because the compared pharmacological properties of racemates and enantiomers remain incompletely understood, unexpected setbacks may occur which only systematic and dedicated research will render less probable.

The book also conveys another, more implicit message, that stereochemistry is an essential dimension in pharmacology and should be understood as such by all drug researchers whatever their background. All too often indeed, investigations are still being conducted and published that ignore the contribution of stereochemistry to data interpretation and extrapolation of results to therapeutic implications (Somogyi 1998). Ariëns's warning (1984) remains as valid now as it was almost two decades ago.

By summarizing in a structured manner our current state of knowledge and ignorance on stereochemical aspects of drug action and disposition, this book aims to guide and inspire drug researchers as they enter the twenty-first century.

Michel Eichelbaum, Bernard Testa and Andrew Somogyi

References

Agranat I, Caner H (1999) Intellectual property and chirality of drugs. Drug Disc Today 4:313–321

Ariëns EJ (1984) Stereochemistry, a basis for sophisticated nonsense in pharmacokinetics and clinical pharmacology. Eur J Clin Pharmacol 26:663–668

Arieë4ns EJ (1986) Stereochemistry: a source of problems in medicinal chemistry. Med Res Rev 6:451–466

Ariëns EJ (1989) Stereoselectivity in the action of drugs. Pharmacol Toxicol 64:319–320

Ariëns EJ (1992) Racemic therapeutics: a source of problems to chemists and physicians. Anal Proceed 29:232–234

Caldwell J (1995) "Chiral pharmacology" and the regulation of new drugs. Chem Ind 176–179

De Camp WH (1989) The FDA perspective on the development of stereoisomers. Chirality 1:2–6

Handley DA (1999) The therapeutic advantages achieved through single-isomer drugs. Pharm News 6:11–15

Millership JS, Fitzpatrick A (1993) Commonly used chiral drugs: a survey. Chirality 5:573–576

Somogyi A (1998) Stereoselectivity in drug disposition. Naunyn-Schmiederberg's Arch Pharmacol 358[Suppl 2]:R760

Stinson SC (1995) Chiral drugs. Chem Eng News 73:44–74

Stinson SC (1997) FDA may confer new status on enantiomers. Chem Eng News 75:28–29

Stinson SC (1998) Counting on chiral drugs. Chem Eng News 76:83–104

Stinson SC (1999) Chiral drug interactions. Chem Eng News 77:101–120

Stinson SC (2000) Chiral drugs. Chem Eng News 78:55–78

Szelenyi I, Geisslinger G, Polymeropoulos E, Paul W, Herbst M, Brune K (1998) The real Gordian knot: racemic mixtures versus pure enantiomers. Drug News Persp 11:139–160

Testa B (1991) Editorial – Stereophilia. Drug Metab Rev 24:1–3

Testa B, Trager WF (1990) Racemates versus enantiomers in drug development: dogmatism or pragmatism? Chirality 2:129–133

Testa B, Carrupt PA, Christiansen LH, Christoffersen P, Reist M (1993) Chirality in drug research: stereomania, stereophobia, or stereophilia? In: Claassen V (ed) Trends in receptor research. Elsevier, Amsterdam, pp 1–8

List of Contributors

BRANCH, S.K., Medicines Control Agency, Market Towers, 1 Nine Elms Lane, Vauxhall, London, SW8 5NQ, UK

BRETT, C.M., Department of Anesthesia, 521 Parnassus Auenue Room C-450, Box-0648 San Francisco California 94143-0648, USA
e-mail: brettc@anesthesia.ucsf.edu

CARRUPT, P.-A., Institut de Chimie Thérapeutique, BEP, Université de Lausanne, 1015 Lausanne-Dorigny, Switzerland

DRESSLER, C., Martin-Luther-Universität Halle-Wittenberg, Department of Pharmaceutical Chemistry, Wolfgang-Langenbeck-Str. 4, 06120 Halle Saale, Germany

EICHELBAUM, M., Dr. Margarete-Fischer-Bosch-Institut für Klinische Pharmakologie, Auerbachstrasse 112, 70376 Stuttgart, Germany
e-mail: michel.eichelbaum@ikp-stuttgart.de

GEISSLINGER, G., Fachbereich Humanmedizin, Zentrum der Pharmakologie, University of Frankfurt, 60596 Frankfurt am Main, Germany
e-mail: geisslinger@em.uni-frankfurt.de

GIACOMINI, K.M., Department of Biopharmaceutical Sciences, UCSF Box 0446, University of California, San Francisco, CA 94143-0446, USA
e-mail: kmg@itsa.ucsf.edu

GRANT, D.J.W., Department of Pharmaceutics, College of Pharmacy, University of Minnesota, Weaver-Densford Hall, 308 Harvard Street SE, Minneapolis, MN 55455-0343, USA
e-mail: grant001@maroon.tc.umn.edu

GROSS, A., Clinical Pharmacology and Discovery Medicine, GSK R&D, James Lance GlaxoSmithkline Medicines Research Unit, Prince of Wales Hospital Randwick, NSW, 2031, Australia
e-mail: asg88523@gsk.com

Gu, C.-H., 1 Squibb Drive, P.O. 191, Bristol-Myers Squibb, New Brunswick, NJ 08903, USA
e-mail: chonghui.gu@bms.com

Hayball P.J., School of Pharmaceutical, Molecular & Biomedical Sciences, City East Campus, Reid Building Frome Road, Adelaide 5000, South Australia
e-mail: peter.hayball@unisa.edu.au

Hostettmann, K., Institute of Pharmacognosy and Phytochemistry, BEP, University of Lausanne, 1015 Lausanne, Switzerland
e-mail: kurt.hostettmann@ipp.unil.ch

Ijzerman, A.P., Leiden/Amsterdam Center for Drug Research, Division of Medicinal Chemistry, Leiden University, PO Box 9502, 2300 RA Leiden, The Netherlands
e-mail: ijzerman@lacdr.leidenuniv.nl

Klebe, G., Philipps-Universität Marburg, Institut für Pharmazeutische Chemie, Marbacher Weg 6, 35032 Marburg, Germany
e-mail: klebe@mailer.uni-marburg.de

Leisen, C., Martin-Luther-Universität Halle-Wittenberg, Department of Pharmaceutical Chemistry, Wolfgang-Langenbeck-Str. 4, 06120 Halle/Saale, Germany

D. Mauleón, D., Development Team Consulting S.L., via Augusta 59, of 302, 08006 Barcelona, Spain

Mayer, J.M., Institut de Chimie Thérapeutique, BEP, Université de Lausanne, 1015 Lausanne-Dorigny, Switzerland

Ott, R.J., GlaxoSmithkline Research and Development, Five Moore Drive, PO Box 13398 Research Triangle Park, NC 27709-3398, USA
e-mail: rjo20389@gsk.com

Reist, M., Institut de Chimie Thérapeutique, BEP, Université de Lausanne, 1015 Lausanne-Dorigny, Switzerland

Rudaz, S., Laboratory of Pharmaceutical Analytical Chemistry, University of Geneva, 20 Bd d'Yvoy, 1211 Geneva 4, Switzerland

Shah, R.R., Medicines Control Agency, Market Towers, 1 Nine Elms Lane, Vauxhall, London, SW8 5NQ, UK
e-mail: clin.safety@lineone.net

Somogyi, A., Department of Clinical and Experimental Pharmacology, University of Adelaide, Adelaide 5005, Australia
e-mail: andrew.somogyi@adelaide.edu.au

Soudijn, W., Leiden/Amsterdam Center for Drug Research, Division of Medicinal Chemistry, Leiden University, PO Box 9502, NL-2300 RA Leiden, The Netherlands

Spahn-Langguth, H., Martin-Luther-University Halle-Wittenberg, Department of Pharmaceutical Chemistry, Wolfgang-Langenbeck-Strasse 4, 06120 Halle/Saale, Germany
e-mail: spahn-langguth@pharmazie.uni-halle.de

Tegeder, I., Universitätsklinikum Frankfurt Haus 75/4. OG Theodor-Stern-Kai 7 60590 Frankfurt/Main, Germany
e-mail: tegeder@em.uni-frankfurt.de

Terreaux, C., Institute of Pharmacognosy and Phytochemistry, BEP, University of Lausanne, 1015 Lausanne, Switzerland

Testa, B., Institut de Chimie Thérapeutique, BEP, Université de Lausanne, 1015 Lausanne-Dorigny, Switzerland
e-mail: bernard.testa@ict.unil.ch

Valenzuela, C., Institute of Pharmacology and Toxicology, CSIC/UCM, School of Medicine, Universidad Complutense, 28040 Madrid, Spain
e-mail: carmenva@eucmax.sim.ucm.es

Vermeulen, N.P.E., Division of Molecular Toxicology, Department of Pharmacochemistry, De Boelelaan 1083, 1081 HV Amsterdam, The Netherlands
e-mail: vermeule@chem.vu.nl

Veuthey, J.-L., Laboratory of Pharmaceutical Analytical Chemistry, University of Geneva, 20 Bd d'Yvoy, 1211 Geneva 4, Switzerland
e-mail: jean-luc.veuthey@pharm.unige.ch

Vogel, P., Section de Chimie de L'Université de Lausanne, BCH, 1015 Lausanne-Dorigny, Switzerland
e-mail: pierre.vogel@epfl.ch

van Wijngaarden, I., Leiden/Amsterdam, Center for Drug Research, Division of Medicinal Chemistry, Leiden University, PO Box 9502, NL-2300 RA Leiden, The Netherlands

Williams, K., Department of Clinical Pharmacology and Toxicology, St. Vincent's Hospital, University of New South Wales, Darlinghurst 2010, Australia
e-mail: ken.williams@unsw.edu.au

Contents

Section I: Chemical Aspects

CHAPTER 9

Stereoselective Drug-Channel Interactions

CHAPTER 10

Stereoselective Bioactivation and Bioinactivation – Toxicological Aspects

CHAPTER 13

Stereoselective Drug Metabolism and Drug Interactions

Section I
Chemical Aspects

CHAPTER 1

Recent Developments in Asymmetric Organic Synthesis: Principles and Examples

P. VOGEL

A. Introduction

Chirality is a fundamental symmetry property in three and other dimensions. A molecule is said to be chiral if it cannot be superimposed upon its mirror image. Putting on one's shoes or shaking hands confronts us with chirality. Although there is no obvious relationship between macroscopic chirality and chirality at the molecular level, it is accepted that homochirality (i.e., molecules with the same chirality; e.g., L-α-amino acids, D-glucose, D-arabinose) is one of the most fundamental aspects of life on Earth. Parity violation discovered in the weak nuclear force (the fourth type of fundamental force, next to gravity, electromagnetism and the strong nuclear force) led to the experimental observation than the β-particles emitted from radioactive nuclei have an intrinsic asymmetry: left-handed (L)-electrons are preferentially formed relative to right-handed (R)-electrons. The major consequence of this finding is that chirality exists at the level of elemental particles, making the two enantiomers of a chiral molecule not to have exactly the same energy (ULBRICHT 1981; MASON 1989; FERINGA and VAN DELDEN 1999).

I. Definitions

Many compounds may be obtained in two different forms in which the molecular structures are constitutionally identical but differ in the three-dimensional arrangement of atoms. For instance L-alanine and D-alanine are called *enantiomers* and are said to be *enantiomeric* with each other. They are related as mirror images. L-alanine ((S)-alanine: according to the *Cahn-Ingold-Prelog* (CIP) rule (CAHN et al. 1966; PRELOG and HELMCHEN 1982)) is a proteinogenic α-amino acid; D-alanine ((R)-alanine) is not. It occurs in some bacterial cell walls (ULBRICHT 1981) and in some peptide antibiotics (Table 1).

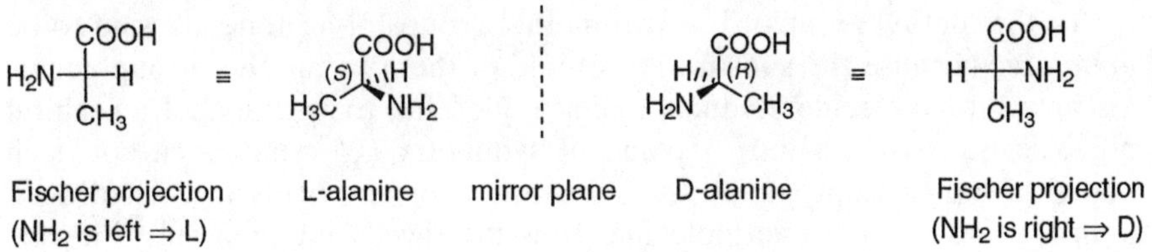

Table 1. Examples of organic stereoisomerisms and definitions

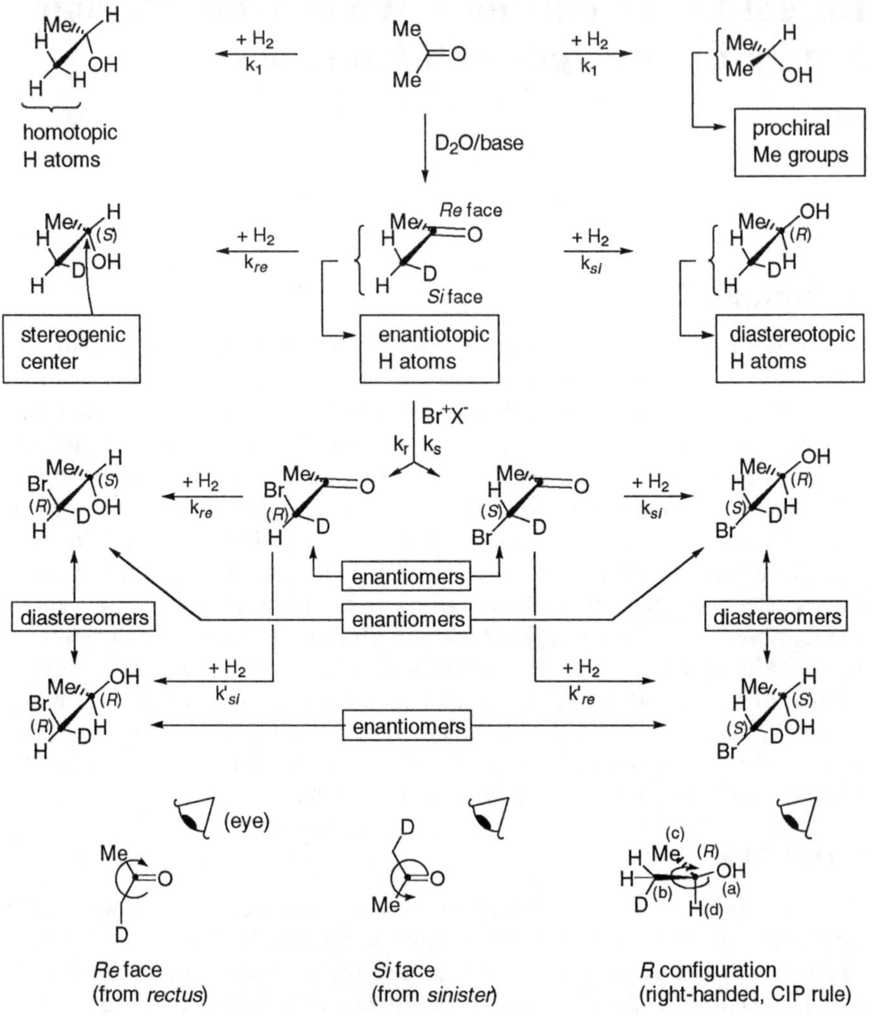

Acetone (CH_3COCH_3) is an *achiral* molecule (C_{2V} symmetry: possesses one C_2 axis of symmetry contained in a mirror plane of symmetry, free rotation of the methyl groups). The two methyl groups of acetone are said to be *homotopic* because the substitution of one of them by another group generates an achiral molecule. Reduction of acetone leads to isopropanol, an achiral molecule possessing a mirror plane of symmetry (C_s symmetry). For each methyl group of isopropanol, the substitution of any of its three hydrogen atoms generates the same molecule. Thus the three hydrogen atoms in one methyl group are *homotopic*. The two methyl groups of isopropanol are said to be *enantiotopic*: when substituting one of them by another group (different from H in this case) two enantiomeric molecules can be obtained. The two methyl groups of isopropanol are said to be *prochiral*, which is not the case in acetone. Monodeuteration of acetone generates CH_2DCOCH_3, a molecule with C_s symmetry in which the three hydrogen atoms of the CH_3 group are homotopic and the two hydrogen atoms of the CH_2D group are enantiotopic. Reduction of CH_2DCOCH_3 generates two enantiomeric alcohols, i.e., (2*S*)-1-deuteropropan-2-ol and (1*R*)-1-deuteropropan-2-ol. The C-2 carbon atom of these alcohols are said to be *stereogenic centers* (the exchange of two ligands attached to these centers gives another stereomer). The two faces of the carbonyl moiety of acetone are not prochiral. The two faces of the carbonyl moiety of 1-deuteroacetone are *prochiral* or *enantiotopic*. The *Re* face corresponds to that for which an observer sees a right-handed sequence for the oxygen → CH_2D → CH_3 moieties. The *Si* face corresponds to that for which an observer sees a left-handed sequence for the oxygen → CH_2D → CH_3 moieties (CIP rule). The monobromation of the CDH_2 group of 1-deuteroacetone generates two enantiomeric compounds, i.e: (1*R*)-1-bromo-1-deuteroacetone and (1*S*)-1-bromo-1-deuteroacetone. Reduction of the carbonyl group of (1*R*)-1-bromo-1-deuteroacetone gives the *diastereomeric* (1*R*,2*R*)- and (1*R*,2*S*)-1-bromo-1-deuteropropan-2-ol. Similarly, the reduction of (1*S*)-1-bromo-1-deuteroacetone generates the *diastereomeric* (1*S*,2*R*)- and (1*S*,2*S*)-1-bromo-1-deuteropropan-2-ol. (1*R*,2*R*)- and (1*S*,2*S*)-1-bromo-1-deuteropropan-2-ol are enantiomers. Similarly, (1*R*,2*S*)- and (1*S*,2*R*)-1-bromo-1-deuteropropan-2-ol are also enantiomers. The *relative configuration* of the (1*R*,2*R*)- and (1*S*,2*S*)-enantiomers is denoted by *l* (like), that of the (1*R*,2*S*) and (1*S*,2*R*) enantiomers by *u* (unlike) (Seebach and Prelog 1982).

The rate constant ratio k_r/k_s for the monobromation of CDH_2COCH_3 defines the *kinetic enantioselectivity* or *selectivity factors* of these reactions. For irreversible reactions following rate laws with the same order for all reactants, it is given by the *enantiomeric ratio* of products: s = k_r/k_s = [(1*R*)-1-bromo-1-deuteroacetone]/[(1*S*)-1-bromo-1-deuteroacetone]. The kinetic enantioselectivity of the reduction of CDH_2COCH_3 giving (2*R*)- and (2*S*)-1-deuteropropan-2-ol is defined by the rate constant ratio k_{re}/k_{si}. For irreversible reactions following rate laws with the same order for all reactants, k_{re}/k_{si} = [(2*R*)-$CH_2DCH(OH)CH_3$]/[(2*S*)-$CH_2DCH(OH)CH_3$]. One defines the enantiomeric excess for these reactions as:

$$\mathrm{ee} = \frac{[\text{enantiomer } (R)] - [\text{enantiomer } (S)]}{[\text{enantiomer } (R)] + [\text{enantiomer } (S)]}$$

The *kinetic diastereoselectivity* of the reduction of (1*S*)-1-bromo-1-deuteroacetone giving (1*S*,2*R*)- and (1*S*,2*S*)-1-bromo-1-deuteroacetone is given by the rate constant ratio k'_{si}/k'_{re}. For irreversible reactions following rate laws with the same order for all reactants k'_{si}/k'_{re} = [(1*S*,2*R*)-1-bromo-1-deuteropropan-2-ol]/[(1*S*,2*S*)-1-bromo-1-deuteropropan-2-ol].

Using enantiomerically enriched chiral reagents, catalysts or medium, the kinetic enantioselectivity can be different from 1/1, i.e., giving products with enantiomeric excess ee >0 for irreversible reactions. Kinetic or thermodynamic diastereoselectivity $\neq 1:1$ can be observed with achiral reagents. The stereogenic center of the starting material controls the diastereoselection of the reaction generating a second, or more stereogenic centers in the products.

II. The Need for Asymmetric Synthesis

Most natural products are chiral. The building-blocks (amino acids, carbohydrates, etc.) making the biological macromolecules of living systems are chiral. Biopolymers are themselves chiral. As demonstrated throughout this book, there is an absolute need for efficient methods generating enantiomerically pure drugs or prodrugs. The majority of natural compounds are formed by stereospecific or stereoselective biological processes (e.g., α-amino acids, carbohydrates, terpenes, alkaloids, steroids, etc.). The preparation of optically active compounds is one of the oldest problems of organic chemistry. New journals dedicated to the topic of chirality or "chiroscience" including *Tetrahedron: Asymmetry*, *Enantiomer*, *Chirality* and *Molecular Asymmetry* complement other journals that are themselves replete with publications on chirality and asymmetric syntheses (STINSON 1995; 1999; RAHMAN and SHAH 1993; AGER and EAST 1996; GAWLEY and AUBÉ 1996). Three main strategies are available:

1. The resolution of racemates to enantiomers either through physical, chemical or biochemical methods
2. The use of a enantiomerically pure starting material obtained from a natural source or by asymmetric synthesis (chiral pool)
3. The conversion of a prochiral precursor into a chiral product applying asymmetric synthesis (chemical or biochemical methods)

With its high yielded (87%) conversion of D-mannose into pure D-glycero-D-talo-heptonic acid EMIL FISCHER realized the first diastereoselective synthesis (1889–1894). The method involves diastereoselective additions of HCN to the carbaldehyde moiety of the hexose, followed by hydrolysis of the intermediate cyanohydrine (JACQUES 1995). In 1904 MARCKWALD claimed to have realized the first asymmetric synthesis by carrying out the decarboxylation of ethylmethylmalonic acid at 170°C in the presence of brucin, the source of enantioselectivity. MARCKWALD defines asymmetric syntheses as those reactions that produce optically active substances from symmetrically constituted compounds, with intermediate use of optically active materials (JACQUES 1995).

B. Resolution of Racemates to Enantiomers

Three successful methods of racemate resolution (spontaneous and initiated resolution, resolution via diastereomers, kinetic resolution) elaborated by LOUIS PASTEUR between 1848 and 1858 have been until recently the only methods of preparation of enantiomerically pure compounds. They are applied on large-scale production of industrial products (CROSBY 1991; SHELDON 1993). The first method, *spontaneous resolution of racemates*, allowed PASTEUR (PASTEUR 1848, 1850) to separate the sodium ammonium salt of tartaric acid as the hemihedric crystals of the enantiomers can be distinguished visually (JACQUES et al. 1994; ELIEL and WILEN 1994; HAGER et al. 1999). The method is not applicable for large-scale production, but may serve to obtain inoculate of enantiomers that can be used in the *entrainment method* which consists of the inoculation of a saturated racemate solution by a crystal of one of the enantiomers. The method, *initiated resolution*), is applied industrially for the resolution of glutamic acid and threonine (PROFIR and MATSUOKA 2000). In the case of other amino acids the resolution is successful after their conversion to salts or other derivatives (IZUMI et al. 1978). If a crystal of pure enantiomer is not available, a crystal of some other enantiomerically pure compound may be used in some cases (e.g., resolution of (±)-chloramphenicol). Scanning tunneling microscopy has been used recently to separate a racemic mixture of two-dimensional molecular clusters (BOHRINGER et al. 1999).

I. Resolution via Diastereomers

This method consists of introducing a new chirality center into the racemate by combining it with an auxiliary optically-active compound. In this way two diastereoisomeric combinations (see e.g., the fractional crystallization of salts obtained by combining cinchonidin with (±)-tartaric acid) are formed from the racemate (e.g., salts, complexes, covalent compounds, etc.) which can be separated by fractional crystallization, distillation, or by chromatographic methods (KAGAN 1999) (Scheme 1).

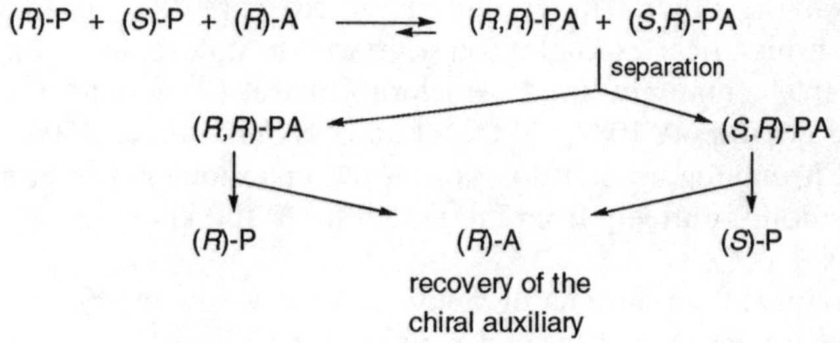

Scheme 1. Resolution via separation of diastereomers

A recent example of this method is the preparation of both "naked sugars of the first generation" (VOGEL et al. 1990; 1999; 2000), (+)-1 and (−)-1, valuable chirons (= enantiomerically pure synthetic intermediates) in the asymmetric synthesis of a great variety of natural products, rare carbohydrates and analogues. Reaction of (*R*,*R*)-1,2-diphenylethylene diamine (*R*,*R*-2) with racemic (±)-1 gave two diastereomeric aminals 3 and 4 that were separated by crystallization and flash chromatography in 46% and 43% yield, respectively. The pure aminals were hydrolyzed by treatment with 0.1 M phosphoric acid and the organic phase was extracted with ether, providing enantiomerically pure (1*S*,4*S*)-(−)-1 (33%) and (1*R*,4*R*)-(+)-1 (30% yield based on (±)-1), respectively. The chiral diamine (*R*,*R*)-2 was recovered in 92% yield from the aqueous phase (FORSTER et al. 1999).

The resolution of racemates can also be carried out with less than stoichiometric amounts of the chiral auxiliary (WILEN 1971; CERVINKA 1995).

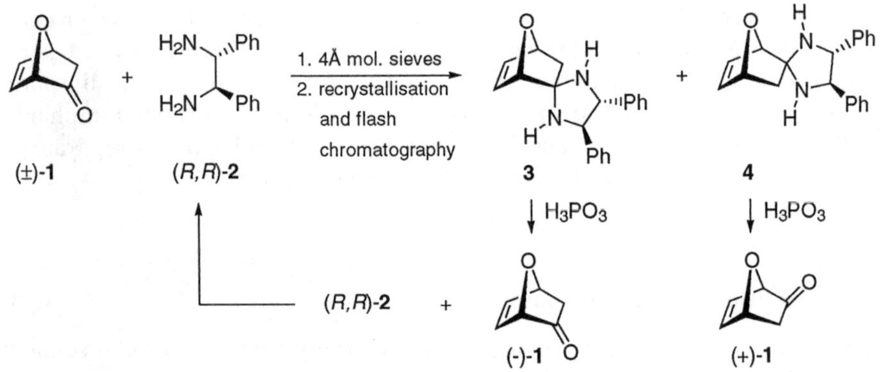

II. Resolution by Means of Chromatography on a Chiral Phase

The chromatographic resolution of a racemate on an optically active phase the resolution of racemates via diastereoisomeric salts or complexes. The complexes formed by adsorption are not equally stable and thus the enantiomer which is absorbed more weakly passes through the column more rapidly then the other. The resolution of Troeger's base on a column of lactose is an historical example. Today, triacetylcellulose and other modified cellulose are commonly used as chiral phase (ZIEF and CRANE 1988; PIRKLE and POCHAPSKY 1989; MILLER et al. 1999; KRIEG et al. 2000; MAGORA et al. 2000). Chromatographic resolution of the enantiomers of pharmaceutical products is done routinely from the milligram to the kilogram scale (MILLER et al. 1999).

Supercritical fluid chromatography (SFC) is a new method for the resolution of racemates (WILLIAMS and SANDER 1997; MEDVEDOVICI et al. 1997). In this method a chiral complexing agent is bonded to a stationary phase and the mobile phase in a supercritical fluid such as sc-CO_2. SFC has several advantages over liquid chromatography, including, highly selective enantiomeric separation, rapid mass transfer, and high-speed separation with high resolution. For example, the racemic levamisole (±)-5 has been separated into (*R*)-(+)-5 and (*S*)-(−)-5 (Keszei et al. 1999) using a chiral solid phase made of (2*R*,3*R*)-O,O′-dibenzoyltartaric acid. Compound (*S*)-(−)-5, called levamisole, is an effective anthelmintic drug with an immunomodulatory activity.

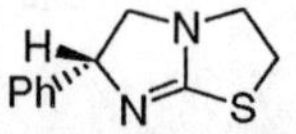

(-)-**5** (levamisole)

Centrifugal partition chromatography using chiral selectors has been used to separate enantiomers (Duret et al. 2000).

III. Simple Kinetic Resolutions

Kinetic resolution of a 1:1 mixture of two enantiomeric substrates (*R*)-P and (*S*)-P with a single enantiomerically pure reagent, or an enantiomerically pure catalyst, or in a chiral medium giving products Q and Q′ is well documented (Brown 1987; Kagan and Fiaud 1988). It can lead to a maximum yield of less than 50% of the slow reacting enantiomer with high enantiomeric excess if the rate constant ratio k_r/k_s = s (selectivity factor) is larger than 200 (Scheme 2).

Products Q and Q′ can be diastereomers, enantiomers or identical depending on the type of substrates P and reactions. For instance, the esterification of a racemic alcohol with a homochiral acylating agent gives two diastereomeric esters; the esterification of a racemic secondary alcohol in which the stereogenic center is the alcoholic center generates two enantiomeric esters whereas the oxidation (or dehydrogenation) of this racemic alcohol furnishes the same achiral ketone from both enantiomeric substrates.

For reactions following first order or pseudo-first order rate laws, the difference is:

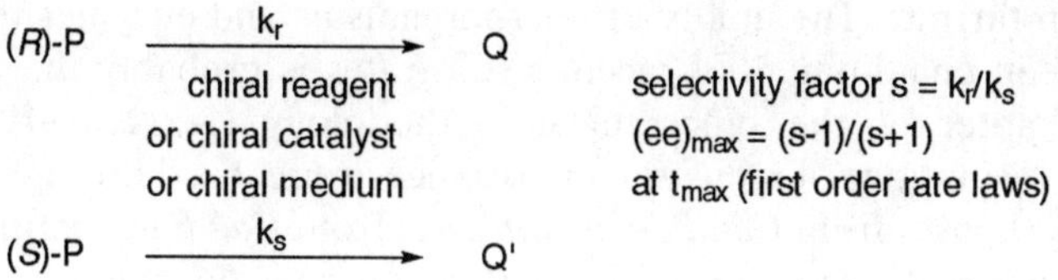

Scheme 2. Kinetic resolution

$$[(S)-\mathrm{P}]-[(R)-\mathrm{P}]=\frac{1}{2}(e^{-k_s t}-e^{k_r t}) \text{ for } k_r/k_s>1$$

At time t = 0 and t = ∞, $[(S)\text{-P}]_{tmax} - [(R)\text{-P}]_{tmax} = 0$ (no enantiomeric excess for both enantiomeric substrates, and for products Q and Q′ if they are enantiomers). There is a value of time t_{max} for which [(*S*)-P] – [(*R*)-P] reaches a maximum. If the reaction is stopped after time t_{max} one obtains [(*S*)-P] – [(*R*)-P] = s and ee_{max} = (s – 1)/(s + 1). For longer reaction time the yield in (*S*)-P diminishes. If products Q and Q′ are enantiomers, their enantiomeric excess ee′ is the highest at the beginning of the reaction: ee'_o = [Q] – [Q′]/[Q] + [Q′] = (s – 1)/(s + 1).

1. An Example of Chemical Kinetic Resolution

Hydrogenation of cyclopentenone (±)-6 with 0.1 mol % of catalyst (*S*)-7 at 4 atmospheres proceeds with k_s/k_r = 11 : 1 and, at 68% conversion (34% maximal yield) gives slow reacting (*R*)-6 in 98% ee together with (*R*)-enriched hydrogenation product 6′. Subsequent silylation ((*t*-Bu)Me_2SiCl, Et_3N, DMAP = 4-Me_2N-pyridine) and treatment with 1,8-diazabicyclo[5.4.0]undec-7-ene (DBU) converted 6′ into achiral cyclopent-2-enone and, after recrystallization homochiral (*R*)-8 was obtained. Enone (*R*)-6 is an important building block for the three-component coupling prostaglandin synthesis (KITAMURA et al. 1988).

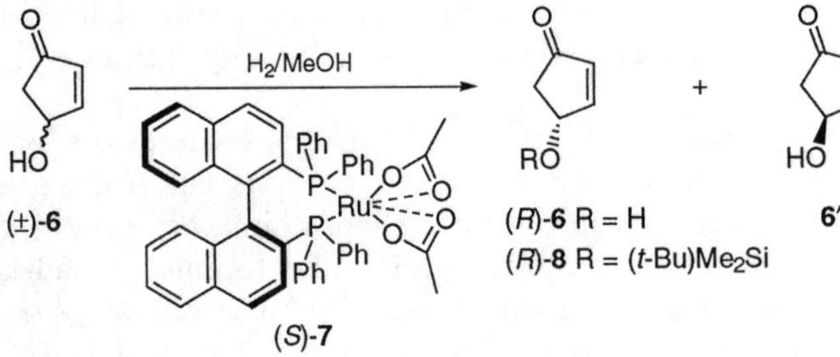

2. Examples of Biochemical Kinetic Resolutions

This method was introduced by PASTEUR in 1858 when he carried out the kinetic resolution of tartaric acid with yeast. The organism *Penicillium glauca* destroyed D-ammonium tartrate more rapidly from a solution of a racemic ammonium tartrate. The ability of microorganisms and enzymes to discriminate between enantiomers of racemic substrates is probably the best documented chapter of the biochemical methodology (AZERAD 1995). Most enzymatic resolutions of industrial importance are realized through the use of hydrolases (lipases from *Candida cylindracea*, from *Candida antartica* B, from *Pseudomonas* sp (= *aeruginosa*), from *Pseudomonas fluorescens* (= *cepacia*), from *Mucor miehei*, from *Humicola lanuginosa*, from *Aspergillus niger*, from

Geotrichum caudidum, from *Rhizopus delemar*, from porcine pancreas; esterases from pig liver, horse liver; acetyl esterase from orange *flavedo*; proteases such as α-chymotrypsin, papain, substilisin A (from *Bacillus licheniformis*), thermolysin (from *B. thermoproteolyticus*) and protease from *Aspergillus oryzae*, amino acylase from porcine kidney, from *Aspergillus sp*; penicillin acylase from *E coli*, D- or L-hydrantoinases (dihydropyrimidinases) from *Agrobacterium radiobacter* or *B. brevis*; nitrilase and nitrile hydratase from *Brevibacterium* or from *Rhodococus* pp). These enzymes are produced in bulk quantities (detergent, food industry) and have the advantage of not requiring expensive coenzymes. They can be active toward a large number of substrates and tolerate organic solvents. A large number of alcohols have been successfully resolved using essentially lipase activity of purified enzymes or whole microbial cells (AZERAD 1995).

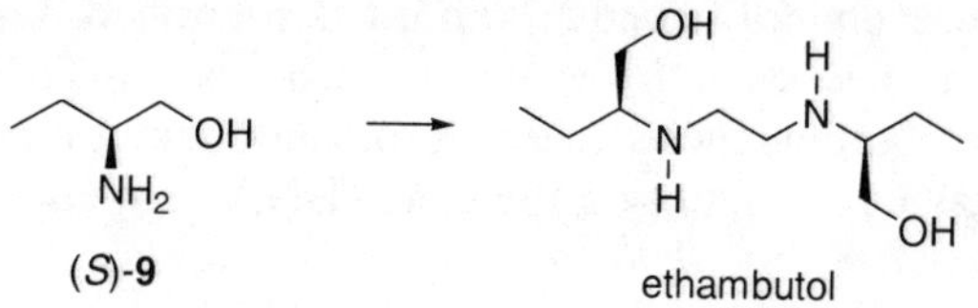

Resolution of racemic 2-aminobutan-1-ol ((±)-9) is an industrially important process since the (*S*)-2-aminobutan-1-ol is used as an intermediate for the production of ethambutol, an antibiotic for the treatment of tuberculosis (WILKINSON et al. 1961). The classical enzymatic resolutions of (±)-9 are based on either an enantioselective hydrolysis of the ester function of an *N,O*-diacetyl derivative with a lipase or porcine pancreatic lipase catalyzed acylation of *N*-acylated derivative (FERNANDEZ et al. 1992) or by hydrolysis of the *N*-benzoyl derivative with a fungus such as *Aspergillus oryzae* where the (*S*) derivative is hydrolyzed due to amino acylase activity (NIKAIDO and KAWADA 1994). A new and simple methodology applies the enantioselective hydrolysis of the racemic *N*-phenylacetyl derivative (±)-10 with penicillin G acylase immobilized on Eupergit C. Penicillin G acylase is used mainly for the production of 6-aminopenicillanic acid; it is commercially available in large quantities. The product (*S*)-(+)-9 is obtained with >99% ee (up to 40% conversion rate) and the enzyme removed by simple filtration can be recycled several times (FADNAVIS et al. 1999a).

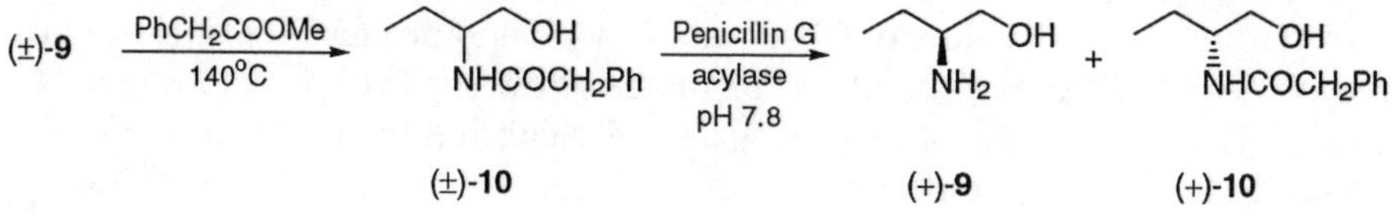

Among the various options in biocatalysis, enzymes are chosen for their high activity, their variety and commercial availability. One of the problems with using enzymes (or microorganisms) is the fact that they are substrate-specific. For a given reaction of interest enantioselectivity may be unacceptably low. In principle, enzymatic activity and enantioselectivity through exchange of a specific amino acid in the enzyme with one of the remaining 19 proteinogenic α-amino acids using site-directed mutagenesis can be envisioned. Alternatively, improved enantioselectivity can be obtained by creating new enzymes by "directed evolution in the test tube" (REETZ and JAEGER 2000). Accordingly, the combination of proper molecular biological methods for random mutagenesis and gene expression coupled with high-through put screening systems for the rapid identification of enantioselective mutant enzymes forms the basis of the approach. One starts with a wild-type enzyme that has an unacceptable selectivity factor $s = k_r/k_s$ (Scheme 2) for a given transformation of interest, $P \rightarrow Q$, to create a library of mutants from which a more enantioselective variant is identified. One repeats the process as often as necessary by using, in each case, an improved mutant for the next round of mutagenesis. In doing so, random mutagenesis is not performed on the enzyme itself, but on the gene (DNA segment) which encodes the protein.

3. Parallel Kinetic Resolutions

For a good yield (maximum 50%) and enantiomeric excess (>99%) in the simple kinetic resolution method the selectivity factor $s = k_r/k_s$ must be larger than 200 (KAGAN and FIAUD 1988). Selectivities such as these are rare for chemical kinetic resolutions, and even above that of some enzymes such as lipase esterases (SIH and WU 1989). Problems do arise depending on this inherent selectivity factor and can have dramatic consequence on the yield and enantiomeric excesses of the remaining reactant P and the product Q (Scheme 2). One way of preventing the concentration effect of the slow reacting enantiomer of the reactant near the end of the resolution is to remove it in a parallel reaction during the course of the resolution. Ideally, its rate of reaction should be similar to that of the other enantiomer. This has led to the strategy termed *parallel kinetic resolution* (EAMES 2000). By far the most elegant application of this strategy has been developed by VEDEJS and CHEN with a chiral DMAP acyl transfer reaction (VEDEJS and CHEN 1997). Activation of the quasienantiomeric pyridines (*R*)-11 and (*S*)-12 with the hindered chloroformate 13 and (+)-fenchyl chloroformate (14) gave the acyl transfer agents 15 and 16, respectively (Scheme 3). The alkyl substitution of these chloroformates (trichlorotertiobutyl and fenchyl) is very important since it is transferred to the resolved alcohol (*S*)-17 and (*R*)-17, and because the substituents are obviously different, the chloroformates allow separation of products 18 and 19. The fact that the fenchyl group in 19 is chiral is irrelevant to the selectivity. Addition of equimolar amounts of 15 and 16 combined with an excess of $MgBr_2$ and Et_3N to (±)-17 gave mixed carbonates 18 and 19. Their separation is made simpler by treatment of the product mixture with zinc in acetic acid which chemoselectively removed the trichlorobutyl protecting group to give (*S*)-17 and 19 that are readily separable. The quasi-enantiomeric chiral DMAP equivalents (*R*)-11 and (*S*)-12 liberated during the acyl transfers are fully recyclable (Scheme 3).

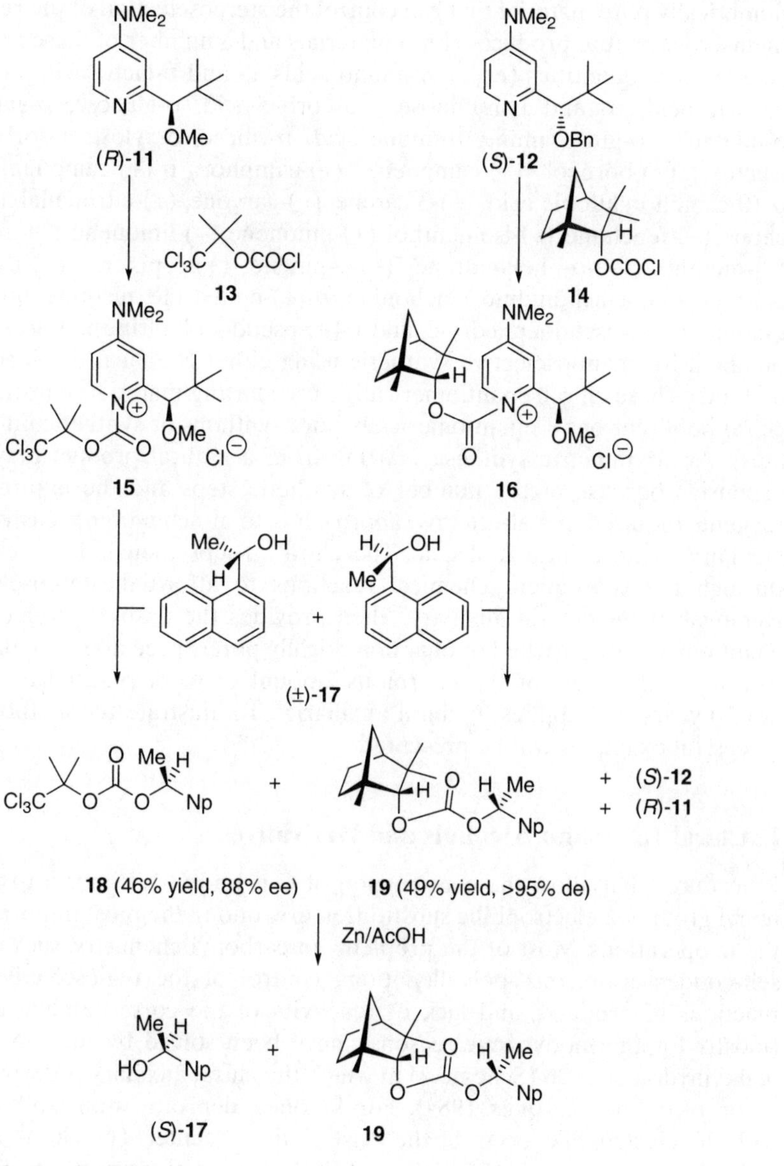

Scheme 3. Parallel kinetic resolution of a secondary alcohol by acyl transfer using quasi-enantiomeric DMAP derivatives

C. The Use of Stoichiometric Chiral Auxiliaries

For reactions that cannot be catalyzed by chemical or biochemical enantiomerically pure catalysts (see Sect. D.), or for racemic products that cannot be resolved readily (Sect. B.), asymmetric synthesis requires that at least one of the starting materials (substrate, reagent, medium) is chiral and enantiomerically enriched. One of the best scenarios is to start with an enantiomerically pure material that can control the stereoselection of the reactions themselves. Nature produces chiral materials and a number of these are available in large quantities (e.g., L-α-amino acids, L- and D-lactic acid, L- and D-tartaric acid, D- and L-arabinose, L-ascorbic acid, D-glucose, D-galactose, D-mannose, D-glucosamine, D-quinic acid, D-ribose, D-xylose, L-sorbose, D-fructose, (–)-borneol, (+)-camphene, (+)-camphor, D-(+)-camphanic acid, D-10-camphorsulfonic acid, (+)-3-carene, (–)-carvone, (+)-citronellal, (+)-fenchone, (–)-fenchone, (+)-isomenthol, (+)-limonene, (–)-limonene, (–)-menthol, (+)-menthol, (–)-α-phellandrene, (–)-α-pinene, (+)-α-pinene, (–)-β-pinene, (+)-pulegone, cinchonidine, cinchonine, D-(+)-ephedrine, nicotine, quinidine, quinine, D-(+)-pseudoephedrine, and L-(–)-pseudoephedrine). Others can be obtained by enantioselective synthesis using either chemical or biochemical methods. These chiral, enantiomerically pure starting materials constitute the chiral pool (chirons = enantiomerically pure synthons or synthetic intermediates). An asymmetric synthesis starting from a natural product may prove expensive because of the number of synthetic steps and the nature of the reagents required. An alternative approach is to attach an inexpensive chiral auxiliary to an achiral molecule. The chiral groups then induce selectivity through the subsequent chemical reactions to afford diastereoselectivity. Removal of the "chiral auxiliary" then provides the desired product either enantiomerically enriched or enantiomerically pure. Space does not allow one to give an overview of the enormous amount of work realized during the last 20 years that applies to chiral auxiliaries. To illustrate the method a few successful examples will be presented.

I. Chiral 1,2-amino Alcohols and Derivatives

Procedures that allow the construction of C-C and C-X bonds a to the carbonyl group via electrophilic substitution are among the most important synthetic operations. Most of the problems in carbonyl chemistry, such as aldol self-condensation, α,α'-polyalkylations, control of the regioselectivity, side reactions of products, and lack of reactivity of the corresponding enolates (mostly for thermodynamic reasons), have been solved by the use of *N,N*-dialkylhydrazones 20 (Scheme 4) in which the chiral auxiliary is derived from L- or D-proline (ENDERS 1984). For ketones, deprotonation, and reaction with the electrophile, occur at the least hindered center. The chiral auxiliary (SAMP if derived from (*S*)-proline, RAMP if derived from (*R*)-proline) can be recovered as shown in Scheme 4 (ENDERS et al. 2000).

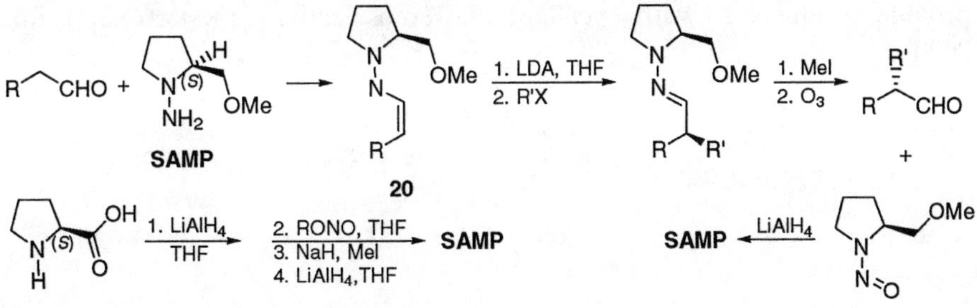

Scheme 4. Asymmetric α-alkylation of aldehydes via homochiral hydrazones

Evans' oxazolidin-2-ones (e.g., 21-*R*) obtained by reaction of enantiomerically pure 1,2-amino alcohols with diethyl carbonate (Evans et al. 1981) have proven particularly effective for controlling variety of reactions of attached acyl fragments, including alkylation, amination, azidation, bromination, hydroxylation, aldol addition, Diels-Alder cycloaddition and conjugate addition (Ager et al. 1996). Davies' 5,5-disubstituted SuperQuat 24 (Davies and Sanganee 1995; Bull et al. 1998) and Gibson's variants 25 (Gibson et al. 1998) appear to be superior to Evans' auxiliaries because of their better recyclability and better crystallinity (important in obtaining diastereomerically pure products).

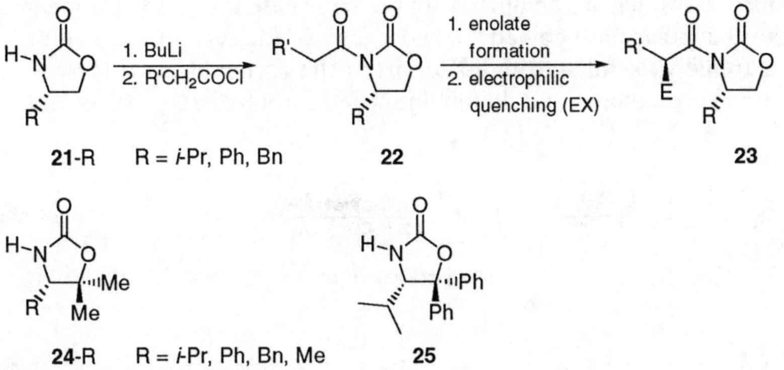

21-R R = *i*-Pr, Ph, Bn **22** **23**

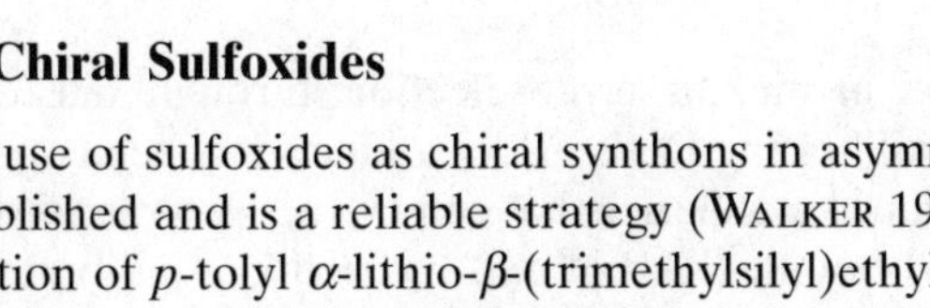

24-R R = *i*-Pr, Ph, Bn, Me **25**

II. Chiral Sulfoxides

The use of sulfoxides as chiral synthons in asymmetric synthesis is now well-established and is a reliable strategy (Walker 1992). A recent example is the reaction of *p*-tolyl α-lithio-β-(trimethylsilyl)ethyl sulfoxide obtained by lithiation of 26, with α,β-unsaturated esters that give the corresponding products of conjugate addition with high diastereoselectivity (>96% de). The intermediate enolates 27 can be trapped with various alkyl halides or aldehydes to provide products 28 with excellent diastereoselectivity (NAKAMURA et al. 2000).

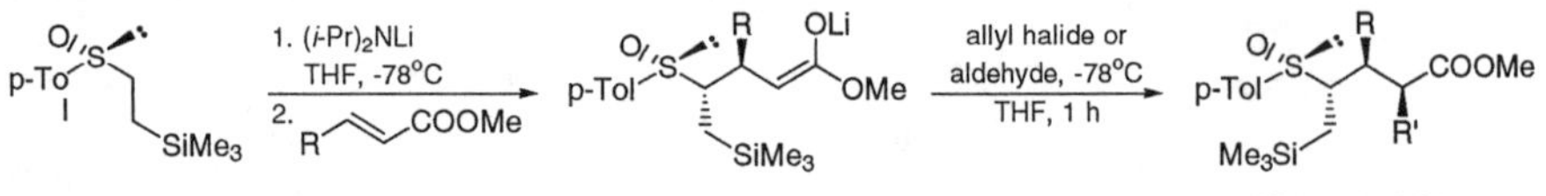

26 **27** **28** (>96% de)

III. Thermodynamic Diastereoselection

The diastereoselectivity of synthetic intermediates bearing a chiral auxiliary is in most cases governed by kinetic factors (e.g., kinetic face selectivity of enolates, enoxysilanes, enoxyboranes, ketones, alkenes, etc.). For reversible reactions, diastereoselectivity could be controlled by the relative stability of the diastereomeric products or by displacing the equilibrium between products of similar stabilities by diastereoselective crystallization. The method has been applied to furan derivatives that undergo reversible Diels-Alder addition (VOGEL et al. 1999). In the presence of ZnI_2 as catalyst, a 1:1 mixture of 1-cyanovinyl (1*R*)-camphanate and 2,4-dimethylfuran (without solvent) is equilibrated with a single adduct (+)-29 that precipitates diastereoselectively (SEVIN and VOGEL 1994). Adduct (+)-29 is converted into enone (+)-30 upon saponification and treatment with formaline. The chiral auxiliary ((1*R*)-camphanic acid) is recovered at this stage. Chirons (+)-29, (+)-30 and their enantiomers which are obtained with the same ease using (1*S*)-camphanic acid as chiral auxiliary, are coined "naked sugars of the second generation." They are extremely useful starting materials in the asymmetric synthesis of long-chain polypropionates and doubly-branched carbohydrates (SEVIN and VOGEL 1994).

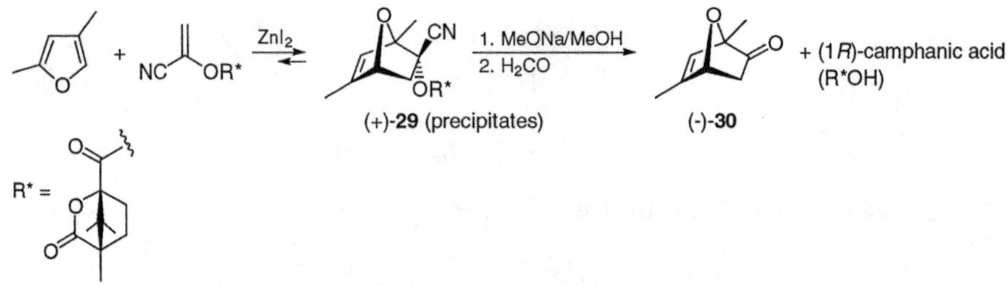

(+)-**29** (precipitates) (-)-**30**

The method of thermodynamic diastereoselection through diastereoselective crystallization of equilibrating Diels-Alder adducts has been applied to furan derivatives bearing chiral auxiliaries that can be recovered easily. For instance, the acetal of (2*S*,3*S*)-butane-2,3-diol and furfural is equilibrated in molten maleic anhydride with one major crystalline product that is a 1:1 complex of maleic anhydride and Diels-Alder adduct (+)-31. Upon treatment with isoprene this complex looses maleic anhydride which gives an adduct that is extracted with hexane, liberating pure (+)-31 (GUIDI et al. 1996). This chiron has been used to generate advanced anthracyclinone precursors (THEURILLAT-MORTIZ et al. 1997). In a similar way, (*S*)-camphanate of furfuryl alcohol undergoes Diels-Alder additions in molten maleic anhydride, giving one major crystalline adduct (+)-32. It has been converted into enantiomerically pure intermediates of the synthesis of taxol (THEURILLAT-MORITZ and VOGEL 1996) and squalestatin analogues (JOTTERAND and VOGEL 1999).

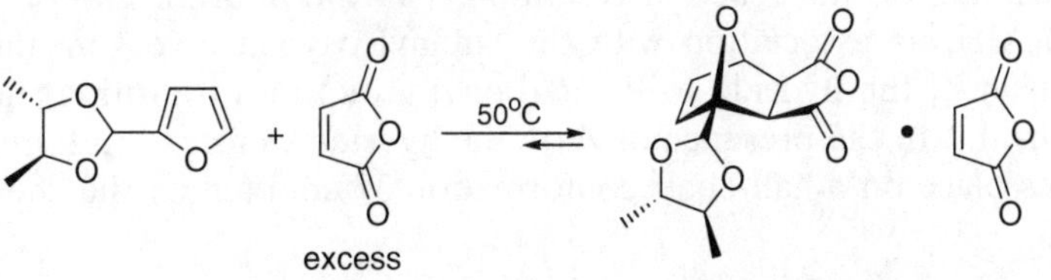

(+)-**31** (crystalline)

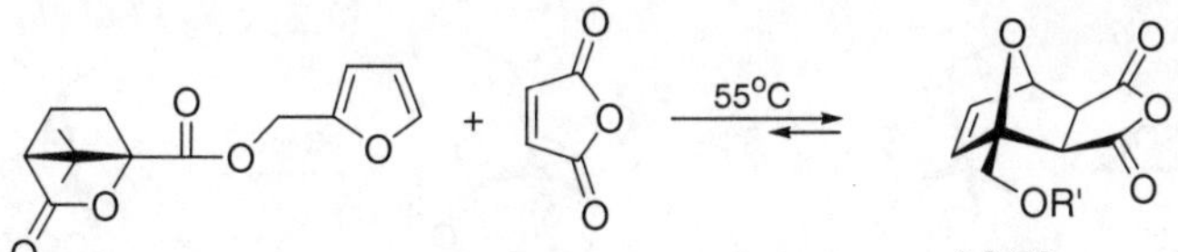

(+)-**32**

IV. One Chiral Auxiliary, Two Enantiomers

The most versatile chiral auxiliaries are those available in both their enantiomeric forms. However, there is a growing number of instances where an enantiomerically pure starting material attached to a chiral auxiliary can undergo different diastereoselective reactions depending on the reagents or/and the reactions conditions. For example, after removal of the chiral auxiliary, the two possible enantiomers of a given targeted compound can be obtained.

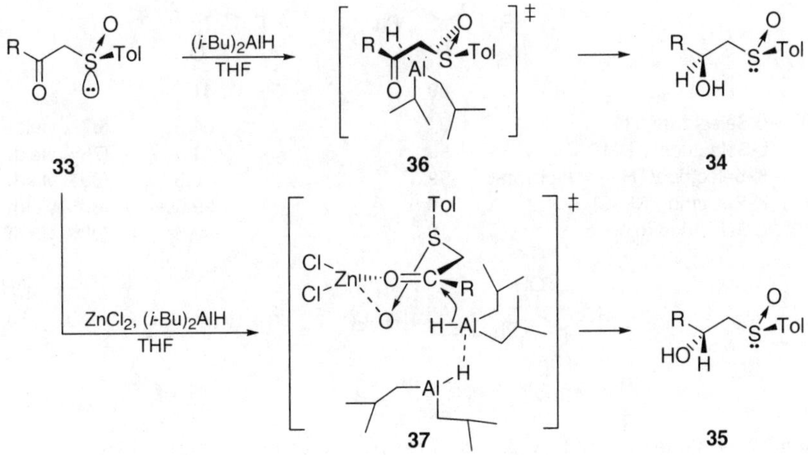

33 **36** **34** **37** **35**

As already mentioned (see Sect. C.II.), homochiral sulfoxides have been used extensively in asymmetric synthesis (SOLLADIÉ 1981). For instance, (+)-(*R*)-methyl *p*-tolylsulfoxide (obtained by addition of methylmagnesium iodide to (−)-(*S*)-menthyl *p*-toluenesulfinate) can be condensed via its lithiated conjugate base, without epimerization to carboxylic esters. This provides the corresponding homochiral 2-ketosulfoxides 33 which can be reduced with high diastereoselectivity into the corresponding 2-hydroxysulfoxides 34 or 35, depending on the reaction conditions (SOLLADIÉ et al. 1982; 1985). With (i-Bu)$_2$AlH, its association with the sulfinyl oxygen favors an intramolecular transfer of the hydride to the carbonyl group in a transition state 36 that is chair-like. In the presence of ZnI_2, the hydride transfer is intermolecular and takes place on a half-chair conformation 37 adopted by the chelated 2-keto-

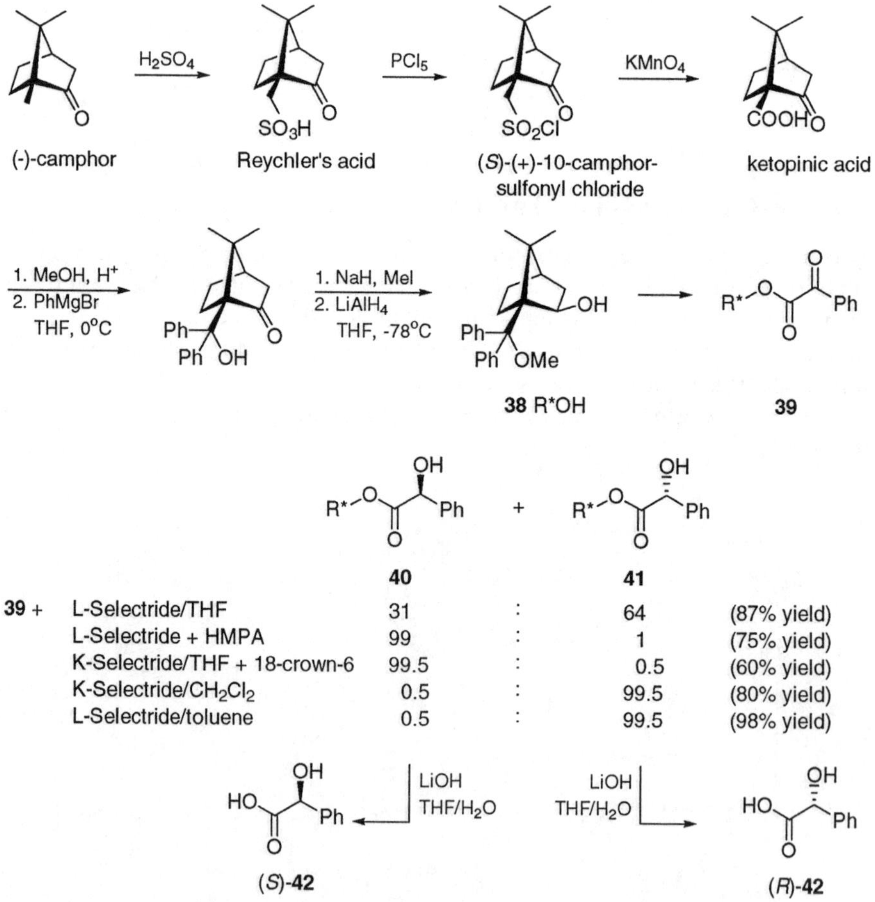

Scheme 5. Enantioselective synthesis of α-hydroxycarboxylic derivatives

sulfoxide with the Lewis acid (CARREÑO et al. 1990). After desulfurization of 34 and 35 with Raney nickel, enantiomeric secondary alcohols are obtained (SOLLADIÉ and ALMARIO 1992).

Many biologically active substances contain homochiral α-hydroxycarboxylic moieties (HANESSIAN 1983). They can be obtained (e.g., (*S*)-42, (*R*)-42 >99% ee) via reduction of α-ketoesters derived from *exo*-10,10-diphenyl-2,10-camphanediol or from *exo*-10,10-diphenyl-10-methoxy-2-camphanol (38) derived readily from (–)-camphor (Scheme 5) (CHU et al. 1999). Ester 39 derived from 38 is reduced with L-Selectride in THF with little diastereoselectivity. Upon adding one equivalent of HMPA (hexamethylphosphoramide), the diastereomeric ratio rises to 99:1. It reaches 99.5/0.5 using K-selectride in THF with one equivalent of 18-crown-6. These results suggest that the counter-ion of the hydride (Li^+, K^+), can interact with the α-ketoester and as consequence affect the face selectivity of its reduction. Very interesting is the observation that solvent can also play a significant role on the diastereoselectivity as shown by the reversal of diastereoselectivity for both K-Selectride and L-Selectride reduction when going from THF, a good coordinating solvent to CH_2Cl_2 or toluene, noncoordinating solvents.

Homochiral α-amino-aldehydes are very useful synthetic intermediates in the asymmetric synthesis of products of biological interest (JURCZAK and GOTÊBIOWSKI 1989). ALEXAKIS and coworkers (ALEXAKIS et al. 1991) have designed an efficient approach for their preparation in both their enantiomeric forms starting with glyoxal, 1,1-dimethylhydrazine, and (1*S*,2*S*)1, 2-diphenyl-*N*,*N*′-dimethylethylenediamine (43). The hydrazone 44 so-obtained reacts with organolithium reagents in THF, giving adducts 45 as single product. Hydrogenolysis, followed by amine protection and acidic treatment liberates the corresponding protected α-amino-aldehydes (*S*)-46 with high enantiomeric purity (>99% ee). Interestingly, when using Grignard reagents in toluene solvent, adducts 45′ are obtained as major products (with diastereomeric ratio 45′/45 4:1 to 99:1), thus allowing preparation of the enantiomers (*R*)-46 with high enantiomeric purity also (ALEXAKIS et al. 1992). In this method the chiral auxiliary (diamine) is readily recovered.

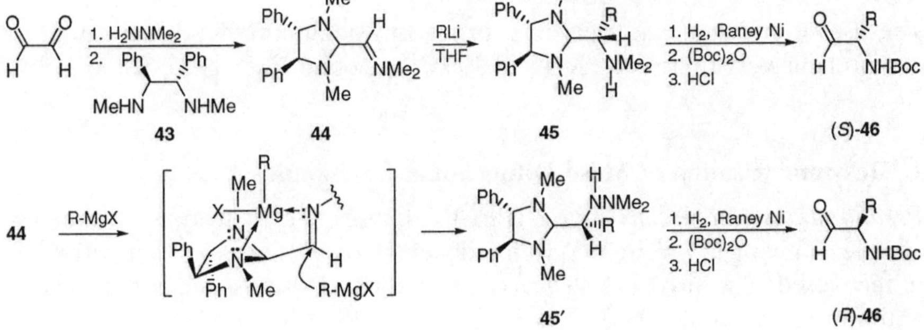

D. Asymmetric Catalysis

EMIL FISCHER recognized that enzymes act as catalysts either in living organisms or as isolated species and proposed the "lock and key" analogy for explaining the stereoselectivity of the enzymes. After PASTEUR's discoveries several attempts were made to generate optically active compounds from inactive precursors. Most of them relied on fermentation in the presence of microorganisms. The synthesis of optically active mandelonitrile by addition of HCN to benzaldehyde catalyzed by an enzyme (emulsin) isolated from almonds was an early example (ROSENTHALER 1908). Another early case of bioorganic enantioselective catalysis was the rearrangement of phenylglyoxal hydrate into optically active (95% ee) mandelic acid in the presence of *B. proteus* (HAYASHI 1908). Mandelonitrile was obtained with less than 10% enantiomeric excess by addition of HCN to benzaldehyde catalyzed by quinine or quinidine (BREDIG and FISKE 1912). Before the 1970s most asymmetric syntheses relying on asymmetric catalysis used enzymes or microorganisms. Today nonenzymatic man-made catalysts can lead to highly enantioselective reactions, with high turnover numbers (TON: moles of substrate produced per moles of catalyst used) and turnover frequencies (TOF: turnover numbers per time unit) and that for a large variety of substrates and reactions under all kinds of conditions. Biochemical methods (catalysis by enzymes or microorganisms) remain valuable for ecological reasons and thus will continue to play an important role in industrial processes. Today, chemical asymmetric catalysis is easier to apply and more flexible than biochemical methods when it deals with the preparation of new compounds with unusual structures, and new leads for drugs, fragrances or pesticides. Enzymes and microorganisms usually are tuned for only one of the two possible enantiomers of a given substrate or product. Man-made catalysts are smaller chemical systems which can be obtained in their two enantiomeric forms (KAGAN 1999). Their tuning in asymmetric synthesis can be realized efficiently by applying combinatorial chemistry methods (JANDELEIT 1999; SHIMIZU et al. 1999).

I. Biochemical Methods

The use of enzymes and microorganisms in asymmetric synthesis has been comprehensively reviewed (AZERAD 1995; AGER and EAST 1996; SUGAI 1999).

1. Desymmetrization of Meso-Difunctional Compounds

Pseudomonas fluorescens lipase (PFL) catalyses the hydrolysis of the bisacetate 47 giving (+)-48 in 33% isolated yield and >99% ee, together with 62% of recovered 47. Chiron (+)-48 was converted into (−)-muscone and polyether antibiotics (XIE et al. 1993).

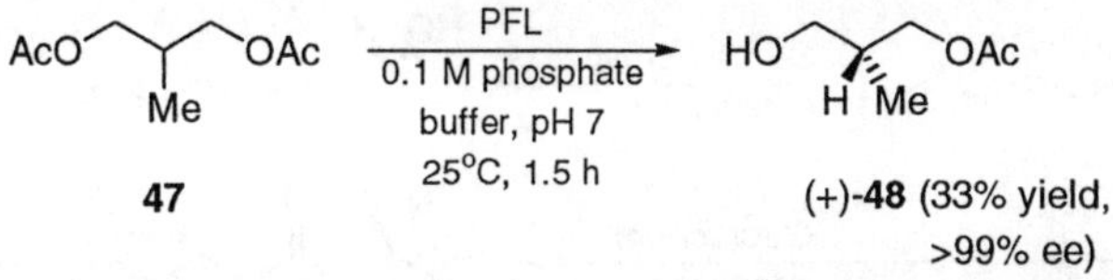

The methodology can be taken one step further through the use of an enzyme for ester hydrolysis or formation. This opens up the possibility to prepare the two enantiomers of a given substrate (Scheme 6) (RAMOS TOMBO et al. 1986).

2. Carbonyl Reductions

Baker's yeast is by far the most common agent used for the enantioselective reduction of achiral ketones (CZUK and GLANZER 1991). A rule as been developed by Prelog that predicts that the hydride transfer from alcohol deshydrogenase of *Curvularia falcata* occurs on the *Re* face of the ketone (AZERAD 1995). Reductions, when performed by an enzymatic system, usually require a cofactor. This is why organisms are invariably used rather than purified enzymes. Isolation of the product from the reaction media may become problematic. Furthermore, because those several enzymes in the microorganisms that can catalyze the reduction, tuning of the reaction conditions is necessary (control of the carbon source, immobilization of the cells, aging, heat treatment, incubation with inhibitors, etc.). In some cases, microorganisms are available that allow for complementary, selective production of either one of the two possible enantiomeric products (Scheme 7) (DAUPHIN et al. 1989).

Sometimes immobilized microorganisms can be used, making the purification procedure simple. For instance, reduction of ethyl 2-hydroxy-3-

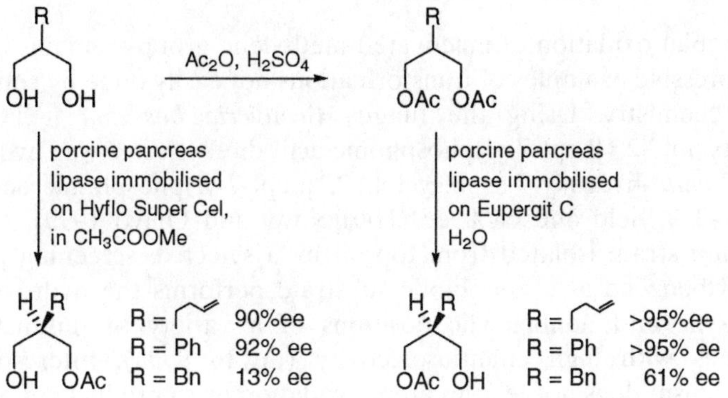

Scheme 6. Lipase-catalyzed acetylation or hydrolysi

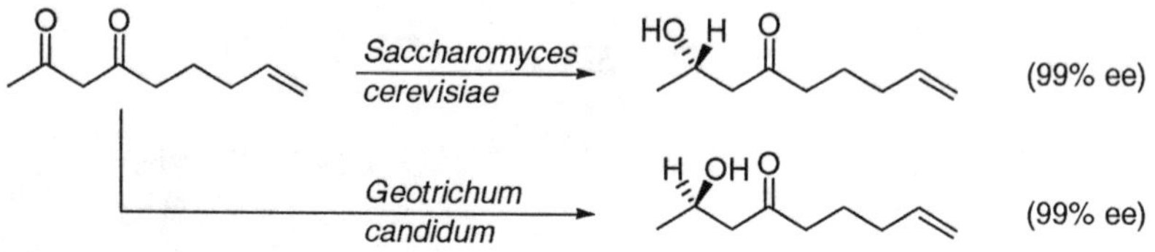

Scheme 7. Stereodivergent microorganism-catalyzed reductions

oxooctanoate with immobilized baker's yeast at pH 4.0 yields the corresponding *anti*-(2*R*,3*R*)-dihydroxy ester with 70% diastereoselectivity and 80% ee (FADNAVIS et al. 1999b).

3. Reductive Amination of α-keto-Acids

Nicotinamide cofactor-dependent L-amino acid dehydrogenases, which catalyze the formation of ketoacids from chiral α-amino acids, have important synthetic applications when used to obtain the reverse formation of natural or unnatural L-amino acids in the presence of ketoacids and ammonium salts. This occurs through the reduction of the imine resulting from the equilibrium with the ketoacid and NH_3 (Scheme 8) (AZERAD 1995).

4. Microbial Oxidations

Since the pioneering work by GIBSON (GIBSON et al. 1968, 1970) cyclohexadiene diols produced by microbial oxidation of substituted benzenes with mutant strains of *Pseudomonas putida* have been used extensively in the asymmetric synthesis of natural products and analogues of biological interest (HUDLICKY 1996; HUDLICKY et al. 1996). Toluene dioxygenase, the active enzyme in these oxidations, has been expressed in recombinant strains of *E. coli* JM109 (pDTG601) for improved efficiency (GIBSON and ZYLSTRA 1989).

Microbial oxidation of inactivated methylene groups remains one of the most impressive examples of transformations not easily done by conventional organic chemistry. Using the fungus *Beauveria bassiana* (ATCC 7159) (7-azabicyclo[2.2.1]hept-7-yl)phosphonic acid diethyl ester was hydroxylated giving (2-*endo*-hydroxy-7-azabicyclo[2.2.1]hept-7-yl)phosphonic acid diethyl ester in 43% yield and 64% ee (HEMENWAY and OLIVO 1999). A *Bacillus megaterium* strain isolated from topsoil by a selective screening procedure with alkylbenzene as a xenobiotic substrate performs the hydroxylation in the benzylic and nonbenzylic positions of a variety of unfunctionalized arylalkanes with high enantioselectivity (up to 99%). Interestingly, the microorganism does not lead to arene oxidation or to products of overoxidation (Scheme 9) (ADAM et al. 2000).

5. Acyloin Condensation

Nature condenses a carbonyl carbon center (via umpolung) to another carbonyl using thiamine pyrophosphate cofactor with yeast pyruvate decarboxylase. With benzaldehyde as a cosubstrate and using yeast whole-cells in the presence of glucose or pyruvate, this constitutes the key step in the industrial process for the manufacture of L-(−)-ephedrine (Scheme 10) (AZERAD 1995).

6. The Use of Catalytic Antibodies

Antibodies, or immunoglobulins, are important components of the mammalian defense mechanism. They are glycosidated proteins (ca. 100 kDa)

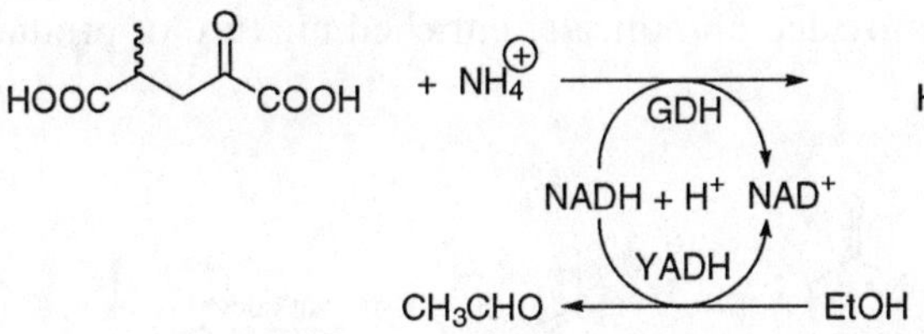

Scheme 8. Asymmetric synthesis of 4-methyl-L-glutamic acids using glutamate dehydrogenase (GDH) and yeast alcohol dehydrogenase (YADH)-ethanol as NADH regenerating system

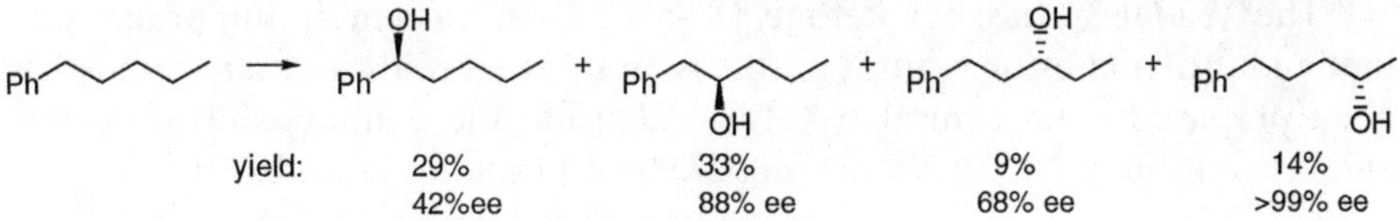

Scheme 9. Nonactivated C-H oxidation by *Bacillus megaterium*

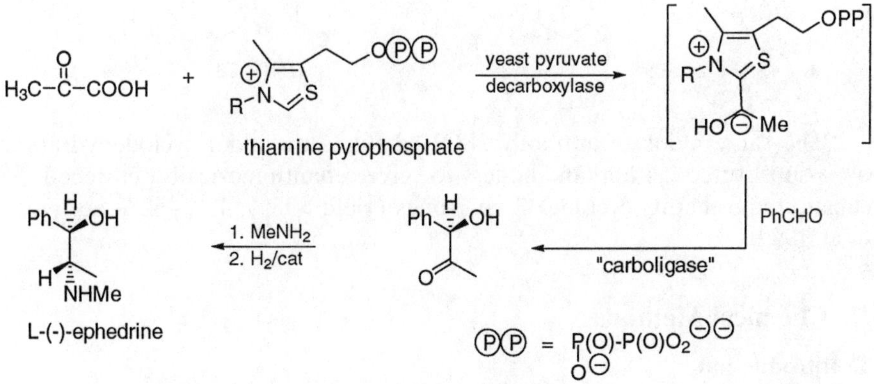

Scheme 10. Industrial synthesis of L-(−)-ephedrine based on a microbial mediated acyloin condensation

that recognize antigens, or haptens, with high specificity, in general through noncovalent interactions leading to affinity constants ranging from 10^{-4} to 10^{-14}, 6 to 20 kcal/mol in terms of the enthalpy of binding. When this energy of binding involves an immoglobulin and an hapten that is a mimic of a transition structure or intermediate of a given organic reaction, the immoglobulin is a catalyst (catalytical antibody = abzyme) of the reaction itself (WENTWORTH and JANDA 1999). One of the first enantiofacially selective processes catalyzed by an antibody involved the hydrolysis of enol esters (FUJI et al. 1991). Hapten 49 (antigen) was used to elicit antibodies for the hydrolysis of enol ester 50. The phosphonate moiety of 49 is a mimic of the tetrahedral transition state anticipated in the hydrolysis. The protein for antibody induction is keyhole limpet hemocyanin (KLH). An antibody called 27B5 was selected and found to catalyze the hydrolysis of 50 with a turnover number of ca. 300 and provided an optically enriched mixture of product 51 with 42% ee.

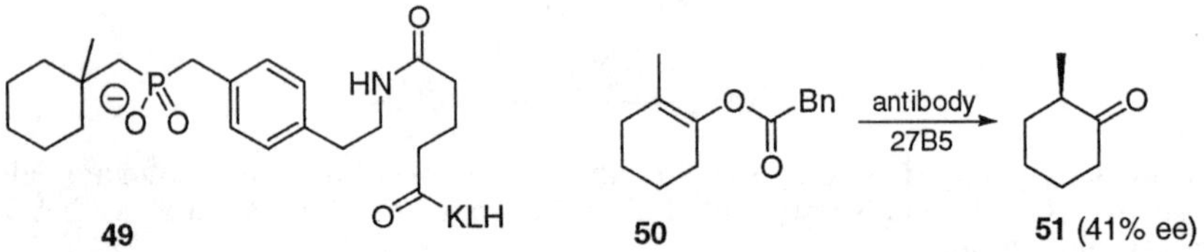

49 **50** **51** (41% ee)

The Wieland-Miescher ketone (*S*)-(+)-53, an extremely important synthetic intermediate for natural products, notably steroids and terpenoids has been prepared from achiral triketone 52 using the commercially available aldolase antibody 38C2 (Aldrich no. 47,995–0) (ZHONG et al. 1997).

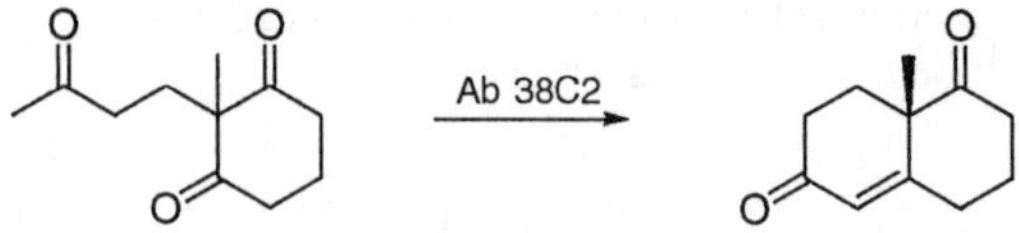

52 (*S*)-(+)-**53**

The same aldolase antibody 38C2 catalyses the aldol cyclodehydration of 4-substituted-2,6-heptanediones to give enantiomerically enriched 5-substituted-3-methyl-cyclohex-2-en-1-ones (yield >95%, 41%–62% ee) (LIST et al. 1999).

II. Chemical Methods

1. Introduction

The first example of asymmetric organometallic catalysis outside the area of polymer chemistry was the cyclopropanation of alkenes by methyl diazoacetate catalyzed by a salen-copper complex, as described by NOZAKI and coworkers in 1966 (NOZAKI et al. 1996). The maximum enantiomeric excess obtained then did not exceed 10%. Better levels of enantioselectivity were reached by the tuning of the structure of the ligands of the copper complex.

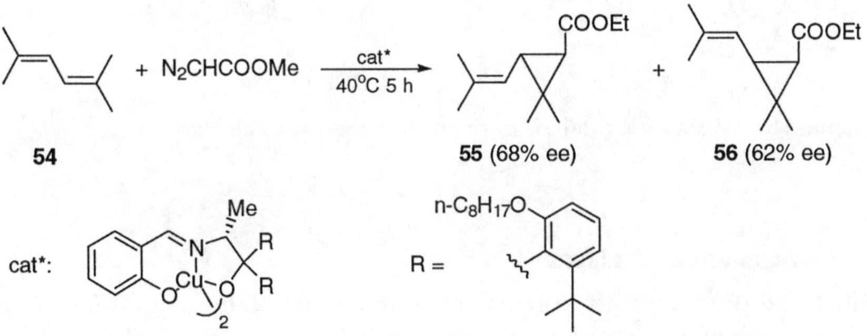

For instance, chrysantemic acid, a naturally occurring insecticide, was obtained this way with a 68% ee. The method implies the asymmetric cyclopropanation of 54 with ethyl diazoacetate giving a mixture of 55 and 56 (ARATANI et al. 1975).

In 1966 WILKINSON established the feasibility of homogenous hydrogenation of alkenes with $RhCl(PPh_3)_3$ as the catalyst precursor (OSBORN et al. 1966). In 1968, HORNER (HORNER et al. 1968) and KNOWLES (KNOWLES and SABACKY 1968) independently selected the enantiomerically pure monophosphine (*S*)-(Ph)(*n*-Bu)(Me)P as chiral auxiliary. This led to a maximum of 15% ee for the catalytical hydrogenation of α-phenylacrylic acid. A giant jump ahead was realized in 1971 and 1972 by KAGAN and DANG who showed that the chelating diphosphine (–)-(*R*,*R*)-DIOP derived from (*R*,*R*)-tartaric acid provided an excellent rhodium catalyst able to lead to enantioselectivities of up to 88% (Scheme 11) (DANG and KAGAN 1971; KAGAN and DANG 1972). This discovery has led to the first industrial application of asymmetric synthesis. At Monsanto, KNOWLES developed a process for the fabrication of L-DOPA, an important drug used in the treatment of the Parkinson's disease (KNOWLES et al. 1975; 1979). This stimulated a huge amount of research and the invention of new and better homochiral catalysts containing transition metals. At the same time other types of homochiral catalysts were developed such as chiral Lewis acids and chiral bases. Impressive results both in catalytic loading (<0.5 mol %) and enantiomeric excess (≥99%) have been achieved. Space does not allow one to survey all the pertinent achievements of the last 20 years. A comprehensive review on asymmetric catalysis has appeared recently (JACOBSEN et al. 1999). Therefore, we shall illustrate the methods with a few selected examples.

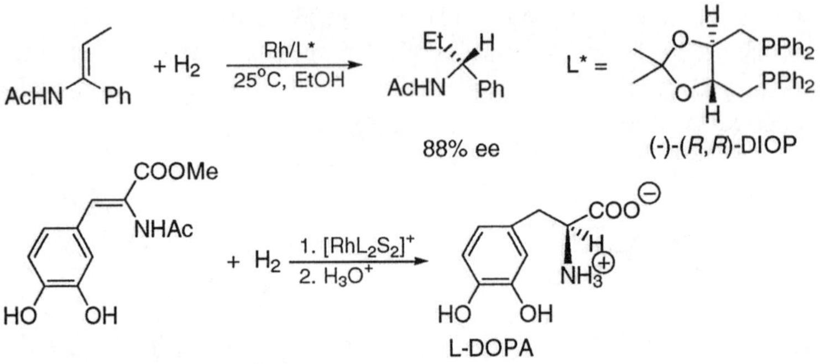

Scheme 11. Asymmetric catalysis of the hydrogenation of alkenes

2. Hydrogenation of Alkenes

Mibefrabil (60), the active ingredient of the antihypertensive *Posicor*, a new type of calcium antagonist, is made industrially starting from 57 according to a 5-step resolution/recycling process. An efficient enantioselective route has been developed which involves a high-yielded two-step conversion of 57 into the trisubstituted acrylic acid 58 and its hydrogenation catalyzed with Ru((*R*)–MeOBIPHEP) $(OAc)_2$. In this process, which constitutes a rare example of asymmetric hydrogenation of a tetrasubstituted alkene, the enantiomeric excess reaches 94% under 180 bar, giving (*S*)-59. Scale-up was performed in a continuous stirred tank reactor at 270 bar and 30°C, with a molar substrate to catalyst ratio of 100 and 93.5% ee. Upgrading of (*S*)-59 to 98% ee is possible by crystallization of the sodium salt. This new process is more economical than the classical process and is ready for implementation in the plant (SCHMID and SCALONE 1999).

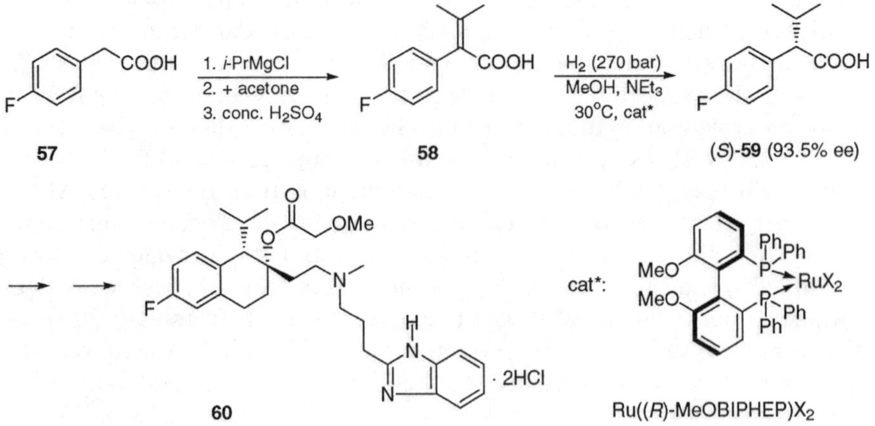

In a total synthesis of α-tocopherol with (*R*,*R*,*R*)-configuration, workers at Roche have found that 61 can be hydrogenated into 62 in the presence of Ru((*S*)–MeOBIPHEP) $(CF_3COO)_2$ with TONs of 20 000 and 100 000 and ee's of 96.5 and 98%, respectively (SCHMID and SCALONE 1999).

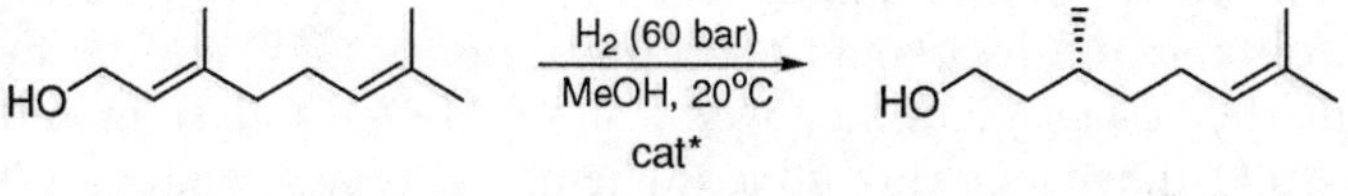

61 cat*: Ru((*S*)-MeOBIPHEP)$(CF_3COO)_2$ **62** (98.5% ee)

3. Hydrogenation of Imines

An important technical application of asymmetric catalytical hydrogenation is the production of (*S*)-metolachlor, an important herbicide invented by Ciba-Geigy AG and commercialized now by Novartis. In this case imine 63 is hydrogenated in the presence of the iridium catalyst Ir-XYLIPHOS iodide into 64 with 80% ee. Amidification of 64 with $ClCH_2COCl$ provides (*S*)-Metolachlor. The relatively low enantioselectivity (80% ee) obtained in this case is tolerable for an agrochemical product. The cost of the catalyst is almost negligible since a substrate to catalyst ratio of 1 000 000 allows the conversion 63 → 64 to be complete within 4 h, with an initial TOF exceeding 1 800 000 h^{-1} (BLASER et al. 1999).

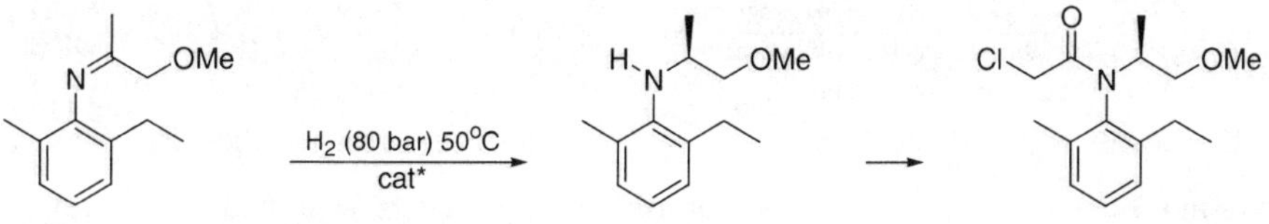

63 cat*: $[Ir(XYLIPHOS)Cl]_2 + NBu_4I$ **64** (80% ee) (*S*)-Metolachlor

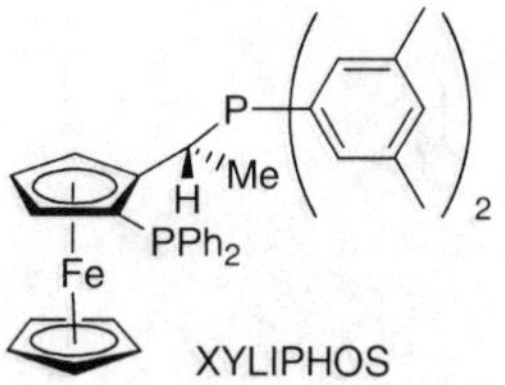

XYLIPHOS

4. Asymmetric Reduction of Ketones

Asymmetric hydrogenation and transfer hydrogenation of ketones are ideal means with which to prepare chiral secondary alcohols in view of the operational simplicity, environmental friendliness, and economics (PALMER and WILLS 1999). High enantioselectivities have been observed for the hydrogenation of α-keto esters and acids to the corresponding alcohols (OHKUMA and NOYORI 1999). Owing to environmental considerations, and safety and cost concerns, heterogeneous catalysts are often preferred to homogeneous catalysts. Cinchona alkaloid-modified platinum surfaces have been found effective in the asymmetric hydrogenation of α-ketoesters, ketoacids, and ketoacetals (BAIKER and BLASER 1997). For instance, ethyl 5-oxotetrahydrofuran-2-carboxylate, an important building block in the synthesis of natural products, has been obtained in high yield and optical purity (96% ee) by hydrogenation of diethyl 2-ketoglutarate over 5% Pt/Al_2O_3 catalyst modified by 9-methoxy-10,11-dihydrocinchonidine (Scheme 12) (BALÁZSIK et al. 2000).

Most nonactivated ketones undergo reduction sluggishly and with low enantioselectivity under asymmetric catalytical hydrogenation conditions. A very powerful alternative is the CBS reduction (COREY et al. 1987) which uses the inexpensive $BH_3 \cdot THF$ (or dialkylboranes) as stoichiometric-reducing agents, and enantiomerically pure oxazoborolidines derived from chiral 1,2-amino alcohols as catalysts. This method has been applied successfully to the enantioselective reduction of a large number of ketones (COREY and HEDAL 1998). For instance, the dopamine D1 agonist A77636 has been obtained in enantiomerically pure form by the CBS reduction of adamantyl bromomethyl ketone (Scheme 13) (DE NINNO et al. 1992). Conversion of the chiral bromohydrine 65 into the corresponding epoxide, ring opening with an arylllithium reagent, intramolecular acetal cyclization, and deprotection gave the desired product.

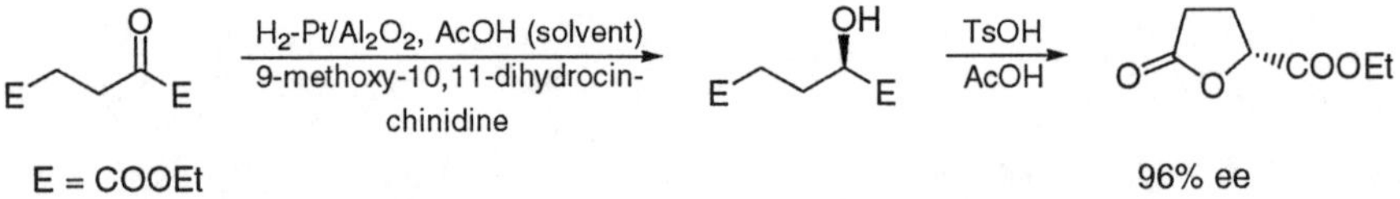

Scheme 12. Asymmetric hydrogenation with a solid catalyst (modified Pt surface)

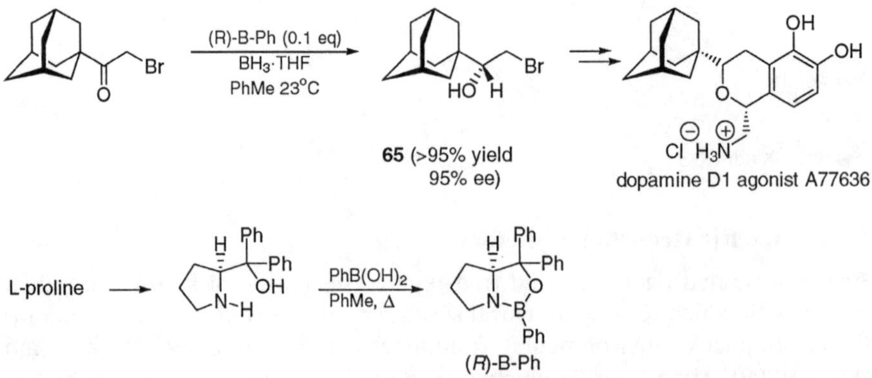

Scheme 13. Asymmetric synthesis of A77636 via the CBS reduction

5. Alkene Isomerization

Among the processes relying on enantioselective catalysis, the asymmetric isomerization of alkenes is certainly one of the most useful as demonstrated by the industrial synthesis of (–)-menthol (1500 tons annually). As shown in Scheme 14, two equivalents of isoprene react with lithium diethylamide to give a C_{10}-allylamine. In the presence of a catalytical amount of Rh[(*R*)-BINAP][COD]ClO_4, it is isomerized into (*R*)-66 in 98% yield and 98% ee. Acidic hydrolysis liberates (+)-citronellal, which undergo a zinc bromide-catalyzed intramolecular ene-reaction to provide (–)-isopulegol, and then, after hydrogenation, (–)-menthol (Akutagawa 1999).

6. Alcohols from Alkenes

Hydroboration of alkenes and alkynes is one of the most valuable synthetic techniques in organic synthesis because the organoboranes formed can be converted into various kind of compounds, and the enantioselective version developed by Brown has greatly expanded the scope of the hydroboration

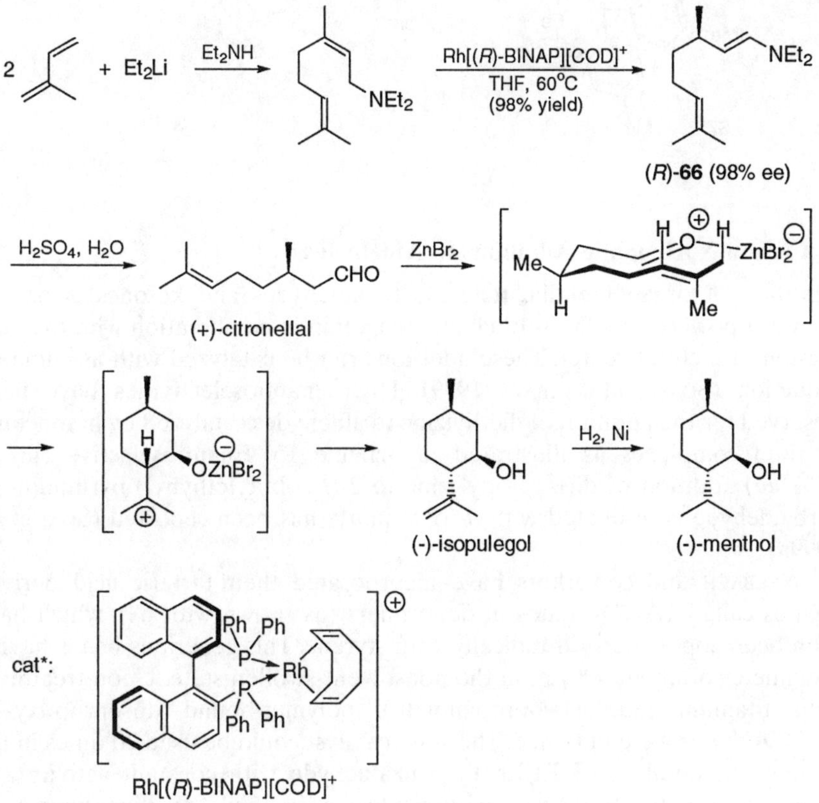

Scheme 14. Industrial asymmetric synthesis of (–)-menthol

(PETTER et al. 1988). Hydroboration of alkenes with enantiomerically pure di(isopinocamphenyl)borane is recognized to be a highly efficient reaction for the synthesis of optically active organoboranes (DHOKTE et al. 1996). In 1985, it was reported that the hydroboration with catecholborane can be catalyzed by rhodium(I)complexes (MÄNNIG and NÖTH 1985). Using homochiral phosphine ligands, the catalyzed hydroboration can be enantioselective. For instance, the reaction of styrene with catecholborane in THF in the presence of a catalytical amount of $[Rh(COD)_2]BF_4$ and pyrrazole-ferrocenylphosphine ligand 67 gives 1-phenylethanol with 98% ee (68% yield) (HAYASHI 1999a).

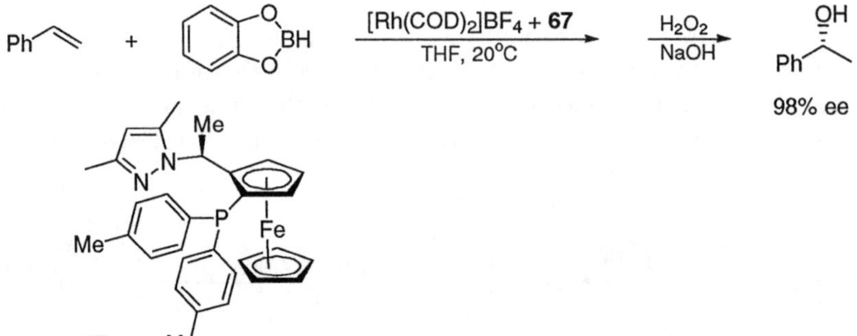

7. Carbon Nucleophile Additions to Aldehydes

Addition of organometallic reagents to aldehydes and ketones is one of the most powerful methods of chain elongation in combination with the generation of a chiral center. These additions can be catalyzed with asymmetric induction (SOAI and SHIBATA 1999). High enantioselectivities have been observed for the addition of diethylzinc to aldehydes catalyzed by homochiral titatium complexes as illustrated in Scheme 15. Enantioselective (up to 97% ee) addition of di(isopropyl)zinc to 2-(*tert*-butylethynyl)-pyrimidine-5-carbaldehyde coordinated with *d*- or *l*-quartz has been reported (SOAI et al. 1999a).

SEEBACH and coworkers have incorporated their tartaric acid derived ligands called TADDOLates in dendrimers (as shown with 68), which have then been copolymerized radically with styrene. This generates beads having a diameter of about 400 μm in the nonsolvent-swollen state. Upon treatment with titanium tetrakis(isopropanolate), polymer-bound diisopropoxy-Ti-TADDOLates were obtained. The solid catalyst could be used 20 times in the asymmetric addition of Et_2Zn to benzaldehyde with constant activity and enantioselectivity (96% ee), without losing its high swelling properties (SELLNER and SEEBACH 1999).

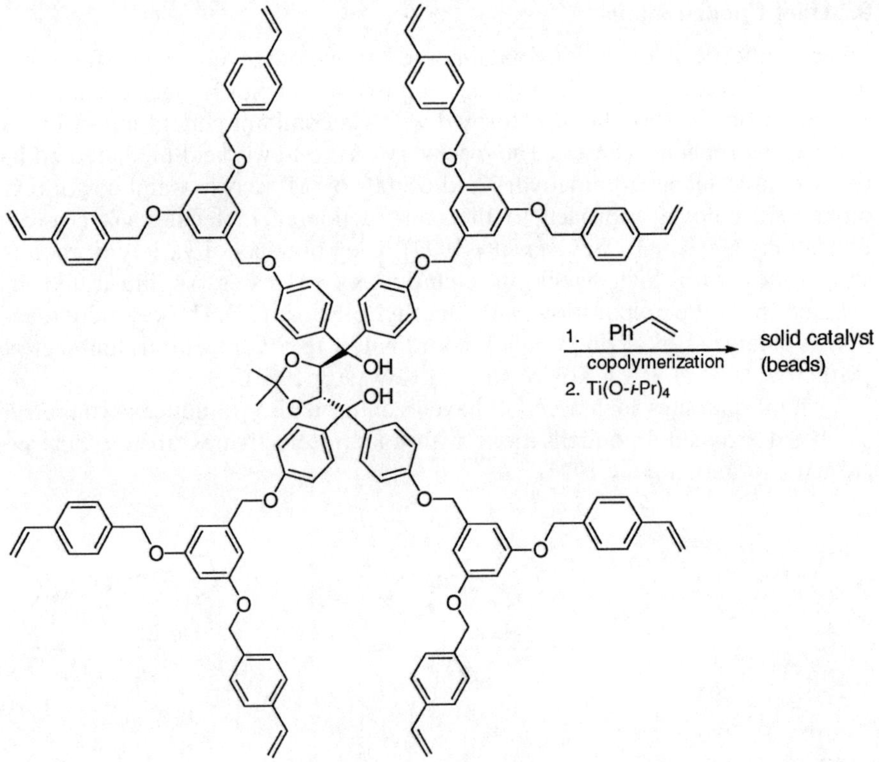

68

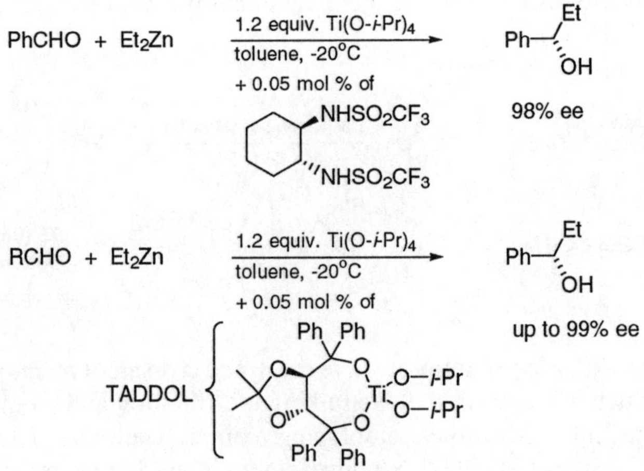

Scheme 15. Asymmetric ethylation of aldehydes

8. Aldol Condensations

The asymmetric aldol addition has emerged as one of the most powerful transformations available to the synthetic chemist for complex molecule synthesis. A new carbon–carbon bond is formed with concomitant generation of up to two new stereogenic centers. The discovery of the Lewis acid-mediated addition of enoxysilanes to aldehydes and acetals by MUKAIYAMA and coworkers pioneered a novel approach to the construction of molecules via crossaldolization (NELSON 1998; CARREIRA 1999). They obtained a variety of monosaccharides with high yield an enantiomeric excesses, as illustrated in Scheme 16 for the preparation of 4-C-methyl-D-ribose (70). The key aldol reaction is catalyzed by a tin complex coordinated to a homochiral diamine 69 (MUKAIYAMA et al. 1990; KOBAYASHI and KAWASUJI 1993).

Chiral diamines such as 71-73 have been used also to induce asymmetric catalyzed crossaldol condensations with enantioselectivities up to >99% ee (KOBAYASHI and HORIBE 1996).

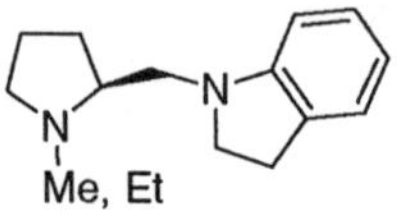

71

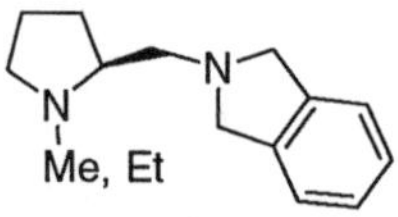

72

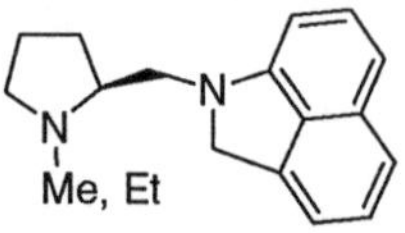

73

9. [2+2]-Cycloadditions

The first efficient catalytic asymmetric [2+2]-cycloaddition was reported in 1989, giving cyclobutane derivative 75 with high yield and enantiomeric excess (HAYASHI and NASARAKA 1989). The catalyst was the Ti-TADDOLate 74, a chiral Lewis acid also capable of inducing enantioselective Diels-Alder additions (Sect. D.II.10.).

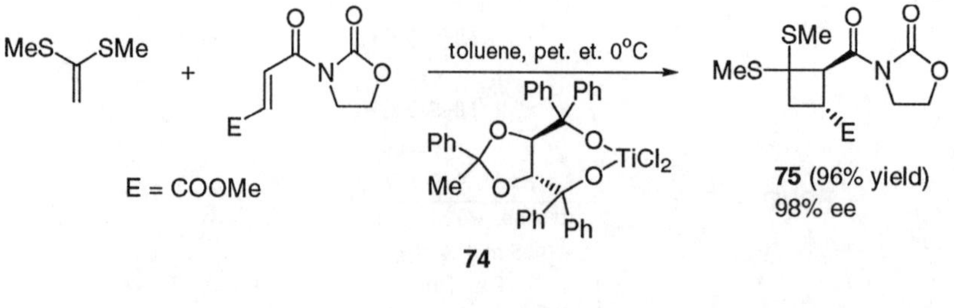

A new strategy for enantioselective catalyzed crossaldol reaction has been proposed which is based on the in situ ketene formation and aldehyde [2+2]-cycloaddition into β-lactones employing commercially available starting materials (NELSON et al. 1999). When straight-chain, β-branched and alkoxy-substituted aldehydes 76 and acetyl bromide were added simultaneously (via syringe) to a solution of 10 mol% Lewis acids 77A or 77B and Hünig's base

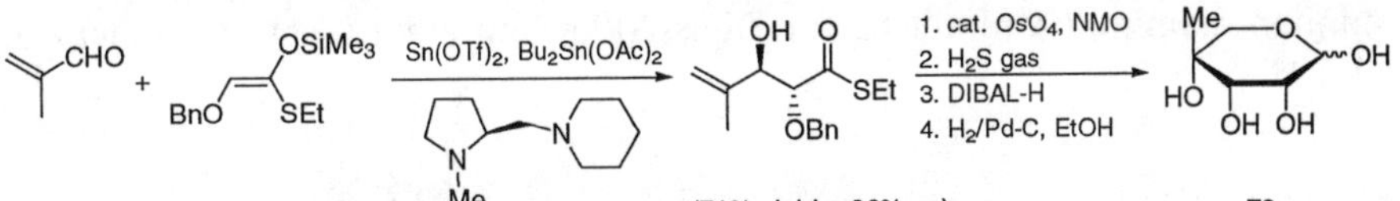

69 (71% yield, >96% ee) 70

Scheme 16. Asymmetric synthesis of 4-C-methyl-D-ribose

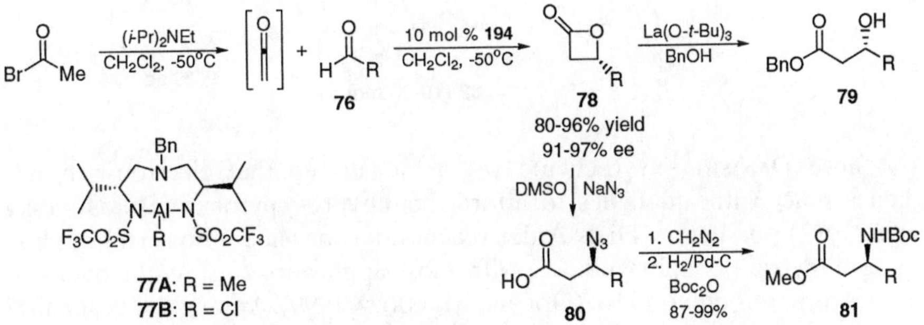

76 78 79

77A: R = Me
77B: R = Cl

80 81

Scheme 17. Asymmetric syntheses of aldols and β-amino acids via catalyzed [2+2]-cycloadditions of ketene and aldehydes

cooled to –50°C, the corresponding β-lactones 78 were obtained in 80%–96% yield and 91%–97% ee (Scheme 17). The La(O*t*-Bu)$_3$-catalyzed ring-opening of 78 with benzyl alcohol afforded the optically active aldols 79 in nearly quantitative yield. Sodium azide (DMSO, 50°C) promoted efficient S_N2 lactone ring opening to give the β-amino acids 80 (78%–95% yield) that could be converted into optically active *N*-protected β-amino acids 81 (NELSON and SPENCER 2000). Optically active β-amino acids have become increasingly prevalent features as chemotherapeutic agents (JUARISTI 1997) and are components of peptidomimetics with unique properties (KOERT 1997; GELLMAN 1998).

10. Diels-Alder and Hetero-Diels-Alder Additions

Since up to four new stereogenic centers can be created in a Diels-Alder reaction giving cyclohexenes with high stereoselectivity (suprafacial, regioselective, Alder *endo* rule), this reaction plays a prominent role in organic synthesis. In 1960, YATES and EATON demonstrated that Lewis acid can accelerate the Diels-Alder additions (YATES and EATON 1960). In 1979, KOGA and coworkers disclosed the first practical example of a catalytic enantioselective Diels-Alder reaction promoted by a Lewis acid complex, presumed to be "menthoxyaluminium dichloride," derived from menthol and ethylaluminium dichloride. This complex catalyzed the cycloaddition of cyclopentadiene to acrolein, methylacrylate, and methacrolein with enantioselectivities as high as 72% ee (HASHIMOTO et al. 1979). A decade later, COREY introduced an effective aluminium-diamine promoter 82 for Diels-Alder and aldol reactions (COREY et al. 1989).

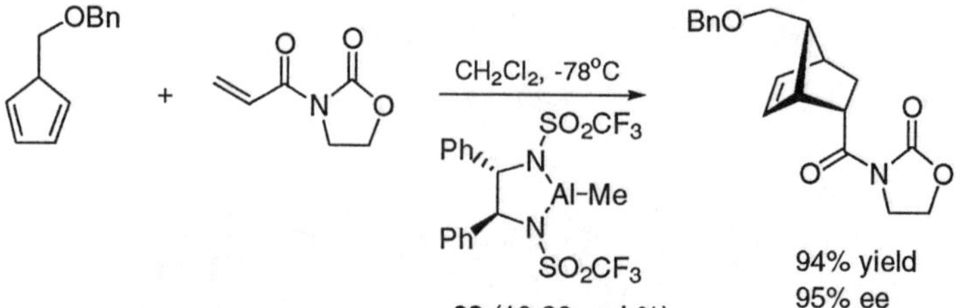

82 (10-20 mol %)

Since DANISHEFSKY (Scheme 18) demonstrated that electron-rich 1,3-dienes react with aldehydes to afford 5,6-dihydro-γ-pyrones (DANISHEFSKY et al. 1982), the hetero-Diels-Alder reaction has became a powerful method in asymmetric organic synthesis, with most applications using aldehydes or imines as heterodienophiles (OOI and MARUOKA 1999). Among the best chiral C_2-symmetrical systems 83 (SIMONSEN et al. 2000) and 84 (EVANS et al. 1998), derived from binaphthols and bis(oxazolines) respectively, must be retained (EVANS et al. 2000).

11. Other Carbon–Carbon Bond-Forming Reactions

Transition metal-catalyzed allylic substitutions are quite useful carbon–carbon bond-forming reactions. They often proceed under milder reaction conditions than ordinary S_N2 or S_N2' substitutions and with different chemo-, regio- and stereoselectivities. Typical leaving groups are acetates or carbonates, rather than the more reactive halides or sulfonates. The first example of enantioselective metal-catalyzed substitution was reported by TROST and STREGE in 1977. It implied tetrakis(triphenylphosphine)Pd and (+)-(*S*,*S*)-DIOP ligand and led to a maximum of 38% ee (TROST and STREGE 1977). More recently, PFALTZ and coworkers prepared palladium complexes derived from phosphinooxazolines (e.g., 85) that turned out to be very reactive, highly selective catalysts for the allylic substitution of 1,3-diphenyl-2-propenylacetate (86) with a range of *N*- and C-nucleophiles (for instance, Fig. 25; VON MATT and PFALTZ 1993).

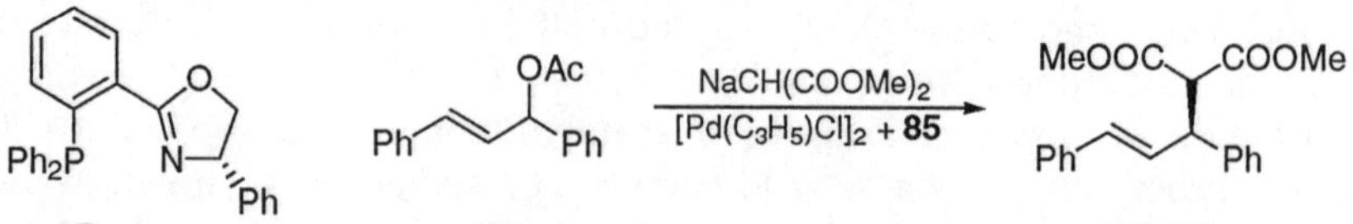

85 **86** **87** 99% ee

Recently, HOVEYDA and coworkers have disclosed the first examples of asymmetric ring opening metathesis of strained disubstituted alkynes, followed by intermolecular crossmetathesis. As illustrated in Scheme 19, when 88 is treated with 2 mol% of homochiral molybdenum catalyst 89, product 91 is obtained with 84% yield and better than 98% ee. The reaction probably involves an intermediate of type 90 (WEATHERHEAD et al. 2000).

There are several other carbon–carbon bond-forming reactions such as the Heck reaction (LOISELEUR et al. 1996; 1997) the Pauson-Khand reaction (BUCHWALD and HICKS 1999), the S_N2' nucleophilic substitution of propargyl

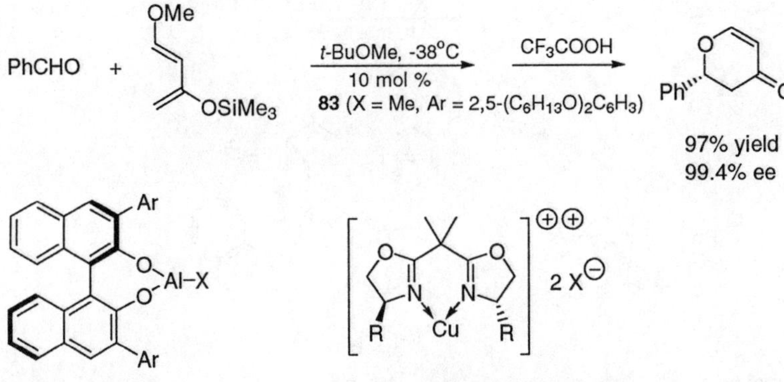

83 (X = Me, Cl)
Ar = Ph, 2,5-(MeO)$_2$C$_6$H$_3$, 2,5-(C$_6$H$_{13}$O)$_2$C$_6$H$_3$

84 (X = OSO$_2$CF$_3$, SbF$_6$, BF$_4$)
R = *t*-Bu, *i*-Pr, Ph, Bn, 4-Cl-C$_6$H$_4$ 4-MeO-C$_6$H$_4$

Scheme 18. Asymmetric hetero-Diels-Alder addition

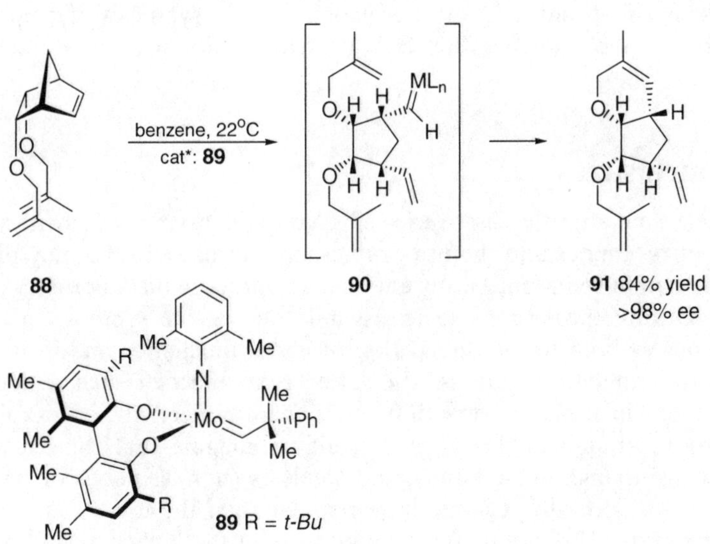

Scheme 19. Tandem asymmetric ring-opening metathesis and ring-closing metathesis

derivatives (OGASAWARA et al. 2000), the alkylation of Schiff bases (HALPERN 1997), alkyne additions to aldehydes (FRANTZ et al. 2000), the ene reaction (MIKAMI and TERADA 1999; DIAS 2000), the conjugate addition reaction (TOMIOKA ET NAGAOKA 1999; YAMAGUCHI 1999), and the nitroaldol reaction (SHIBAZAKI and GRÖGER 1999) for which useful asymmetric versions have been proposed.

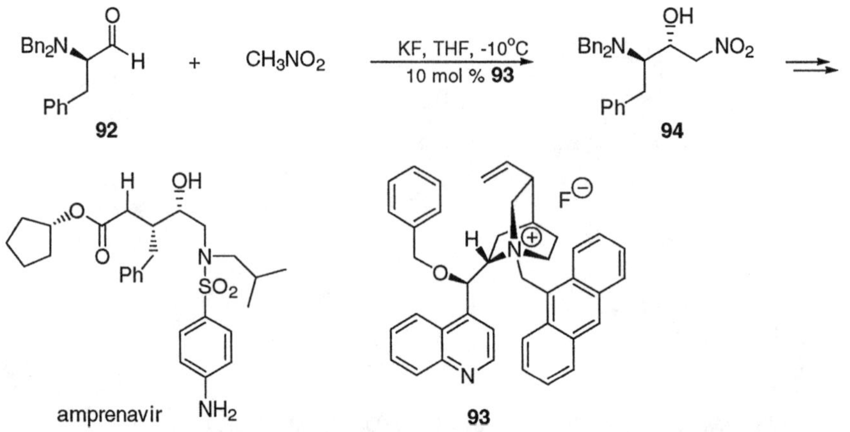

The nitroaldol reactions can be catalyzed by enantiomerically pure quaternary salts. An example is the condensation of nitromethane to aldehyde *92* in the presence of salt *93* which generates the nitro-alcohol *94* in 86% yield. This product is an intermediate in the synthesis of amprenavir, an important second generation HIV protease inhibitor (COREY and ZHANG 1999).

E. Conclusion

For a long time after the discoveries of PASTEUR on asymmetric synthesis, resolution of racemates and the use of microorganisms were the only practical means to obtain enantiomerically enriched materials in the laboratory and the factory. During the last 20 years, asymmetric organic synthesis has made huge progress both in the application of chiral auxiliaries and in inventing asymmetric catalytic reactions. The latter have reached such a degree of development in terms of versatility, high enantioselectivity, chemical yields and turnover numbers that it is difficult to imagine that the asymmetric synthesis using man-made homochiral catalysts (homogeneous or heterogeneous) is not going to become more popular than the use of enzymes and microorganisms. This can be foreseen because of the higher flexibility of the chemical catalysts compared with the biocatalysts, at least in the hands of synthetic chemists searching rapidly for new structures, and new enantiomerically pure compounds. Enzymatic and microbiological processes will remain preferable to "pure" chemical processes if they are cost effective. For a time, they will be better tolerated in the public view than "polluting" chemical processes (heavy metals, expensive solvents and toxic reagents, waste, energy consumption).

References

Adam W, Lukacs Z, Harmsen D, Saha-Möller CR, Schreier P (2000) Biocatalytic asymmetric hydroxylation of hydrocarbons with the top soil microorganism *Bacillus megaterium*. J Org Chem 65:878–882

Ager DJ, East MB (1996) Asymmetric Synthetic methodology. CRC Press, Boca Raton, FL, USA

Akutagawa S (1999) Isomerization of carbon – carbon double bonds. In: Jacobsen EN, Pfaltz A, Yamamoto H (eds) Comprehensive asymmetric catalysis, Springer, Berlin, pp 813–830

Alexakis A, Lensen N, Mangeney P (1991) Chiral aminal templates. Diastereoselectivity of hydrazone alkylation. Asymmetric synthesis of α-aminoaldehydes. Tetrahedron Lett 32:1171–1174

Alexakis A, Lensen N, Tranchier J-P, Mangeney P (1992) Reactivity and Diastereoselectivity of Grignard Reagents toward the hydrazone functionality in toluene solvent. J Org Chem 57:4563–4565

Aratani T, Yoneyoshi Y, Nagase T (1975) Asymmetric synthesis of chrysanthemic acid. An application of copper carbenoid reaction Tetrahedron Lett 1707–1710

Azerad R (1995) Application of biocatalysts in organic synthesis. Bull Soc Chim Fr 132:17–51 and references cited

Baiker A, Blaser H-U (1997) Enantioselective catalysts and reactions vol 5. In: Ertl G, Knözinger H, Weilkamp J (eds) Handbook of Heterogeneous Catalysis, Wiley-VCH, New York 4:2422–2436

Balázsik K, Szöri K, Felföldi K, Török B, Bartók M (2000) Asymmetric synthesis of alkyl 5-oxotetrahydrofuran-2-carboxylates by enantioselective hydrogenation of dialkyl 2-oxoglutarates over cinchona modified Pt/Al_2O_3 catalysts. J Chem Soc Chem Commun 555–556

Blaser H-U, Buser H-P, Jalett H-P, Pugin B, Spindler F (1999) Iridium ferrocenyl diphosphine catalyzed enantioselective reductive alkylation of a hindered aniline. Synlett S1:867–868

Bohringer M, Morgenstern K, Schneider WD, Berndt R (1999) Separation of a racemic mixture of two-dimensional molecular clusters by scanning tunneling microscopy. Angew Chem Int Ed Engl 38:821–823

Brown JM (1987) Directed homogeneous hydrogenation. Angew Chem Int Ed Engl 26:190–203

Buchwald SL, Hicks FA (1999) Pauson-Khand type reactions. In: Jacobsen EN, Pfaltz A, Yamamoto H (eds), Springer, Berlin, pp 491–510

Bull SD, Davies SG, Jones S, Polywka MEC, Prasad RS, Sanganee HJ (1998) A pratical procedure for the multigram synthesis of SuperQuat chiral auxiliaries. Synlett 519–521

Cahn RS, Ingold CK, Prelog V (1966) Specification of molecular chirality. Angew Chem Int Ed Engl 5:385

Carreira EM (1999) Mukaiyama aldol reaction. In: Jacobsen EN, Pfaltz A, Yamamoto H (eds) Comprehensive asymmetric catalysis. Springer, Berlin, pp 997–1065

Carreño MC, García Ruano JL, Martín AM, Pedregal C, Rodríguez JH, Rubio A, Sanchez J, Solladié G (1990) Stereoselective reductions of 2-keto sulfoxides with hydrides. J Org Chem 55:2120–2128

Cervinka O (1995) Resolution of racemates to enantiomers. In: Enantioselective Reactions in Organic Chemistry, Ellis Horwood Series in Organic Chemistry, London, Chapter 3, pp 5–16

Chu Y-Y, Yu C-S, Chen C-J, Yang K-S, Lain J-C, Lin C-H, Chen K (1999) Novel camphor-derived chiral auxiliaries: significant solvent and additive effects on asymmetric reduction of chiral α-keto esters. J Org Chem 64:6993–6998

Corey EJ, Hedal CL (1998) Reduction of carbonyl compounds with chiral oxazoborolidine catalysts: a new paradigm for enantioselective catalysis and a powerful new synthetic method. Angew Chem Int Ed Engl 37:1986–2012

Corey EJ, Zhang F-Y (1999) *re* and *si*-face-selective nitroaldol reactions catalyzed by a rigid chiral quaternary ammonium salt: a highly stereoselective synthesis of the HIV protease inhibitor Amprenavir (Vertex 478). Angew Chem Int Ed 38: 1931–1934

Corey EJ, Bakshi RK, Shibata S (1987) Highly enantioselective borane reduction of ketones catalyzed by chiral oxazaborolidines. Mechanism and synthetic implications. J Am Chem Soc 109:5551–5553

Corey EJ, Imwinkelried R, Pikul S, Xiang YB (1989) Practical enantioselective Diels-Alder and aldol reactions using a new chiral controller system. J Am Chem Soc 111:5493–5495

Crosby J (1991) Synthesis of optically active compounds: a large scale perspective. Tetrahedron 47:4789–4846

Czuk R, Glanzer BI (1991) Baker's yeast mediated transformations in organic chemistry. Chem Rev 91:49–97

Dang TP, Kagan HB (1971) The asymmetric synthesis of hydratropic acid and aminoacids by homogeneous catalytic hydrogenation. J Chem Soc Chem Commun 481

Danishefsky SJ, Kerwin JF, Kobayashi S (1982) Lewis acid catalyzed cyclocondensations of functionalized dienes with aldehydes. J Am Chem Soc 104:358–360

Dauphin G, Fauve A, Veschambre H (1989) Preparation of Stereoisomeric 2,4-diols: synthesis and conformational study of bicyclo derivatives, isomeric components of a pheromone of *trypodendrun lineatom.* J Org Chem 54:2238–2242

Davies SG, Sanganee HJ (1995) 4-Substituted-5,5-dimethyl-oxazolidin-2-ones as effective chiral auxiliaries for enolates alkylations and Michael additions. Tetrahedron Asymmetry 6:671–674

De Ninno MP, Perner RJ, Morton HE, Di Domenico Jr. S (1992) The enantioselective synthesis of the potent dopamine D1 agonist (1*R*,3*S*)-3-(1′-adamantyl)-1-(aminomethyl)-3,4-dihydro-5,6-dihydroxy-1*H*-2-benzopyran (A77636). J Org Chem 57:7115–7118

Dhokte UP, Soundararajan R, Ramachandran PV, Brown HC (1996) A general, efficient, convenient synthesis of chiral bis(terpenyl)haloborane reagents, valuable for asymmetric synthesis via organoboranes. Tetrahedron Lett 37:8345–8348

Dias LC (2000) Chiral Lewis acid catalyzed ene-reactions. Curr Org Chem 4:305–342

Duret P, Foucault A, Margraff R (2000) Vancomycin as a chiral selector in centrifugal partition chromatography. J Liq Chromatogr Mel Technol 23:295–312

Eames J (2000) Parallel kinetic resolution. Angew Chem Int Ed Engl 39:885–888 and references cited

Eliel EL, Wilen SH (1994) Separation of Stereoisomers. Resolution. Racemization. In: Stereochemistry of Organic Compounds. Wiley J & Sons, Chapter 7, pp 297–464

Enders D (1984) Alkylation of chiral hydrazones. In: Asymmetric Synthesis, Morisson JD (ed), Academic Press, Orlando FL, USA, Vol 3, pp 275–339

Enders D, Wortmann L, Peters R (2000) Recovery of Carbonyl Compounds from N,*N*-dialkylhydrazones. Acc Chem Res 33:157–169

Evans DA, Johnson JS (1999) Diels-Alder reactions. In: Jacobsen EN, Pfaltz A, Yamamoto H (eds) Comprehensive asymmetric catalysis, Springer, Berlin, pp 1177–1234

Evans DA, Johnson JS, Olhava EJ (2000) Enantioselective synthesis of dihydropyrans. Catalysis of hetero-Diels-Alder reactions by bis(osazoline) copper(II) complexes. J Am Chem Soc 122:1635–1649
Evans DA, Olhava EJ, Johnson JS, Janey JM (1998) Chiral C_2-symmetric Cu(II) complexes as catalysts for enantioselective hetero-Diels-Alder reactions. Angew Chem Int Ed 37:3372–3375
Evans DA, Takacs JM, McGee LR, Ennis MD, Mathre DJ, Bartoli J (1981) Chiral enolate design. Pure Appl Chem 53:1109–1127
Fadnavis NW, Sharfuddin M, Vadivel SK (1999a) Resolution of racemic 2-amino-1-butanol with immobilized penicillin G acylase. Tetrahedron Asymmetry 10:4495–4500
Fadnavis NW, Vadivel SK, Sharfuddin M (1999b) Chemoenzymatic synthesis of (4*S*,5*R*)-5-hydroxy-g-decalactone. Tetrahedron Asymmetry 10:3675–3680
Feringa BL, van Delden RA (1999) Absolute asymmetric synthesis: the origin, control and amplification of chirality. Angew Chem Int Ed Engl 38:3418–3438
Fernandez S, Brieva R, Rebolledo P, Gotor V (1992) Lipase-catalysed enantioselective acylation of *N*-protected and unprotected 2-aminoalkan-1-ols. J Chem Soc Perkin Trans I: 2885–2889
Forster A, Kovac T, Mosimann H, Renaud P, Vogel, P (1999) Resolution of 7-oxabicyclo[2.2.1]hept-5-en-2-one via cyclic aminals. Tetrahedron Asymmetry 40:567–571
Frantz DE, Fässler R, Carreira EM (2000) Facile enantioselective synthesis of propary-gylic alcohols by direct addition of terminal alkynes to aldehydes. J Am Chem Soc 122:1806–1807
Fuji I, Lerner RA, Janda KD (1991) Enantiofacial protonation by catalytic antibodies. J Am Chem Soc 113:8528–8529
Gawley RE, Aubé J (1996) Principles of asymmetric synthesis. Elsevier Sciences Ldt, Oxford, UK
Gellman SH (1998) Foldamers: a manifesto. Acc Chem Res 31:173–180
Gibson CL, Gillon K, Cook S (1998) A study of 4-substituted 5,5-diaryl-oxazolidin-2-ones as efficacious chiral auxiliaries. Tetrahedron Lett 39:6733–6736
Gibson DT, Zylstra GJ (1989) Toluene degradation by *Pseudomonas putida* F1. 264:14940–14946
Gibson DT, Hensley M, Yoshioka H, Mabry J (1970) Formation of (+)-*cis*-2,3-dihydroxy-1-methylcyclohexa-4,6-diene from toluene by *Pseudomonas putida*. J Biochem 9:1626–1630
Gibson DT, Koch JR, Schuld CL, Kallio RE (1968) Oxidative degradation of aromatic hydrocarbons by microorganisms. II. Metabolism of halogenated aromatic hydrocarbons. Biochem 7:3795–3802
Guidi A, Theurillat-Mortiz V, Vogel P, Pinkerton AA (1996) Enantiomerically pure Diels-Alder adducts of maleic anhydride to furfural acetals through thermodynamic control. Single crystal and molecular structure of (1*S*,4*R*,4′*S*,5′*S*)-1-(4′,5′-dimethyldioxolan-2′-yl)-5,6-dimethylidene-7-oxabicyclo[2.2.1]hept-2-ene. Tetrahedron Asymmetry 7:3153–3162
Hager O, Llamas-Saiz AL, Foces-Foces C, Claramont RM, López C, Elguero J (1999) Complexes between 1,1′-binaphthyl-2,2′-dicarboxylic acid and pyrazoles: a case of manual sorting of conglomerate crystals (triage). Helv Chim Acta 82:2213–2230
Halpern ME (1997) Phase-transfer catalysis: mechanisms and synthesis ACS Symp Ser 659, (ed) American Chemical Society, Washington, DC
Hanessian S (1983) Total Synthesis of Natural Products: the chiron approach. Pergmanon Press, New York
Hashimoto *S*-I, Komeshima N, Koga K (1979) Asymmetric Diels-Alder reaction catalysed by chiral alkoxyaluminium dichloride. J Chem Soc Chem Commun 437–438
Hayashi S (1908) Experimentelle Untersuchungen über die sterischen Verhältnisse bei der Dismutation von Phenylglyoxal-hydrat durch verschiedene Bakterien (*B.*

proteus, *B. fluorescens*, *B. pyocyaneum*, *B. prodigiosum* und *B. coli*). Biochem Z 46:7–15

Hayashi T (1999) Hydroboration of carbon – carbon bonds. In: Jacobsen EN, Pfaltz A, Yamamoto H (eds) Comprehensive asymmetric catalysis, Springer, Berlin, pp 351–364

Hayashi Y, Nasaraka K (1989) Asymmetric [2+2]-cycloaddition reaction catalyzed by a chiral titanium reagent. Chem Lett 793–796

Hemenway MS, Olivo HF (1999) Syntheses of new phosphorous-containing azabicycloalkanes and their microbial hydroxylation using *Beauveria bassiana*. J Org Chem 64:6312–6318

Horner L, Siegel H, Büthe H (1968) Aymmetric catalytical hydrogenation with an optically active homogenous and soluble phosphine rhodium complex Angew Chem Int Ed Engl 7:942–943

Hudlicky T (1996) Design constraints in pratical syntheses of complex molecules: current status, case studies with carbohydrates and alkaloids, and future perspectives. Chem Rev 96:3–30

Hudlicky T, Entwistl DA, Pitzer KK, Thorpe AH (1996) Modern Methods of monosaccharide synthesis from noncarbohydrate sources. Chem Rev 96:1195–1220

Izumi Y, Chibata I, Hoh T (1978) Herstellung und Verwendung von Aminosäuren. Angew Chem 90:187–194

Jacobsen EN, Pfalz A, Yamamoto H (1999) Compehensive asymmetric catalysis I, II, III, Springer, Berlin.

Jacques J (1995) Brève préhistoire de la synthèse asymétrique. Bull Soc Chim Fr 132:352–359

Jacques J, Collet A, Wilen SH (1994) Resolution by direct crystallization. In: Enantiomers, Racemates, and Resolutions, Krieger Publishing Co., Malabar, Florida, Chapter 4, pp 217–250

Jandeleit B, Schaefer DJ, Powers TS, Turner HW, Weinberg WH (1999) Combinatorial Material Science and Catalysis. Angew Chem Int Ed Engl 38:2494–2532

Jotterand N, Vogel P (1999) En route toward squalestatins and analogues from furfuryl alcohol and maleic anhydride. Tetrahedron Lett 40:5499–5502

Juaristi J (1997) Enantioselective synthesis of β-amino acids, Wiley-VCH, Weinheim

Jurczak J, Gołêbiosky A (1989) Optically active *N*-protected α-aminoaldehydes in organic synthesis. Chem Rev 89:149–164

Kagan HB (1999) Historical Perspective. In: Jacobsen EN, Pfaltz A, Yamamoto H (eds) Comprehensive Asymmetric Catalysis. Springer, Berlin pp 10–30

Kagan HB, Dang TP (1972) Asymmetric catalytic reduction with transition metal complexes. I. A catalytic system of rhodium(I) with (–)-2,3-*O*-isopropylidene-2,3-dihydroxy-1,4-bis(diphenylphosphino)butane, a new chiral diphosphine. J Am Chem Soc 94:6429–6433

Kagan HB, Fiaud JC (1988) Kinetic resolution. Top Stereochem 18:249–330

Keszei S, Simandi B, Szekely E, Fogassy E, Sawinsky J, Kemeny S (1999) Supercritical fluid extraction: a novel method for the resolution of tetramisole. Tetrahedron Asymmetry 10:1275–1281

Kitamura M, Kasahara I, Manabe K, Noyori R, Takaya H (1988) Kinetic resolution of racemic allylic alcohols by BINAP-ruthenium(II)-catalyzed hydrogenation. J Org Chem 53:708–710

Knowles WS, Sabacky MJ (1968) Catalytic asymmetric hydrogenation employing a soluble, optically active rhodium complex. J Chem Soc Chem Commun 1445–1446

Knowles WS, Sabacky MJ, Wineyard BD, Weinkarft DJ (1975) Asymmetric hydrogenation with a complex of rhodium and a chiral bisphosphine. J Am Chem Soc 97:2567–2566

Knowles WS, Wineyard BD, Sabacky MJ, Stults BB (1979) Fundamental Research. In: Ishii Y, Tsutsui M (eds) Homogeneous Catalysis. Plenum, New York

Kobayashi S, Horibe M (1996) Preparation of both enantiomers of 2-methyl-3-hydroxythioesters based on chiral Lewis acid-controlled synthesis. Tetrahedron 52: 7277–7286

Kobayashi S, Kawasuji T (1993) A new synthetic route to monosaccharides from simple achiral compounds by using a catalytic asymmetric aldol reaction as a key step. Synlett 911–913

Koert U (1997) β-Peptides: Novel secondary structures take shape. Angew Chem Int Ed Engl 36:1836–1837

Kriaules WS, Sabacky ML (1968) Catalytical Asymmetric hydrogenation employing a soluble, optically active, rhodium complex. J. Chem Soc Chem Commun 1445–1446

Krieg HM, Breytenbach JC, Keizer K (2000) Chiral resolution by β-cyclodextrin polymer-impregnated ceramic membranes. J Membrane Science 164:177–185

List B, Lerner RA, Barbas III CF (1999) Enantioselective aldol cyclodehydrations catalyzed by antibody 38C2. Org Lett 1:59–61

Loiseleur O, Hayashi M, Schmees N, Pfaltz A (1997) Enantioselective Heck reactions catalyzed by chiral phsphinooxazoline-palladium complexes. Synthesis 1338–1345

Loiseleur O, Meier P, Pfaltz A (1996) Chiral phosphanyldihydooxazoles in asymmetric catalysis: enantioselective Heck reactions. Angew Chem Int Ed 35:200–202

Magora A, Abu-Lafi S, Levin S (2000) Comparison of the enantioseparation of racemic uridine analogs on Welk-01 and ChiralPak-AD columns. J Chromatogr 866: 183–194

Männig D, Nöth H (1985) Catalytic hydroboration with rhodium complexes. Angew Chem Int Ed Engl 24:878–879

Mason S (1989) The origin of biomolecular chirality in nature. In: Krstuloviá AM (ed) Chiral separations by HPLC, applications to pharmaceutical compounds, Ellis Horwood Ltd, Chichester, pp 13–30

Medvedovici A, Sandra P, Toribio L, David F (1997) Chiral packed column subcritical fluid chromatography on polysaccharide and macrocyclic antibiotic chiral stationary phases. J Chromatogr 785:159–171

Mikami K, Terada M (1999) Ene-type reactions. In: Jacobsen EN, Pfaltz A, Yamamoto H (eds), Springer, Berlin, pp 1143–1174

Miller L, Orihuela C, Fronek R, Honda D, Dapremont O (1999) Chromatographic resolution of the enantiomers of a pharmaceutical intermediate from the milligram to the kilogram scale. J Chromatogr 849:309–317

Miller L, Orihuela C, Fronek R, Murphy J (1999) Preparative chromatographic resolution of enantiomers using polar organic solvents with polysaccharide chiral stationary phases. Angew Chem Int Ed Engl 865:211–226

Mukaiyama T, Shiina I, Kobayashi S (1990) A convenient and versatile route for the stereoselective synthesis of monosaccharides via key chiral synthons prepared from achiral source. Chem Lett 2201–2204

Nakamura S, Watanabe Y, Toru T (2000) Extremely efficient chiral induction in conjugate additions of tolyl α-lithio-β-(trimethylsilyl)ethyl sulfoxide and subsequent electrophilic trapping reactions. J Org Chem 65:1758–1766

Nelson SG (1998) Catalyzed enantioselective aldol additions of latent enolate equivalents. Tetrahedron Asymmetry 9:357–389

Nelson SG, Spencer KL (2000) Enantioselective β-amino acid synthesis base on catalyzed asymmetric acyl halide-aldehyde cyclocondensation reactions. Angew Chem Int Ed Engl 39:1323–1325

Nelson SG, Peelen TJ, Wan Z (1999) Catalytic asymmetric acyl halide-aldehyde cyclocondensations. A strategy for enantioselective catalyzed cross aldol reactions. J Am Chem Soc 121:9742–9743

Nikaido T, Kawada N (1994) Kokai Tokkyo Koho JP 6,209,781

Nozaki H, Moroiti S, Takaya H, Noyori R (1966) Asymmetric inductions in carbenoid reaction by means of a dissymmetric copper chelate. Tetrahedron Lett 2:5239–5244

Ogasawara M, Ikeda H, Hayashi T (2000) π-Allylpalladium-mediated catalytic synthesis of functionalized allenes. Angew Chem Int Ed 39:1042–1044

Ohkuma T, Noyori R (1999) Hydrogenation of carbonyl groups. In: Jacobsen EN, Pfaltz A, Yamamoto H (eds) Comprehensive asymmetric catalysis, Springer, Berlin, pp 199–246

Ooi T, Maruoka K (1999) Hetero-Diels-Alder and related reactions. In: Jacobsen EN, Pfaltz A, Yamamoto H (eds) Comprehensive asymmetric catalysis, Springer, Berlin, pp 1237–1254

Osborn JA, Jardine FH, Young JF, Wilkinson G (1966) The preparation and properties of tris(phenylphosphine)halogenorhodium (I) and some reactions thereof including catalytical homogneous hydrogenation of olefins and acetylenes and their derivatives. J Chem Soc (A) 1711–1732

Palmer MJ, Wills M (1999) Asymmetric transfer hydrogenation of C = Oand C = N bonds. Tetrahedron Asymmetry 10:2045–2061

Pelter A, Smith K, Brown HC (1988). Borane reagents. Academic Press, New York

Pirkle WH, Pochapsky TC (1989) Considerations of chiral recognition relevant to the liquid chromatographic separation of enantiomers. 89:347–362

Prelog V, Helmchen G (1982) Basic Principles of the CIP-System and Proposals for a Revision. Angew Chem Int Ed Engl 21:567–583

Profir VM, Matsuoka M (2000) Processes and phenomena of purity decrease during the optical resolution of DL-threonine by preferential crystallization. Colloids & Surfaces A-Physicochemical & Engineering Aspects 164:315–324

Rahman A, Shah Z (1993) Stereoselective synthesis in organic chemistry. Springer-Verlag, New York

Ramos Tombo GM, Schär H-P, Fernandez i Busquets X, Ghisalba O (1986) Synthesis of both enantiomeric forms of 2-substituted-1,3-propanediol monoacetates starting from a common prochiral precursor, using enzymatic transformations in aqueous and in organic media. Tetrahedron Lett 27:5707–5710

Reetz MT, Jaeger KE (2000) Enantioselective enzymes for organic synthesis created by directed evolution. Chem Eur J 6:407–412

Ronan B, Kagan HB (1992) Highly enantioselective synthesis of a Corey prostaglandin intermediate. Tetrahedron Asymmetry 3:115–122

Rosenthaler L (1908) Biochem Z 14:232, 365

Schmid R, Scalone M (1999) Process R&D of Pharmaceutical, Vitamins, and fine chemicals. In: Jacobsen EN, Pfaltz A, Yamamoto H (eds) Comprehensive asymmetric catalysis, Springer, Berlin, pp 1439–1449

Schurig V (1984) Gas chromatographic separation of enantiomers on optically active metal-complex-free stationary phases. Angew Chem Int Ed Engl 23:747–830

Seebach D, Prelog V (1982) The unambigous specification of the steric course of asymmetric syntheses. Angew Chem Int Ed Engl 21:654–660

Sellner H, Seebach D (1999) Dentritically cross-linking chiral ligands: high stability of a polystyrene-bound Ti-TADDOLate catalyst with diffusion control. Angew Chem Int Ed Engl 38:1918–1920

Sevin A-F, Vogel P (1994) A new stereoselective and convergent approach to the syntheis of long-chain polypropionate fragments. J Org Chem 59:5920–5926

Seyden-Penne J (1994) Chiral auxiliaries and ligands in asymmetric synthesis. John Wiley & Sons, Inc, New York

Sheldon RA (1993) Chirotechnology, Dekker M, New York

Shibazaki M, Gröger H (1999) Nitroaldol reaction. In: Jacobsen EN, Pfaltz A, Yamamoto H (eds), Springer, Berlin, pp 1075–1090

Shimizu KD, Snapper ML, Hoveyda AH (1999) Combinatorial approaches. In: Jacobsen EN, Pfaltz A, Yamamoto H (eds) Comprehensive Asymmetric Catalysis, Springer, Berlin, pp 1389–1399

Sih CJ, Wu SS (1989) Resolution of Enantiomers via biocatalysis. Top Stereochem 19:63–125

Simonsen KB, Svenstrup N, Roberson M, Jørgensen KA (2000) Development of an unusually highly enantioselective hetero-Diels-Alder reaction of benzaldehyde with activated dienes catalyzed by hyper-coordinating chiral aluminium complexes. Chem Eur J 6:123–128

Soai K, Osanai S, Kadowaki K, Yonekudo S, Shibata T, Sato I (1999a) *d*- and *l*-Quartz-promoted highly enantioselective synthesis of a chiral organic compound. J Am Chem Soc 121:11235–11236

Soai K, Shibata T (1999b) Alkylation of carbonyl groups. In: Jacobsen EN, Pfaltz A, Yamamoto H (eds) Comprehensive asymmetric catalysis, Springer, Berlin, pp 911–922

Solladié G (1981) Asymmetric synthesis using nucleophilic reagents containing a chiral sulfoxide group. Synthesis 185–196

Solladié G, Almario A (1992) Asymmetric synthesis of both enantiomers of methyl and *t*-butyl 3-hydroxybutyrates monitored by optically active sulfoxides. Tetrahedron Lett 33:2477–2480

Solladié G, Demailly G, Greck C (1985) Reduction of *β*-hydroxysulfoxides: application to the synthesis of optically active epoxides. Tetrahedron Lett 26:435–438

Solladié G, Greck C, Demailly G, Solladié-Cavallo A (1982) Reduction of *β*-ketosulfoxides: a highly efficient asymmetric synthesis of both enantiomers of methyl carbinols from the corresponding esters. Tetrahedron Lett 23:5047–5050

Stinson SC (1995) Chiral drugs: Market growth in single-isomer forms spurs research advances. Chem Ing News (Oct 5) 44–77

Stinson SC (1999) Chiral drug interactions. Chem Ing News (Oct 11) 101–120

Sugai T (1999) Application of enzyme- and microorganism-catalyzed reactions to organic synthesis. Curr Org Chem 3:373–406

Theurillat-Mortiz V, Vogel P (1996) Synthesis of enantiomerically pure 7-oxabicyclo[2.2.1]hept-2-enes precursors in the preparation of taxol analogues. Tetrahedron Asymmetry 7:3163–3168

Theurillat-Mortiz V, Guidi A, Vogel P (1997) Remote control of Diels-Alder additions. Enantioselective synthesis of (2*R*)-1,2,3,4-tetrahydro-2-hydroxy-5,8-dimethoxy-napththalen-2-yl methyl ketone (Wong's anthracycline intermediate) from furfural. Tetrahedron Asymmetry 8:3497–3501

Tomioka K, Nagaoka Y (1999) Conjugate addition of organometallic reagents. In: Jacobsen EN, Pfaltz A, Yamamoto H (eds), Springer, Berlin, pp 1105–1120

Trost BM, Strege PE (1977) Asymmetric inductions in catalytic allylic alkylation. J Am Chem Soc 99:1649–1651

Ulbricht TLV (1981) Reflexion on the origin of optical asymmetry on Earth. Origins Life Evol Biosphere 11:55–70

Vedejs E, Chen E (1997) Parallel kinetic resolution. J Am Chem Soc 119:2584–2585

Vince R, Brownell J (1990) Resolution of racemic carbovir and selective inhibition of human immunodeficiency virus by the (−)-enantiomer. Bioch Bioph Res Comm 168:912–916

Vogel P (2000) Synthesis of rare carbohydrates and analogues starting from enantiomerically pure 7-oxabicyclo[2.2.1]heptyl derivatives ("naked sugars"). Contempory Organic Chemistry 4:455–480

Vogel P, Cossy J, Plumet J, Arjona O (1999) Derivatives of 7-oxabicyclo[2.2.1]heptane in nature and as useful synthetic intermediates. Tetrahedron 55:13521–13642

Vogel P, Fattori D, Gasparini F, Le Drian L (1990) Optically pure 7-oxabicyclo[2.2.1]hept-5-en-2-yl derivatives ("naked sugars") as new chirons. Synlett 173–184

Vogl EM, Gröger H, Shibasaki M (1999) Towards Perfect asymmetric catalysis: additives and cocatalysts. Angew Chem Int Ed Engl 38:1570–1577

Von Matt P, Pfaltz A (1993) Chiral Phosphinoaryldihydrooxazoles as ligands in asymmetric catalysis: Pd-catalyzed allylic substitution. Angew Chem Int Ed 32: 566–568

Walker AJ (1992) Asymmetric Carbon–carbon bond formation using sulfoxide-stabilized carbanions. Tetrahedron Asymmetry 3:961–998

Warm A, Vogel P (1987) Synthesis of (+)- and (–)-methyl 8-epinonactate and (+)- and (–)- methyl nonactate. Helv Chem Acta 70:690–700
Weatherhead GS, Ford JG, Alexanian EJ, Schrock RR, Hoveyda AH (2000) Tandem catalytic asymmetric ring-opening metathesis/ring-closing metathesis. J Am Chem Soc 122:1828–1829
Wentworth Jr P, Janda KD (1999) Catalytic Antibodies. In: Jacobsen EN, Pfaltz A, Yamamoto H (eds) Comprehensive Asymmetric Catalysis. Springer, Berlin, pp 1403–1423
Wilen SH (1971) Resolving agents and resolution in organic chemistry. In: Allinger NL, Eliel EL (eds) Topics in Stereochemistry 6:107–176
Wilkinson TG, Shepherd RG, Thomas JP, Baughn C (1961) Stereospecificity in a new type of synthetic antituberculous agent. J Am Chem Soc 83:2212–2213
Williams KL, Sander LC (1997) Enantiomer separation on chiral stationary phases in supercritical fluid chromatography. J Chromatogr 785:149–158
Xie Z-F, Suemune H, Sakai K (1993) Synthesis of chiral building blocks using *Pseudomonas fluorescens* lipase-catalyzed asymmetric hydrolysis of meso diacetates. Tetrahedron Asymmetry 4:973–980
Yamaguchi M (1999) Conjugate additions of stabilized carbanions. In: Jacobsen EN, Pfaltz A, Yamamoto H (eds), Springer, Berlin, pp 1121–1139
Yates P, Eaton P (1960) Accelerations of the Diels-Alder reaction by aluminium chloride. J Am Chem Soc 82:4436–4437
Zhong G, Hoffmann T, Lerner RA, Danishefsky S, Barbas III CF (1997) Antibody-catalyzed enantioselective Robinson annulation. J Am Chem Soc 119:8131–8132
Zief M, Crane, LJ (1988) Editors Of: Chromatographic Chiral Separation, Chromatographic Sciences Series, Vol 40, Marcel Dekker, New York

CHAPTER 2

Stereoselective Separations: Recent Advances in Capillary Electrophoresis and High-Performance Liquid Chromatography

S. RUDAZ and J.-L. VEUTHEY

A. Introduction

In 1848, Louis Pasteur discovered, for the first time, a method to resolve a racemate into its two enantiomers by crystallization (WAINER and DRAYER 1993). Since then, this kind of separation has remained a challenging task in analytical chemistry as enantiomers possess identical properties except when in the presence of a chiral handle (plane-polarized light, chiral solvent, chiral reagent, etc.). As Pasteur proposed, enantiomers are separated as diastereomeric derivatives after a chemical reaction with a chiral selector which adds another asymmetric center. Indeed, diastereomers possess different physical-chemical properties which is why the technique of fractional recrystallization remains widely used after so many years (COLLET 1989). However, other analytical techniques are now available for enantiomer analysis, which can be divided into two different categories: methods that do not involve separation such as polarimetry, nuclear magnetic resonance, calorimetry and enzymatic techniques, and methods based on separation. In this chapter, only the latter category will be discussed; description of the former may be found in different textbooks (ALLENMARK 1991; KRUSTULOVIC 1989; WAINER and DRAYER 1988).

Since the tragic accident of thalidomide in the early 1960s (see Chap. 4, this volume), a great number of pharmacological and toxicological studies have shown that enantiomers can have different activities and have therefore to be considered as two distinct substances. Recently, several countries have decided to impose toxicity studies on both enantiomers as well as on the racemate (FDA 1992; GROSS et al. 1993; SHINDO and CALDWELL 1995). Therefore, the growing awareness of drug stereochemistry has initiated a tremendous development of analytical methods for determining enantiomeric proportions in pure substances as well as in biological matrices. Separation techniques such as liquid, thin layer, supercritical fluid and gas chromatography, as well as more recent techniques, such as capillary electrophoresis and capillary electrochromatography, are now recognized as methods of choice for the separation of enantiomers in the pharmaceutical industry.

Three strategies are generally adopted for enantiomer resolution by chromatographic or electrophoretic techniques:

1. The indirect method, which separates enantiomers via diastereomeric derivatives, after a prederivatization step with an optically pure reagent.
2. The dynamic mode, where resolution of enantiomers is carried out by addition of a chiral selector directly in the mobile phase or in the buffer electrolyte solution for in situ formation of diastereomeric derivatives.
3. The direct resolution of enantiomers in a chiral stationary phase immobilized on a chromatographic support. In this case, labile diastereomers are formed on the support.

As mentioned above, different chromatographic and electrophoretic techniques can be used for the stereoselective separation of enantiomers. However, liquid chromatography is considered the method of choice for the analysis of pharmaceutical compounds, especially in the presence of biological fluids. More recently, capillary electrophoresis has gained importance in pharmaceutical analysis because of its low cost and great efficiency. This chapter will therefore focus on the description of these two techniques only. For information regarding chiral separations with gas and supercritical fluid chromatography, textbooks (Allenmark 1991; Krustulovic 1989; Wainer and Drayer 1988) and review articles can be consulted (Markides and Juvancz 1992; Macaudière et al. 1989; Anton et al. 1994; Schurig 1994; König 1987).

B. Liquid Chromatography

Liquid chromatography (LC), and more specifically high-performance liquid chromatography (HPLC), were developed to a considerable extent in the 1970s and 1980s and are now recognized as powerful methods for the analysis of pharmaceutical compounds at every stage of drug development. Indeed, HPLC is valuable for the quality control of an active substance from its initial synthesis through to the finished product as well as in pharmacological, toxicological, and pharmacokinetic studies. The success of this technique resides principally in its versatility, since it can be applied to the analysis of inorganic, organic, polar, and nonpolar compounds, as well as large biomolecules such as oligosaccharides and proteins in organic or aqueous media. Furthermore, several detectors such as the universal nonsensitive refractive index detector or the selective and sensitive mass spectrometer can be coupled to the chromatographic column and allow qualitative and quantitative analyses. Finally, liquid chromatography can be entirely automated, which increases its throughput and enhances its precision and accuracy. A description of this technique (Poole and Poole 1994; Rosset et al. 1991; Willoughby et al. 1998) and several pharmaceutical applications can be found in the literature (Riley et al. 1994; Adamovics 1997; Gosh 1992).

LC is widely used to determine the enantiomeric purity of a drug substance, as well as the enantiomeric ratio in biological matrices such as serum and urine. In the last 10 years, several textbooks (Allenmark 1991;

KRUSTULOVIC 1989; WAINER and DRAYER 1988) and review articles (SIRET et al. 1992; CECCATO et al. 1999; BOJARSKI and ABOUL-ENEIN 1996) have been published on this topic. As mentioned in the Introduction, three strategies can be used to resolve enantiomers:

1. Precolumn formation of diastereomers
2. Addition of chiral selectors in the mobile phase
3. Use of chiral stationary phases (CSP)

I. Precolumn Formation of Diastereomers

This approach consists forming diastereomeric derivatives prior to the chromatographic separation. For this purpose, a pure optically active reagent is added to the sample and the enantiomers can be separated into two diastereomers with different chromatographic properties. The separation can then be conducted on a conventional achiral and cost-effective stationary phase. However, this procedure is limited to molecules which have a suitable group for derivatization, such as amino and carboxylic acid functions. Furthermore, a reagent with a known and high degree of optical purity is necessary. This parameter is of utmost importance, since it directly affects the accuracy of the results. Finally, the derivatization reaction should be quantitative and should avoid racemization or epimerization.

Several derivatization reagents are now commercially available for LC such as *o*-phthalaldehyde (OPA), 1-(9-fluorenyl)ethyl chloroformate (FLEC), 1-(*p*-nitrophenyl)ethylamine and 2,3,4,6-tetra-*O*-acetyl-b-D-glucopyranosyl isothiocyanate (GITC). Generally, these reagents contain a chromophore or a fluorophore to enhance method sensitivity and possess a stereogenic center which is as close as possible to the reaction site. This strategy, very successful in the 1980s, is now mainly used for the stereoselective analysis of amino acids and peptides, which lack strong chromophores (ALLENMARK 1991).

II. Addition of Chiral Selectors in the Mobile Phase

Here, a chiral selector is added to the mobile phase and forms in situ, with the compounds of interest, complexes that can be separated on a conventional column (LINDNER and PETTERSON 1985). In this case, a prederivatization step is not necessary but a large amount of expensive chiral selector compatible with the detection system is generally required. The metal complex of a chiral ligand, a chiral counter-ion or a cyclodextrin derivative is added in the mobile phase and separation is conducted by ligand exchange chromatography, normal phase or reversed phase ion-pair chromatography, or by reversed-phase chromatography, respectively (KRUSTULOVIC 1989). Several papers have been published on the separation of amino acids, hydroxy acids, amino alcohols, and carboxylic and sulfonic acids, as well as of numerous drugs. This method, which is used less now because of the high consumption of chiral selector and because reproducible results are difficult to obtain, is mainly applied in capillary electrophoresis (see Sect. C.).

III. Chiral Stationary Phases

Chiral stationary phases (CSPs) are certainly the method of choice for the separation of pharmaceutical compounds by LC. Since the pioneering works of PIRKLE in 1981 (PIRKLE et al. 1981), over a hundred CSPs have been made commercially available. However, a universal stationary phase for resolution of all types of enantiomers does not exist yet. Therefore, the chemist has to choose the appropriate column depending on the nature of the analytes. In 1987, WAINER (WAINER 1987) and CAUDE et al. (LIENNE et al. 1987) proposed a classification of chiral stationary phases based upon the formation of solute-CSP complexes and on the nature of the chiral selector. Chiral stationary phases can be thus classified into five or four groups according to WAINER or CAUDE, respectively.

As shown in Fig. 1, CSPs can be divided into two main groups, the independent and the cooperative stationary phases, according to their chiral recognition mechanism. With the former, a chiral selector is grafted on a chromatographic support and each selector forms in situ; with both enantiomers, labile diastereomeric complexes with the stoechiometric ratio 1:1. The model established by DALGLIESH in 1952 allows selection of the appropriate chiral selector (DALGLIESH 1952). He postulated that chiral recognition requires a minimum of three simultaneous interactions between the selector and the analyte, which can include hydrogen bonding, dipolar, and charge transfer interactions, as shown in Fig. 2. Enantioselectivity results from the difference in the interaction energy of the complexes thus formed.

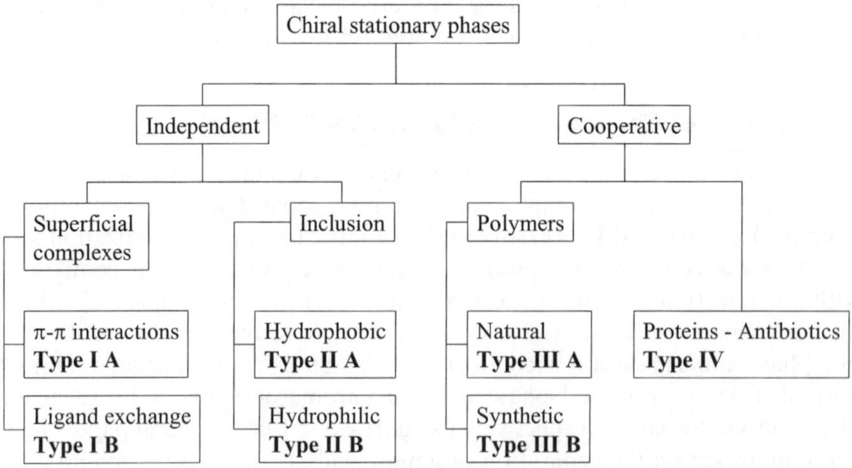

Fig. 1. Chiral stationary phases: classification according to chiral recognition mechanisms and chemical structures. (Adapted from SIRET et al. 1992)

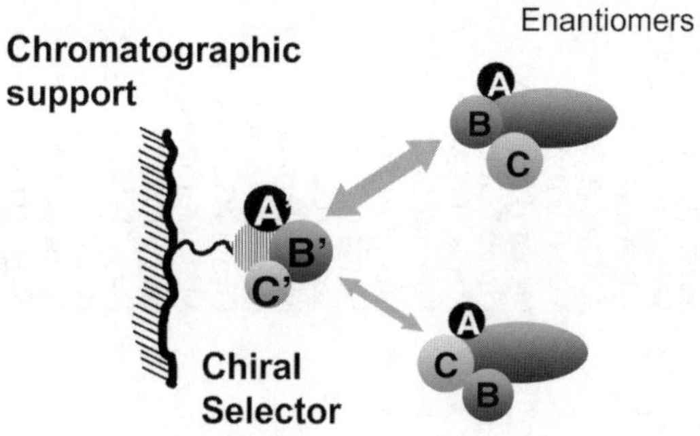

Fig. 2. Three-point interactions model. (Adapted from Rosset et al. 1991)

This model cannot be applied to polymeric chiral selectors which have many asymmetric centers and cavities. In this case, the chiral selector acts cooperatively and inclusion phenomena generate chiral recognition ability.

1. Type I A (Pirkle Type)

Following the concept established by Dalgliesh, Pirkle developed numerous CSPs (Pirkle et al. 1984). The chiral selector contains a π-donor or π-acceptor aromatic moiety bonded to a silica-based material (Table 1). Enantiomers with a π-electron-acceptor or π-electron-donor group form labile diastereomeric complexes with the chiral selector. π-π-Interaction, as well as hydrogen bonding and dipole–dipole stacking, can be involved in the recognition mechanism. These CSPs present good stability and efficiency and are generally used in normal phase mode as well as in supercritical fluid chromatography (SFC) (Caude and Bargmann-Leyder 1993). With the latter, selectivity can be modified and separations are fast. However, these stationary phases are limited to compounds of low or medium polarity and it is often necessary to add a derivatization step to introduce an aromatic moiety with π-electron-acceptor or π-electron-donor groups.

2. Type I B (Ligand Exchange Chromatography)

These selectors contain an amino acid grafted either onto a polymer (polystyrene-divinylbenzene) (Kurganov et al. 1983) or a silica (Gubitz and Juffmann 1987) matrix and a transition metal (Cu^{2+}, Ni^{2+}, Zn^{2+} or Cd^{2+}) loaded by percolation before enantioselective separation (Table 1). Hence, during chromatography with an aqueous mobile phase, a ternary labile complex is formed with the metal at the center. These selectors have been mainly used for the chiral separation of amino acids and derivatives as well as for amino alcohols and hydroxyacids. The major drawback of this kind of selector is its poor efficiency.

Table 1. Type I CSP: (adapted from Ceccato et al. 1999)

Selector	Commercial name	Provider
(*R*)-(dinitro-3,5-benzoyl)phenylglycine	Covalent D-phenylglycin	Regis (USA)
	Bakerbond DNBPG (cov.)	Baker (Netherlands)
	Bakerbond DNBPG (ionic)	Baker (Netherlands)
	Chiral D-DPG	Serva (Germany)
	ChiraSep DNBPG	Merck (Germany)
	ChiRSil I	Alltech (USA)
	Chi-RoSil	Alltech (USA)
	Chiral-2	MN (Germany)
	Grom-Chiral-*R*-DNBPG-C	Grom (Germany)
	Grom-Chiral-*R*-DNBPG-I	Grom (Germany)
	Ionic D-Phenylglycine	Regis (USA)
	Sumichiral OA-2000	Sumimoto (Japan)
	Sumichiral OA-2000A	Sumimoto (Japan)
(*S*)-(dinitro-3,5-benzoyl)phenylglycine	Covalent L-Phenylglycin	Regis (USA)
	Chiral L-DPB	Serva (Germany)
	ES D-DNB-PHGLY	MN (Germany)
	Grom-Chiral-*S*-DNBPG-C	Grom (Germany)
	Grom-Chiral-*S*-DNBPG-I	Grom (Germany)
	Chiral-3	MN (Germany)
(*R*,*S*)-(dinitro-3,5-benzoyl)phenylglycine	Covalent D.L-Phenylglycin	Regis (USA)
(3*R*,4*R*)-4-(3.5-dinitrobenzamido)-1,2,3,4-THP	(*R.R*)-Whelk-O I	Regis (USA)
(3*S*,4*S*)-4-(3.5-dinitrobenzamido)-1,2,3,4-THP	(*S.S*)-Whelk-O I	Regis (USA)
(3*R*,4*R*)-4-(3.5-dinitrobenzamido)-1,2,3,4-THP-trifunct.	(*R.R*)-Whelk-O 2	Regis (USA)
(3*S*,4*S*)-4-(3.5-dinitrobenzamido)-1,2,3,4-THP-trifunct.	(*S.S*)-Whelk-O 2	Regis (USA)
(*S*)-(DNB) leucine	Covalent L-Leucine	Regis (USA)
	Ionic L-Leucine	Regis (USA)
	ES L-DNB-LEU	MN (Germany)
	Grom-Chiral-*S*-DNBL-C	Grom (Germany)
	Grom-Chiral-*S*-DNBL-I	Grom (Germany)
	Chiral L-DL	Serva (Germany)
	Bakerbond DNBLeu	Baker (Netherlands)

(*R*)-(DNB) leucine	Covalent D-Leucine	Regis (USA)
	ES D-DNB-LEU	MN (Germany)
	Grom-Chiral-*R*-DNBL-C	Grom (Germany)
	Grom-Chiral-*R*-DNBL-I	Grom (Germany)
	Chiral L-DDL	Serva (Germany)
(*S*)-valyl-phenylurea	Apex Prepsil L-valyl phenyl-urea	Jones (Switzerland)
(*R*)-naphtylalanine	Covalent D-naphtylalanine	Regis (USA)
(*S*)-naphtylalanine	Covalent L-naphtylalanine	Regis (USA)
rac-naphtylalanine	Covalent D,L-naphtylalanine	Regis (USA)
(*R*)-ou (*S*)-*N*-(DNB)tyrosine *n*-butyl-amide	ChyRoSine A	Serva (Germany)
(*S*)-ou (*S*)-*N*-(DNB)tyrosine [(naphtyl-1)-I ethyl]-amide	ChyRoSine A	Serva (Germany)
(*S*)-(DNB) phenylalanine	Chiraline	SFCC (France)
(*R*)-α-methylbenzylurea	Supelcosil-LC-®-Urea	Supelco (USA)
R-*N*-α-phenylethylurea	Grom-Chiral-U	Grom (Germany)
	Spherisorb Chiral-1	PhaseSep
Dimethyl *N*-3,5-DNB-amino-2,2-dimethyl-4-pentenyl phosp.	(*R*)-α-Burke 2	Regis (USA)
Dimethyl *N*-3,5-DNB-amino-2,2-dimethyl-4-pentenyl phosp.	(*S*)-α-Burke 2	Regis (USA)
Ruthenium complex	Ceramosphsere RU-2 Chiral	Phenomenex (USA)
3,5 dinitrobenzamide of β-lactam	Pirkle I-J	Regis (USA)
O,*O′*-bis(3,5-dimethylbenzoyl)-*N*,*N′*-diallyl-L-tartardiamide	Kromasil DMB	Eka Chem. (Sweden)
O,*O′*-bis(4-*tert*-butylbenzoyl)-*N*,*N′*-diallyl-L-tartardiamide	Kromasil TBB	Eka Chem. (Sweden)
N-3,5-DNB-3-amino-3-phenyl-2-(1,1-dimethylethyl)-prop.	(*R*,*R*)-β-Gem 1	Regis (USA)
N-3,5-DNB-3-amino-3-phenyl-2-(1,1-dimethylethyl)-prop.	(*S*,*S*)-β-Gem 1	Regis (USA)
(*S*)-(α-naphtyl-1)-ethylamine	Sumichiral OA-1000	Sumimoto (Japan)
(*R*)-phenylglycine and (*S*)-(Cl-4-phenyl)-4-isovaleric acid derivat.	Sumichiral OA-2100	Sumimoto (Japan)
(*R*)-phenylglycine and (1*R*,3*R*)-chrysanthemic acid derivat.	Sumichiral OA-2200	Sumimoto (Japan)
(*R*) ou (*S*)-(naphtyl-1) glycine (DNB) derivative	Sumichiral OA-2500	Sumimoto (Japan)
Tert-butylamine and (*S*)-valine derivatives	Sumichiral OA-3000	Sumimoto (Japan)
Dinitro-3,5 aniline and (*S*)-valine derivatives	Sumichiral OA-3100	Sumimoto (Japan)
Dinitro-3,5 aniline and (*S*)-*tert*-leucine derivatives	Sumichiral OA-3200	Sumimoto (Japan)
(*S*)-(α-naphtyl-1)-ethylamine and (*S*)-valine derivatives	Sumichiral OA-4000	Sumimoto (Japan)
(*R*)-(α-naphtyl-1)-ethylamine and (*S*)-valine derivatives	Sumichiral OA-4100	Sumimoto (Japan)
(*R*)-phenylglycine and (*R*)-(α-naphtyl-1)-ethylamine	Sumichiral OA-4200	Sumimoto (Japan)
(*R*)-phenylglycine and (*R*)-(α-naphtyl-1)-ethylamine	Sumichiral OA-4300	Sumimoto (Japan)
(*S*)-proline and (*S*)-(α-naphtyl-1)-ethylamine derivatives	Sumichiral OA-4400	Sumimoto (Japan)
(*S*)-proline and (*R*)-(α-naphtyl-1)-ethylamine derivatives	Sumichiral OA-4500	Sumimoto (Japan)
(*S*)-*tert*-leucine and (*S*)-(α-naphtyl-1)-ethylamine derivatives	Sumichiral OA-4600	Sumimoto (Japan)
(*S*)-*tert*-leucine and (*R*)-(α-naphtyl-1)-ethylamine derivatives	Sumichiral OA-4700	Sumimoto (Japan)

3. Type II A (Cyclodextrin-Bonded Phases)

Commercially available cyclodextrin-bonded phases are made of α-, β-, or γ-cyclodextrins grafted to silica (Table 2). Cyclodextrins (CD) are cyclic oligosaccharides of 6, 7, or 8 D-(+)-glucopyranose units, respectively, with the shape of a truncated cone. The hydrophobic cavity of cyclodextrins enables formation of inclusion complexes with apolar or moderately polar compounds of the correct size (host–guest chemistry). A secondary interaction can occur between the secondary hydroxyl groups at the entrance of the cavity and the solute. Stability of the inclusion complex is dependent on the hydrophobic and steric character of the analyte. These CSPs were developed by Armstrong and operate with hydroorganic mobile phases in a reversed-phase mode (Armstrong et al. 1988). Derivation of the hydroxyl groups of the CD enhances the number of stationary phases commercially available. Derivatives, such as acetyl-, hydroxypropyl-, and alkylcarbamate-cyclodextrins can be used with apolar or slightly polar mobile phases. Furthermore, functions located at the rim of the cavity induce different types of interactions with the analyte and/or can modify the cavity size entrance, thus altering enantioselectivity. These stationary phases present good stability and are cost effective. Several applications have been published in the literature for the stereoselective analysis of drugs in pharmaceutical formulations and biological matrices (Cyclobond 1992).

4. Type II B (Crown Ether)

Crown ethers are macrocyclic polyethers which can form selective complexes with inorganic cations such as alkali and alkaline-earth metal ions, as well as with organic cations such as protonated amino compounds. With the latter, complex stability depends on the molecular structure around the amino group. It is noteworthy that in 1974, Cram developed an LC method for the chiral separation of amino compounds with crown ethers (Cram and Cram 1974). Three types of stationary phases can be used: polymer matrices, bonded, and coated crown ethers on silica material. However, stationary phases with crown ethers bearing atropisomers (bi-napthyl derivatives) coated on silica are the most effective CSPs. These columns used with aqueous mobile phases are now commercially available (Table 2), but are limited to the separation of compounds with primary protonated amines such as amino acid enantiomers.

5. Type III A (Natural Polymers)

Polysaccharides derivatives of cellulose and amylose are certainly the most appropriate CSPs for the enantioselective separation of pharmaceutical compounds. Natural cellulose was recognized in 1948 as a potent chiral selector for the chiral separation of amino acids in paper chromatography. Since then, natural polymers have been replaced by derivatives with better mechanical

Table 2. Type II CSP (adapted from CECCATO et al. 1999)

Selector	Commercial name	Provider
α-Cyclodextrin	Cyclobond III	Astec (USA)
	α-Cyclodextrin	Serva (Germany)
Permethylated α-CD	Nucleodex α-PM	MN (Germany)
Acetylated α-cyclodextrin	Cyclobond III AC	Astec (USA)
α-Cyclodextrin	Chiradex	Merck (Germany)
	Cyclobond I 2000	Astec (USA)
	β-Cyclodextrin	Serva (Germany)
	Grom-chiral-Beta-CD	Grom (Germany)
	Chiral β-dex = Si 100	Serva (Germany)
	Nucleodex β-OH	MN (Germany)
	Sumichiral OA-7000	Sumimoto (Japan)
Acetylated β-cyclodextrin	Cyclobond I 2000 Ac	Astec (USA)
β-Cyclodextrin derived (*S*)-hydroxypropyl	Cyclobond I 2000 SP	Astec (USA)
β-CD derived hydroxypropyl (racemic)	Cyclobond I 2000 RSP	Astec (USA)
β-CD derived (*S*)-[1-(1-naphtyl)-1)-ethyl]carbamate	Cyclobond I 2000 SN	Astec (USA)
β-CD derived (*R*)-[1-(1-naphtyl)-1)-ethyl]carbamate	Cyclobond I 2000 RN	Astec (USA)
β-CD derived (Rac)-[1-(1-naphtyl)-1)-ethyl]carbamate	Cyclobond I RSN	Astec (USA)
β-CD derived 3,5-dimethylphenylcarbamate	Cyclobond I 2000 DMP	Astec (USA)
β-CD derived paratoluoyl	Cyclobond I PT	Astec (USA)
Permethylated β-CD	Nucleodex β-PM	MN (Germany)
Sulfonated β-CD	Cyclobond I 2000 SO	Astec (USA)
Carboxymethyl β-CD	Shodex Orpak CDBS	Showa Denko (Japan)
γ-Cyclodextrin	Cyclobond II	Astec (USA)
	ChiraDex	Merck (Germany)
γ-CD derived (*S*)-(α-naphtyl)-1 ethylcarbamate	Cyclobond II SN	Astec (USA)
Permethylated γ-CD	Nucleodex γ-PM	MN (Germany)
Acetylated γ-cyclodextrin	Cyclobond II AC	Astec (USA)
Crown ether with chiral barriers	Crownpak CR(+)	Daicel (Japan)

properties and different selectivity. The first derivative was triacetylated microcrystalline cellulose (HESSE and HAGEL 1973), which includes the solute into molecular cavities of the polymer network (cooperative stationary phases). Beside the inclusion process, the solute can also form other interactions with the selector. Later, in order to improve the mechanical properties of cellulose-based supports, ester and carbamate derivatives were prepared, coated on silica (Table 3) and commercialized under the trade names of ChiralCel (cellulose) and ChiralPak (amylose). Generally, these stationary phases are used with nonpolar mobile phases (e.g., hexane) modified with an alcohol. For example, isopropanol was demonstrated to be a modifier of choice to modulate selectivity. Numerous applications with this kind of selector in supercritical fluid chromatography have also been published (ROSSET et al. 1991). Recently, new cellulose- and amylose-coated silica materials have been commercialized for enantioselective separations in reversed-phase mode (Table 3). The application domain of these CSPs for pharmaceutical compounds is vast (DAICEL 1989), and can be extended to the quantitative determination of enantiomers in biological matrices (CECCATO et al. 1997).

6. Type III B (Synthetic Polymers)

Polyacrylamides and polymethacrylamides synthesized by BLASCHKE (BLASCHKE 1986) as well as poly(triphenylmethylmethacrylate) described by OKAMOTO (OKAMOTO et al. 1981) are particularly suitable for the separation of atropisomers (Table 3). These selectors are impregnated on silica and, as demonstrated with natural polymers, chiral recognition is mainly governed by an inclusion mechanism into asymmetric cavities. Such stationary phases are used in normal phase mode but offer only a limited scope of applications for pharmaceutical compounds.

Applications of molecularly imprinted polymers as chiral stationary phases have recently appeared in the literature (MAGES and MOSBACH 1997). This material involves arranging polymerizable functional monomers around a template and incorporating them by polymerization into a highly cross-linked macroporous polymer matrix. The template is finally eliminated. The polymer thus possesses binding sites with a specific shape and functional groups that can recognize a particular molecule. However, even if impressive enantioselectivities have been reported, the use of molecularly imprinted polymers as CSPs is relatively limited since efficiency is very low. This drawback will certainly be overcome in the near future and will allow for powerful chiral stationary phases.

7. Type IV (Proteins-Antibiotics)

Different proteins and enzymes such as bovine and human serum albumin (BSA and HSA, respectively), α_1-acid glycoprotein (AGP), ovomucoid (OVM), avidine and cellobiohydrolase (CBH), as well as trypsin, α-chymotrypsin and pepsin (ALLENMARK and ANDERSSON 1994) have been

Table 3. Type III CSP (adapted from Ceccato et al. 1999)

Selector	Commercial name	Provider
Microcrystalline cellulose triacetate (crude polymer)	Cellulose triacetate	Merck (Germany)
	Chiral Triacel	MN (Germany)
	Chiralcel CA-1	Daicel (Japan)
Cellulose triacetate	Chiralcel OA	Daicel (Japan)
Cellulose tribenzoate	Chiralcel OB	Daicel (Japan)
	Chiralcel OB-H	Daicel (Japan)
Cellulose triphenylcarbamate	Chiralcel OC	Daicel (Japan)
Cellulose tri-(3,5-dimethylphenylcarbamate)	Chiralcel OD	Daicel (Japan)
	Chiralcel OD-H	Daicel (Japan)
	Chiralcel OD-R	Daicel (Japan)
	Chiralcel OD-RH	Daicel (Japan)
Cellulose tri-(-4-chlorophenyl)carbamate	Chiralcel OF	Daicel (Japan)
Cellulose tri-(-4-méthylphenyl)carbamate	Chiralcel OG	Daicel (Japan)
Cellulose tri-(-4-methylbenzoate)	Chiralcel OJ	Daicel (Japan)
	Chiralcel OJ-R	Daicel (Japan)
Cellulose tri-cinnamate	Chiralcel OK	Daicel (Japan)
Amylose tri-(phenylethylamine)carbamate	Chiralcel AS	Daicel (Japan)
Amylose tri-(-3,5-dimethylphenyl)carbamate	Chiralcel AD	Daicel (Japan)
	Chiralcel AD-R	Daicel (Japan)
Poly(*N*-acryloyl-1-phenylalanine ethylester)	Chiraspher	Merck (Germany)
Poly(triphenylmethylmethacrylate)	Chiralpak OT (+)	Daicel (Japan)
Poly(pyridyl-2 diphenylmethylmethacrylate)	Chiralpak OP (+)	Daicel (Japan)
Proline	Chiral ProCu = Si 100	Serva (Germany)
	Grom-Chiral-PC	Grom (Germany)
Hydroxyproline	Chiral HyproCu = Si 100	Serva (Germany)
	Grom-Chiral-HP	Grom (Germany)
Valine	Chiral Val Cu-Si 100	Serva (Germany)
	Grom-Chiral-VC	Grom (Germany)
N-propyl hydroxyproline	Nucleosil chiral-1	MN (Germany)
	Chiralpak WM	Daicel (Japan)
	Chiralpak WH	Daicel (Japan)
(1*R*,2*S*)-2-carboxymethylamino-1,2-diphenyl-ethanol	Chiralpak WE	Daicel (Japan)
Proline derivatives	Chiralgel L-prolinamide	MN (Germany)
	Grom-Chiral-P	Grom (Germany)
Valine derivatives	Chiralgel L-valinamide	MN (Germany)
Phenylalanine derivatives	Chiralgel phenylalinamide	MN (Germany)
1,2-Diphenylethyldiamine	ULMO	Régis (USA)
D-Penicillamine	Sumichiral OA-5000	Sumimoto (Japan)

immobilized on silica (Table 4). It is well known that proteins form very strong and stereoselective complexes with several compounds of pharmaceutical interest where different interactions such as ion exchange, hydrogen bonding, host–guest inclusion, and steric contributions are involved. However, the importance of these interactions depends on the protein structure, which varies depending on pH and mobile phase. Therefore, the chiral recognition mechanism is very complex and enantioselectivity is difficult to predict.

Protein-based columns are used in reversed-phase mode to separate a very broad range of chiral compounds (acidic, basic, and neutral compounds). Enantioselectivity is regulated by the pH of mobile phase buffer concentration, and the nature and content of the organic modifier. In comparison with polysaccharide-based columns, protein-based columns lack ruggedness when biological matrices are injected. Thus, special precautions have to be taken to periodically regenerate the column. Also, capacity of these CSPs is limited for separative purposes because of the low surface coverage of proteins on supports.

The use of macrocyclic antibiotics as a new class of chiral selectors was introduced by ARMSTRONG et al. in 1994 (ARMSTRONG et al. 1994a). These selectors are covalently bonded to silica and two phases have been made commercially available, Chirobiotic V (vancomycin) and Chirobiotic T (teicoplanine), as reported in Table 4. Vancomycin is an amphoteric glycopeptide which contains 18 chiral centers surrounding three cavities while teicoplanine contains 23 chiral centers surrounding four cavities. These selectors with hydroxyl, amine and amido groups, aromatic moieties, and sugar units have a recognition mechanism that has not yet been elucidated. However, π-π complexation, hydrogen bonding, dipole stacking, and electrostatic interactions are involved in the retention mechanism as already reported for protein-based columns. These CSPs can be used in normal- and reversed-phase modes as well as in polar organic mode. These three different modes have different selectivity and a large number of applications have been reported with these two columns (CHIROBIOTIK 1996).

C. Capillary Electrophoresis

Since its introduction in the 1980s, capillary electrophoresis (CE) has rapidly become a powerful separation technique and has found applications in a number of different fields such as environment, clinical chemistry, and biochemistry. In particular, CE has revealed great potential for the analysis of pharmaceutical compounds. High efficiency, short analysis time, rapid method development, simple instrumentation and low sample requirement are the main reasons for this success. In addition, exotic and expensive background electrolyte solutions can be used with fewer economical and disposal problems.

Separation by electrophoresis is obtained by differential migration of solutes in an electrical field. In CE, separation is performed in narrow-bore

Table 4. Type IV CSP (adapted from Ceccato et al. 1999)

Selector	Commercial name	Provider
Bovine serum albumin (BSA)	Resolvosil BSA-7	MN (Germany)
	Resolvosil BSA-7PX	MN (Germany)
	BSA Column	SFCC (France)
Human serum albumin (HSA)	Chiral protein-2	SFCC (France)
		Astec (USA)
	Chiral HSA	Chromtech AB (Sweden)
	Chiral HSA HPLC column	Astec (USA)
α_1-Glycoprotein acid	Enantiopac	Pharmacia LKB (Sweden)
	Chiral AGP	Chromtech AB (Sweden)
	Chiral AGP HPLC column	Astec (USA)
Ovomucoïd (ovoglycoproteine)	Ultron E*S*-OVM	Shinwa (Japan)
Cellobiohydrolase	Chiral CBH	Chromtech AB (Sweden)
	Chiral CBH HPLC column	Astec (USA)
Avidine	Bioptic AV-1	GL Sciences (USA)
Vancomycin	Chirobiotic V	Astec (USA)
Teicoplanin	Chirobiotic T	Astec (USA)
Teicoplanin aglycone	Chirobiotic TAG	Astec (USA)
Ristocetin	Chirobiotic R	Astec (USA)
Avoparcin	Chirobiotic A[a]	Astec (USA)

[a] To be commercialized.

capillaries which are usually filled only with a buffer solution. In contrast to conventional gel electrophoresis, CE presents numerous advantages, particularly with respect to the detrimental effect of Joule heating. The high electrical resistance of the capillary enables application of very high electrical fields (100–500 V/cm) with only minimal heat generation. Moreover, the large surface-to-volume ratio of the capillary efficiently dissipates the generated heat. Application of high electrical fields results in short analysis time as well as high efficiency and resolution. In addition, minimal sample requirements (1–10 nl) and the overall simplicity of instrumentation (Fig. 3) give CE numerous advantages vis-à-vis conventional separation techniques such as chromatography (Grossman and Colburn 1992; Heiger 1992; Landers and Spelsberg 1994).

In CE, transport of analytes is due to two phenomena, electrophoretic migration and electrophoresis which occur in the so-called electroosmotic flow (EOF). A unique feature of EOF in the capillary is the flat profile of the flow. Since the driving force of the flow is uniformly distributed along the capillary there is no pressure drop within the capillary and the flow remains nearly constant throughout. The flat flow profile is beneficial since it avoids contributions directly to the dispersion of the solute zone in contrast to the laminar flow profile observed in pressure-driven techniques such as liquid chromatography. As a result of the EOF and as a function of electrophoretic and electroosmotic mobilities, all sample components, be they cations, neutrals, or anions, are drawn towards the cathode and the detector, although with different velocities. The major limitation of CE is perhaps the fact that relatively high analyte concentrations are required. It follows that the major improvements to exist-

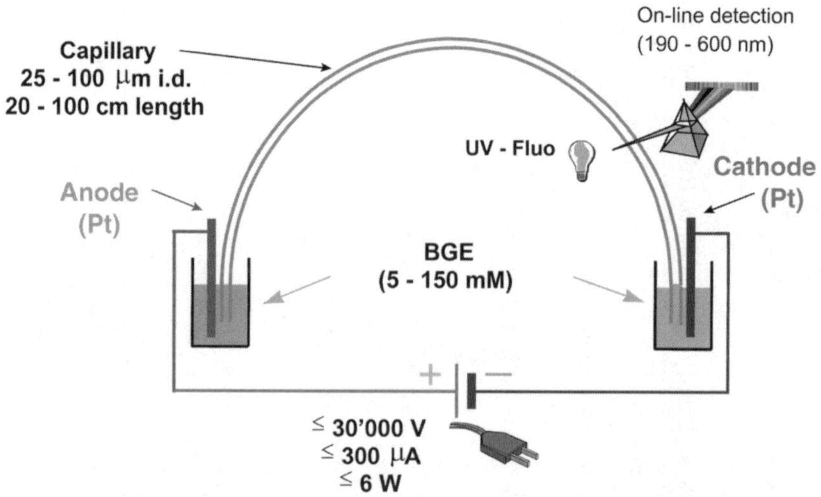

Fig. 3. Instrumental set-up of CE

ing methodologies will reside in improving the detection systems, e.g., by interfacing CE and mass spectrometry (KEMP 1998).

I. Introduction

As mentioned above, CE enantioseparation is generally performed in the direct separation mode, requiring addition of a chiral selector to the background electrolyte (BGE). The most frequently applied CE techniques for chiral separations are capillary zone electrophoresis (CZE), micellar electrokinetic chromatography (MEKC) with addition of a chiral selector, and capillary electrochromatography (CEC) (GÜBITZ and SCHMID 1997; AMINI 2001). In all cases, the chiral selector creates a stereoselective environment which enables the separation of the two enantiomers. One of the most attractive advantages of CZE and MEKC for enantiomer separation is that the separation media can easily be modified during method development to screen the best stereoselective environment and find optimum operating conditions at low cost according to the minimal amount of additives and solvents required.

II. Capillary Zone Electrophoresis

1. Cyclodextrins

Among the great number of chiral selectors reported in the literature, cyclodextrins (CDs) are now the most widely used in CZE. Neutral and charged CD derivatives, with various functional groups, have been developed to induce different stereoselective interactions and enhance enantioselectivity and solubility. The use of CDs in CE was first reported by TERABE for the separation of positional isomers (TERABE et al. 1985). Later, GUTTMAN reported on chiral CE separations via CD incorporated within polyacrilamide gel-filled columns (GUTTMAN et al. 1988) and FANALI obtained a chiral separation in CZE by simply adding CD to the BGE (FANALI 1989).

More than 700 analytes have been resolved using CDs as chiral selectors, including various pharmaceutical compounds such as β-blockers, sympathomimetics, antipsychotics, antidepressants, hypnotics, barbiturates, locals anaesthetics, bronchodilators, and anticoagulants (GÜBITZ and SCHMID 1997). Several parameters such as the chiral selector type and its concentration, pH, nature of the background electrolyte and its ionic strength, as well as the addition of organic modifiers or other additives have an important influence on resolution. WREN and ROWE postulated a separation model to show that an optimum CD concentration exists (WREN and ROWE 1992a,b, 1993). Other separation models were proposed by RAWJEE (RAWJEE et al. 1993a,b; RAWJEE and VIGH 1994). Most commercially available CDs are a mixture of randomly substituted components with different degrees of substitution and regiochemistry. The degree of substitution is an important aspect in chiral selectivity and thus in the resolution of chiral compounds (VALKÓ et al. 1994; FANALI and ATURKI 1995; WESELOH et al. 1995). A detailed characterization of these derivatives is of key importance for the validation, optimization, and standardization of chiral separations (CHANKVETADZE et al. 1996a)

Charged CDs have been used for both charged and uncharged analytes. The advantages of charged CD derivatives are better solubility and ability to display additional electrostatic interactions. In the ionized state, they migrate with their own electrophoretic mobility. In fact, as already shown by several authors (DETTE et al. 1994; QUANG and KHALEDI 1995; DESIDERIO and FANALI 1995), a chiral selector with a mobility opposite to that of the analyte shows relevant resolving power even at very low concentrations. The use of a charged CD at the lowest possible concentration with low analyte concentration and high buffer concentration is recommended to avoid electrophoretic dispersion of the peaks. However, the higher conductivity of the separation medium (higher Joule heating), with charged CDs instead of neutral CDs, may limit their application in specific approaches. Cationic CDs such as mono(6-β-aminoethylamino-6-deoxy)-β-CD were first introduced by TERABE (TERABE 1989). Different amino substituted-CDs were screened by NARDI et al. (NARDI et al. 1993) for the separation of several acidic compounds. The introduction of mono(6-amino-6-deoxy)-β-CD by LELIÈVRE et al. (LELIÈVRE et al. 1997) demonstrated the potential of this selector compared to neutral CDs for the separation of solutes bearing a carboxylic function. Quaternary ammonium β-CDs were also chosen as chiral selectors with the advantage that these derivatives are positively charged whatever the pH (JAKUBETZ et al. 1997; CHANKVETADZE et al. 1996b). Anionic CDs such as carboxymethyl-β-CD (CM-β-CD), carboxyethyl-β-CD (CE-β-CD), succinyl-β-CD (Succ-β-CD), sulfobutyl-β-CD (SBE-β-CD), sulfoethyl-β-CD (SEE-β-CD), phosphated and sulfated CD have been applied as chiral selector to resolve cationic, neutral, and anionic compounds. As mentioned above, for positively charged compounds, mobility difference between the free and complexed analyte is increased with negatively charged chiral selectors, often leading to higher resolution than with neutral cyclodextrins, according to the theoretical approach of WREN and ROWE (WREN and ROWE 1992a). Even if the number of known anionic CD derivatives is lower than that of cationic CDs, a great number of anionic CDs are now commercially available.

2. Macrocyclic Antibiotics

Recently ARMSTRONG and coworkers first demonstrated macrocyclic antibiotics to be a broad, useful new class of chiral selectors in CE, HPLC, and thin-layer chromatography for the separation of a wide variety of enantiomers (ARMSTRONG et al. 1994b,c; ARMSTRONG and ZHOU 1994d; ARMSTRONG and NAIR 1997). These compounds contain multiple stereogenic centers and a variety of functional groups. Most macrocyclic antibiotics are ionisable (they can be used in either charged or uncharged states) and are sufficiently soluble in aqueous buffers and solvents such as those used in CE. These selectors are structurally diverse and complementary in the type of compounds they can resolve. The glycopeptide antibiotics ristocetin A, teicoplanine, and vancomycin consist of an aglyconportion of fused macrocyclic rings which form a characteristic "basket" shape, and several carbohydrate moieties. Although they have structural similarities, glycopeptide antibiotics are complementary to one another in their enantioselectivities. Armstrong suggested that if only a partial enantioresolution can be obtained with one glycopeptide, there is a high probability that a baseline or better separation may be obtained with another glycopeptide (GASPER et al. 1996). The most versatile seems to be ristocetin A, followed by vancomycin and teicoplanin, respectively (WARD and OSWALD 1997). Ansamycins, rifamycin B, and rifamycin SV have a characteristic ansa structure, which includes a ring structure or chromophore spanned by an aliphatic chain. Rifamycin B is normally negatively charged and can resolve cationic compounds which makes it complementary to glycopeptide antibiotics which are better suited to resolve anionic compounds. Recently, avoparcin, a new macrocyclic glycopeptide antibiotic, was introduced for the CE separation of many *N*-blocked amino acids, as well as several nonsteroidal antiinflammatory drugs (EKBORG-OTT et al. 1999). The strongly absorbing nature of these selectors in the UV region of the spectrum does not allow a direct detection of most solutes. In order to achieve separations with this type of chiral selector, indirect detection or the partial filling counter-current technique can be used.

3. Crown Ethers

Crown ethers are macrocyclic polyethers, synthesized in 1967, which form stable inclusion complexes with alkali, alkaline soil and primary ammonium cations (PEDERSEN 1967). A more advanced generation of crown ethers derivatized with carboxylic groups (18-crown-6-tetracarboxylic acid) was introduced by CRAM et al. (KYBA et al. 1978). The cyclic polyether forms a cavity where the analyte can penetrate with its hydrophilic part and interact with oxygen atoms. Because this complexation does not exhibit chiral recognition, additional lateral interactions between the four carboxylic acids and the analyte, as well as electrostatic interactions such as hydrogen bonds or coulometric attraction or repulsion forces with polar substituents of the analytes, are necessary for the discrimination of enantiomers. To date, more than 100 different chiral amines have been separated with crown ethers (KUHN 1999). Interesting works have been undertaken on chiral separation by dual complex formation with these selectors (KUHN et al. 1994; ARMSTRONG et al. 1998) (see Sect. C.II.7., "Dual System").

4. Proteins

A wide variety of protein selectors have been employed in CE. The vast majority has demonstrated selectivity towards enantiomers of small molecules, mostly drugs. pH is a fundamental parameter to optimize, because proteins are composed of several amino acids either positively or negatively charged according to their p*I*. pH modulates the charge of both analytes and chiral selector involving attractive and/or repulsive interactions. The major drawback of this chiral selector is its absorption in the low UV range and the adsorption onto the fused silica capillary wall. Both problems can be circumvented by applying the partial filling method and a coated capillary. To date, several proteins, including BSA, AGP, avidin, human albumin, ovomucoid, and cellobiohydrolase I have been used in CE for enantiomeric separation (LLOYD et al. 1997).

5. Polysaccharides

Recently, oligosaccharides and polysaccharides have been found effective as chiral selectors in CE. When electrically neutral polysaccharides such as dextrin, dextran, or amylose are chosen, only ionic enantiomers can be separated (CZE mode). On the other hand, electrically neutral enantiomers as well as ionic compounds can be separated with charged polysaccharides such as heparin, condroitin sulfate, or dextran sulfate. The chiral recognition mechanism of this type of selector is still not clear, but previous investigations by HPLC with polysaccharide-coated chiral stationary phases have shown that hydrophobic interactions, hydrogen bonding, and dipole–dipole interactions play an important role (YASHIMA and OKAMOTO 1995). Macroscopic structure (helical) or polymer network also contribute to enantio-recognition (OKAMOTO and KAIDA 1994). In particular, mucopolysaccharides such as heparin and chondroitin sulfates (which are natural ionic polysaccharides) and dextrins showed a wide enantioselectivity. Enantioseparation was also successful with CD polymers which can be regarded as a type of polysaccharide. In CE, two CD polymers have been employed, namely, CM-β-CD and neutral β-CD (INGELSE et al. 1995; ATURKI and FANALI 1994; NISHI et al. 1994; FANALI 1994).

6. Other Chiral Selectors

Ergot alkaloids were first used as chiral selectors by INGELSE et al. (INGELSE et al. 1996) for a number of racemic acidic compounds such as mandelic acid and racemic herbicides. Other chiral separations employ particular selectors such as Calix-4-arene (PENA et al. 1997) (SHOHAT and GRUSHKA 1994) or ligand-exchange complexation (GASSMANN et al. 1985) but an extensive review of the numerous selectors of CE is outside the scope of the present chapter.

7. Dual System

The simultaneous use of different selectors in BGE is an attractive feature of CE. As recently reported, the combination of two CDs, e.g., charged–

uncharged, may be advantageous in CE in order to achieve and/or improve chiral resolution. Separation selectivity in chiral CE with mixed CDs was recently reviewed (LURIE 1997). Dual systems other than the CDs, such as a mixture of taurodeoxycholic acid and β-CD (TERABE et al. 1993), polymerized chiral micelle and γ-CD (WANG and WARNER 1995), and chiral crown ether and hydroxypropyl-β CD (HP-β-CD) (KUHN et al. 1994) have also been successful for the enantiomeric separation of several drugs. Due to the additional interaction processes involved in enantioselective discrimination, the optimization of chiral separation in a dual system is generally a complex process, but can solve a number of particularly different stereoselective separations.

8. Migration Order Reversal

The possibility of reversing the migration order of enantiomers is an important development in determining enantiomeric impurities. This reversal can be achieved in different ways. Firstly, as in LC, the chiral environment can be reversed but this method is only possible with synthetic products and not with CDs which naturally occur in one enantiomeric form only. The most widely used method in CE is therefore EOF reversal by addition of long chain cationic surfactants, cationic polyamides, or immobilized cationic coating. The third method to reverse migration order is selecting different CDs as chiral selectors, because of the different separation and complexation mechanisms that can occur (VERLEYSEN and SANDRA 1998). Another possibility is to use chargeable CDs, such as CMB-β-CD. Enantiomer migration time can also be inverted simply by varying the pH of the buffer, as demonstrated by SCHMITT and ENGELHART for the separation of ephedrine (SCHMITT and ENGELHARDT 1995).

9. Chiral Separation in NACE

Because of its high efficiency and extremely high selectivity, nonaqueous capillary electrophoresis (NACE) has evolved as an interesting alternative to aqueous CE for the separation of closely related compounds (e.g., in drug metabolism as well as in drug purity studies). In fact, because of their different physical and chemical properties (dielectric constant, viscosity, autoprotolysis constant, polarity, volatility, etc.), organic solvents allow easy manipulation of separation performances, including selectivity, efficiency, and separation time. By selecting a suitable medium, a wide range of solutes as well as chiral selectors such as β-CD become soluble. Applications using quaternary ammonium β-CD have demonstrated the potential of nonaqueous media to separate profens (WANG and KHALEDI 1998). Heptakis-β-CD was used for a variety of weak base analytes with limited solubility in aqueous background electrolytes (BJORNSDOTTIR et al. 1996). Recently, quinine and carbamoylated quinine were found to be highly stereoselective for separation of *N*-protected amino acids (PIETTE et al. 1999).

10. Micellar Electrokinetic Chromatography

Micellar electrokinetic chromatography (MECK or MECC) was first introduced by TERABE for the separation of uncharged as well as charged compounds and is based on the repartition of analytes between two phases (TERABE et al. 1984). The MECK separation principle consists of adding surfactants in BGE which play the role of a pseudo-stationary phase. At concentrations above the critical micelle concentration (CMC), aggregates of individual surfactant molecules, called micelles, are formed. Surfactants and thus micelles are usually charged and migrate electrophoretically as a solute in CZE. During migration, micelles can interact with solutes in a chromatographic manner through both hydrophobic and electrostatic interactions.

Many surfactants are commercially available but only some are chiral. Amino acid-derived synthetic surfactants, bile salts, an abundant source of chiral surfactants, and a number of others have been successfully used for enantiomer separation, including electrically neutral compounds (TERABE et al. 1994). Another possibility is to employ CD together with ionic micelles, such as sodium dodecyl sulfate (SDS). This feature was originally developed to separate highly hydrophobic compounds which are almost totally incorporated into the micelles and cannot be separated (TERABE et al. 1990). Addition of CD to MECK solution changes the apparent distribution coefficients between micellar and nonmicellar phases, because CD can include hydrophobic compounds in its cavity, thus increasing the apparent solubility of the aqueous phase. This approach, applied to chiral separations in microfabricated channels etched in a glass substrate, demonstrated that stereoselective analyses can be achieved in microchip devices with high speed and high resolution (RODRIGUEZ et al. 2000).

11. Capillary Electrochromatography

Capillary electrochromatography (CEC) is a hybrid of electrophoresis and high-performance liquid chromatography. Silica-based particles are packed into capillaries generally used in CZE, and CEC separation results from both the effect of electrophoresis and partitioning (DITTMANN et al. 1995). This technique, which combines the advantages of CE (i.e., separation efficiency) with those of LC (i.e., well-characterized retention), has recently generated a great interest. The main operating problem with CEC is the preparation of frit, required to prevent the packed bed movement under the effect of the EOF or pressure. The frit introduces some inhomogeneities, inducing an additional source of band broadening and favors bubble formation within the column, leading to the breakdown of current and EOF (ALTRIA et al. 1997). Applications of CEC appears to be similar to LC and CE, including determination of impurities, main components assay and, of course, chiral separation. For the latter, this technique can be realized in two basic modes: CEC in wall-coated open tubular capillaries (Mayer and Schurig 1992; Francotte and Jung 1996; Vindevogel and Sandra 1994) and CEC in packed with CSPs (Li and Lloyd 1993; Lelièvre et al. 1996). Brush-type, proteins, macrocyclic antibiotic, CDs, natural and molecular imprint-based polymer have already been immobilized as chiral selectors in CEC. One important advantage of CEC over CE is the possibility of using chiral selectors insoluble in common BGE or which possess a detector response.

12. CE–MS Coupling

UV-VIS spectrophotometry is often chosen for the on-line detection of compounds separated by CE. However, the CE-UV bottleneck is its relatively low sensitivity due to the short optical path length afforded by the small internal diameter of capillaries. Additionally, many compounds do not possess a chromophore and, therefore, their UV detection requires a derivatization procedure. Moreover, with spectroscopic detectors, peak identity is generally confirmed by migration times only. However, this information is often insufficient to identify unequivocally compounds of interest because the electroosmotic flow can vary. In this context, the coupling of capillary electrophoresis with mass spectrometry is a powerful technique which opens new development perspectives in chemical analysis (Banks 1997; Johansson et al. 1991; Smith et al. 1993; Cai and Henion 1995).

The on-line coupling of CE with MS is a promising combination of two analytical techniques. Among the available interfaces, electrospray ionization (ESI) is the most widely used for on-line coupling of CE with MS (Wheat et al. 1997; Bruins 1998; Wahl and Schmith 1994). Because of its instrumental simplicity at the capillary outlet and of its possibility to enhance ionization by chemical reaction, the coaxial sheath-flow interface is by far the most popular and suitable configuration for CE-ESI-MS. It consists of adding a make-up flow to attain optimal conditions of electrospray ionization (1–10 μl/min). This coupling induces some limitations concerning the choice of the background electrolyte solution. Indeed, nonvolatile buffers commonly used in CE, such as phosphate, borate and citrate are not compatible with MS, although the use of a make-up flow containing 50%–80% organic solvent may encompass this incompatibility. These nonvolatile buffers may have detrimental effects on the performance of the mass spectrometer, since they can enhance contamination risk of the ionization chamber and suppress the analyte signal. Therefore, the use of volatile buffers such as ammonium acetate and ammonium formate or nonaqueous media are often recommended to achieve CE-ESI-MS experiments (Tomlinson et al. 1994). In addition, additives such as surfactants, cyclodextrins or ion-pairing agents, necessary for selectivity and stereoselective separations in CE, are not suitable for ESI-MS. Therefore, the partial filling counter-current technique is generally used to overcome this problem.

13. Partial Filling Counter-Current Technique

This technique was first introduced by Valtcheva to perform separations with CBH as chiral selector (Valtcheva et al. 1993) and was largely used with other proteins (Tanaka and Terabe 1997), macrocyclic antibiotics (Fanali and Desiderio 1996), and CDs (Amini et al. 1997). More recently, this method was applied in chiral CE-MS determination to avoid the presence of a chiral selector in the MS detector ion source, which can have a detrimental effect on method sensitivity (Schulte et al. 1998; Tanaka et al. 1998; Jäverfalk et al. 1998). This technique involves filling a discrete portion of the capillary with a background electrolyte containing a suitable amount of chiral selector for enantiomeric separation (partial filling). Generally, a coated capillary is used to reduce the EOF (Cherkaoui et al. 2001; Rudaz et al. 2001). As mentioned before, enhanced enantioselectivities can be achieved when electromigrations of the chiral selector and chiral analytes are opposite to one another (counter-current) (Ward et al. 1995) (Fig. 4).

Negatively charged CDs are used for basic compounds. The electrical field then makes the chiral selector and the enantiomers migrate in opposite directions. The potential of CE-ESI-MS combined with this technique was demonstrated in the stereoselective analysis of chiral drugs in biological matrices, such as tramadol and its phase I metabolites, as shown in Fig. 5. In spite of the peak overlapping observed in the reconstructed ion electropherogram (RIC), the recording of selected masses allowed unambiguous determination of each analyte, which demonstrates the high selectivity of MS compared to conven-

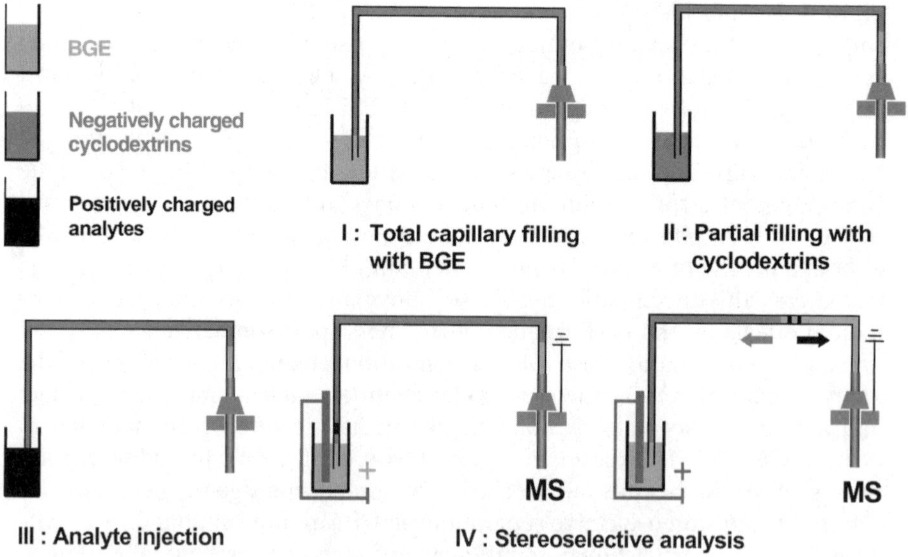

Fig. 4. Schematic presentation of the partial filling method

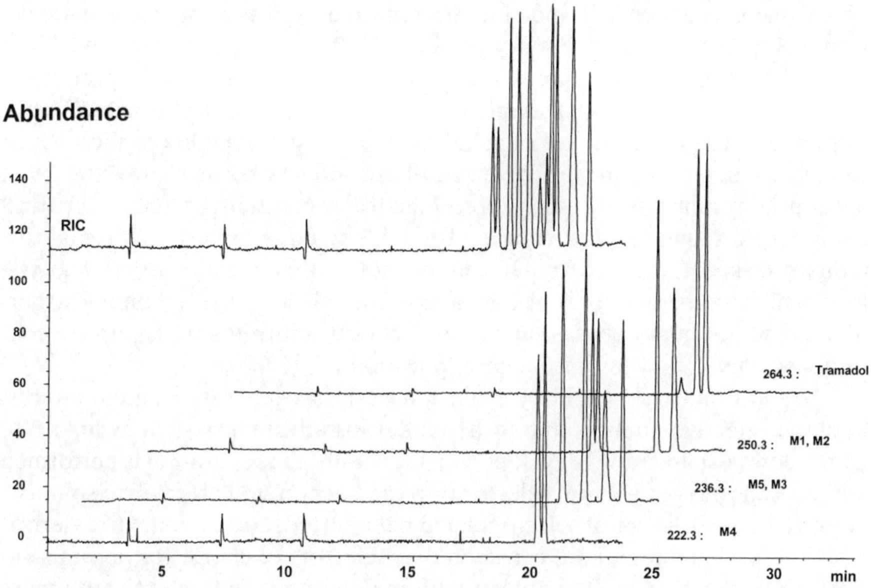

Fig. 5. CE-ESI-MS enantioseparation of tramadol and its phase I metabolites (*M1–M5*) in the presence of a negatively charged chiral selector. Experimental conditions – CE conditions: running buffer, 40 mM ammonium acetate at pH 4 in the presence of SBE-β-CD (2.5 mg/ml); PVA-coated capillary 70 cm × 50 μm ID.; partial filling of the capillary (90%); sample concentration, 100 ng/ml; pressure injection, 50 mbar for 10 s; applied voltage, 30 kV; temperature, 20°C. MS conditions: SIM positive ion mode (4 ions); capillary voltage, 3 kV; fragmentor, 70 V; drying gas N_2 flow and temperature, 6 l/min and 150°C; nebulizer pressure 4 p.s.i; sheath liquid, 0.5% formic acid in water-isopropanol (50/50, v/v); sheath flow, 3 μl/min. *RIC*, reconstructed electropherogram. (Adapted from RUDAZ et al. 2000)

tional detectors. Acquisition in the SIM mode allowed enhancement of both sensitivity and selectivity. All compounds were baseline-resolved within reasonable analysis time (RUDAZ et al. 2000).

D. Discussion and Conclusion

The increased need of stereoselective analyses has induced a tremendous development of analytical techniques for resolving enantiomers. Among these techniques, liquid chromatography (LC) and more recently capillary electrophoresis (CE) are recognized as methods of choice for the chiral separation of pharmaceutical compounds.

LC is particularly suited to determine the enantiomeric purity of a drug substance as well as its enantiomeric ratio in biological matrices for pharmacological studies. Indeed, LC is mature, reliable, efficient, and easily automated, and affords good sensitivity since several detectors can be used such as UV-visible, fluorimetric and electrochemical detectors, as well as mass spectrometry (MS). Chiral stationary phases (CSP) are generally employed and more than a hundred columns are now commercially available permitting direct resolution of a great number of enantiomers. Natural polymers such as polysaccharide derivatives of cellulose and amylose coated on silica, which can be used in normal and reversed-phase modes, are certainly the most appropriate supports for the stereoselective separation of pharmaceutical compounds. However, the choice of a CSP is not obvious and remains a tedious task for the analyst. It can be noted that a database (CHIRBASE; http://chirbase.u-3mrs.fr/) is also available for selecting the column. Furthermore, new stationary phases such as molecularly imprinted polymers are continuously developed for particular applications.

In addition to LC, capillary electrophoresis has generated a great interest in stereoselective analysis due to its numerous advantages such as high efficiency, low cost and rapid development. Generally, the separation is performed with the addition of a chiral selector into the background electrolyte solution. A large number of chiral selectors have been tested, and cyclodextrin derivatives (neutral and ionic) are considered as selectors of choice for pharmaceutical compounds. Capillary electrochromatography with chiral stationary phases is also developed but to a lesser extent. It can be noted that its lack of sensitivity remains the methods bottleneck of CE for biological matrices. However, coupling of CE with mass spectrometry can overcome this drawback and the partial filling technique is particularly adapted with chiral selectors added into the electrolyte solution. There is still considerable interest in the development of new chiral selectors to enlarge the field of stereoselective separation in capillary electrophoresis. With the success of cyclodextrins as chiral selector in CE, it is difficult to predict the future role of these new additives.

List of Abbreviations

ACN	acetonitrile
AGP	α_1-acid glycoprotein
BGE	background electrolyte
BSA	bovine serum albumin
CBH-I	cellobiohydrolase I
CD	cyclodextrins
CD-MEKC	cyclodextrin-modified electrokinetic micellar chromatography
CE	capillary electrophoresis
CEC	capillary electrochromatography
CE-β-CD	carboxyethyl-β-cyclodextrin
CM-β-CD	carboxymethyl-β-cyclodextrin
CMC	critical micelle concentration
CSP	chiral stationary phase
CZE	capillary zone electrophoresis
EOF	electroosmotic flow
ESI	electrospray ionization
HPLC	high-performance liquid chromatography

HP-β-CD	hydroxypropyl-β cyclodextrin
HE-β-CD	hydroxyethyl-β-cyclodextrin
HSA	human serum albumin
LC	liquid chromatography
MEKC	micellar electrokinetic chromatography
NSAI	non-steroids anti-inflammatory
OVM	ovomucoid
RIC	reconstructed ion monitoring
SBE-β-CD	sulfobutyl-β-cyclodextrine
SDS	sodium dodecyl sulfate
SEE-β-CD	sulfoethyl-β-CD
SFC	supercritical fluid chromatography
SIM	single ion monitoring
Succ-β-CD	succinyl-β-cyclodextrin

References

Adamovics JA (1997) Chromatographic analysis of pharmaceuticals. Marcel Dekker, New York

Allenmark S (1991) Chromatographic Enantioseparation, methods and applications. 2, Ellis Horwood, New York

Allenmark S, Andersson S (1994) Proteins and peptides as chiral selectors in liquid chromatography. J Chromatogr 666:167–179

Altria KD, Smith NW, Turnbull CH (1997) A review of the current status of CEC technology and applications. Chromatographia 46:664–674

Amini A, Pettersson C, Westerlund D (1997) Enantioresolution of disopyramide by capillary affinity electrokinetic chromatography with human α1-acid glycoprotein (AGP) as chiral selector applying a partial filling technique. Electrophoresis 18: 950–957

Amini A (2001) Recent developments in chiral capillary electrophoresis and applications of this technique to parmaceutical and biomedical anaéysis. Electrophoresis 22:310–3130

Anton K, Eppinger J, Frederiksen L, Francotte E, Berger TA, Wilson WH (1994) Chiral separations by packed-column super- and subcritical fluid chromatography. J Chromatogr A 666:395–401

Armstrong DW, Han YI, Han SM (1988) LC resolution of enantiomers containing single aromatic rings with β-cyclodextrin-bonded phases. Anal Chim Acta 208: 275–281

Armstrong DW, Tang YB, Chen SS, Zhou YW, Bagwill C, Chen JR (1994a) Macrocyclic antibiotics as a new class of chiral selectors for liquid chromatography. Anal Chem 66:1473–1484

Armstrong DW, Rundlett K, Reid GL (1994b) Use of a macrocyclic antibiotic, rifamycin b, and indirect detection for the resolution of racemic amino alcohols by CE. Anal Chem 66:1690–1695

Armstrong DW, Rundlett KL, Chen J-R (1994c) Evaluation of the macrocyclic antibiotic vancomycin as a chiral selector for CE. Chirality 6:496–509

Armstrong DW, Zhou Y (1994d) Use of a macrocyclic antibiotic as chiral selector for enantiomeric separations by TLC. J Liq Chromatogr 17:1695–1707

Armstrong DW, Nair UB (1997) Capillary electrophoretic enantioseparations using macrocyclic antibiotics as chiral selectors. Electrophoresis 18:2331–2342

Armstrong DW, Chang LW, Chang SS (1998) Mechanism of capillary electrophoresis enantioseparations using a combination of an achiral crown ether plus cyclodextrins. J Chromatogr A 793:115–134

Aturki Z, Fanali S (1994) Use of β-CD polymer as a chiral selector in CE. J Chromatogr A 680(1):137–146

Banks JF (1997) Recent advances in CE / ESI / MS. Electrophoresis 18:2255–2266
Bjornsdottir I, Hansson SH, Terabe S (1996) Chiral separation in nonaqueous media by capillary electrophoresis using the ion-pair principle J Chromatogr A 475:37–44
Blaschke G (1986) Chromatographic resolution of chiral drugs on polyamides and cellulose triacetate. J Liq Chromatogr 9:341–368
Bojarski J, Aboul-Enein H (1996) Recent application of chromatographic resolution of enantiomers in pharmaceuticals analysis. Biomed Chromatogr 10:297–302
Bruins AP (1998) Mechanistic aspects of electrospray ionization. J Chromatogr A 794: 345–357
Cai J, Henion JD (1995) Capillary electrophoresis-mass spectrometry. J Chromatogr A 703:667–692
Caude M, Bargmann-Leyder N (1993) Séparations chirales par chromatographie en phases liquide, supercritique et gazeuse. Techniques de l'ingénieur 10:1470/ 1–1470/25
Ceccato A, Chiap P, Hubert Ph, Crommen J (1997) Automated determination of tramadol enantiomers in human plasma using solid phase extraction in combination with chiral liquid chromatography J Chromatogr B 698:161–170
Ceccato A, Hubert Ph, Crommen J (1999) Analyse de médicaments enantiomériques en chromatographie liquide STP Pharma Pratiques 9:295–310
Chankvetadze B, Endresz G, Blaschke G, Juza M, Jakubetz H, Schurig V (1996a) Analysis of charged cyclomalto-oligosaccharides (CD) derivatives by ion-spray, matrix-assisted laser-desorption/ionization time-of-flight and fast atom bombardment mass spectrometry, and by capillary electrophoresis. Carbohydr Res 287: 139–155
Chankvetadze B, Endresz G, Schulte G, Bergenthal D, Blaschke G (1996b) Capillary electrophoresis and H/1 nmr studies on chiral recognition of atropisomeric bunaphthyl derivatives by cyclodextrin hosts. J Chromatogr A 732:143–150
Cherkaoui S, Rudaz S, Varesio E, Veuthey J-L (2001) On-line capillary electrophoresis-electrospray mass spectrometry for the stereoselective analysis of drugs and metabolites. Electrophoresis 22:3308–3315
Chirobiotik Handbook (1996) A guide to using macrocyclic antiobiotic bonded phases for chiral LC separations. ASTEC. U.S.A.
Collet A (1989) Optical resolution by crystallisation methods. in: Krstulovic AM (Ed) Chiral separations by HPLC. Application to pharmaceuticals compounds. Ellis Horwood Ltd., Chichester, p 81
Cram DJ, Cram JM (1974) Host-guest chemistry. Science 183:803–809
Cyclobond Handbook (1992) A Guide to Using Cyclodextrin Bonded Phases. ASTEC U.S.A.
Daicel (1989) Application Guide for Chiral Column Selection. Daicel Chemical Industries, Tokyo
Dalgliesh CE (1952) Optical resolution of amino acids on paper chromatograms. J Chem Soc 137:3940–3940
Desiderio C, Fanali S (1995) Use of negatively charged sulfobutyl ether-b-cyclodextrin for enantiomeric separation by capillary electrophoresis. J Chromatogr A 716: 183–196
Dette C, Ebel S, Terabe S (1994) Neutral and anionic CDs in CZE: Enantiomeric separation of ephedrine and related compounds. Electrophoresis 15:799–803
Dittmann M, Wienand K, Bek F, Rozing G (1995) Theory and practice of capillary electrochromatography. LC-GC 13:800–814
Ekborg-Ott KH, Zientara GA, Schneiderheinze JM, Gahm K, Armstrong DW (1999) Avoparcin, a new macrocyclic antibiotic chiral run buffer additive for capillary electrophoresis. Electrophoresis 20:2438–2457
Fanali S (1989) Separation of optical isomers by capillary zone electrophoresis based on host-guest complexation. J Chromatogr 474:441–446
Fanali S (1994) Use of β-cyclodextrin polymer as a chiral selector in capillary electrophoresis. J Chromatogr A 680:137–146

Fanali S (1996) Identification of chiral drug isomers by capillary electrophoresis. J Chromatogr A 735:77–121
Fanali S, Aturki Z (1995) Use of cyclodextrins in capillary electrophoresis for the chiral resolution of some 2-arylpropionic acid nonsteroidal antiinflammatory drugs. J Chromatogr A 694:297–305
Fanali S, Desiderio C (1996) Use of vancomycin as chiral selector in capillary electrophoresis. Optimisation and quantitation of loxiglumide enantiomers in pharmaceuticals. J High Res Chromatogr 19:322–326
FDA (1992) FDA's Policy Statement for the Development of New Stereoisomeric Drugs. Chirality 4:338–340
Francotte E, Jung M (1996) Enantiomer separation by open-tubular LC and electrochromatography in cellulose-coated capillaries. Chromatographia 42:521–527
Gasper MP, Berthod A, Nair UB, Armstrong DW (1996) Comparison and modeling study of vancomicin, ristocetin A, and teicoplanin for CE. Anal Chem 68: 2501–2514
Gassmann E, Kuo JE, Zare RN (1985) Electrokinetic separation of chiral compounds. Science 230:813–814
Gosh MK (1992) HPLC Methods on Drug Analysis. Springer Verlag, Berlin
Gross M, Cartwrigth A, Campbell DB, Bolton R, Holmes K, Roberts JL (1993) Regulatory requirements for chiral drugs. Drug Info J 27:453–457
Grossman PD, Colburn JC (1992) Capillary Electrophoresis Theory & Practice. 1, Academic Press, Inc., San Diego
Gubitz G, Juffmann F (1987) Resolution of the enantiomers of thyroid hormones by high-performance ligand exchange chromatography using a chemically bonded chiral stationary phase. J Chromatogr 404:391–393
Guttman A, Paulus A, Cohen AS, Grinberg N, Karger BL (1988) Use of complexing agents for selective separation in high performance capillary electrophoresis, chiral resolution via cyclodextrins incorporated within polyacrylamide gel column. J Chromatogr 448:41–53.
Gübitz G, Schmid MG (1997) Chiral separation principles in capillary electrophoresis. J Chromatogr A 792:179–225
Heiger D (1992) High Performance Capillary Electrophoresis – An Introduction. Hewlett-Packard Company, France
Hesse G, Hagel R (1973) Eine vollständige Racemattrennung durch Elutions-Chromatographie an Cellulose-tri-acetate. Chromatographia 6:277–277
Ingelse BA, Everaerts FM, Desiderio C, Fanali S (1995) A study on the enantiomeric separation by capillary electrophoresis using a soluble neutral β-cyclodextrin polymer. J Chromatogr A 709:89–98
Ingelse BA, Reijenga JC, Claessens HA, Everaerts F, Flieger M (1996) Ergot alkaloids as novel chiral selectors in capillary electrophoresis. J High Resolut Chromatogr 19:225–226
Jakubetz H, Juza M, Schurig V (1997) Electrokinetic chromatography employing an anionic and cationic β-cyclodextrin derivative. Electrophoresis 18:897–904
Jäverfalk EM, Amini A, Westerlund D, Andrén PE (1998) Chiral separation of local anesthetics by capillary electrophoresis/partial filling technique coupled on-line to micro-electrospray Mass Spectrometry. J Mass Spectrom 33:183–186
Johansson IM, Pavelka R, Henion JD (1991) Determination of small drug molecules by capillary electrophoresis atmospheric pressure ionisation mass spectometry. J Chromatogr A 559:515–525
Kemp G (1998) Capillary electrophoresis: a versatile family of analytical techniques. Biotechnol Appl Biochem 27:9–17
König WA (1987) Practice of enantiomer separation by capillary gas chromatography. Huetig Verlag, Heidelberg
Krustulovic AM (1989) Introduction to chromatographic resolution of enantiomers. 8. Ellis Horwood, Chichester

Kuhn R (1999) Enantiomeric separation by capillary electrophoresis using a crown ether as chiral selector. Electrophoresis 20:2613–2620

Kuhn R, Steinmetz C, Bereuter T, Haas P, Erni F (1994a) Enantiomeric separations in capillary zone electrophoresis using a chiral crown ether. J Chromatogr A 666: 367–373

Kurganov AA, Tevlin AB, Davankov V (1983) High performance ligand-exchange chromatography of enantiomers. J Chromatogr 261:223–233

Kyba EP, Timko JM, Kaplan LJ, de Jong F, Gokel GW, Cram D (1978) Host-guest complexation: II. Survey of chiral recognition of amine and amino esters. J Am Chem Soc 100:4555–4565

Landers JP, Spelsberg TC (1994) Future prospects for capillary electrophoresis. Handbook of Capillary Electrophoresis

Lelièvre F, Gareil P, Jardy A (1997) Selectivity in CE: application to chiral separations with CDs. Anal Chem 69:385–392

Lelièvre F, Yan C, Zare R, Gareil P (1996) Capillary electrochromatography: operating characteristics and enantiomeric separations. J Chromatogr A 723: 145–156

Li S, Lloyd DK (1993) Direct chiral separations by capillary electrophoresis using capillaries packed with an α_1-acid glycoprotein chiral stationary phase. Anal Chem 65:3684–3690

Lienne M, Caude M, Tambute A, Rosset R (1987) Séparation d'énantiomères par chromatographie en phase liquide sur phases stationnaires chirales. Analusis 15: 431–476

Lindner W, Petterson K (1985) in: Wainer IW (Ed) Liquid chromatography in pharmaceutical development: an introduction. Asters Publishing, Springfield

Lloyd DK, Aubry AF, De Lorenzi E (1997) Selectivity in capillary electrophoresis: the use of proteins. J Chromatogr A 792:349–369

Lurie IS (1997) Separation selectivity in chiral and achiral capillary electrophoresis with mixed cyclodextrins. J Chromatogr A 792:297–307

Macaudière P, Caude M, Rosset R, Tambute A (1989) CO_2 SFC with chiral stationary phases: A promising coupling for the resolution of various racemates. J Chromatogr Sci 27:383–392

Mages AG, Mosbach K (1997) Recent progress in open tubular liquid chromatography. Trends Anal Chem 16:332–342

Markides K, Juvancz Z (1992) Enantiomer separation using supercritical fluid chromatography, a promising possibility. LC-GC Intl 5:44–56

Mayer S, Schurig V (1992) Enantiomer separation by electrochromatography on capillaries coated with chirasil-dex. J High Resolut Chromatogr 15:129–131.

Nardi A, Eliseev A, Bocek P, Fanali S (1993) Use of charged and neutral cyclodextrins in CZE: enantiomeric resolution of some 2-hydroxy acids. J Chromatogr 638: 247–254

Nishi H, Nakamura K, Nakai H, Sato T (1994) Chiral separation of drugs by capillary electrophoresis using β-cyclodextrin polymer. J Chromatogr A 678: 333–342

Okamoto Y, Honda S, Yuki H, Murata S, Noyori R, Takaya H (1981) Novel packing material for optical resolution: (+)-Poly(trimethylmethacrylate) coated on macroporous silica gel. J Am Chem Soc 103:6971–6973

Okamoto Y, Kaida Y (1994) Resolution by HPLC using polysaccharide carbamates and benzoates as chiral stationary phases. J Chromatogr A 666:403–419

Pedersen CJ (1967) Cyclic polyethers and their complexes with metal salts. J Am Chem Soc 89:7017–7036

Pena MS, Zhang YL, Warner IM (1997) Enantiomeric separations by use of calixarene electrokinetic chromatography. Anal Chem 69:3239–3242

Piette V, Lammerhofer M, Lindner W, Crommen J (1999) Enantiomeric separation of *N*-protected amino acides by nonaqueous capillary electrophoresis using quinine or *tert*-butyl carbamoylated quinine as chiral additive. Chirality 11:622–630

Pirkle WH, Finn JM, Schreiner JL, Humper BC (1981) A widely useful chiral stationary phase for the high performance liquid separation of enantiomers. J Am Chem Soc 103:3964–3966

Pirkle WH, Hyun MH, Banks F (1984) A rational approach to the design of highly-effective chiral stationary phase. J Chromatogr 316:581–585

Poole CF, Poole SK (1994) Chromatography today. 3th Ed., Elsevier, Amsterdam

Quang C, Khaledi MG (1995) Extending the scope of chiral separation of basic compounds by cyclodextrin-mediated capillary zone electrophoresis. J Chromatogr A 692:253–265

Rawjee YY, Staerk DU, Vigh G (1993a) CE chiral separations with cyclodextrin additives I. Acids: chiral selectivity as a function of pH and the concentration of β-cyclodextrin for fenoprofen and ibuprofen. J Chromatogr 635:291–306

Rawjee YY, Williams RL, Vigh G (1993b) Capillary electrophoretic chiral separations using β-cyclodextrin as resolving agent. II. Bases: Chiral selectivity as a function of pH and the concentration of β-cyclodextrin. J Chromatogr A 652:233–245

Rawjee YY, Vigh G (1994) A peak resolution model for the CE separation of the enantiomers of weak acids with hydroxypropyl β-cyclodextrin containing background electrolytes. Anal Chem 66:619–627

Riley CM, Lough WJ, Wainer IW (1994) Pharmaceutical and biomedical applications of liquid chromatography. Pergamon, Oxford

Rodriguez I, Jin LJ, Li SFY (2000) High-speed chiral separations on microchip electrophoresis devices. Electrophoresis 21:211–219

Rosset R, Caude M, Jardy A (1991) Chromatographies en phases liquide et supercritique. 3, Masson, Paris

Rudaz S, Cherkaoui S, Dayer P, Fanali S, Veuthey J-L (2000) Simultaneous stereoselective analysis of tramadol and its main phase I metabolites by on-line capillary zone electrophoresis-electrospray ionization mass spectrometry. J Chromatogr A 868:295–303

Rudaz S, Cherkaoui S, Gauvrit J-Y, Lantéri P, Veuthey J-L (2001) Experimental designs to investigate capillary electrophoresis-electrospray mass spectrometry enantioseparation with the partial-filling technique. Electrophoresis 22:3316–3326

Schmitt T, Engelhardt H (1995) Optimization of enantiomeric separations in capillary electrophoresis by reversal of the migration order and using different cyclodextrins. J Chromatogr A 697:561–570

Schulte G, Heitmeier S, Chankvetadze B, Blaschke G (1998) Chiral capillary electrophoresis-electrospray mass spectrometry coupling with charged cyclodextrin derivatives as chiral selectors. J Chromatogr A 800:77–82

Schurig V (1994) Enantiomer separation by gas chromatography on chiral stationary phases. J Chromatogr A 666:111–130

Shindo H, Caldwell J (1995) Development of chiral drugs in Japan: an update on regulatory and industrial opinion. Chirality 7:349–352

Shohat D, Grushka E (1994) Use of calixarenes to modify selectivities in capillary electrophoresis. Anal Chem 66:747–750

Siret L, Bargamann-Leyder N, Tambute A, Caude M (1992) Choice and use of chiral stationary phases in liquid chromatography. Analusis 20:427–435

Smith RD, Goodlett DR, Hofstadler SA (1993) Capillary electrophoresis-mass spectrometry. Anal Chem 65:574–586

Tanaka Y, Kishimoto Y, Terabe S (1998) Separation of acidic enantiomers by capillary electrophoresis-mass spectrometry employing partial filling technique. J Chromatogr A 802:83–88

Tanaka Y, Terabe S (1997) Separation of the enantiomers of basic drugs by affinity capillary electrophoresis using a partial filling technique and a_1-acid glycoprotein as chiral selector. Chromatographia 44:119–128

Terabe S (1989) EKC: an interface between electrophoresis and chromatography. Trends Anal Chem 8(4):129–134

Terabe S, Miyashita Y, Ishihama Y, Shibata O (1993) Cyclodextrin-modified micellar electrokinetic chromatography – Separation of hydrophobic and enantiomeric compounds. J Chromatogr A 636:47–55
Terabe S, Miyashita Y, Shibata O (1990) Separation of highly hydrophobic compounds by cyclodextrin- modified MEKC. J Chromatogr 516:23–31
Terabe S, Otsuka K, Ichikama K, Tsuchiya A, Ando T (1984) Electrokinetic separations with micellar solutions and open-tubular capillaries. Anal Chem 56: 111–113
Terabe S, Otsuka K, Nishi H (1994) Separation of enantiomers by capillary electrophoretic techniques. J Chromatogr A 666:295–319
Terabe S, Ozaki H, Otsuka K, Ando T (1985) Electrokinetic chromatography with 2-*O*-carboxymethyl-β-cyclodextrin as a moving "stationnary" phase. J Chromatogr 332:211–217
Tomlinson AJ, Benson LM, Naylor S (1994) On-line capillary electrophoresis mass spectrometry for the analysis of drug metabolite mixture: practical considerations. J Cap Elec 1:127–135
Valkó I, Billiet H, Frank J, Luyben KCAM (1994) Effect of the degree of substitution of (2-hydroxy)propyl-β-cyclodextrin on the enantioseparation of organic acids by capillary electrophoresis. J Chromatogr A 678:139–144
Valtcheva L, Mohammad J, Pettersson G, Hjerten S (1993) Chiral separation of beta-blockers by HPCE based on non immobilized cellulase as enantioselective protein. J Chromatogr 638:263–268
Verleysen K, Sandra P (1998) Separation of chiral compounds by capillary electrophoresis. Electrophoresis 19:2798–2833
Vindevogel J, Sandra P (1994) On the possibility of performing chiral wall-coated open-tubular electrochromatography in 50 μm internal diameter capillaries. Electrophoresis 15:842–847
Wahl JH, Schmith RD (1994) Comparison of buffer systems and interface designs for capillary electrophorcsis-mass spectrometry. J Cap Elec 1:62–70
Wainer IW (1987) Proposal for the high-performance chromatographic chiral stationary phases: how to choose the right column. TRAC 6:125–134
Wainer IW, Drayer DE (1988) Drug Stereochemistry. Analytical Methods and Pharmacology. Marcel Dekker, New York
Wainer IW, Drayer DE (1993) The early history of stereochemistry. In. Drug stereochemistry. Analytical Methods and Pharmacology. Marcel Dekker, New York
Wang F, Khaledi MG (1998) Nonaqueous capillary electrophoresis chiral separations with quaternary ammonium β-cyclodextrin. J Chromatogr A 817:121–128
Wang J, Warner IM (1995) Combined polymerized chiral micelle and g-cyclodextrin for chiral separation in capillary electrophoresis. J Chromatogr A 711:297–304
Ward TJ, Dann CI, Blaylock A (1995) Enantiomeric resolution using the macrocyclic antibiotics rifamycin B and rifamycin SV as chiral selectors for capillary electrophoresis. J Chromatogr A 715:337–344
Ward TJ, Oswald TM (1997) Enantioselectivity in capillary electrophoresis using the macrocyclic antibiotics. J Chromatogr A 792:309–325
Weseloh G, Bartsch H, Koenig WA (1995) Separation of basic drugs by capillary electrophoresis using selectively modified cyclodextrins as chiral selectors. J Microcol Sep 7:355–363
Wheat TE, Lilley KA, Banks JF (1997) Capillary electrophoresis with electrospray mass spectrometry detection for low-molecular-mass compounds. J Chromatogr A 781:99–105
Willoughby R, Sheehan E, Mitrovich S (1998) A global view of LC-MS. Global View Publishing, Pittsburgh
Wren SAC, Rowe RC (1992a) Theoretical aspects of chiral separation in CE. I: initial evaluation of a model. J Chromatogr 603:235–241

Wren SAC, Rowe RC (1992b) Theoretical aspects of chiral separation in CE. II: the role of organic solvent. J Chromatogr 609:363–367
Wren SAC, Rowe RC (1993) Theoretical aspects of chiral separation in CE. III: application to beta-blockers. J Chromatogr 635:113–118
Yashima E, Okamoto Y (1995) Chiral discrimination on polysaccharides derivatives. Bull Chem Soc Jpn 68:3289–3307

CHAPTER 3

Stereochemical Issues in Bioactive Natural Products

C. Terreaux and K. Hostettmann

A. Introduction

Beside molecules produced by classical organic synthesis, by combinatorial chemistry or designed by molecular modeling, natural products represent a large source of new chemical entities, potential lead compounds or novel drugs. These compounds originate from higher plants, marine organisms, fungi and animals, or are produced by microorganisms or yeasts using biotechnology. This chapter will focus on secondary metabolites from higher plants.

Among the main plant biosynthetic pathways, i.e., the shikimate, acetate-malonate, acetate-mevalonate and amino acids metabolic routes, the first two usually lead to polyphenolic derivatives, while terpenoids and alkaloids are produced by the last ones. Compounds from these two chemical classes always contain at least one chiral center. Biosynthetic and enzymatic pathways in higher plants generally allow production of single enantiomers in a definite plant. However, another species might possess other enzymatic functionalities with opposite enantiospecificities. Despite these considerations, the characterization of opposite stereoisomers in one plant is quite common and is most probably due to the various processing steps of the plant material such as drying, exposure to light, extraction with various solvents and fractionation by chromatography.

Numerous natural products have important pharmacological or toxicological properties. As receptor interactions are specific and, in many cases, even stereospecific, chirality in natural products is of prime importance. Therefore, chiral pharmacological specificities such as those observed with the nonsteroidal antiinflammatory profens might also occur with natural products.

Some examples of chiral plant secondary metabolites in relation with their biological and pharmacological properties are presented here. Since many compounds contain more than one chiral center, the discussion will be expanded to diastereomerization or epimerization processes.

B. Natural Products Occurring in Opposite Enantiomeric Forms

Usually, natural products occur in one single isomeric form in a plant, due to the presence of specific enzymatic systems. This stereospecific biosynthesis can generally be extended to a definite plant genus or family. However, it is not rare to detect the presence of opposite enantiomers in nature.

I. Occurrence of Camphor

Camphor is a well-known example for the occurrence of opposite enantiomers in plants. The European Pharmacopoea describes the natural (+)-(1*R*;4*R*)-camphor (1a), as well as the racemic mixture produced by organic synthesis. The pure enantiomeric form is a crystalline powder obtained by sublimation from the essential oil of the camphor tree [*Cinnamomum camphora* (L.) Siebold, Lauraceae], a tree native from the coasts of Eastern Asia.

The camphor molecule has two chiral centers at positions C(1) and C(4), so that four stereoisomers would be expected. However, steric constraints (bonds for the bridge over carbon 7 from positions 1 and 4 are in *cis* configuration) allow only two enantiomers, (+)-camphor (1a) and (–)-camphor (1b) (HÄNSEL et al. 1999). While the dextrorotatory isomer (1a) is present in the camphor tree, levorotatory camphor is found in feverfew (*Tanacetum parthenium* Schultz Bip., Asteraceae) and tansy (*Tanacetum vulgare* L.) and follows different biosynthetic pathways compared to (+)-camphor (CROTEAU et al. 1990). The presence of racemic mixtures of camphor has also been described in the essential oil of some Lamiaceae species: (+)-camphor is predominant in rosemary oil (*Rosmarinus officinalis* L.), while in sage oil (*Salvia officinalis* L.) either enantiomer can predominate (RAVID et al. 1993).

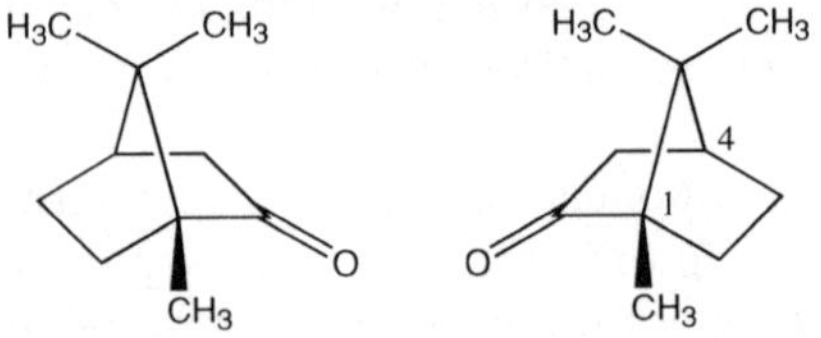

1a **1b**

In external use, camphor is widely applied as a hyperemic drug and enters in the composition of numerous oils and ointments for massage or for the treatment of rheumatic pains. It is also applied in cases of respiratory difficulty due to a cold (HÄNSEL et al. 1999). As many camphor-containing preparations are commercially available, acute toxicity of camphor is of importance, particularly to newborns and young children. Moreover, camphor has shown a high epileptogenic potential as observed with other monoterpenic ketones (BURKHARD et al. 1999).

II. Organoleptic Properties of Menthol and Carvone

Besides camphor, various monoterpenic essential oil constituents are naturally encountered in different isomeric forms. Unlike camphor, some of them display distinct organoleptic properties, as exemplified by menthol or carvone.

Naturally occurring (–)-menthol (2a) is obtained from the essential oil of peppermint (*Mentha piperita* L., Lamiaceae) or cornmint (*Mentha arvensis* L.). It represents the generally used form, although the racemic mixture is also accepted in the European Pharmacopoea. Menthol has three chiral centers, allowing four pairs of optical isomers (2a–5b). The complete mixture of isomers can derive from the partial synthesis of menthol by hydration of thymol. Racemic menthol is then obtained after separation of the diastereomeric pairs of enantiomers (ECCLES 1994). (–)-Menthol and (+)-menthol have similar physical properties. However, they differ in their organoleptic properties: levomenthol has typical strong peppermint odor, while dextromenthol has a weak peppermint and musty smell (HÄNSEL et al. 1999). Both enantiomers also strongly differ in their cooling and refreshing effect. (+)-Menthol is almost inactive, but (–)-menthol is responsible for the cold sensation observed when applied on skin or mucosa (ECCLES 1994).

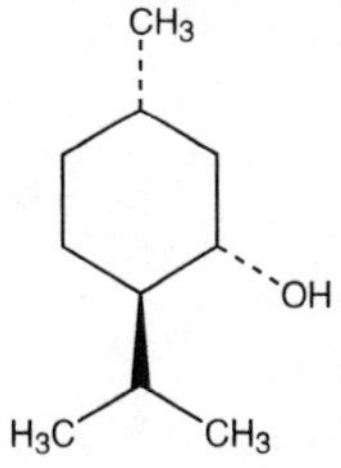

2a (+)-menthol

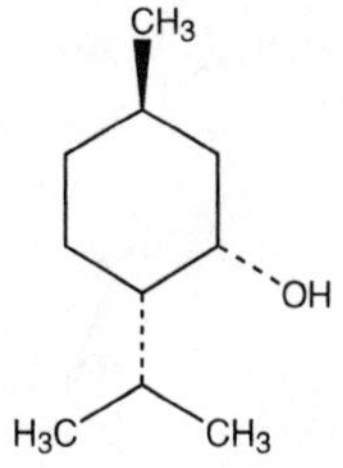

3a (+)-neomenthol

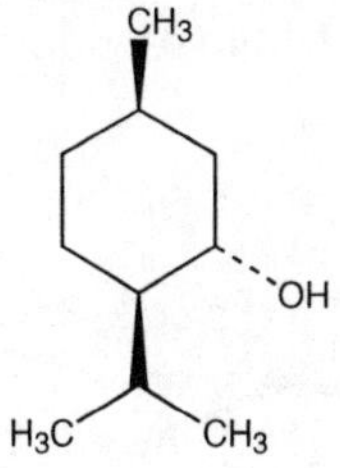

4a (+)-isomenthol

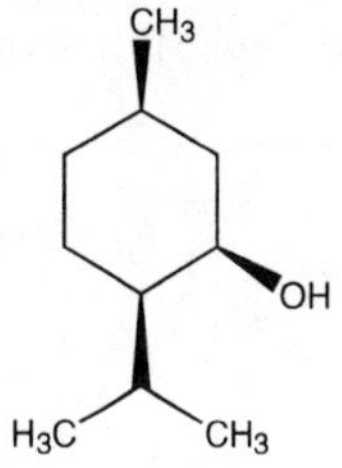

5a (+)-neoisomenthol

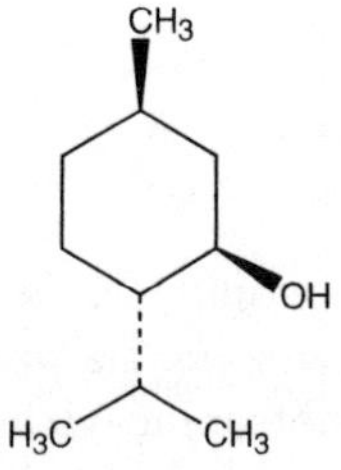

2b (-)-menthol

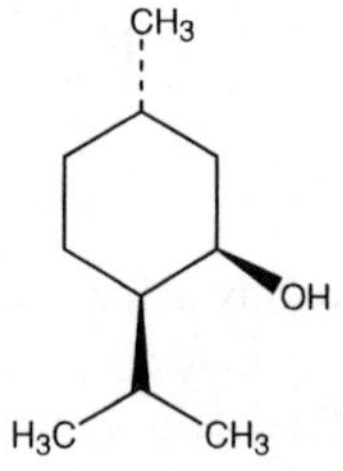

3b (-)-neomenthol

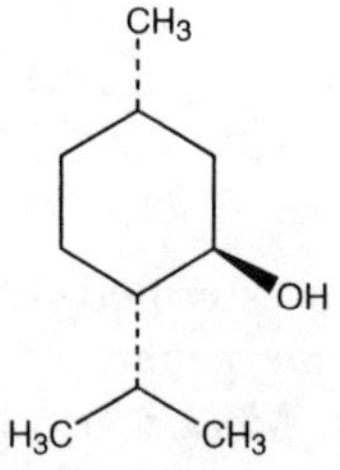

4b (-)-isomenthol

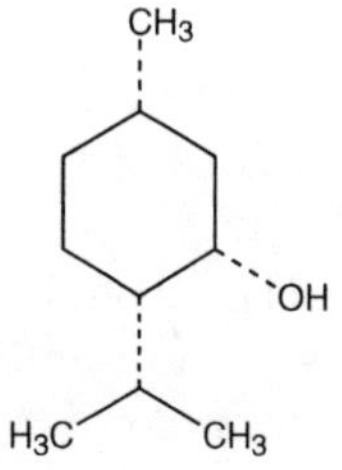

5b (-)-neoisomenthol

Another example of the differing odor of two enantiomers is given by the main caraway oil constituent, (+)-(*S*)-carvone (6a), and its optical isomer, (–)-(*R*)-carvone (6b) (LEITEREG et al. 1971). (+)-Carvone is extracted from the essential oil of caraway (*Carum carvi* L., Apiaceae) and is responsible for the strong characteristic caraway odor. On the other hand, (–)-carvone is the major component of curled mint oil [*Mentha spicata* L. var. *crispa* (Benth.) Danert] and possesses a typical mint odor.

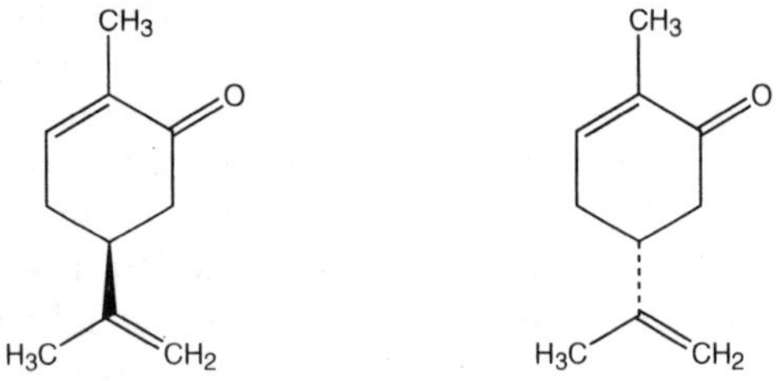

6a **6b**

C. Stereochemistry and Pharmacological Implications

I. Tropane Alkaloids

Tropane alkaloids derive from the amino acid ornithine, which is transformed into tropanone (7). This intermediate is then reduced to either tropan-3α-ol (tropine, 8) or tropan-3β-ol (pseudotropine, 9) (DRÄGER 1996).

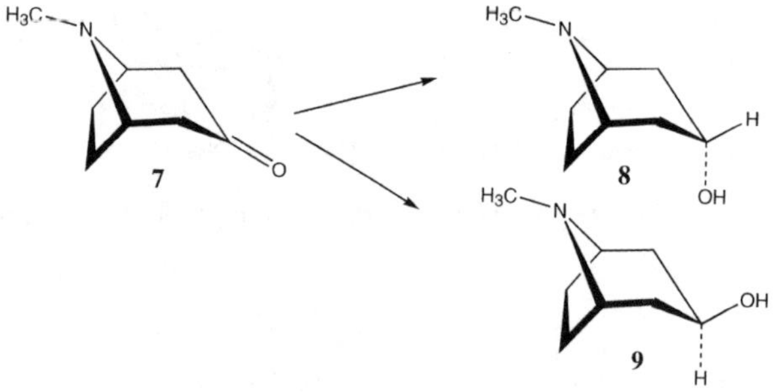

The alcohol is then esterified with various acids and the tropine derivatives are present in some genera of the Solanaceae family (*Atropa*, *Datura*, *Hyoscyamus*, *Scopolia*, *Mandragora*). Esterification of pseudotropine yields the alkaloids found in the genus *Erythroxylum* (Erythroxylaceae), the leader in this category being cocaine.

The medicinal plants night shade (*Atropa belladonna* L.), thorn apple (*Datura stramonium* L.) and black henbane (*Hyoscyamus niger* L.) have been included in most pharmacopoeias since ancient times. They contain large amounts of the alkaloids (–)-(*S*)-hyoscyamine (10) and (–)-(*S*)-scopolamine (11), in various proportions. These substances are strong antagonists of the muscarinic acetylcholine receptors and thus act as parasympatholytic drugs.

The active (–)-(*S*)-hyoscyamine racemizes very easily to (+)-(*R*)-hyoscyamine, which possesses almost no affinity for the muscarinic receptors. This racemic mixture is called atropine. Racemization can occur during extraction from the plant material or can be induced under weak alcohol/base conditions in order to produce atropine, which is used for its therapeutic properties. The pharmacological action will appear either as peripheral or central effects, such as dry mouth, mydriasis, tachycardia, constipation or, at high dosage, respiratory depression. Central effects are memory loss, sedation, then hallucinations and delirium (Rang et al. 1997). The pharmacological difference between hyoscyamine or atropine and scopolamine is mainly due to their lipophilicity, since the epoxide function in scopolamine considerably increases its lipophilic character. Thus, hyoscyamine and atropine are used preferably as spasmolytics for peripheral disorders, while scopolamine crosses the brain–blood barrier better and its main indication is the treatment and prevention of nausea and travel sickness, applied as a transdermal therapeutic system. These considerations also allowed Mann (1992) to offer a tentative explanation for the common representation of witches flying on their broomsticks. In the Middle Ages, witches yielded their secret under torture: they prepared an ointment with leaves of Solanaceae plants and then applied it to the armpits or covered their broomstick, which they rode, in order to rub the ointment on their genital mucosa. This way of administration hindered the absorption of hyoscyamine and, consequently, its peripheral toxic effects, while scopolamine easily reached the CNS and induced hallucinations and flying sensations.

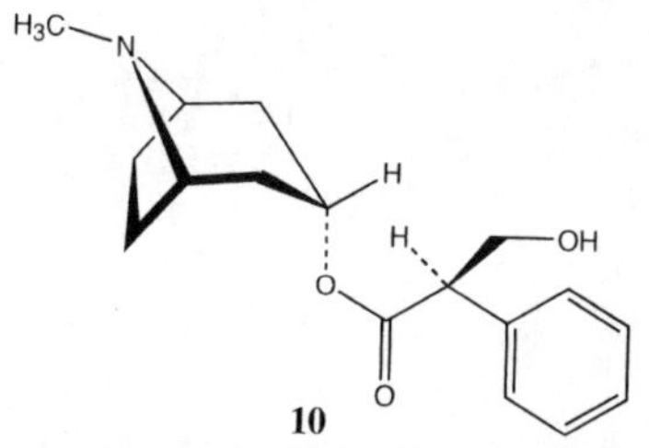

10

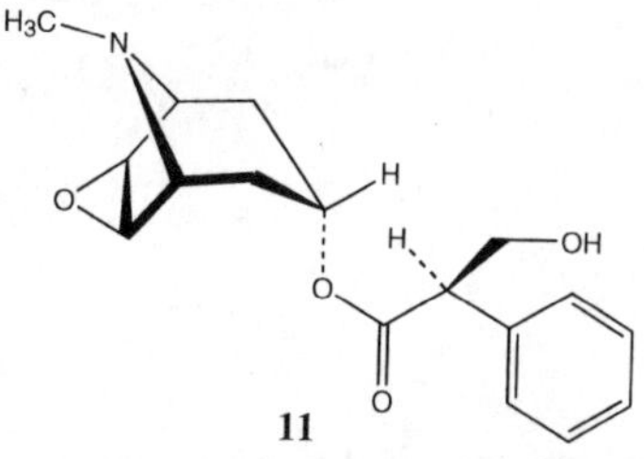

11

Angel's trumpet, *Datura suaveolens* Humb. and Bonpl. ex Willd., is a widespread ornamental plant of the Solanaceae family, but recently became notorious through its abuse as a hallucinogenic drug. The World Wide Web is diffusing recipes for hallucinogenic beverages prepared by maceration of angel's trumpet leaves and flowers in alcoholic drinks (wine, beer). In that case, scopolamine as well as hyoscyamine are extracted. Therefore, besides the hallucinogenic effect of scopolamine, an anticholinergic syndrome can occur through hyoscyamine. Many cases of intoxication have been reported (Francis and Clarke 1992), even the death of a 20-year-old woman (Rauber et al. 1999).

II. Khat and Cathinone Derivatives

Khat designates the fresh leaves or branch endings of *Catha edulis* Forsk., a shrub of the Celastraceae family, which are chewed for their stimulating effect. Khat cultures look similar to tea cultures and the main producers are Yemen, Ethiopia, Somalia, Kenya and Madagascar. Khat chewers usually take 100–400 g fresh leaves and form balls in their mouth, which are either ingested or spat (PALLENBACH 1996). The active substance of khat is (–)-(*S*)-cathinone (12), a pseudoalkaloid showing structural but also pharmacological similarities with amphetamine. Thus, (–)-(*S*)-cathinone acts as an indirect sympathomimetic and enhances the release of adrenergic neurotransmitters such as noradrenaline (inducing a higher activity, anorexy, stimulation) and dopamine (stimulation, stereotypical behavior). The difference between cathinone and amphetamine resides in their relative potency, the latter being around eight times more potent (HÄNSEL et al. 1999). Fresh leaves of khat are sold in traditional markets and are usually chewed, as the stimulating effect of dried leaves is around five times lower. During the drying process, cathinone undergoes an enzymatic reduction to (1*R*;2*S*)-norephedrine (13) and (1*S*;2*S*)-norpseudoephedrine (14). This enzymatic reaction is highly stereospecific as only traces of the former can be found in the dried plant.

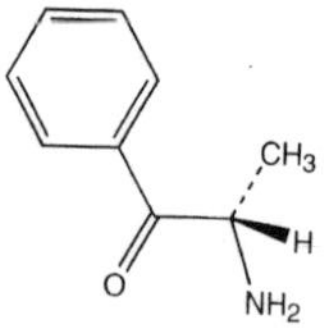

12

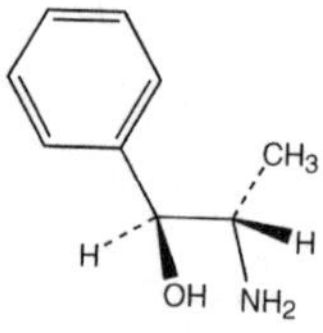

13

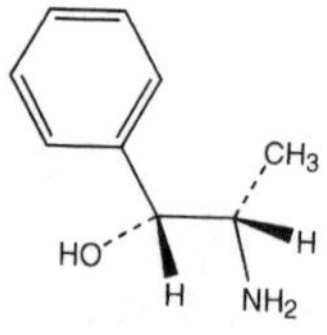

14

The effects of khat chewing can be described as follows: decreased tiredness, higher resistance to stressful jobs, slight excitation and improved sociability, while intoxication signs are mydriasis, extrasystoles, increased blood pressure, headache, hyperthermia and insomnia (KALIX 1996). The maximal plasmatic concentration following khat chewing is reached after 1.5–3 h (HALKET et al. 1995). Khat addiction leads to a poor physical but strong psychological dependance. Khat use is not restricted to the countries of origin of the plant, but is also widespread over Europe and is submitted to the legislation of drugs of abuse. Recently, a pharmacological study has evaluated the analgesic effect of a khat extract compared to amphetamine and ibuprofen (CONNOR et al. 2000). All three were active, although to a different degree.

III. Kawalactones

Kava (also called kawa or kava-kava) is a traditional beverage of the South Pacific Islands (Samoa, Tonga, Fiji, Vanuatu, etc.), as well as the plant it is made from. The kava "ceremony" is very common in these regions and starts with the mastication of kava rhizomes (*Piper methysticum* Forst., Piperaceae) by men only. The mixture, including saliva, is then spat into water, macerated and filtered: the kava beverage is ready. Kava drinking induces a feeling of well-being and an increased resistance to fatigue. A higher dosage will have a tranquilizing effect and cause sleepiness (SINGH 1992). This fascinating plant has attracted the curiosity of chemists and pharmacologists for more than 100 years. Anxiolytic (HOLM et al. 1991), but also anticonvulsive, analgesic and central muscle relaxing properties have been demonstrated for a kava extract. Phytomedicines from kava rhizomes are now commercialized in Europe as a natural anxiolytic, and clinical studies have assessed such preparations as an alternative to benzodiazepines for the treatment of anxiety (VOLZ and KIESER 1997). The active principles of kava appear to be kawapyrones, the main constituents of kava rhizomes. Although little is known about their mode of action, pharmacological studies are investigating the effects of this chemical class on various targets such as γ-aminobutyric acid (GABA) receptors (BOONEN and HÄBERLEIN 1998), benzodiazepine receptors (DAVIES et al. 1992), neurotransmitter systems (BOONEN et al. 1998), and direct neuronal interactions (BOONEN et al. 2000).

The main kawapyrones from *P. methysticum* are (+)-kavain (15) and (+)-methysticin (16), their dihydrogenated derivatives (+)-dihydrokavain (17) and (+)-dihydromethysticin (18), together with yangonin (19) and desmethoxyyangonin (20). According to their degree of insaturation in the lactone ring, they are classified into enolides and dienolides. The enolides are 5,6-dihydro-α-pyrones with a chiral center at position 6. As yet, no study has reported differences in the pharmacological activities of the corresponding isomers. However, the stereochemical issue of kawapyrones is still current: ($\pm$)-kavain is commonly produced by synthesis and is often added to kava preparations in order to increase their kavain content. An analytical HPLC method has been developed to separate (*R*)-(+)- and (*S*)-(−)-kavain (HÄBERLEIN et al. 1997), and analysis of the presence of both enantiomers in a kava extract can indicate a falsification of the original plant extract with synthetic kavain.

Kava is gaining increasing importance in the market of nonprescription drugs and health foods. Further studies are required regarding its mode of action, but also its potential liver toxicity, as severe hepatic troubles have recently been described following long-term kava ingestion (CSPV 2000).

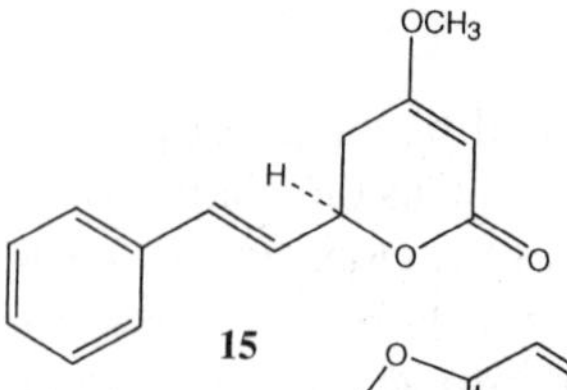

15

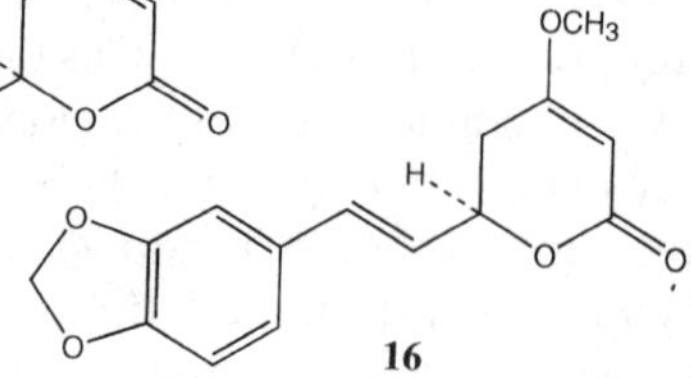

16

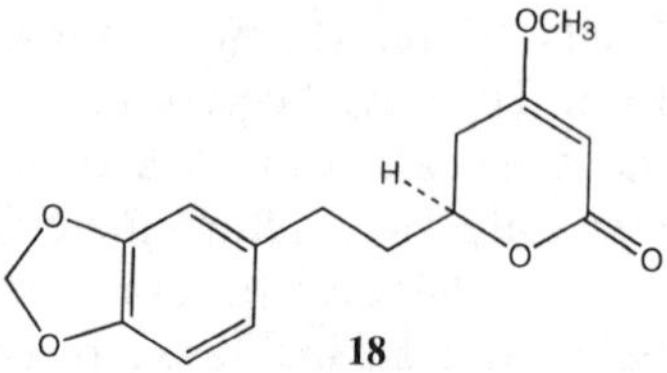

18

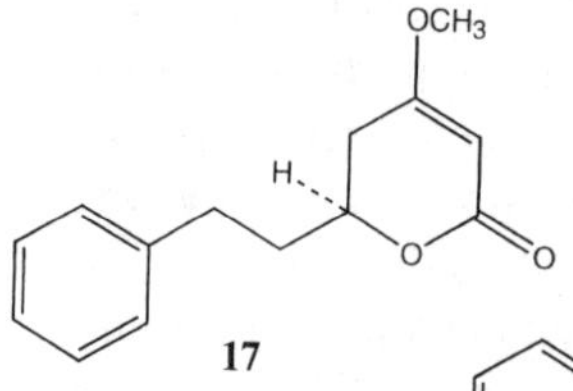

17

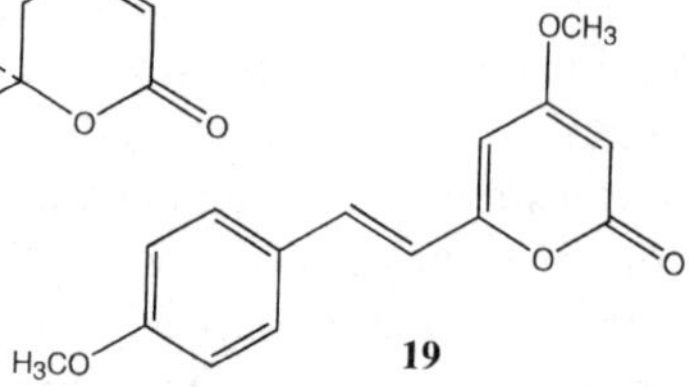

19

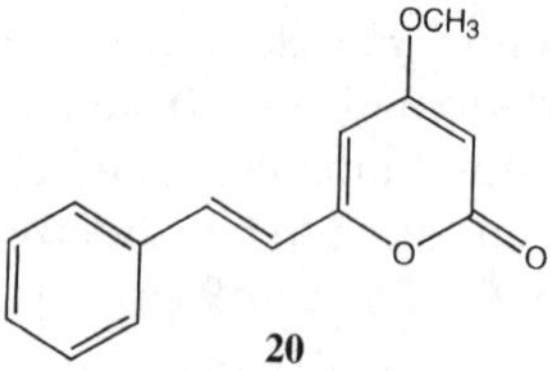

20

IV. Gossypol

Gossypol (21) (2,2′-bis-(8-formyl-1,6,7-trihydroxy-3-methylnaphthalene) is a natural polyphenolic dialdehyde occurring at high concentration in the pigment glands of the cotton plant *Gossypium herbaceum* L. or *G. hirsutum* L. (Malvaceae). The discovery of its male antifertility properties in the early 1970s raised considerable interest in family planning programs. A large-scale study with more than 10000 patients assessed the male contraceptive activity and provided extensive information on the toxicology and pharmacokinetics of gossypol (WU 1989). However, the slow onset of action, the risks of sterility and hypokalemia and other adverse effects are major disadvantages for its use as an antifertility agent (WU 1987).

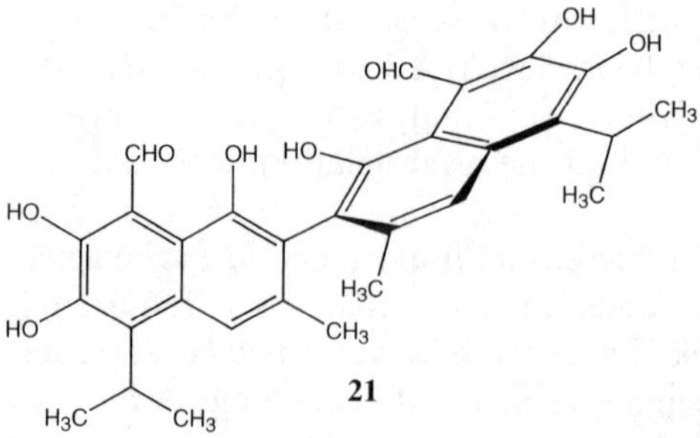

21

Though possessing no chiral center, gossypol exists in the cotton plant as a racemate of two enantiomers as a result of the axial dissymmetry of the molecule (atropisomerism). The rotation around the C–C bond linking the two naphthyl rings is hindered by the presence of –OH and –CH_3 substituents in *ortho* position, which results in (+)- (21) and (–)-isomers (22).

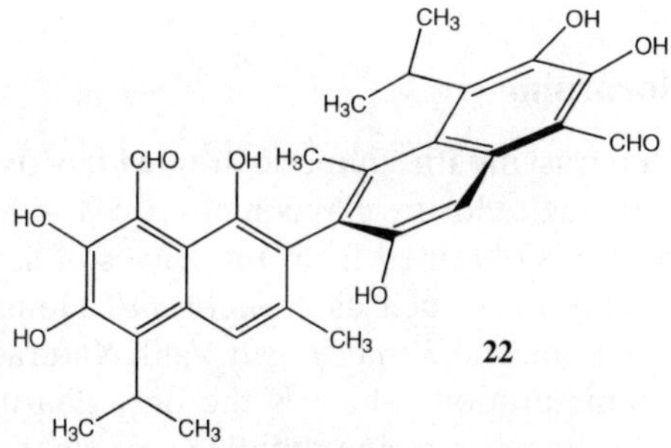

22

There are strong differences in the pharmacokinetic and pharmacological profiles of both enantiomers. (–)-Gossypol is responsible for the antifertility action, while the (+)-isomer does not affect sperm counts (no antispermatogenic activity) and does not cause testis atrophy (Wu 1987). This stereospecificity has also been shown for the cytotoxic activity of gossypol, with a significant higher cytotoxicity of (–)-gossypol towards various cell lines (Shelley et al. 1999). On the other hand, the active enantiomer is very rapidly eliminated compared to the inactive form. In humans, (+)-gossypol and (–)-gossypol have elimination half-lives of approximately 133 h and 4.5 h, respectively. Thus, long-term administration of racemic gossypol leads to (+)-gossypol accumulation in the body and may increase toxicity. The half-life differences may be due to a lower binding capacity of (–)-gossypol to tissue proteins (Wu 1987). In addition to the high elimination rate of the active enantiomer, the antifertility action appears after 2–4 months treatment with the cumulative effect of single doses. As mentioned above, this slow onset of action was one of the major drawbacks in the use of gossypol as a male contraceptive.

D. Epimerization

Epimers are diastereomers possessing an opposite configuration at only one of several chiral centers. The most common epimers in the field of natural products are the sugars. They occur either as free units, oligosaccharides or long-chain polysaccharides, or in the glycosidic part of secondary metabolites. For example, glucose and mannose are C(2)-epimers, while glucose and galactose are C(4)-epimers. Sugars are usually not encountered as open-chain tautomers, but with their aldehyde and an alcoholic group cyclized to an hemiacetal. This ring formation generates a new asymmetric carbon at position C(1), which can have either the α- or the β-configuration. The α- and β-forms of a sugar are diastereomers called anomers and C(1) is described as the anomeric carbon. Naturally occurring α- and β-glucose can be mentioned as examples. Since many plant secondary metabolites contain more than one asymmetric carbon, epimerization phenomena might also be of importance.

I. Stability of Pilocarpine

Pilocarpine (23) is a parasympathomimetic drug widely used externally in the treatment of the acute angle closure glaucoma, as 0.5%–4% solutions (HÄNSEL et al. 1999). Pilocarpine is obtained from the leaves of several species of the genus *Pilocarpus* (Rutaceae), such as *P. jaborandi* Holmes, *P. microphyllus* Stapf, *P. pennatifolius* Lem. and *P. racemosus* Vahl. Natural pilocarpine occurs in the (+)-(3*S*;4*R*)-configuration, which is the only pharmacologically active form. In aqueous solution, pilocarpine rapidly converts to the inactive epimer (+)-(3*R*;4*R*)-isopilocarpine, and to the lactone-opened pilocarpic acid through base catalysis (NUNES and BROCHMANN-HANSSEN 1974). Further mechanistic investigations have shown that pilocarpine hydrolysis forms pilocarpic acid and isopilocarpic acid, but the latter with its higher thermodynamic stability leads only to isopilocarpine (SZEPESI et al. 1983).

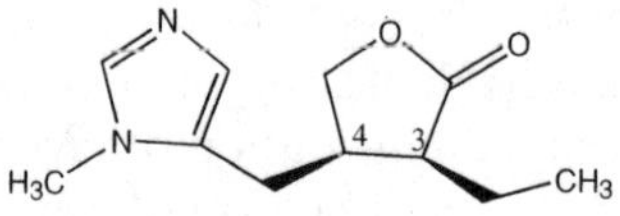

23

As the degradation rate is quite slow (half-life of about 36 days, at pH 7.4 and 35°C), the issue of pilocarpine stability has a low pharmacological relevance, but is of prime pharmaceutical importance (TESTA et al. 1993). Therefore, one major challenge of companies producing pilocarpine eye drops is to ensure stability of the active principle in the drug formulation. Thus, a pilocarpine salt is usually conditioned in a dry form and the solution is prepared just before use.

II. Lysergic Acid and Derivatives

(–)-(5*R*;8*R*)-Lysergic acid (24) and its derivatives are the pharmacologically active alkaloids from the ergot of rye, *Claviceps purpurea* (Fries) Tulasne (Clavicipitaceae). Ergot is an ascomycete fungus contaminating rye (*Secale cornutum* Offic. ex Nees, Gramineae), which caused many intoxications since the Middle Ages. The strong vasoconstrictory effects of the lysergic acid alkaloids induced gangrene epidemics when ingested in contaminated cereals.

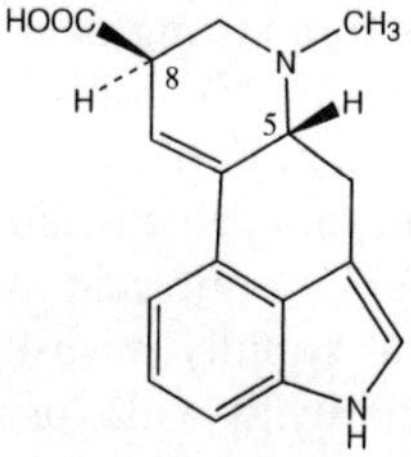

24

Lysergic acid possesses two asymmetric carbons at positions C(5) and C(8). Native lysergic acid and its derivatives have an *R*,*R*-configuration, but handling of these substances during extraction, liquid–liquid partition or solubilization can easily lead to epimerization at C(8) and yield isolysergic acid (25). Epimerization occurs through a keto-enol equilibrium with the carbonyl function attached at C(8) (HÄNSEL et al. 1999). Epimerization affects not only lysergic acid, but also its natural and semisynthetic derivatives, such as amides (ergometrin, 26) and the peptidic alkaloids (ergotamin, 27). Dihydrogenated semisynthetic derivatives (e.g., dihydroergotamine) are less likely to epimerize, the lack of the double bond Δ-9,10 resulting in a lower stability of the enol intermediate.

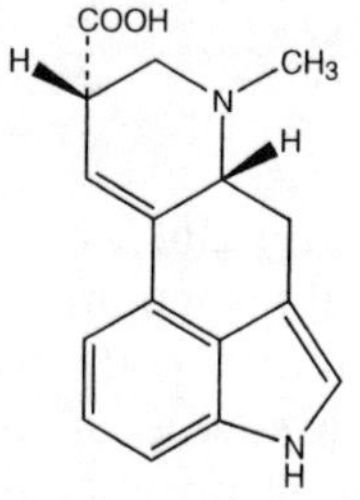

25

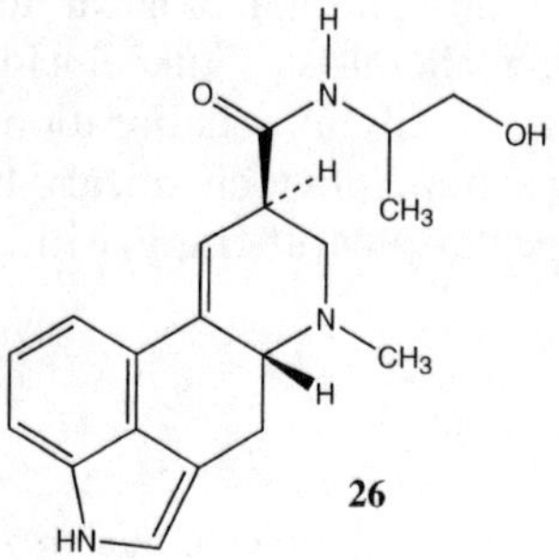

26

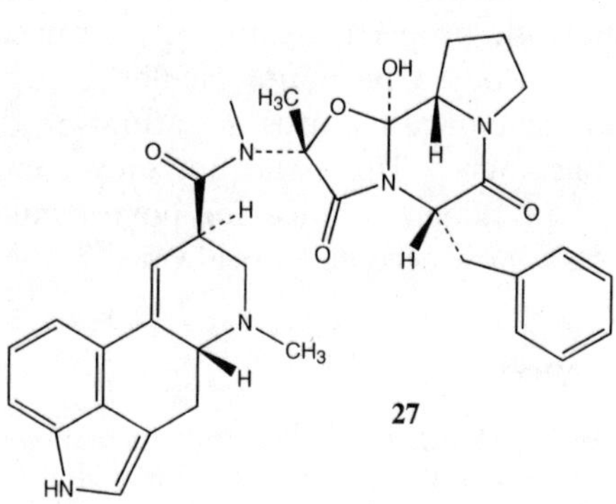

27

C(8)-Epimerization leads to a strong decrease in pharmacological activity of the natural or semisynthetic alkaloids, which are still used in various therapies. Ergotamine is a component of many preparations used in the treatment of acute migraine, while dihydroergotamine finds a place in the long-term management of migraine. Lysergic acid diethylamide (LSD, 28) was first synthesized as a drug, but was rapidly misused because of its strong hallucinogenic properties. The relatively wide use of these prescription drugs or drugs of abuse can of course lead to intoxications due to overdosage or interactions. Recently, severe ergotism (intoxication by ergot alkaloids) was reported following an interaction between ergotamine and ritonavir, a component of HIV tritherapies (Liaudet et al. 1999).

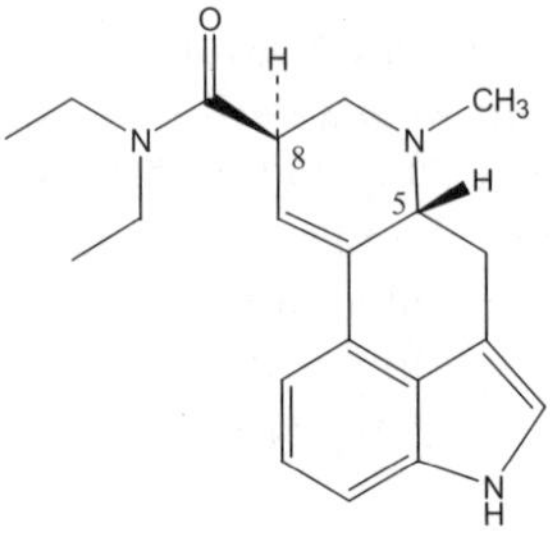

28

E. Conclusion

Stereochemistry has a prominent influence on various steps of drug development, from synthesis to therapeutic action, including drug formulation, distribution and pharmacological interactions (Jamali et al. 1989; Reist et al. 1995). Higher plants represent a rich source of new drugs and lead compounds (Hostettmann et al. 1998) and in many cases, these potential new chemical entities are chiral molecules. As shown by several examples, stereochemistry can have implications at different levels, namely isolation procedures, organoleptic properties, stability, pharmacokinetics, pharmacological activity and toxicology. One major challenge for phytochemists is the elucidation of stereochemical features in natural products, mainly in collaboration with pharmacologists. This issue not only concerns pure phytochemicals, but is also of prime importance for phytomedicines, as phytotherapy with plant extracts is now accepted worldwide.

References

Boonen G, Häberlein H (1998) Influence of genuine kavapyrone enantiomers on the GABA-A binding site. Planta Med 64:504–506

Boonen G, Ferger B, Kuschinsky K, Häberlein H (1998) In vivo effects of the kavapyrones (+)-dihydromethysticin and (+/–)-kavain on dopamine,

3,4-dihydroxyphenylacetic acid, serotonin and 5-hydroxyindoleacetic acid levels in striatal and cortical brain regions. Planta Med 64:507–510
Boonen G, Pramanik A, Rigler R, Häberlein H (2000) Evidence for specific interactions between kavain and human cortical neurons monitored by fluorescence correlation spectroscopy. Planta Med 66:7–10
Burkhard PR, Burkhardt K, Haenggeli CA, Landis T (1999) Plant-induced seizures: reappearance of an old problem. J Neurol 246:667–670
Connor J, Makonnen E, Rostom A (2000) Comparison of analgesic effects of khat (*Catha edulis* Forsk.) extract, D-amphetamine and ibuprofen in mice. J Pharm Pharmacol 52:107–110
Croteau R, Gershenzon J, Wheeler CJ, Satterwhite DM (1990) Biosynthesis of monoterpenes: stereochemistry of the coupled isomerization and cyclization of geranyl pyrophosphate to camphane and isocamphane monoterpenes. Arch Biochem Biophys 277:374–381
CSPV (2000) Atteintes hépatiques liées aux extraits de kawa. J Suisse Pharm 138: 437–438
Davies LP, Drew CA, Duffield P, Johnston GA, Jamieson DD (1992) Kava pyrones and resin: studies on GABA-A, GABA-B and benzodiazepine binding sites in rodent brain. Pharmacol Toxicol 71:120–126
Dräger B (1996) Wie kommt Atropin in die Tollkirsche? Neues zur Regulation des pflanzlichen Sekundärstoffwechsels. Pharm Uns Zeit 25:242–249
Eccles R (1994) Menthol and related cooling compounds. J Pharm Pharmacol 46:618–630
Francis PD, Clarke CF (1999) Angel trumpet lily poisoning in five adolescents: clinical findings and management. J Paediatr Child Health 35:93–95
Häberlein H, Boonen G, Beck M-A (1997) *Piper methysticum*: enantiomeric separation of kavapyrones by high performance liquid chromatography. Planta Med 63:63–65
Halket JM, Karasu Z, Murray-Lyon IM (1995) Plasma cathinone levels following chewing khat leaves (*Catha edulis* Forsk.). J Ethnopharmacol 49:111–113
Hänsel R, Sticher O, Steinegger E (1999) Pharmakognosie – Phytopharmazie. Springer, Berlin
Holm E, Staedt U, Heep J, Kortsik C, Behne F, Kaske A, Mennicke I (1991) The action profile of D,L-kavain. Cerebral sites and sleep-wakefulness-rhythm in animals. Arzneim-Forsch / Drug Res 41:673–683
Hostettmann K, Potterat O, Wolfender J-L (1998) The potential of higher plants as a source of new drugs. Chimia 52:10–17
Jamali F, Mehvar R, Pasutto FM (1989) Enantioselective aspects of drug action and disposition: therapeutic pitfalls. J Pharm Sci 78:695–715
Kalix P (1996) *Catha edulis*, a plant that has amphetamine effects. Pharm World Sci 18:69–73
Leitereg TJ, Guadagni DG, Harris J, Mon TR, Teranishi R (1971) Evidence for the difference between the odours of the optical isomers (+)- and (–)-carvone. Nature 230:455–456
Liaudet L, Buclin T, Jaccard C, Eckert P (1999) Drug points: severe ergotism associated with interaction between ritonavir and ergotamine. BMJ 318:771
Mann J (1992) Murder, Magic and Medicine. Oxford University Press, Oxford, UK
Nunes MA, Brochmann-Hanssen E (1974) Hydrolysis and epimerization kinetics of pilocarpine in aqueous solution. J Pharm Sci 63:716–721
Pallenbach E (1996) Die Männer mit der dicken Backe. Khat im Jemen. Dtsch Apoth Ztg 136:3399–3410
Rang HP, Ritter JM, Dale MM (1995) Pharmacology. Churchill Livingstone, Edinburgh
Rauber C, Guirguis M, Meier-Abt AS, Gossweiler B, Meier PJ (1999) Lethal poisoning after ingestion of a tea prepared from the angel's trumpet (*Datura suaveolens*). XIX International Congress, EAPCCT, Dublin, Eire, June 22–25

Ravid U, Putievsky E, Katzir I (1993) Determination of the enantiomeric composition of (1*R*) (+)- and (1*S*) (–)-camphor in essential oils of some Lamiaceae and Compositae herbs. Flav Fragr J 8:225–228

Reist M, Testa B, Carrupt P-A, Jung M, Schurig V (1995) Racemization, enantiomerization, diastereomerization, and epimerization: their meaning and pharmacological significance. Chirality 7:396–400

Shelley MD, Hartley L, Fish RG, Groundwater P, Morgan JJ, Mort D, Mason M, Evans A (1999) Stereo-specific cytotoxic effects of gossypol enantiomers and gossypolone in tumour cell lines. Cancer Lett 135:171–180

Singh YN (1992) Kava: an overview. J Ethnopharmacol 37:13–45

Szepesi G, Gazdag M, Ivancsics R, Mihalyfi K, Kovacs P (1983) Investigation of degradation mechanism of pilocarpine by HPLC. Pharmazie 38:94–98

Testa B, Carrupt PA, Gal J (1993) The so-called "interconversion" of stereoisomeric drugs: an attempt at clarification. Chirality 5:105–111

Volz H-P, Kieser M (1997) Kava-kava extract WS 1490 versus placebo in anxiety disorders – a randomized placebo-controlled 25-week outpatient trial. Pharmacopsychiatry 30:1–5

Wu YW (1987) Probing into the mechanism of action, metabolism and toxicity of gossypol by studying its (+)- and (–)- stereoisomers. J Ethnopharmacol 20:65–78

Wu D (1989) An overview of the clinical pharmacology and therapeutic potential of gossypol as a male contraceptive agent and in gynaecological disease. Drugs 38: 333–341

CHAPTER 4

Drug Racemization and Its Significance in Pharmaceutical Research

M. REIST, B. TESTA, and P.-A. CARRUPT

A. Introduction

The presence of one or more elements of chirality (i.e., centers, axes or planes of chirality, and generally helicity) (CAHN et al. 1966; TESTA 1979) in drug molecules generates specific properties which may be advantageous in some cases, but inevitably require special consideration and studies. Examples of advantages include the possibility of increased selectivity and the fact that chirality per se is an invaluable probe in molecular pharmacology and biochemistry (TESTA 1989, 1990; TESTA and TRAGER 1990). In contrast, problems generated by stereoisomerism include the need for stereospecific synthetic and analytical methods, the influence of the degree of resolution on activity (BARLOW et al. 1972), and the increased complexity of metabolic, pharmacological, and clinical studies (ARIËNS 1986; TESTA et al. 1993a).

One additional complicating factor in such studies is the possibility of low configurational stability, which results in nonnegligible racemization or epimerization (TESTA et al. 1993a,b). Ignoring the question of possible chiral inversion can cause misleading or ambiguous results in the pharmacodynamic and pharmacokinetic study of pure enantiomers. Thus, special attention to this problem is required. In this chapter, we examine the problem of low configurational stability as encountered for a number of chiral drugs and resulting in racemization or epimerization. The focus is on implications in pharmaceutical research and drug development. After offering a clarification of some relevant terms and concepts, methods available to screen compounds for their configurational stability are briefly reviewed. Prediction of low configurational stability and chemical factors influencing it are discussed in Sects. D and E. Finally, the case of thalidomide is presented as a most relevant and timely example of the interplay of biological activities, metabolism, and stereoselectivity of drugs with low configurational stability.

B. Background and Concepts

I. Racemization, Enantiomerization, Diastereomerization and Epimerization

The configurational lability of stereoisomeric drugs can be studied at two levels, namely macroscopic and microscopic. At the macroscopic and statistical level, the process considered is the formation of an equilibrium mixture of two stereoisomers when starting from a single stereoisomer. At the microscopic and molecular level, we consider the reversible conversion of one stereoisomer into the other (REIST et al. 1995a).

In the case of enantiomers, i.e., compounds containing only one element of chirality, the irreversible process leading to equilibrium is known as *racemization*, and at equilibrium a racemic mixture (i.e., a mixture of enantiomers in equal amounts) is obtained (Fig. 1). The rate constant of racemization (k_{rac}) applies to the rate of formation of the racemic mixture, starting from a single enantiomer or an enantiomerically enriched mixture. In contrast to racemization, the term *enantiomerization* describes the microscopic process occurring at the level of individual molecules and is defined as the reversible conversion of one enantiomer into the other. The microscopic chemical model in Fig. 2 shows that one enantiomer inverts its configuration, from *R* to *S* or from *S* to *R*, by passing through a transition state or an intermediate product. The rate constant of enantiomerization (k_{enant}) thus applies to this process of *R*-to-*S*, or *S*-to-*R*, conversion. A comparison of Figs. 1 and 2 makes it evident that the rate of enantiomerization is half that of racemization, since the interconversion of one molecule reduces the enantiomeric excess by two molecules. Thus, when discussing the configurational lability of chiral compounds, it is obviously of importance to discern whether the reported rate constants refer to the interconversion of enantiomers (i.e., enantiomerization) or to the formation of the racemic mixture (i.e., racemization).

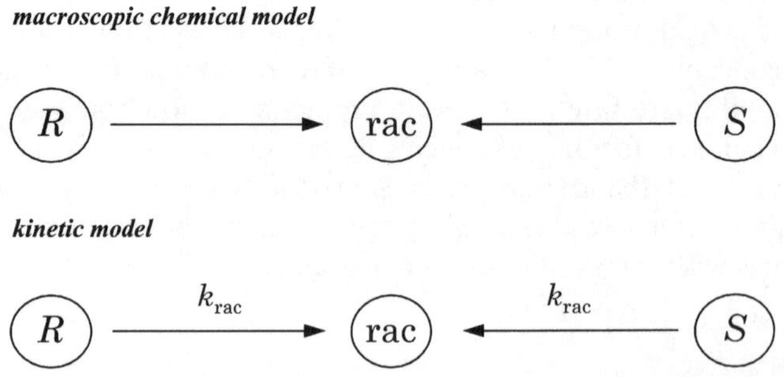

Fig. 1. Chemical and kinetic models of racemization. *R*, (*R*)-enantiomer; *S*, (*S*)-enantiomer; *rac*, racemic mixture; k_{rac}, rate constant of racemization. (Reproduced with permission from REIST et al. 1995a)

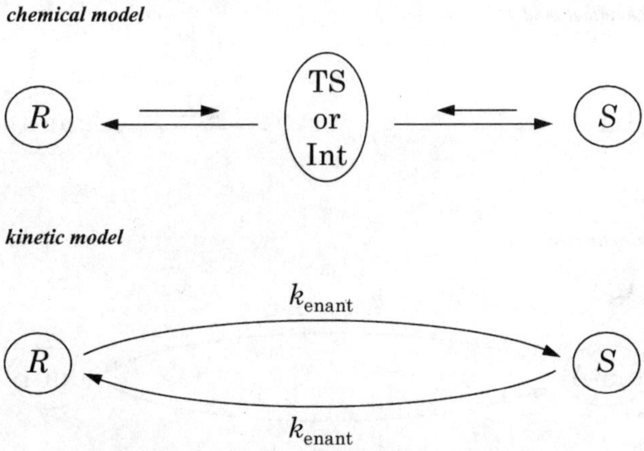

Fig. 2. Chemical and kinetic models of enantiomerization. *R*, (*R*)-enantiomer; *S*, (*S*)-enantiomer; *TS*, transition state; *Int*, intermediate product; k_{enant}, rate constant of enantiomerization. (Reproduced with permission from Reist et al. 1995a)

A number of cases are known of drugs that can exist as two or more diastereomers and which contain one stereogenic element of poor configurational stability. There are some examples in the literature of interconversion of geometric isomers, and more numerous examples of interconversion of epimers (Testa and Trager 1990). Since epimers are but a particular case of diastereomers, the class and its subclass will be discussed together here. In contrast to enantiomers, two diastereomers obligatorily differ in their internal energy (heat of formation), however small the difference. It follows that when the interconversion of two diastereomers is left to proceed to equilibrium, an exact 50/50 ratio cannot be reached. As a result, the rate constant of the conversion of diastereomer A to diastereomer B can only be different from that of the reverse reaction, the difference being small or large depending on the relative heat of formation of the two isomers. The conversion of one diastereomer (or one epimer) into the other, observed at the microscopic and molecular level, is designated as a reaction of *diastereomerization* (or *epimerization*) (Fig. 3). The two rate constants are designated as $k_{diast/A\text{-}to\text{-}B} \neq k_{diast/B\text{-}to\text{-}A}$ (or $k_{epi/A\text{-}to\text{-}B} \neq k_{epi/B\text{-}to\text{-}A}$). In the case of diastereomers, the only term that exists to describe the macroscopic process of formation of an equilibrium mixture of two isomers when starting from a single diastereomer, is that of mutarotation applicable only to optically active epimers (Reist et al. 1995a).

In brief and as summarized in Table 1, the following terms are recommended to discuss the inversion of stereoisomers. Depending on the nature of the stereoisomers (i.e., enantiomers, diastereomers, or epimers), the terms of enantiomerization, diastereomerization, and epimerization, respectively, describe the microscopic process. The terms of racemization (in the case of enantiomers) and mutarotation (in the case of optically active epimers) refer to the macroscopic process.

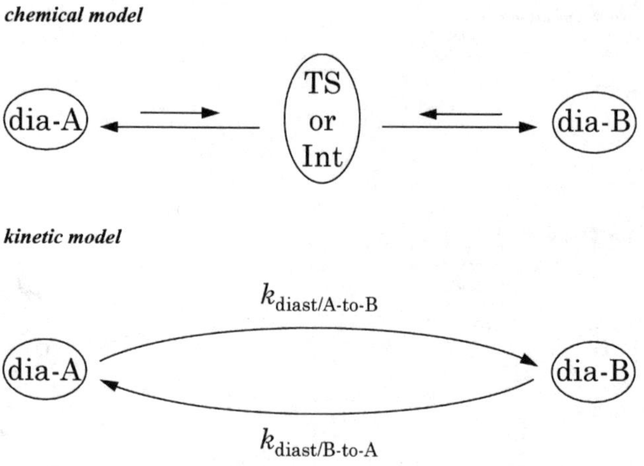

Fig. 3. Chemical and kinetic models of diastereomerization (or epimerization), i.e., of interconversion between diastereomer A and diastereomer B. *TS*, transition state; *Int*, intermediate product; k_{diast}, rate constants of diastereomerization. (Reproduced with permission from REIST et al. 1995a)

Table 1. Processes of isomerization of stereoisomers

	Enantiomers	Diastereomers	Epimers
Microscopic process (conversion of one stereoisomer into the other)	Enantiomerization	Diastereomerization	Epimerization (a particular case of diastereomerization)
Macroscopic process (leading to an equilibrium mixture)	Racemization		Mutarotation[a]
Relations	$k_{enant} = 1/2\ k_{rac}$	$k_{diast/A\text{-}to\text{-}B} \neq k_{diast/B\text{-}to\text{-}A}$	$k_{epi/A\text{-}to\text{-}B}$ $\neq k_{epi/B\text{-}to\text{-}A}$ $\neq 1/2\ k_{muta/A\text{-}to\text{-}eq}$ $\neq 1/2\ k_{muta/B\text{-}to\text{-}eq}$

[a]Applies only to optically active epimers.

II. Enzymatic and Nonenzymatic Inversion of Chiral Compounds

Beside nonenzymatic racemization reactions, which are the main object of this chapter, a limited variety of enzymatic (i.e., metabolic) reactions can operate to interconvert stereoisomers (TESTA 1995; TESTA et al. 1993b). Several racemases, such as α-methylacyl-CoA-racemase (SCHMITZ et al. 1995), glutamate-racemase (YAGASAKI et al. 1995), *N*-acylamino acid-racemase (TOKUYAMA and HATANO 1995), mandelate-racemase (MITRA et al. 1995), alanine-racemase (HOFFMANN et al. 1994), aspartate-racemase (YAMAUCHI et al. 1992), and hydantoin racemase (LICKEFETT et al. 1993), have been isolated, identified and characterized. Also, reactions of oxidoreduction (TESTA 1995; TESTA and JENNER 1976), sulfoxide/sulfide and *N*-oxide/tertiary amine equilibria, as well as reactions of conjugation, offer some interesting cases of enzymatic chiral inversion (TESTA et al. 1993b). A well-known example is that of anti-inflammatory 2-arylpropionates (MAYER 1990, see also Chap. 14, this volume). The inactive (*R*)-enantiomer of ibuprofen and a few analogues are enantioselectively conjugated with coenzyme A to yield the (*R*)-acyl-CoA thioester. This conjugate is then epimerized to a mixture of the (*S*)-and (*R*)-acyl-CoA, followed by hydrolysis to yield a profen enriched in the (*S*)-form. However, such enzymatic reactions should not be confused with nonenzymatic inversion of chiral compounds, which will be discussed in the remainder of this chapter.

Cases indeed exist of drugs whose enantiomers interconvert readily due to the configurational instability of their element of chirality. A representative example is the drug *oxazepam*, whose asymmetric carbon-3 undergoes rapid inversion (ASO et al. 1988). Its kinetics of racemization is interesting and relevant from a pharmacological viewpoint, since it indicates that oxazepam racemizes at ambient temperature and in the neutral pH range with a pseudo-first-order rate constant of $0.1 \pm 0.05\,\text{min}^{-1}$, suggesting a half-life of racemization of 1–4 min at 37°C. This rate of *racemization* is extremely fast compared to the duration of action of the drug, indicating that oxazepam is correctly viewed as a single compound existing in two very rapidly interconverting chiral states.

The case of oxazepam is obviously an extreme example of a racemization that is so rapid that the separation and pharmacological testing of the separate enantiomers would be pointless and impossible, respectively. However, between apparently complete configurational stability and this rapid chiral inversion, all possibilities exist a priori, forming a continuum of high to intermediate to low stability. A case in point is that of the anticholinergic alkaloid (*S*)-(−)-hyoscyamine, whose asymmetric carbon can be partly racemized during isolation and storage in solution (RAMA SASTRY 1981), offering a rationale for the therapeutic use of *rac*-hyoscyamine, i.e., *atropine*. Unfortunately, studies appear to be lacking on the configurational lability of (*S*)-(−)-hyoscyamine under physiological conditions of temperature and pH, perhaps because the reaction is not detectable under these conditions.

The two above examples deal with the configurational stability of compounds having one center of chirality and undergoing racemization. Compounds having two or more elements of chirality and undergoing epimerization are also of relevance in pharmaceutical research since epimerization is documented for a number of drugs. *Pilocarpine*, whose absolute configuration is 2*S*;3*R*, thus epimerizes to (2*R*;3*R*)-isopilocarpine by inversion of the C(2)-chiral center. A detailed mechanistic and kinetic study has demonstrated hydroxide ion-catalyzed epimerization and hydrolytic ring opening (NUNES and BROCHMANN-HANSSEN 1974). From the second-order rate constant of the two reactions, activation energies of 28.5 and 25.0 kcal/mol, respectively, were calculated. The data provided allow one to estimate a pseudo-first-order rate constant of pilocarpine disappearance (epimerization to isopilocarpine plus hydrolysis to pilocarpic acid) of about $8 \times 10^{-4}\,h^{-1}$ at pH 7.4 and 35°C, i.e., a $t_{1/2}$ of about 36 days. At 35°C, about 26% of pilocarpine undergoes epimerization. Of mechanistic importance is the fact that epimerization does not occur in pilocarpic acid due to the presence of the free carboxylic group that acts as a strong configurational stabilizer (see Table 3).

III. Relevant Time Scales for the Configurational Instability of Drugs

An important aspect to take into account when considering the configurational lability of stereoisomers is its significance in drug research and development. Indeed, one should always bear in mind that configurational stability and lability are relative phenomena. Given proper conditions (e.g., temperature and pH), no stereoisomer is configurationally stable. However, only two time scales (and their related sets of conditions) are of relevance as far as drug research is concerned (REIST et al. 1995a). As schematized in Fig. 4, the *pharmacological time scale* applies to the time of residence of a drug in the body and under physiological conditions (37°C, pH 7.4; e.g., oxazepam). Very fast rates of isomerization (in the order of minutes) are of interest only for drug-receptor or drug-enzyme recognition, whereas a half-life of isomerization of months is no longer of importance in pharmacology and therapy.

Half-lives of isomerization in the order of months or a few years are important in a pharmaceutical perspective, i.e., compared to the duration of the manufacturing process and the shelf-life of drugs. This is the *pharmaceutical time scale*, which applies to the manufacturing process and to the shelf-life of drugs (e.g., hyoscyamine and pilocarpine).

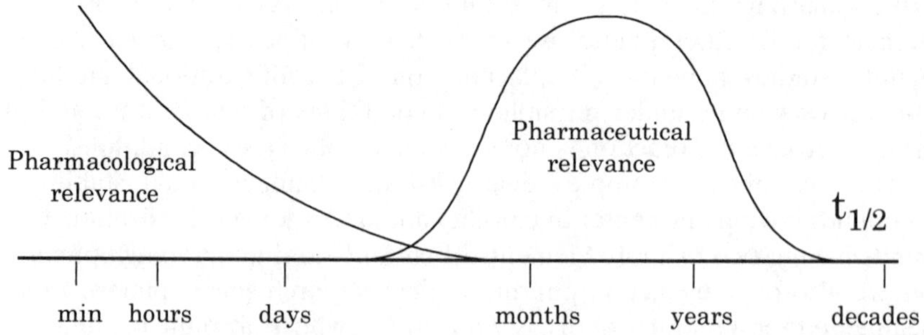

Fig. 4. Time scales of relevance for the configurational instability of drugs (indicative $t_{1/2}$ of interconversion). (Reproduced with permission from REIST et al. 1995a)

C. Methods to Assess Configurational Stability

I. Direct and Indirect Methods

A brief overview of the main methods available to study the configurational stability (or lack thereof) of chiral compounds is given here (TESTA 1979; TESTA and JENNER 1978, see also Chap. 2, this volume). According to Table 2, these methods can be divided into direct and indirect ones. Direct methods investigate the chiral inversion (e.g., racemization, enantiomerization or epimerization) of an optically active compound. They follow directly the increase or decrease of the concentration of an enantiomer or the loss of optical activity of a substance over a period of time. As shown in Table 2, the direct methods may be further subdivided into two different categories depending on whether they involve physical separation of the enantiomers or not (ALLENMARK 1991). *Direct methods* are traditional ones and will not be discussed, except to mention that their implementation calls for the availability of the separate enantiomers.

On the other hand, *indirect methods* take advantage of the fact that inversion of chiral centers and substitution reactions share a common mechanism (COWARD and BRUICE 1969; CRAM et al. 1961; INGOLD and WILSON 1934). What is being monitored is not the chiral inversion itself, but the substitution reaction. Almost a century ago, a number of authors noted that substitution reactions are accompanied by inversion of configuration (e.g., HOLMBERG 1913; WERNER 1911). Various studies investigating the steric course of substitution reactions have been published since, and it is well accepted that a number of chemicals experience the same mechanism for substitution and chiral inversion. For example, base-catalyzed racemizations, halogenations and deuterations of carbon atoms α to a carbonyl group share a common mechanism of

Table 2. Methods for the determination of the configurational stability of chiral compounds (REIST et al. 1997)

Direct (chiral) methods		Indirect methods
Methods involving physical separation (resolution)	Methods not involving physical separation	
LC (HPLC, SFC, CPC)	Polarimetry	Substitution reactions (proton-deuterium substitution monitored by ^{1}H-NMR)
GC	CD	
CE	ORD	
	^{1}H-NMR with chiral solvents, chiral shift reagents or inclusion complexes	

the S_E1 type, with a rate-limiting formation of a carbanion intermediate stabilized by resonance (HSÜ and WILSON 1936; HSÜ et al. 1938; INGOLD and WILSON 1934). Hence the configurational stability of a chiral compound can be determined by monitoring a substitution reaction. A condition for a quantitative interpretation of results, however, is that previous studies must have established the existence of a common mechanism of racemization and substitution in analogs.

The indirect methods have the major advantage of not needing separate enantiomers, and are thus of marked interest as a means of screening the configurational stability of chiral drugs prior to the synthesis or resolution of their enantiomers. An indirect method of interest is proton-deuterium substitution (MATSUO et al. 1967; MAYER et al. 1989; TESTA 1973), discussed here in some detail.

II. Screening of Compounds for Configurational Stability by ^{1}H-NMR

A substitution reaction that can easily be monitored by ^{1}H-NMR is *proton-deuterium substitution* (KAWAZOE and OHNISHI 1964; MATSUO et al. 1967). When a compound having a chirally labile carbon atom is dissolved in deuterium oxide (or an other solvent with exchangeable deuterium), the proton-deuterium exchange occurs irreversibly due to the absence of protium, and the deuteration (i.e., racemization or enantiomerization) can be followed by integrating the decreasing signal of the proton connected to the chiral carbon.

Four limiting ratios of k_{deut} (rate constant of deuteration) over k_{rac} (rate constant of racemization) can be envisaged (CRAM and CRAM 1973; MARCH 1985; ROITMAN and CRAM 1971):

1. If deuteration occurs with complete retention of configuration, the ratio k_{deut}/k_{rac} approaches infinity (isoinversion). Such a reaction was seen in nondissociating nonpolar solvents and appears to have little relevance in a pharmacological and pharmaceutical context.
2. If deuteration takes place with complete racemization, each carbanion is captured from either side with equal probability, and k_{deut}/k_{rac} equals unity (i.e., the rate of deuteration would be identical to that of racemization).
3. If deuteration occurs with inversion of configuration (i.e., enantiomerization), its rate is half that of racemization (i.e., equal to that of enantiomerization) and the ratio k_{deut}/k_{rac} equals 0.5.
4. If racemization takes place without deuteration, the ratio k_{deut}/k_{rac} approaches zero (isoracemization). Such reactions were observed in aprotic solvents with aprotic bases such as tertiary amines and are of little relevance in a pharmacological and pharmaceutical context.

Thus only cases 2 and 3 appear of relevance in pharmaceutical research and drug design.

A few years ago, we validated ^{1}H-NMR spectroscopy as an indirect method to monitor rates of racemization using 5-substituted hydantoins as model compounds (Reist et al. 1996). The comparison between ^{1}H/^{2}H substitution monitored by ^{1}H-NMR and racemization investigated by chiral HPLC was performed for (*S*)-5-phenyl- and (*S*)-5-benzylhydantoin because of their good resolution on the chiral stationary phase (compounds 1 and 4 in Fig. 5). Identical solutions (solvent mixture, pH or pD, ionic strength, concentrations of test compound and of sodium 3-trimethylsilylpropionate-d_4) were investigated under identical conditions (four temperatures between 26°C and 80°C, sealed tubes) by both methods. To detect a possible solvent isotope effect, chiral HPLC experiments were performed in deuterated and in nondeuterated solvents. The experiments yielded the activation energies of deuteration, racemization in nondeuterated solvents, and racemization in deuterated solvents, as calculated from Arrhenius plots. Within experimental errors, each compound had the same activation energy in all three reactions. This confirms a common mechanism of ^{1}H/^{2}H substitution and racemization for 5-substituted hydantoins. This mechanism was of the S_E2 push-pull type, with the reaction of deuteration seemingly occurring with inversion of configuration (case 3 above).

As already mentioned, the main advantage of an indirect ^{1}H-NMR spectroscopic method is the fact that the configurational stability of a product can be checked with the racemate. The disadvantages of the method are a poorer sensitivity and inferior precision compared to chiral HPLC. The sensitivity of magnetic resonance methods has improved following the introduction of high-frequency spectrometers, but is still below that of HPLC applications. As for

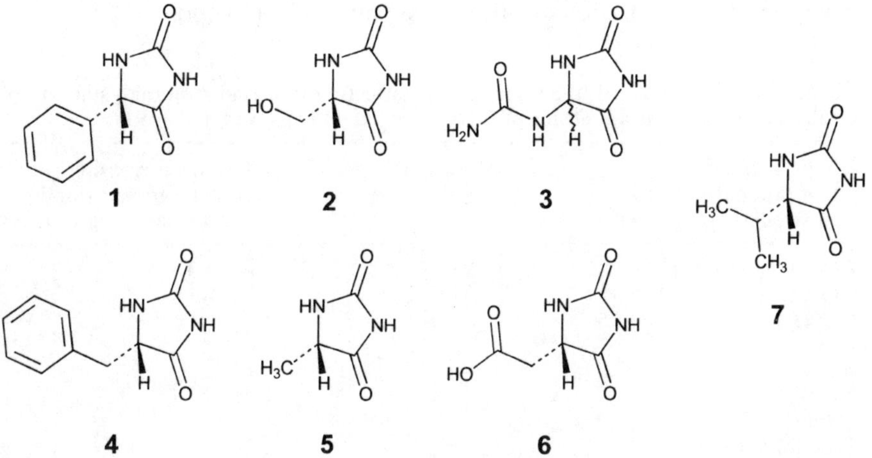

Fig. 5. Molecular structures of the investigated 5-monosubstituted hydantoins. *1*, (*S*)-5-phenyl-; *2*, (*S*)-5-(hydroxymethyl)-; *3*, (*RS*)-5-ureido-; *4*, (*S*)-5-benzyl-; *5*, (*S*)-5-methyl-; *6*, (*S*)-5-(carboxymethyl)-; *7*, (*S*)-5-isopropylhydantoin

precision, it is markedly lower when the signal of interest (i.e., the signal of the proton attached to the chiral center) is near a large solvent peak or even masked by the spinning side-bands of the solvent (SANDERS and HUNTER 1993). Nevertheless, ^{1}H-NMR spectroscopy is a valuable alternative method to chiral approaches.

D. Predicting Configurational Stability

Based on a literature survey, it was possible some years ago to propose a preliminary generalization of the structural factors decreasing the configurational stability of a chirally substituted carbon atom of the type R″R′RC–H (TESTA et al. 1993b). Configurational inversion at such a center is generally catalyzed by bases, and involves the deprotonated form (i.e., the carbanion R″R′RC^{-}) as the intermediate or transition state. It thus appears convenient to distinguish between acid-strengthening, neutral, and acid-weakening groups (GU and STRICKLEY 1987).

Such functional groups usually act by stabilizing the carbanion, but effects on the acid or the transition state should not be neglected. To be of pharmaceutical or even pharmacological relevance, low configurational stability appears to require the presence of either (a) three carbanion-stabilizing groups, or (b) two carbanion-stabilizing groups (one of which must be strong) and one neutral group. A provisional list of such groups, based on available evidence, is given in Table 3 (TESTA et al. 1993b). A number of question marks were added to indicate suspected cases. Much additional quantitative work is needed to verify and complete the list in Table 3, and to assess the relative importance of mesomeric, inductive, and solvation contributions. More distant objectives, perhaps difficult or impossible to reach, are a ranking of the listed groups and the elaboration of rules for quantitative prediction.

Table 3. Provisional list of functional groups affecting the configurational stability of chirally substituted carbon atoms of the type R″R′–H (TESTA et al. 1993b)

Groups decreasing configurational stability (acid-strengthening groups)	Neutral groups	Groups increasing configurational stability (acid-weakening groups)
–CO–O–*R* (strong)	$-CH_3$	–COO–
–CO–aryl (strong)	$-CH_2CH_3$	$-SO_3^-$
–CONRR′		
–OH		
–NRR′		
–N=R		
–Halogens		
–Pseudohalogens		
–Aryl		
$-CH_2$–aryl		
$-CH_2OH$		

In an effort to gain insight into structure-reactivity relationships and to confirm some of the rules in Table 3, we examined the configurational stability of the seven chiral 5-substituted hydantoins shown in Fig. 5 (REIST et al. 1996). $^{1}H/^{2}H$ Substitution rates in a solvent mixture of phosphate buffer (pD 7.4, 0.1 M, ionic strength 0.22) and d_6-DMSO in the proportion 1:1 (v/v) were determined for all compounds by ^{1}H-NMR at four or five different temperatures between 26° and 80°C. The addition of 50% d_6-DMSO as cosolvent was necessary to solubilize some of the compounds. No hydrolysis was observed under these conditions. Semilogarithmic rate plots of the rates of deuteration at 50°C are shown in Fig. 6 and demonstrate the very large differences in reactivity. Half-lives of deuteration at 50°C varied from 3.8 min for (*S*)-5-phenylhydantoin to 115.5 h for (*S*)-5-isopropylhydantoin, a 1800-fold difference. The rates of deuteration decreased in the order phenyl > hydroxymethyl > ureido > benzyl > methyl > carboxymethyl > isopropyl-hydantoin.

Enthalpies and entropies of activation ($\Delta H^{\ddagger}$ and $\Delta S^{\ddagger}$, respectively) were also determined, but did not correlate with the order of reactivity. Hence the configurational stability of the investigated hydantoins depends on both the enthalpy and entropy of activation. In an attempt to rationalize the above reactivity sequence, energies of deprotonation (i.e., the sum of the heats of formation of the carbanion and proton, minus the heat of formation of the neutral molecule) were obtained by semiempirical molecular orbital calculations performed at an AM1 Hamiltonian level. No correlation between the activation

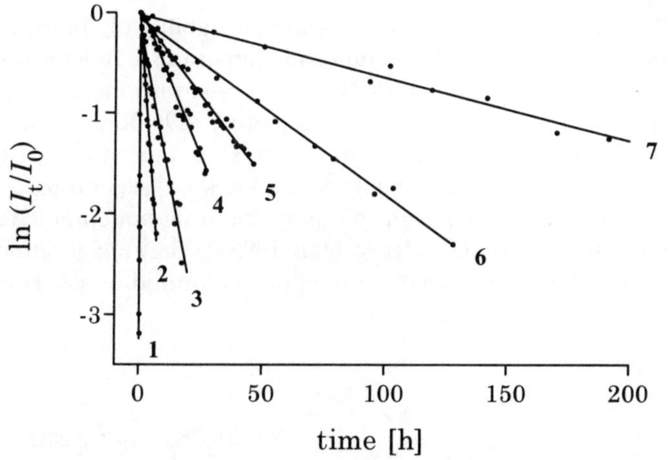

Fig. 6. Pseudo-first-order rate plots for the deuteration of the seven investigated 5-monosubstituted hydantoins at 50°C. *1*, (*S*)-5-phenyl-; *2*, (*S*)-5-(hydroxymethyl)-; *3*, (*RS*)-5-ureido-; *4*, (*S*)-5-benzyl-; *5*, (*S*)-5-methyl-; *6*, (*S*)-5-(carboxymethyl)-; *7*, (*S*)-5-isopropylhydantoin; I_t, integral of exchanging methine proton at time *t*; I_0, integral of exchanging methine proton at time 0. (Reproduced with permission from REIST et al. 1996)

energy $E_{a(deut)}$ and the calculated energy E_{calc} was seen. One of the implications of these data is that inductive and resonance effects on the transition state (which are reflected in a change of $\Delta H^{\ddagger}$) are only partially responsible for the observed influences of substituents on the rates of deuteration. Solvation and steric effects (which alter $\Delta H^{\ddagger}$ and $\Delta S^{\ddagger}$) must be of similar importance.

In conclusion, a qualitative estimate of configurational stability is feasible. Indeed, carbonyl-, amine-, amide-, phenyl-, hydroxymethyl-; ureido-, and benzyl groups (all acid-strengthening) were found to decrease, and the acid-weakening carboxy- and carboxymethyl groups to increase, the configurational stability of chiral carbons of the type R″R′RC-H. Hence, the substituents around the chiral carbon play a determining role whose magnitude depends on their inductive and resonance effects. Given enough experimental evidence, such substituents are classifiable as having a modest or strong effect in stabilizing or destabilizing the configurational stability of chiral carbons.

In contrast, a quantitative prediction of rates of racemization is very difficult in view of the complexity of the intramolecular and intermolecular factors involved. Indeed, solvation and steric effects appear to affect so markedly the entropy of the reaction that their accurate computation is beyond reach at present, and no general rules for a quantitative prediction of substituent effects on the configurational stability of chiral compounds can be derived.

E. Conditions Influencing Configurational Stability

Like all chemical reactions, the configurational stability of chiral compounds depends on conditions such as medium and temperature. Thus, inversion rates of chiral centers of the type R″R′RC-H increase with increasing temperature, pH, and polarity of the solvent (Reist et al. 1996, Reist et al. 2000). An interesting and relevant aspect in this context is the nature of the base-catalyzed mechanism, i.e., specific-base catalysis or general-base catalysis. Here, we summarize studies with the anorectic drug amfepramone and the alkaloid *rac*-cathinone (Fig. 7) (Reist et al. 1995b). The configurational stability of amfepramone in phosphate buffers (as monitored by $^{1}H/^{2}H$-exchange)

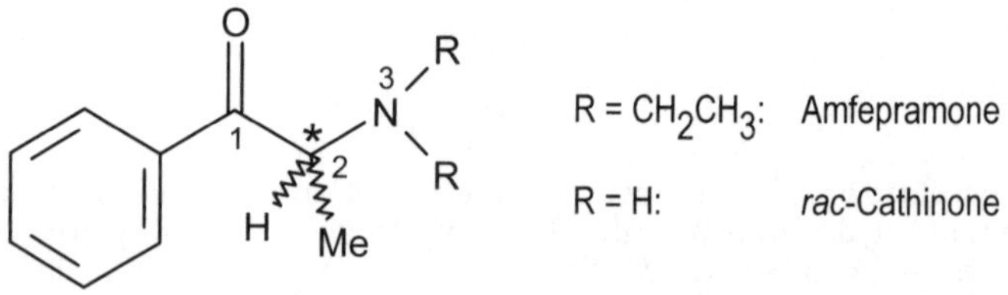

Fig. 7. Molecular structures of amfepramone (diethylpropion) and *rac*-cathinone (α-aminopropiophenone). (Reproduced with permission from Reist et al. 1995b)

was found to decrease with increasing pD (range 2.3–7.5), indicating a base-catalyzed process. Further investigations showed that the proton-deuterium exchange of amfepramone and *rac*-cathinone is subject to general-base catalysis and obeys an S_E1 mechanism (case 2 above). As Fig. 8 illustrates, the observed rate constants for both compounds were a linear function of phosphate concentration. The racemization of amfepramone was also found to follow general-base catalysis by hydroxylamine. Preliminary investigations of the general-base catalyzed racemization of amfepramone by carbonate showed similar results.

The racemization of amfepramone at pD 7.4 was about 5 to 6 times faster in a phosphate buffer than in a hydroxylamine buffer of identical molarity. This demonstrates that the nature of the buffer base has an important influence on the rate of chiral inversion. Hence, comparing literature data and drawing conclusions from them can only be done with great caution. Only when rate constants are obtained under identical conditions (i.e., identical buffer bases and molarities) can they be validly compared and discussed.

A mechanism of general-base catalysis has also important pharmacological implications. Indeed, the racemization of chiral compounds in biological media can be expected to be catalyzed by a variety of endogenous buffers such as plasma proteins, amines, and perhaps anions such as thiolates, phosphate and bicarbonate. Thus, it will not be possible to deduce rates of in vivo racemization or epimerization of chiral drugs from results obtained in nonbiological media. In other words, only experiments conducted in vivo or in biological

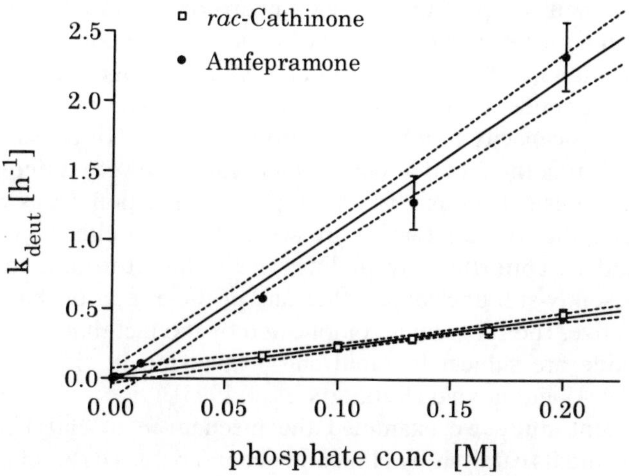

Fig. 8. Influence of phosphate concentration on the rate of proton-deuterium exchange of amfepramone and *rac*-cathinone at 37°C (phosphate buffers of pD 7.4, ionic strength 0.43). (Reproduced with permission from REIST et al. 1995b)

media such as plasma will yield clinically useful figures for racemization or epimerization of configurationally labile chiral drugs.

The above discussions focus on base-catalyzed inversions via a carbanion, but acid-catalyzed inversions have also been seen in a very limited number of cases. Another possibility is acid-catalyzed enolization involving an acyl group. However, systematic studies of medicinal relevance do not appear to be available.

F. The Case of Thalidomide: Interplay of Configurational Stability, Biological Activities, Metabolism, and Stereoselectivity

Its sad history made thalidomide one of the most (in)famous chiral drugs, to become synonymous with tragedy following the discovery of its catastrophic teratogenic effects (MELLIN and KATZENSTEIN 1962). Despite this justifiable, negative accreditation, clinical interest in thalidomide began to rise again when SHESKIN reported that thalidomide could cause a dramatic improvement of the inflammatory reactions in leprosy patients (SHESKIN 1965). The clinical benefit exhibited by the drug was confirmed, and it is now widely employed for the treatment of erythema nodosum leprosum. Thalidomide has also been successfully used to treat various inflammatory conditions and autoimmune diseases (STEVENS et al. 1997; VOGELSANG et al. 1992).

Here, the case of thalidomide is discussed as an example of the interplay of biological activities, metabolism, and stereoselectivity of drugs having low configurational stability. The biological activities of thalidomide and the complexity of its metabolism and stereoselectivity are outlined in Fig. 9. Beside its sedative, neurotropic, teratogenic, neurotoxic, antiinflammatory, and immunomodulatory effects, thalidomide has also been reported to reduce HIV type 1 replication in vitro (MOREIRA et al. 1997). Although the mechanisms of action of thalidomide are only poorly understood, some of its activities may be related to its capacity to inhibit the production of TNF-α (SHANNON et al. 1997). Considering the fact that enantiomers can have very different pharmacological activities (TESTA and TRAGER 1990), the question arises whether the biological activities of thalidomide are associated with either one or both of the enantiomers. Unfortunately, findings in the literature are contradictory, and it is presently still unclear whether any of the effects of thalidomide are enantioselective. The problem is complicated by the fact that the enantiomers of thalidomide are subject to rapid chiral inversion in vitro (KNOCHE and BLASCHKE 1994) and in vivo (ERIKSSON et al. 1995).

In a recent study, we examined the mechanism of chiral inversion of thalidomide and its catalysis by HSA and general bases (REIST et al. 1998). All experiments were performed at 37°C using a stereoselective HPLC assay. As expected (KNOCHE and BLASCHKE 1994), human serum albumin catalyzed the chiral inversion of thalidomide. The observed pseudo-first-order rate constants

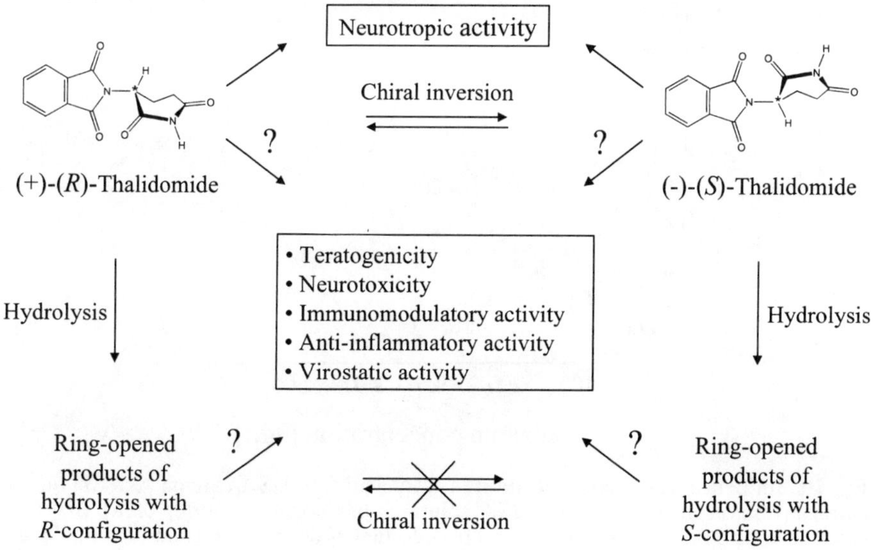

Fig. 9. Interplay of biological activities, metabolism and stereoselectivity of thalidomide enantiomers and their ring-opened metabolites. (Reproduced with permission from REIST et al. 1998)

of chiral inversion increased with increasing HSA concentration (Fig. 10). To obtain further insights into the mechanism of chiral inversion of thalidomide and its catalysis by HSA, the influence of pH (2.0–11.0), phosphate concentration (0.05–0.3 M) and various acidic, neutral and basic amino acids (0.1 M solutions of Arg, Asp, Gln, Glu, Gly, Lys, and Ser) was investigated. Our results showed that the chiral inversion of thalidomide was pH-dependent. At acidic pH the rate of chiral inversion was about zero, and it increased with increasing pH, suggesting a base-catalyzed process. It was also shown that at pH 7.4 the rate of inversion was linearly dependent on phosphate concentration. Thus, chiral inversion is deduced to be subject to general-base catalysis. The influence of various acidic, neutral and basic amino acids on the chiral inversion of thalidomide showed that the basic amino acids Arg and Lys (which possess two basic groups) had a superior catalytic potency than neutral and acidic amino acids with only one basic group. This also implies a general-base catalysis in the chiral inversion of thalidomide. As albumin has many reactive ε-amino groups and almost all of its ligand binding sites involve lysine and/or arginine residues (PETERS 1996), it is suggested that the ability of albumin to catalyze the chiral inversion of thalidomide is due to its Arg and Lys residues. No single catalytic site on the albumin molecule is thought to exist. This suggestion is also consistent with the fact that the chiral inversion of thalidomide in vivo, in plasma and in blood was found to be slower than in HSA solutions of physiological concentrations (ERIKSSON et al. 1995). Endogenous albumin

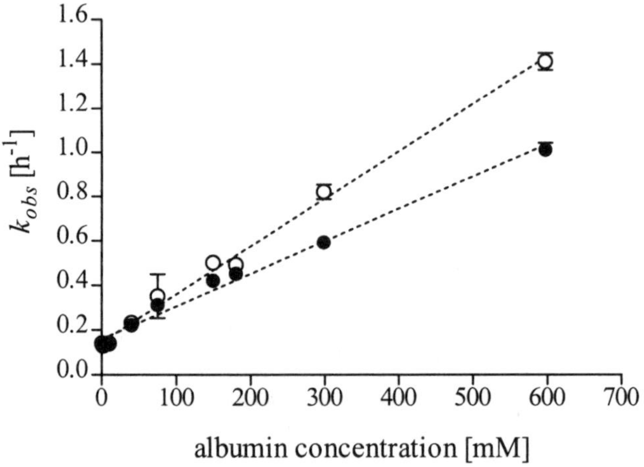

Fig. 10. Influence of the concentration of fatty acid-free HSA (Sigma A-1887) on the chiral inversion of thalidomide (HSA solutions in phosphate buffer 0.1 M, pH 7.40, ionic strength 0.3, 37°C). k_{obs}, Observed pseudo-first-order rate constant of chiral inversion; *open circles*, inversion from *S* to *R*; *filled circles*, inversion from *R* to *S*. Values are means (±standard deviation) of triplicate determinations. Some standard deviations were smaller than the symbols

ligands such as fatty acids and bilirubin may block the catalytic basic amino acid residues of HSA and as a result reduce the catalysis of the chiral inversion. Furthermore, the rates of chiral inversion of (*R*)- and (*S*)-thalidomide showed a slight HSA concentration-dependent stereoselectivity (Fig. 10). At low albumin concentrations no significant difference between the rates of inversion of the two enantiomers could be observed, whereas at physiological concentration the inversion from (*S*)- to (*R*)-thalidomide was about 1.4 times faster than that from the (*R*)- to the (*S*)-enantiomer. This may be due to the fact that albumin is subject to concentration-dependent polymerization caused by various factors such as heating, freeze-drying and others (ZINI et al. 1981). For the commercial fatty-acid free HSA used in the present study, which is freeze-dried, it is known that at low albumin concentrations no polymerization occurs whereas at a physiological concentration up to 50% of the HSA can be polymerized (ZINI et al. 1981). This polymerization might lead to conformational changes in the HSA molecule, inducing stereoselectivity by an unknown mechanism. This hypothesis is in agreement with the finding that in freshly drawn blood no significant difference was observable between the rates of chiral inversion of the enantiomers (ERIKSSON et al. 1995). No albumin polymerization is believed to exist in untreated human blood. In vivo in humans, it was even found that (*R*)-thalidomide inverses its configuration faster than (*S*)-thalidomide (ERIKSSON et al. 1995). Hence the stereoselectivity of the chiral inversion of thalidomide is dependent on the medium and it would be misleading to draw conclusions on in vivo stereoselectivity from in vitro studies. In summary, the chiral inversion of thalidomide is suggested to occur by electrophilic substitution with protons as incoming and leaving groups. Specific and general-base catalysis accelerates the reaction by facilitating the abstraction of the proton bound to the chiral center (REIST et al. 1998).

Parallel to inverting its configuration, thalidomide is hydrolyzed to ring-opened products (Fig. 9). The drug has been found to be eliminated mainly by spontaneous hydrolysis in blood and tissues of humans and animals (CHEN et al. 1989; SCHUMACHER et al. 1965a). All four amide bonds of the molecule are susceptible to hydrolytic cleavage at pH > 6 (SCHUMACHER et al. 1965b) (Fig. 11). The main urinary metabolites in humans are 2-phthalimidoglutaramic acid (3) to about 50% and α-(*o*-carboxybenzamido)-glutarimide (4) to about 30% (BECKMANN 1963). Of the 12 hydrolysis products of thalidomide only those three which contain the intact phthalimide moiety showed teratogenic activity, i.e., 2- (3) and 4-phthalimidoglutaramic acid (2), and 2-phthalimidoglutaric acid (5) (KOCH 1981) (Fig. 11). An investigation of the teratogenic potency of the optical antipodes of 2-phthalimidoglutaric acid (5) in pregnant mice showed that the (*S*)-enantiomer caused dose-dependent

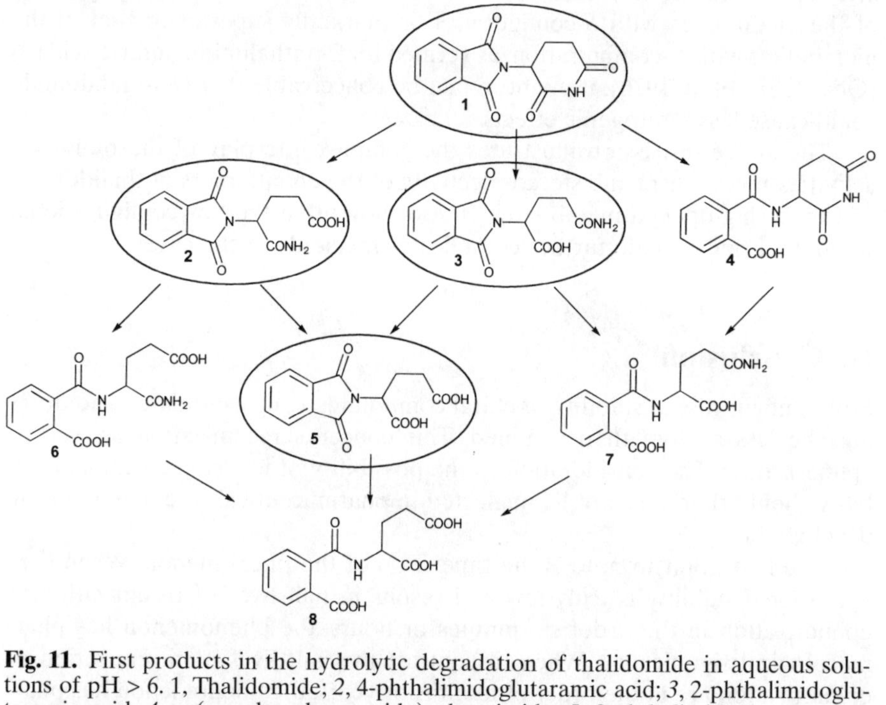

Fig. 11. First products in the hydrolytic degradation of thalidomide in aqueous solutions of pH > 6. 1, Thalidomide; *2*, 4-phthalimidoglutaramic acid; *3*, 2-phthalimidoglutaramic acid; *4*, α-(*o*-carboxybenzamido) glutarimide; *5*, 2-phthalimidoglutaric acid; *6*, 4-(*o*-carboxybenzamido)glutaramic acid; *7*, 2-(*o*-carboxybenzamido) glutaramic acid; *8*, 2-(*o*-carboxybenzamido)glutaric acid. Teratogenic compounds are *circled*. (Reproduced with permission from REIST et al. 1998)

teratogenicity, whereas the (*R*)-enantiomer was devoid of effect even at four times higher doses (OCKENFELS et al. 1976).

The discussion of enantioselective pharmacological activities of the ring-opened products of hydrolysis implies the consideration of their configurational stability. To address this question, we investigated the configurational stability of the three teratogenic metabolites of thalidomide 2, 3, and 5 (Fig. 11) by using the indirect method of proton-deuterium substitution monitored by ^{1}H-NMR (REIST et al. 1998). In a 0.3-M phosphate buffer of pD 7.4 at 37°C, no deuteration was observed for any of the three metabolites in up to 7 days of incubation. This is in agreement with the fact that a carboxy group is known to stabilize chiral carbon atoms of the type R″R′-H (see Sect. D. and Table 3). The complete stability of their configuration for at least 7 days suggests that the pharmacological testing of the separate enantiomers produces unbiased results. Further, the configurational stability of the teratogenic metabolites of thalidomide invites to reflect on a possible enantioselectivity on the teratogenicity of thalidomide. Given the stable configuration of the three teratogenic metabolites, inversion of the chiral center must stop with the hydrolysis of thalidomide. Thus, after administration of (*S*)-thalidomide the concentration of teratogenic metabolites with *S*-configuration is postulated to be higher than after application of (*R*)-thalidomide. Assuming that the teratogenic potency of the metabolites with *S*-configuration is markedly superior to that of the metabolites with *R*-configuration, as verified for 2-phthalimidoglutaric acid (5) (OCKENFELS et al. 1976), it might in fact be conceivable that (*R*)-thalidomide could cause less teratogenic effects.

The above discussion illustrates the complex interplay of the biological activities, metabolism and stereoselectivity of the enantiomers of thalidomide and their ring-opened metabolites. It also shows that the low configurational stability of thalidomide further complicates the study of the case.

G. Conclusion

Low configurational stability of chiral compounds, drug candidates, and drugs may be less seldom than assumed. This concerns racemization as well as epimerization. The consideration of the possibility of low configurational stability should therefore not be neglected in pharmaceutical research and drug development.

The first point to note is the time scale of the phenomenon. When configurational stability is truly low and results in half-lives of racemization or epimerization in the order of minutes or hours, the phenomenon has pharmacological significance. When in contrast the half-lives of racemization or epimerization are in the order of months or years, the phenomenon has pharmaceutical significance and may shorten the shelf-live of the drug. Presumably the majority of reactions of nonenzymatic racemization or epimerization have pharmaceutical but not pharmacological relevance.

The second message emerging from the above examples and discussions is that a fast method exists to screen the configurational stability of drug candidates having a chiral center of the type R″R′-H, namely proton-deuterium exchange. As an alternative method to chiral approaches it has the considerable advantage in that it can be performed with the racemate. In other words, it is a convenient method to screen chiral drug candidates prior to their resolution, which can be especially useful in the early stages of drug development. This advantage more than compensates the main disadvantage of the method, which is its lower precision compared to chiral HPLC.

The third point made here is that a qualitative prediction of configurational stability is feasible. The substituents around the chiral carbon play a determining role whose magnitude depends on their inductive and resonance effects. Thus, given enough experimental evidence, such substituents should be classifiable as having a modest or strong effect in stabilizing or destabilizing the configurational stability of chiral carbons. However, the complexity of the intramolecular and intermolecular factors involved renders a quantitative prediction of rates of racemization very difficult. Indeed, solvation and steric effects appear to affect so markedly the entropy of the reaction that their accurate computation is beyond reach at present.

Another message to keep in mind is that a mechanism of general-base catalysis has important pharmacological implications. Indeed, the racemization of chiral compounds in biological media is expected to be catalyzed by a variety of endogenous buffers such as plasma proteins, amines, and perhaps thiolates, phosphate and bicarbonate. It will therefore not be possible to deduce rates of in vivo racemization or epimerization of chiral drugs from results obtained in nonbiological media. Only experiments conducted in vivo or in biological media such as blood or plasma will yield clinically useful figures for racemization or epimerization of configurationally labile chiral drugs.

Finally, the case of thalidomide demonstrates that the presence of an element of chirality with low configurational stability in drug molecules requires a thorough investigation of the drug's pharmacological activities, metabolism, and stereoselectivity.

References

Allenmark S (1991) Chromatographic Enantioseparation: Methods and Applications. Ellis Horwood, London

Ariëns EJ (1986) Stereochemistry: a source of problems in medicinal chemistry. Med Res Rev 6:451–466

Aso Y, Yoshioka S, Shibazaki T, Uchiyama M (1988) The kinetics of the racemization of oxazepam in aqueous solution. Chem Pharm Bull 36:1834–1840

Barlow RB, Franks FM, Pearson JDM (1972) The relation between biological activity and the degree of optical resolution of optical isomers. J Pharm Pharmacol 24: 753–761

Beckmann R (1963) Ueber das Verhalten von Thalidomid im Organismus. Arzneim-Forsch 13:185–191

Cahn RS, Ingold C, Prelog V (1966) Specification of molecular chirality. Angew Chem Int Ed 5:385–415
Chen TL, Vogelsang GB, Petty BG, Brundrett RB, Noe DA, Santos GW, Colvin OM (1989) Plasma pharmacokinetics and urinary excretion of thalidomide after oral dosing in healthy male volunteers. Drug Metab Dispos 17:402–405
Coward JK, Bruice TC (1969) Intramolecular amine-catalyzed ketone enolization. A search for concerted intramolecular general-base, general-acid catalysis. J Am Chem Soc 91:5339–5345
Cram DJ, Cram JM (1973) Carbanions and carbonium ions – stereochemical analogs. Intra Science Chem Rep 7:1–17
Cram DJ, Kingsbury CA, Rickborn B (1961) Electrophilic substitution at saturated carbon. XIV. Asymmetric solvation of carbanions in stereospecific hydrogen-deuterium exchange reactions. J Am Chem Soc 83:3688–3696
Eriksson T, Björkman S, Roth B, Fyge A, Höglund P (1995) Stereospecific determination, chiral inversion in vitro and pharmacokinetics in humans of the enantiomers of thalidomide. Chirality 7:44–52
Gu L, Strickley RG (1987) Diketopiperazine formation, hydrolysis, and epimerization of the new dipeptide angiotensin-converting enzyme inhibitor R*S*-10085. Pharm Res 4:392–397
Hoffmann K, Schneider-Scherzer E, Kleinkauf H, Zocher R (1994) Purification and characterization of eucaryotic alanine racemase acting as key enzyme in cyclosporin biosynthesis. J Biol Chem 269:12710–12714
Holmberg B (1913) Stereochemie der halogensubstituierten Bernsteinsäuren. J Prakt Chem 88:553–603
Hsü SK, Wilson CL (1936) Optical activity in relation to tautomeric change. Part VI. Comparison of the rates of racemization and of bromination of a ketone. A Further study under conditions of basic catalysis. J Chem Soc 623–625
Hsü SK, Ingold CK, Wilson CL (1938) Prototropy in relation to the exchange of hydrogen isotopes. Part III. Comparison of the rates of racemization and of hydrogen exchange in α-acidic ketone. J Chem Soc 78–81
Ingold CK, Wilson CL (1934) Optical activity in relation to tautomeric change. Part IV. Comparison of the rates of racemization and of bromination of a ketone. J Chem Soc 773–777
Kawazoe Y, Ohnishi M (1964) Quantitative analysis of active hydrogen by nuclear magnetic resonance. Chem Pharm Bull 11:846–848
Knoche B, Blaschke G (1994) Investigations on the in vitro racemization of thalidomide by high-performance liquid chromatography. J Chromatogr 666:235–240
Koch H (1981) Die Arenoxid-Hypothese der Thalidomid-Wirkung. Ueberlegungen zum molekularen Wirkungsmechanismus des "klassischen" Teratogens. Sci Pharm 49:67–99
Lickefett H, Krohn K, König WA, Gehrcke B, Syldatk C (1993) Enantioseparation of 5-monosubstituted hydantoins by capillary gas chromatography – investigation of chemical and enzymatic racemization. Tetrahedron Asym 4:1129–1135
March J (1985) Advanced Organic Chemistry. Wiley, New York, pp 512–575
Matsuo H, Kawazoe Y, Sato M, Ohnishi M, Tatsuno T (1967) Studies on the racemization of amino acids and their derivatives. I. On the deuterium-hydrogen exchange reaction of amino acid derivatives in basic media. Chem Pharm Bull 15:391–398
Mayer JM (1990) Stereoselective metabolism of antiinflammatory 2-arylpropionates. Acta Pharm Nord 2:197–216
Mayer JM, Young M, Testa B, Etter JC (1989) Modelling the metabolic epimerization of antiinflammatory 2-arylpropanoyl-coenzyme-A conjugates: solvent effects on the $^{1}H/^{2}H$ exchange in *S*-[2-(dimethylamino)ethyl] 2-phenylpropanethioate. Helv Chim Acta 72:1225–1232
Mellin GW, Katzenstein M (1962) The saga of thalidomide: neuropathy to embryopathy, with case reports and congenital anomalies. N Eng J Med 267:1184–1244

Mitra B, Kallarakal AT, Kozarich JW, Gerlt JA, Clifton JG, Petsko GA, Kenyon GL (1995) Mechanism of the reaction catalyzed by mandelate racemase: importance of electrophilic catalysis by glutamic acid 317. Biochemistry 34:2777–2787

Moreira AL, Corral LG, Ye W, Johnson B, Stirling D, Muller GW, Freedman VH, Kaplan G (1997) Thalidomide and thalidomide analogs reduce HIV type 1 replication in human macrophages in vitro. AIDS Res Hum Retrov 13:857–863

Nunes MA, Brochmann-Hanssen E (1974) Hydrolysis and epimerization kinetics of pilocarpine in aqueous solution. J Pharm Sci 63:716–721

Ockenfels H, Köhler F, Meise W (1976) Teratogenic effect and stereospecificity of a thalidomide metabolite. Pharmazie 31:492–493

Peters Jr T (1996) All about Albumin. Biochemistry, Genetics, and Medical Applications. Academic Press, San Diego, pp 9–75

Rama Sastry BV (1981) Anticholinergics: antispasmodic and antiulcer drugs. In: Wolff ME (ed) Burger's Medicinal Chemistry, Fourth Edition, Part III. Wiley, New York, p 371

Reist M, Testa B, Carrupt PA, Jung M, Schurig V (1995a) Racemization, enantiomerization, diastereomerization, and epimerization: their meaning and pharmacological significance. Chirality 7:396–400

Reist M, Christiansen LH, Christoffersen P, Carrupt PA, Testa B (1995b) Low configurational stability of amfepramone and cathinone: mechanism and kinetics of chiral inversion. Chirality 7:469–473

Reist M, Carrupt PA, Testa B, Lehmann S, Hansen JJ (1996) Kinetics and mechanisms of racemization: 5-substituted hydantoins (= imidazoline-2,4-diones) as models of chiral drugs. Helv Chim Acta 79:767–778

Reist M, Testa B, Carrupt PA (1997) The racemization of enantiopure drugs: helping medicinal chemists to approach the problem. Enantiomer 2:147–155

Reist M, Carrupt PA, Francotte E, Testa B (1998) Chiral inversion and hydrolysis of thalidomide: mechanisms and catalysis by bases and serum albumin, and chiral stability of teratogenic metabolites. Chem Res Toxicol 11:1521–1528

Reist M, Roy-de Vos M, Montseny JP, Mayer JM, Carrupt PA, Berger Y, Testa B (2000) Very slow chiral inversion of clopidogrel in rats: a pharmacokinetic and mechanistic investigation. Drug Metal Disposit 28:1405–1410

Roitman JN, Cram DJ (1971) Electrophilic substitution at saturated carbon. XLV. Dissection of mechanisms of base catalyzed hydrogen-deuterium exchange of carbon acids into inversion, isoinversion, and racemization pathways. J Am Chem Soc 93:2225–2243

Sanders JKM, Hunter BK (1993) Modern NMR Spectroscopy: a Guide for Chemists. Oxford University Press, Oxford, pp 33–52

Schmitz W, Albers C, Fingerhut R, Conzelmann E (1995) Purification and characterization of an alpha-methylacyl-CoA racemase from human liver. Eur J Biochem 231:815–822

Schumacher H, Smith RL, Williams RT (1965a) The metabolism of thalidomide: the fate of thalidomide and some of its hydrolysis products in various species. Br J Pharmacol 25:338–351

Schumacher H, Smith RL, Williams RT (1965b) The metabolism of thalidomide: the spontaneous hydrolysis of thalidomide in solution. Br J Pharmacol 25:324–337

Shannon EJ, Morales EJ, Sandoval F (1997) Immunomodulatory assays to study structure-activity relationships of thalidomide. Immunopharmacology 35:203–212

Sheskin J (1965) Thalidomide in the treatment of lepra reactions. Clin Pharmacol Ther 6:303–306

Stevens RJ, Andujar C, Edwards CJ, Ames PR, Barwick AR, Khamashta MA, Hughes GR (1997) Thalidomide in the treatment of the cutaneous manifestations of lupus erythematosus: experience in sixteen consecutive patients. Br J Rheumatol 36:353–359

Testa B (1973) Some chemical and stereochemical aspects of diethylpropion metabolism in man. Acta Pharm Suec 10:441–454

Testa B (1979) Principles of Organic Stereochemistry. Dekker, New York
Testa B (1989) Mechanisms of chiral recognition in xenobiotic metabolism and drug-receptor interactions. Chirality 1:7–9
Testa B (1990) Definitions and concepts in biochirality. In: Holmstedt B, Frank H, Testa B (eds) Chirality and Biological Activity. Liss, New York, pp 15–32
Testa B (1995) The Metabolism of Drugs and other Xenobiotics – Biochemistry of Redox Reactions. Academic Press, London
Testa B, Jenner P (1976) Drug Metabolism: Chemical and Biochemical Aspects. Dekker, New York
Testa B, Jenner P (1978) Stereochemical methodology. In: Garrett ER, Hirtz JL (eds) Drug Fate and Metabolism: Methods and Techniques. Dekker, New York, pp 143–193
Testa B, Trager WF (1990) Racemates versus enantiomers in drug development: dogmatism or pragmatism? Chirality 2:129–133
Testa B, Carrupt PA, Christiansen LH, Christoffersen P, Reist M (1993a) Chirality in drug research: stereomania, stereophobia, or stereophilia? In: Claassen V (ed) Trends in Receptor Research. Elsevier, Amsterdam, pp 1–8
Testa B, Carrupt PA, Gal J (1993b) The so-called "interconversion" of stereoisomeric drugs: an attempt at clarification. Chirality 5:105–111
Tokuyama S, Hatano K (1995) Purification and properties of thermostable *N*-acylamino acid racemase from Amycolatopsis sp. T*S*-1–60. Applied Microbiol Biotechnol 42:853–859
Vogelsang GB, Farmer ER, Hess AD, Altamonte V, Beschorner WE, Jabs DA, Corio RL, Levin LS, Colvin OM, Wingard JR (1992) Thalidomide for the treatment of chronic graft-versus-host disease. N Eng J Med 326:1055–1058
Werner A (1911) Ueber den räumlichen Stellungswechsel bei Umsetzungen von raumisomeren Verbindungen. Berichte 44:873–882
Yagasaki M, Iwata K, Ishino S, Azuma M, Ozaki A (1995) Cloning, purification, and properties of a cofactor-independent glutamate racemase from Lactobacillus brevis ATCC 8287. Biosci Biotechnol Biochem 59:610–614
Yamauchi T, Choi SY, Okada H, Yohda M, Kumagai H, Esaki N, Soda K (1992) Properties of aspartate racemase, a pyridoxal 5′-phosphate-independent amino acid racemase. J Biol Chem 267:18361–18364
Zini R, Barré J, Brée F, Tillement JP, Sébille B (1981) Evidence for a concentration-dependent polymerization of a commercial human serum albumin. J Chromatogr 216:191–198

CHAPTER 5

Physical Properties and Crystal Structures of Chiral Drugs

C.-H. Gu and D.J.W. Grant

A. Introduction

More than half of all marketed drugs contain one or more chiral centers and are therefore members of the category of chiral drugs (Millership and Fitzpatrick 1993). Because the biochemical processes of life are either stereospecific or stereoselective, opposite enantiomers and racemates of chiral drugs may differ markedly in the following properties: pharmacological (Islam 1997), toxicological (Wainer 1993), pharmacodynamic and pharmacokinetic (Drayer 1986; Midha et al. 1998). Some chiral drugs must therefore be marketed as pure enantiomers from both the practical and regulatory points of view (Pifferi and Perucca 1995; Testa and Trager 1990). However, from a commercial viewpoint, racemates can usually be manufactured more economically than enantiomers; hence, many chiral drugs are still marketed as racemates. However, a recent trend is towards marketing more single-enantiomer drugs (Stinson 1998).

No matter whether a racemate or an enantiomer is chosen for administration, the physical properties of the chiral drug must be thoroughly characterized in both the racemic and enantiomeric forms in the development of the drug for a safe, efficacious, and reliable pharmaceutical formulation. Furthermore, the physical properties of a chiral drug also provide the basic information needed to achieve the purification of enantiomers via crystallization, either directly or via a diastereomer. Because most properties are governed by the structure of the crystals, knowledge of the crystal structure of a chiral drug should, in principle, provide an in-depth understanding of its properties. In this chapter, emphasis is placed on the characterization of the physical properties of chiral drugs and the relationship between physical properties and crystal structure.

B. Nature of Racemates

The majority of chiral drugs are marketed as racemates (Brocks and Jamali 1995). Three types of crystalline racemates are known as shown schematically in Fig. 1 (Roozeboom 1899):

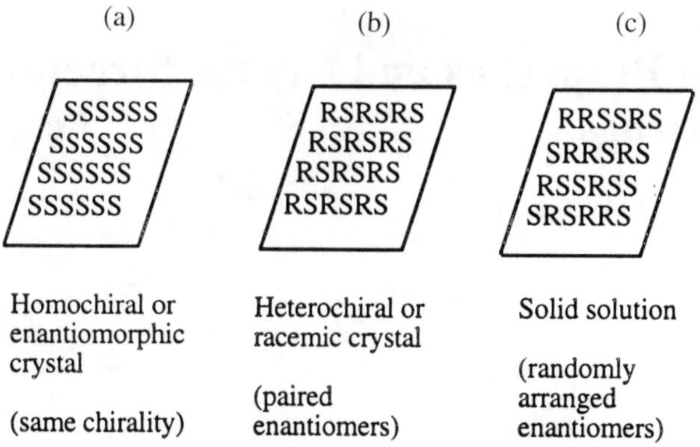

Fig. 1a–c. Schematic representation of the molecular arrangements in three types of racemic species. **a** Homochiral or enantiomorphic crystal (same chirality); **b** heterochiral or racemic crystal (paired enantiomers); **c** pseudoracemate or solid solution (randomly arranged enantiomers). (Collet et al. 1995, reproduced by permission of Kruwer Academic)

1. Conglomerate, which is a 1:1 physical mixture of homochiral crystals of the two enantiomers. Conglomerates belong to the class of homochiral crystals, because individual crystals in a conglomerate are themselves homochiral (Fig. 1a), such as asparagine and sodium ibuprofen. Approximately 5%–10% of crystalline racemates are of this type.
2. Racemic compound, which is a homogeneous solid phase with the two enantiomers present in equal amounts in each unit cell of the crystal lattice, such as ibuprofen free acid and ephedrine free base. A racemic compound consists of heterochiral crystals (Fig. 1b). This type of crystalline racemate is the most common, representing 90%–95% of racemates.
3. Pseudoracemate, or solid solution, which is a homogeneous solid phase containing equimolar amounts of the opposite enantiomers more or less unordered in the crystal, such as camphor. A pseudoracemate consists of heterochiral crystals (Fig. 1c). This type is rather rare, representing less than 1% of racemates.

The nature of a racemate can be determined by reference to the binary phase diagram (Fig. 2), or by spectroscopic techniques to be discussed in Sect. C. A racemic chiral drug may, in principle, crystallize in any one of the three types summarized above, but only one type is the most stable under given conditions. This phenomenon is discussed in detail in Sect. D.

The differences in properties between the two enantiomers in a conglomerate and a racemic compound arise from the different interactions between the homochiral or heterochiral molecules and from the different packing arrangements in the crystal structures. The homochiral interactions in a crystal of an enantiomer, (R . . . R), are defined as intermolecular nonbonded attractions or repulsions in assemblies of molecules of the same chirality. On

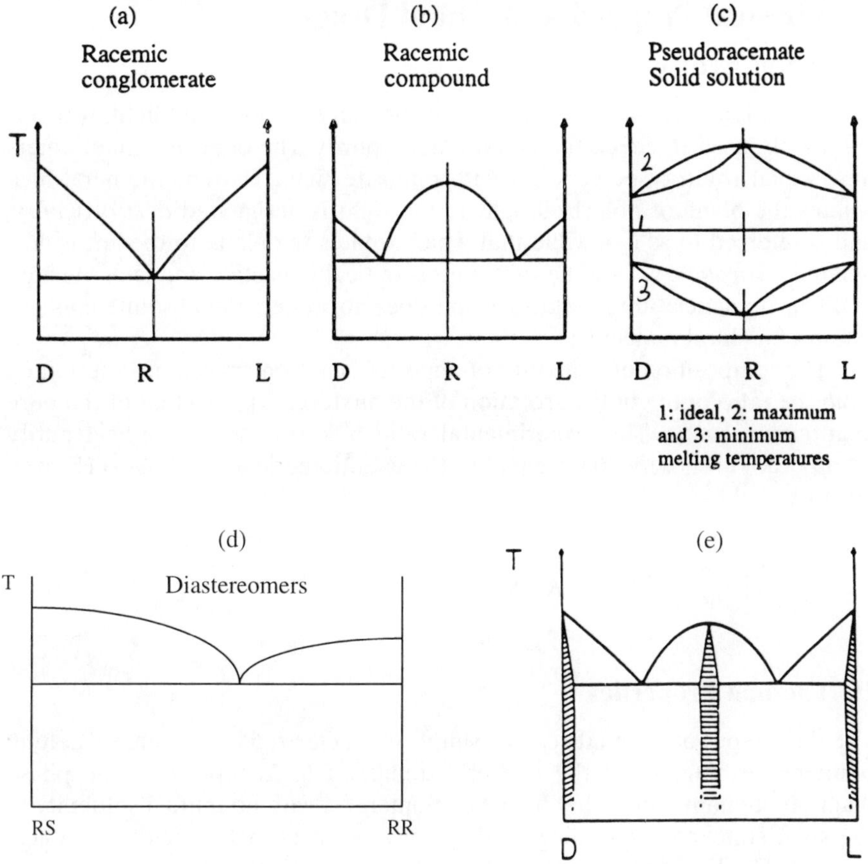

Fig. 2a–e. Typical phase diagrams of melting point against composition (**a–c**) for three types of racemic species, and (**d**) for a pair of diastereomers. In some systems, terminal solid solutions may exist, which is illustrated in **e** for the racemic compound and its enantiomers. (Jacques et al. 1981, reproduced by permission of John Wiley and Sons)

the other hand, the heterochiral interactions (R . . . S) in the racemate are interactions between molecules of different chirality (Eliel et al. 1994). These two interactions are unlikely to be the same, because they are diastereomerically related. However, the difference between these interactions is small enough to be neglected in the gaseous or liquid state, or in an achiral solution. In the solid state or in a chiral environment, the difference between these interactions is significant enough to result in different physical properties between the racemic compound and the corresponding pure enantiomer (Eliel et al. 1994). The differences between the homochiral and heterochiral interactions lead to different crystal structures which, in turn, lead to different physical properties and/or biological activities (Li and Grant 1997).

C. Physical Properties of Chiral Drugs

I. Optical Activity

Optical activity refers to the rotation of the plane of polarized light, which is commonly used to check the enantiomeric purity. The opposite enantiomers have equal rotatory powers but with opposite signs. The enantiomer which rotates the plane of polarized light to the right is designated *dextrorotatory*, and is referred to as (+), while that which rotates the plane to the left is designated *levorotatory*, (−). The racemate is optically inactive and is designated (±). This nomenclature is arbitrary and does not reflect the absolute configuration of a chiral molecule.

The composition of a mixture of opposite enantiomers can be determined from the ratio of the optical rotation of the mixture, $[\alpha]_{exp}$, to that of the pure enantiomer, $[\alpha]_{pure}$. This experimental ratio is known as the optical purity (%*op*, Eq. 1), whose value is equal to the enantiomeric excess (%*ee*) (KAGAN 1995).

$$\%op = \frac{[\alpha]_{exp}}{[\alpha]_{pure}} \times 100\% = \frac{R-S}{R+S} \times 100 = \%ee \tag{1}$$

II. Thermal Properties

The three types of racemate can be simply characterized by the binary melting point phase diagram of the two enantiomers. Fig. 2a represents the phase diagram of enantiomers that form conglomerates with no mutual solubility in the solid state, corresponding to the typical eutectic behavior of a physical mixture. The liquidus line in the diagram can be calculated by the following Schröder-van Laar equation, which gives the relationship between the composition of mixtures and the terminal melting temperature,

$$\ln x = \frac{\Delta H_A^f}{R}\left(\frac{1}{T_A^f} - \frac{1}{T^f}\right) \tag{2}$$

where x is the mole fraction of the more abundant enantiomer, ΔH_A^f is the molar enthalpy of fusion of the pure enantiomer, R is the gas constant, and T_A^f and T^f are the melting point of the pure enantiomer and the mixture, respectively (JACQUES et al. 1981).

The phase diagram of enantiomers that form a racemic compound is represented by Fig. 2b, which is similar to the typical phase diagram of two solids forming a solid complex. Figure 2b indicates that the racemic compound is a new solid phase, which shows eutectic behavior when mixed with either enantiomer. The liquidus line between the pure enantiomer and the eutectic point can be calculated by the Schröder-Van Laar equation, whereas the liquidus line between the two eutectic points can be calculated by the Prigogine-Defay equation which relates the solid composition to the melting temperature,

$$\ln 4x(1-x) = \frac{(2\Delta H_{\mathrm{R}}^{\mathrm{f}})}{R}\left(\frac{1}{T_{\mathrm{R}}^{\mathrm{f}}} - \frac{1}{T^{\mathrm{f}}}\right) \tag{3}$$

where x is the mole fraction of more abundant enantiomer and $\Delta H_{\mathrm{R}}^{\mathrm{f}}$ is the enthalpy of fusion of the racemic compound, and $T_{\mathrm{R}}^{\mathrm{f}}$ and T^{f} are the melting point of the racemic compound and of the solid mixture, respectively. The chiral purity can be estimated by applying either the Schröder-Van Laar or the Prigogine-Defay equations or by measuring the area of the eutectic peak (Jacques et al. 1981).

As mentioned above, relatively few racemic species (less than 1%) are pseudoracemates (Jacques et al. 1981), for which the two enantiomers exhibit complete solid solubility. Figure 2c represents the phase diagrams of pseudoracemates, which include an ideal solid solution (Fig. 2c, 1), and those giving positive deviations (Fig. 2c, 2) and negative deviations (Fig. 2c, 3) from the ideal solid solubility behavior.

When two chiral drugs are diastereomers, their binary melting phase diagrams can be analyzed similarly to those of enantiomers discussed above. The most common phase diagram of diastereomers is eutectic (Fig. 2d; Jacques et al. 1981). Phase diagrams of diastereomers do not exhibit the symmetry shown by enantiomers.

Besides the melting phase diagram, the phase diagram based on the enthalpy of fusion, which is usually determined by differential scanning calorimetry (DSC), can also be used to infer the type of racemate (Elsabee and Prankerd 1992a). The DSC curve may be analyzed by a multiple nonlinear regression program to deconvolute the overlapping endothermic peaks (Elsabee and Prankerd 1992b). The enthalpy of fusion so obtained may be plotted against the composition. It is easier to determine the eutectic composition from the phase diagram based on the enthalpy of fusion than from that based on the melting point (Elsabee and Prankerd 1992a).

The thermodynamics of racemates may be quantified from the thermal data using thermodynamic cycles, as developed by Jacques et al. (1981). Recently, Li et al. (1999) employed this approach and derived the enthalpy, entropy, and Gibbs free energy of racemates with respect to the individual enantiomers from thermal data. The formation of a racemic compound is given by:

$$\mathrm{D_S} + \mathrm{L_S} = \mathrm{R_S} \tag{4}$$

where the left-hand-side represents the racemic conglomerate which is an equimolar mixture of opposite enantiomers, D and L, in the solid state, subscripts, and the right-hand-side represents the racemic compound, R, in the solid state, subscript. If the free energy of formation of the racemic compound is negative, the racemic compound is formed spontaneously and is more stable than the conglomerate. By assuming that enantiomers behave ideally in the liquid mixture and that the heat capacity difference between the solid and the melt is small enough to be neglected, the free energy of formation of the racemic compound at the melting point is given by:

$$\Delta G^0_{T^f_A} = -\Delta S^f_R (T^f_R - T^f_A) - T^f_A R \ln 2 \tag{5}$$

when $T^f_A < T^f_R$

$$\Delta G^0_{T^f_R} = -\Delta S^f_A (T^f_R - T^f_A) - T^f_R R \ln 2 \tag{6}$$

when $T^f_A < T^f_R$

In Eqs. 5 and 6, ΔS^f is the entropy of fusion, T is the melting temperature, and subscripts A and R refer to enantiomer and racemic compound, respectively, as mentioned above. By extrapolating the free energy at the melting temperature with knowledge of the heat capacity, the free energy of formation at lower, practical temperatures, e.g., ambient, can be derived (Li et al. 1999). The sign of the Gibbs free energy of formation at a certain temperature may be different from that at the melting temperature, which reflects the change of relative stability between the racemic compound and the conglomerate.

Examination of 23 chiral drugs that form racemic compounds indicates that the negative free energy of formation at the melting temperature is proportional to the difference in melting point, ΔT^f, between the racemic compound and enantiomer, as suggested by Eqs. 5 and 6 (Li et al. 1999). This linearity suggests that the greater the melting temperature of a racemic compound, as compared with that of its enantiomer, the more stable it is. When the melting point of the racemic compound and that of an enantiomer are equal, the racemic compound is still stabilized by the entropy of mixing term in Eqs. 5 and 6.

III. Solubility

Although construction of the melting point phase diagram is a simple technique for determining the nature of the racemate, it is suitable only for a thermally stable compound. A ternary solubility phase diagram can overcome this limitation for thermally labile drugs. Furthermore, knowledge of the solubility behavior of racemates and enantiomers is essential in the resolution of racemates via crystallization (Jacques et al. 1981). The theoretical phase diagram of a conglomerate is represented by Fig. 3a, which corresponds to eutectic behavior. Most solubility diagrams of diastereomers also show eutectic behavior.

A representative ternary phase diagram showing the solubility of a racemic compound is presented in Fig. 3b. The solubility of the racemic compound can be either greater or less than that of the enantiomer. If the solubility of the enantiomer is less than that of the corresponding racemic compound, a rapid crystallization in the presence of seed crystals may provide pure enantiomer without allowing the system to reach its solubility equilib-

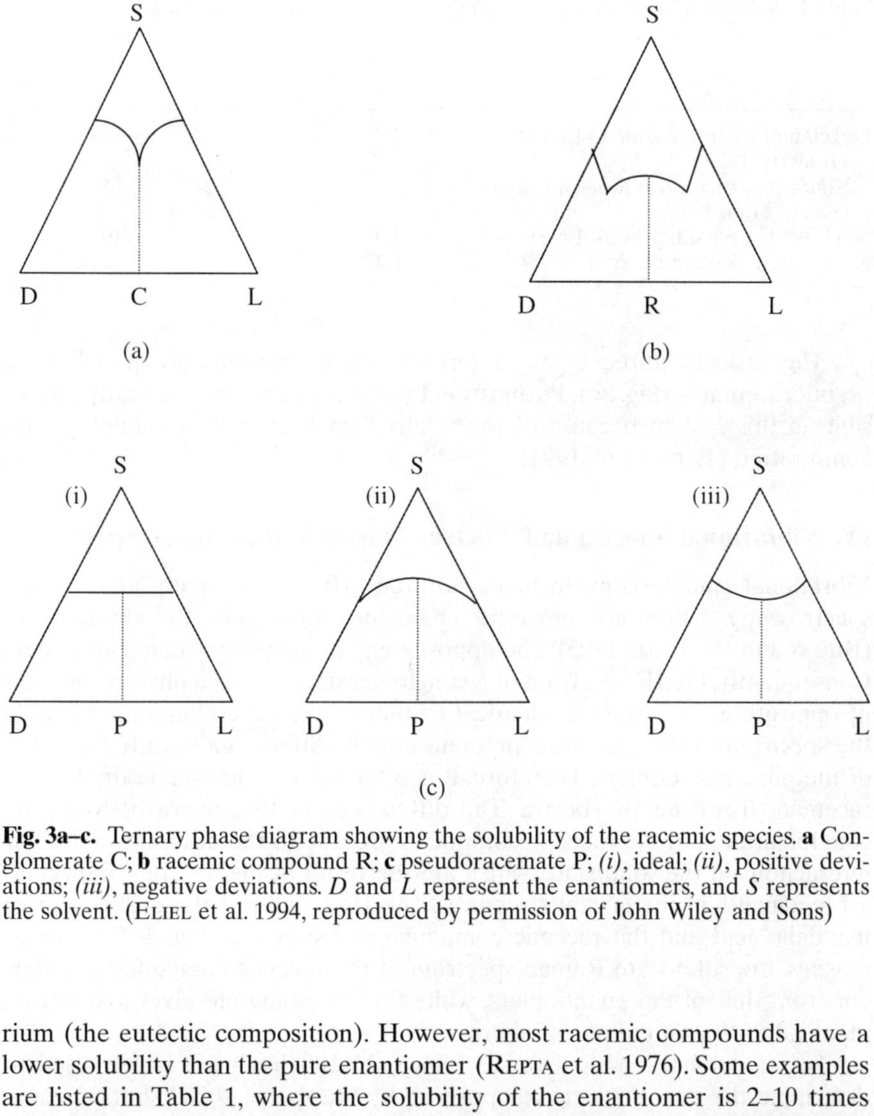

Fig. 3a–c. Ternary phase diagram showing the solubility of the racemic species. **a** Conglomerate C; **b** racemic compound R; **c** pseudoracemate P; *(i)*, ideal; *(ii)*, positive deviations; *(iii)*, negative deviations. *D* and *L* represent the enantiomers, and *S* represents the solvent. (Eliel et al. 1994, reproduced by permission of John Wiley and Sons)

rium (the eutectic composition). However, most racemic compounds have a lower solubility than the pure enantiomer (Repta et al. 1976). Some examples are listed in Table 1, where the solubility of the enantiomer is 2–10 times that of the racemic compound. The different solubilities of the racemate and enantiomer will lead to different dissolution rates. Regardless of the rate-controlling step, the dissolution rate increases with an increase in the solubility (Grant and Higuchi 1990). When drug absorption is controlled by the dissolution rate, the greater dissolution rate of the enantiomer may lead to significantly greater bioavailability over the racemic drug, which is another advantage of administering the enantiomeric drug. It was observed that the enantiomeric drug in a tropical formulation, being more soluble, exhibits a significantly greater skin permeation rate than the corresponding racemic drug (Wearley et al. 1993).

Table 1. Solubility of some racemic compounds and the corresponding enantiomers

	Solubility (mg/ml) Racemic compound	Enantiomer
Dexclamol hydrochloride (Liu and Hurwitz 1978)	3.3	16.4
1,2-Di(4-piperazine-2,6-dione)propane (Repta et al. 1976)	3	15
SCH-39304 (Wearley et al. 1993)	3.6	39.2
Ketoprofen (Kommuru et al. 1998)	1.42	2.32

Theoretically, three types of ternary phase diagrams are possible for pseudoracemates (Fig. 3c). Purification by crystallization is practically impossible in this system because of the relative small change of solubility with composition (Eliel et al. 1994).

IV. Vibrational Spectra and Nuclear Magnetic Resonance Spectra

Vibrational spectroscopy includes infrared (IR) spectroscopy and Raman spectroscopy, which are powerful characterization tools for chiral drugs (Bugay and Williams 1995). The opposite enantiomers give identical spectra. Consequently, the IR spectrum of a conglomerate, which is a physical mixture of opposite enantiomers, is identical to that of the pure enantiomers, while the spectrum of the racemic compound usually differs significantly from that of the pure enantiomers. Therefore, it is often easy to determine the type of racemate from the IR spectra. The differences in IR spectra between the conglomerate and the racemic compound are due to the different molecular interactions in the solid state, which are the result of the different molecular arrangements in the crystal structures. The IR spectra of the enantiomer of mandelic acid and the racemic compound are shown in Fig. 4. For similar reasons, the solid-state Raman spectrum of the racemic compound is different from that of the enantiomers, while the conglomerate gives a spectrum identical to that of the enantiomers.

Solution nuclear magnetic resonance (NMR) has been widely used to determine the enantiomeric composition by the use of chiral shift reagents. For characterizing solid chiral drugs, solid-state ^{13}C NMR spectroscopy can be used to probe the short-range intermolecular interactions in the solid state, and is especially useful for identifying the conformational differences (Bugay 1995). The type of racemate can be identified by the solid-state NMR spectra because the racemic compound gives a spectrum that is different from that of either enantiomer (Fig. 5), whereas the spectrum of the conglomerate is identical to that of either enantiomer. Solid-state ^{13}C NMR spectroscopy is also a quantitative method that can be applied to determine the enantiomeric composition in a solid sample, provided that the racemate is a racemic compound. The distinctive peaks are marked with *asterisks*.

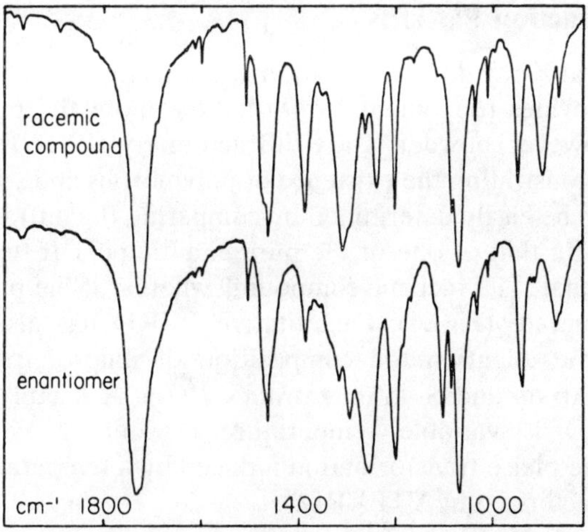

Fig. 4. Infrared spectra of the racemic compound and enantiomer of mandelic acid in Nujol mulls. (Jacques et al. 1981, reproduced by permission of the copyright owner, John Wiley and Sons)

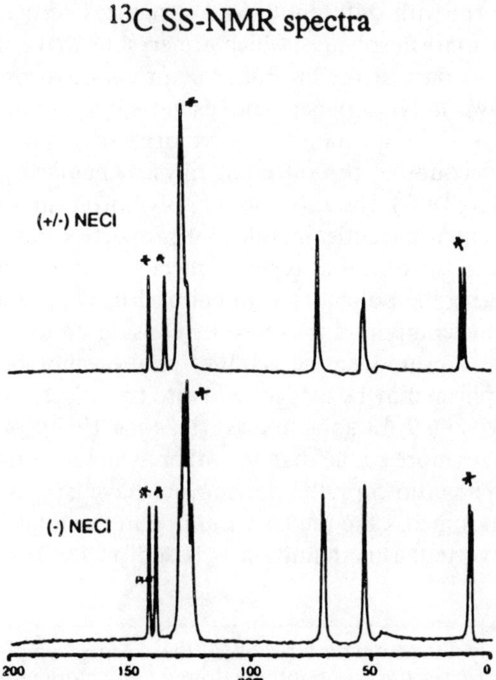

Fig. 5. ^{13}C solid-state NMR spectra of the racemic compound and an enantiomer of norephedrine hydrochloride. (Li et al. 1999, reproduced by permission of American Pharmaceutical Association)

V. X-ray Diffraction Patterns

The single crystal X-ray diffraction pattern reveals the crystal structure of the solid, which provides fundamental information about the properties of the chiral drug. However, powder X-ray diffractometry (PXRD) is the primary method for demonstrating the existence of polymorphs and solvates. The type of racemate can be easily determined by comparing the diffraction pattern of the racemate with that of one of the pure enantiomers. If the pattern is different, the racemate is a racemic compound, whereas if the pattern is similar, it is a conglomerate (Fig. 6). Quantitative PXRD has also been applied to determine the enantiomeric composition of ibuprofen in the racemic compound (Phadnis and Suryanarayanan 1997). A useful complement to ordinary PXRD is variable temperature powder X-ray diffractometry (VTPXRD). The phase transformation induced by a temperature change can be readily characterized by VTPXRD.

D. Polymorphism and Pseudopolymorphism of Chiral Drugs

Polymorphism is a phenomenon whereby a chemical substance crystallizes in more than one form with different crystal structures, termed polymorphs. A compound may also form solvates, which are crystals that contain the solvent of crystallization as part of the crystal structure, termed pseudopolymorphs (Byrn et al. 1999). Polymorphism and pseudopolymorphism are common among chiral drugs. The importance of polymorphism in pharmaceutics is well documented on account of the different physicochemical properties of the polymorphs (Grant 1999). The existence of polymorphism of a chiral drug will not only affect its pharmaceutically relevant properties, but will also result in conversion between the different types of racemate. It may be rather complicated to characterize the polymorphs of chiral drugs because the conglomerate and the racemic compound may also be considered to be polymorphs.

Based on the thermodynamic stability relationships between the polymorphs, polymorphism may be categorized into two types, namely, monotropy and enantiotropy (Fig. 7; Burger and Ramberger 1979). In monotropy, one polymorph is always more stable than the other at all temperatures below their melting points. In enantiotropy, the polymorphs have a transition point below the melting points; on crossing the transition point the relative stability of the polymorphs is inverted. This definition is based on the assumption that the

→

Fig. 6. Powder X-ray diffraction patterns of (**a**) the racemic compound (*upper traces*) and an enantiomer (*lower traces*) of norephedrine hydrochloride, and (**b**) the pseudoracemate (+/–) and an enantiomer (–) of atenolol. The distinctive peaks are marked with *asterisks*. Shift of peak position is indicated by an *asterisk* and is characteristic of a solid solution. (Li et al. 1999, reproduced by permission of American Pharmaceutical Association)

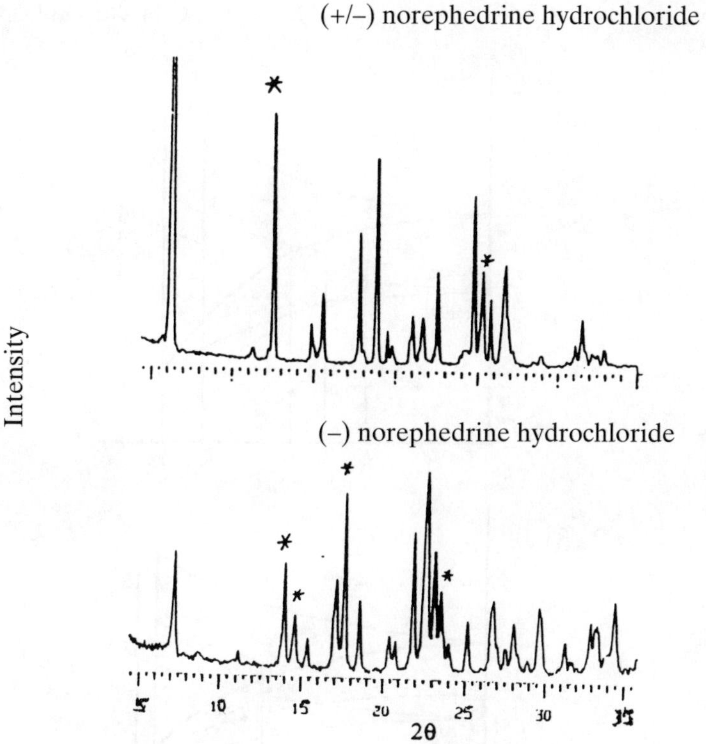
(+/–) norephedrine hydrochloride
(–) norephedrine hydrochloride
Intensity
5
10
15
20
25
30
35
2θ

(a)

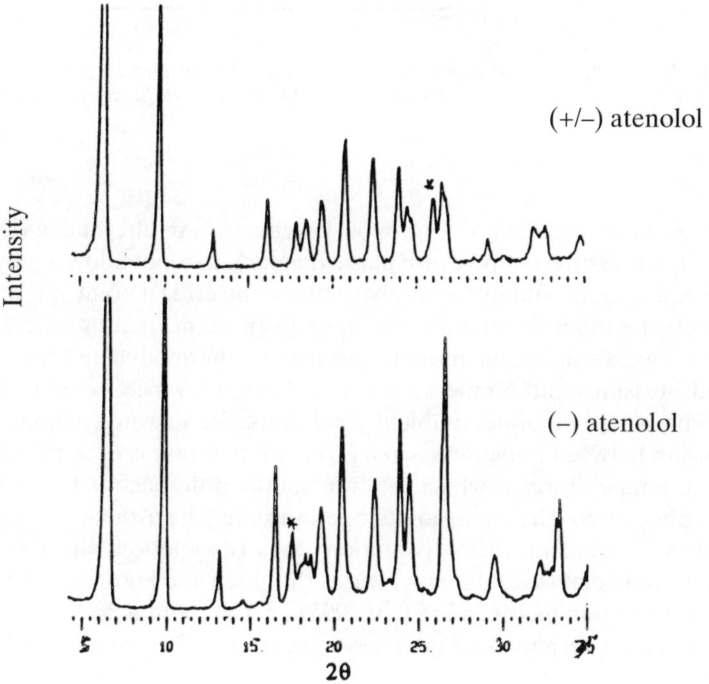
(+/–) atenolol
(–) atenolol
Intensity
5
10
15
20
25
30
35
2θ

(b)

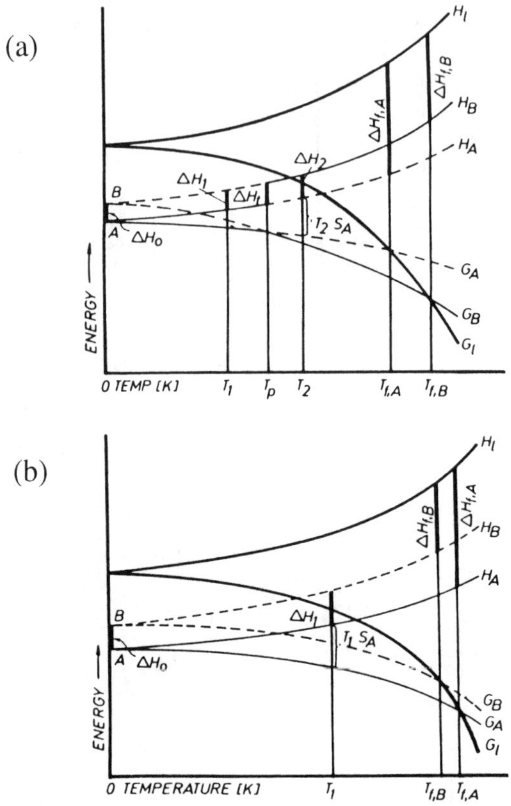

Fig. 7a, b. Energy/temperature diagrams of polymorphic systems. **a** An enantiotropic system, **b** a monotropic system. (Burger and Ramburger 1979, reproduced by permission of Springer)

pressure remains constant at atmospheric pressure. An alternative definition is that, if the pressure-temperature phase diagram does not allow a polymorph to be in equilibrium with its vapor phase below the critical point, it is an unstable monotrope, otherwise it is a stable monotrope or an enantiotrope. This definition recognizes that some monotropes may be thermodynamically stable at elevated pressures and temperatures, e.g., diamond, which is the metastable polymorph of carbon under ambient conditions. The thermodynamic stability relationship between polymorphs at a given temperature is determined by the transition temperature, at which the free energy difference between the two polymorphs is zero. The transition temperature may be estimated either from solubilities or from intrinsic dissolution rates (dissolution rate per surface area) determined at several temperatures (Shefter and Higuchi 1963; Grant et al. 1984), or from melting data (Yu 1995) or from both heat of solution and solubility data determined at any one temperature (Gu and Grant 2001).

Table 2. Several theoretically possible polymorphic systems of enantiomers and of the corresponding racemic species, with examples ("=" represents the coexistence of different species)

Enantiomers	Racemic species	Examples
Individual enantiomers exhibit polymorphism	Conglomerate I = Conglomerate II	Carvoxime (Jacques et al. 1981)
	Racemic compound I = Racemic compound II	Nicotine derivatives (Langhammer 1975)
	Solid solution I = Solid solution II	
	Conglomerate = Racemic compound	Nitrendipine (Burger et al. 1997)
		Nimodipine (Grunenberg et al. 1995)
	Racemic compound with no polymorph	Difficult to prove absence of polymorphism
Individual enantiomers have no polymorphs	Racemic compound I = Racemic compound II	Mandelic acid (Kuhnert-Brandstätter and Ulmer 1974)
	Racemic compound = Conglomerate	Sodium ibuprofen (Zhang 1998)
		α-Bromocamphor binaphthyl (Collet et al. 1972)
	Racemic compound = Solid solution	Camphoroxime (Jacques and Gabard 1972)
	Conglomerate = Solid solution	cis-π-Camphanic acid (Brienne and Jacques 1970)
	Solid solution I = Solid solution II	No example available

Several theoretically possible polymorphs of chiral drugs are listed in Table 2. The most common are polymorphic enantiomers, polymorphic racemic compounds, and the existence of both a conglomerate and a racemic compound of a chiral drug. Polymorphs may be discovered by crystallization under various conditions or by thermally induced transformations (Guillory 1999). The ultimate evidence of polymorphism is a comparison of the single crystal structure. DSC, hot-stage microscopy, FTIR, solid state NMR, PXRD, and VTPXRD are routinely used to characterize polymorphs, while thermogravimetric analysis (TGA) provides the stoichiometric number of the solvent molecules in the crystal lattice of solvates (Brittain 1999). The complication of polymorphism in chiral systems makes the DSC curve of a chiral drug sometimes difficult to interpret. It is therefore necessary to apply several techniques to characterize both the enantiomer and the racemate and to identify the origin of the polymorphism. For example, three monotropically related modifications were found for (*RS*)-nitrendipine (Burger et al. 1997). The melting phase diagram showed that the thermodynamically stable form (I) is a racemic compound, while the other polymorphs, form II and III, are both conglomerates. These conclusions were confirmed by the IR spectra and PXRD patterns. Study of the pure enantiomer also revealed that three modifications exist in which enantiomeric form I corresponds to the metastable conglomerate form II, enantiomeric form II corresponds to conglomerate form III, while enantiomeric form III is not related to the racemic modifications. Therefore, both the enantiomer and the racemate are polymorphic in this system. When the designations of the enantiomeric polymorphs and the conglomerate polymorphs do not correspond, careful attention to the nomenclature is necessary.

Solvates may be formed by certain chiral drugs. When the incorporated solvent is water, the solvates are known as hydrates, which are the most abundant. The incorporated solvent may be present either in stoichiometric amounts, e.g., histidine monohydrate (Jacques and Gabard 1972), or in non-stoichiometric amounts, e.g., cromolyn sodium hydrates (Chen et al. 1999). The relative stability of the anhydrate (unsolvated crystal) and the hydrate is determined by the thermodynamic activity of the solvent and the temperature (Shefter and Higuchi 1963). The activity of a solvent can be reduced to the desired value by dilution with a miscible cosolvent. The resulting solvent mixtures can then be used to prepare solvates with different degrees of solvation (Zhu et al. 1996).

In chiral systems, it is observed that the racemic compound and the enantiomer undergo different degrees of solvation under given conditions. There are many examples of a change of degree of solvation as a result of a change of the nature of the racemate (Jacques et al. 1981). For example, enantiomeric histidine hydrochloride forms a monohydrate when crystallized from water, whereas racemic histidine hydrochloride forms a dihydrate. The racemic dihydrate transforms to the conglomerate monohydrate above 45°C (Fig. 8; Jacques and Gabard 1972). It would be informative to perform similar studies on chiral drugs to determine whether the formation of a solvate can be used to transform a racemic compound into a conglomerate and thereby facilitate resolution.

Diastereomeric pairs also form solvates with different degrees of solvation. This phenomenon is useful for achieving a high efficiency of resolution. The solubility of a solvate with a higher degree of solvation is usually lower than that with a lower degree of solvation in the corresponding solvent. On the other hand, in a solvent that is miscible with the solvating solvent, the solubility of the more highly solvated solid is usually higher than that of the solid with a lower solvation (Shefter and Higuchi 1963). Therefore, by choosing the appropriate solvent, the solubility ranking of a diastereomeric

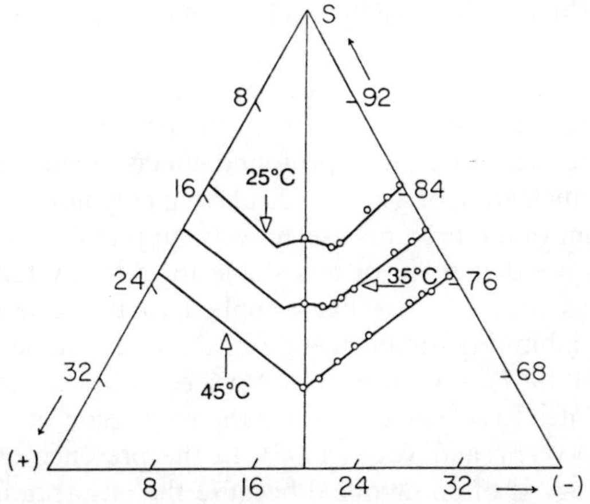

Fig. 8. Ternary phase diagram showing the solubility of histidine hydrochloride enantiomers [(+) and (–)] in water (*S*) at various temperatures. The maximum at 25°C and 35°C corresponds to the solubility of the racemic compound. The sharp minimum at 45°C corresponds to the equal solubilities of the two enantiomers in the racemic conglomerate. (JACQUES and GABARD 1972, reproduced by permission of Société Française de Chimie)

pair can be changed to obtain the desired enantiomer. For example, in the resolution of DL-leucine by forming salts with (*S*)-(–)-phenylethanesulfonic acid (PES), L-leucine-(–)-PES (LS) is less soluble in acetonitrile-methanol mixture than its diastereomer, D-leucine-(–)-PES (DS). However, in acetonitrile-water mixture, DS monohydrate is formed and is less soluble than LS, which does not form a hydrate (YOSHIOKA et al. 1998).

It is important to be aware that polymorphic transitions may occur during pharmaceutical processing, such as grinding, milling, and wet granulation. Both mechanical forces, especially shear, and relative humidity (RH) are believed to exert influences. The external mechanical energy may transform to structural energy, facilitating the transition, while the water content of a drug may increase under higher RH and will significantly increase the molecular mobility (ZOGRAFI 1988). Hydration and dehydration may occur during these processes. It was found that physical mixtures of the enantiomers of malic acid, tartaric acid and serine transformed to racemic compounds after grinding or during storage at 53% RH and at 40°C (PIYAROM et al. 1997).

E. Influence of Impurities on the Physical Properties of Chiral Drugs

Impurities exert profound impacts on the acceptability, quality, and properties of drug substances. For chiral drugs, it is especially important to consider the influence of impurities because it is often difficult to separate chiral isomers completely, such that commercially available chiral compounds often contain trace amounts of their chiral isomers (Duddu et al. 1991). Crystallization from solution is one of the most important techniques for separation and purification of chiral drugs. During crystallization, the presence of trace amounts of structurally related impurity exerts profound effects on nucleation and crystal growth. During nucleation of a system displaying polymorphism, the presence of a small amount of impurity may selectively suppress the nucleation of one form permitting another more or less stable form to crystallize (Weissbuch et al. 1995). This principle has been applied to the kinetic resolution of enantiomers by inhibiting the nucleation of the more stable racemic crystals (Weissbuch et al. 1987). Additives may also serve as nucleation promoters by providing templates to selectively induce the nucleation of desired polymorphic forms (Bonafede and Ward 1995). In the presence of impurities, the crystal morphology is often modified because the adsorption of an impurity onto a certain face during crystal growth may alter the growth rate of that specific face (Mullin 1993). The morphological change may lead to changes in particle size distribution, in specific surface area, and in true density. These changes can further alter the mechanical properties, such as the flowability and tablet properties, and the bulk dissolution rate.

The impurities present in the crystallization medium may be incorporated into the lattice of the host crystals to form terminal solid solutions during crystallization (Fig. 2e). In addition to their possible toxicological effects, the impurity molecules will interact with the neighboring host molecules in a way which is different from the interaction between host molecules themselves. Thus, the incorporation of the impurity may disrupt the crystal lattice, thereby changing the nature and the concentration of the crystal defects. Such defects, created by the incorporation of impurity, are found to significantly alter the solid-state properties of crystalline drugs, such as the thermodynamic properties (Li and Grant 1996; Gu and Grant 2000a), mechanical properties (Law 1994), dissolution behavior (Li and Grant 1996; Gu and Grant 2000a), and stability (Byrn et al. 1999). These changes may be the source of batch-to-batch or lot-to-lot variations among pharmaceutical compounds, which may result in bioinequivalence and instability of the solid drug in its final dosage form (York 1983).

Studies of the ephedrine and pseudoephedrine systems have demonstrated that appreciable amount of chiral impurities (guests), including the opposite enantiomer, excess enantiomer of a racemate, and a diastereomer, can be incorporated into the lattice of the host chiral crystals. The incorporation of impurities leads to decreases in the enthalpy and entropy of fusion, suggesting increases of lattice energy and disorder of the chiral host crystals (Duddu et al. 1993, 1996; Li and Grant 1996; Gu and Grant 2000a). The effects of the incorporation of the impurity on the intrinsic dissolution rate (IDR) of the host are profound and vary from system to system. Depending on whether the mole fraction of the guest in the host crystal lattice is less than or greater than the solid solubility limit, either a stable or a metastable solid solution may be formed. When a stable solid solution is formed, in which case the guest and the host in the impure crystals have lower chemical potentials than in the respective pure crystals, the impure crystals will have a lower solubility than the pure crystals and therefore a lower IDR under constant hydrodynamic conditions (GIORDANO et al. 1999). Conversely, if the concentration of incorporated impurity exceeds the solid solubility, a metastable solid solution is formed, in which case the impure crystals are expected to have a higher IDR than the pure crystals (DUDDU et al. 1996). Furthermore, the impurity itself may exert an inhibitory effect on the dissolution rate of the host, either by segregating at dislocations (GILMAN et al. 1958), where the dissolution process often begins, or by entering the dissolution medium and acting as an inhibitor (BUNDGAARD 1974).

Considering the significant effects of impurities on the crystal properties of chiral drugs, the impurity profile of a chiral drug must be carefully controlled to limit the batch-to-batch variations and to maintain the quality of the drug. The extent of impurity incorporation is greatly dependent on the crystallization conditions. A general rule is that the faster the crystal growth, the greater the mole fraction of impurities that are incorporated. GU and GRANT (2002) observed that larger crystals incorporate more impurity than smaller crystals in the same batch. It is therefore possible to minimize the impurity incorporation by optimizing the crystallization conditions. On the other hand, crystallization in the presence of tailor-made guest molecules may be used as a technique of crystal engineering to modify the crystal properties, such as habit, mechanical properties, and dissolution rate.

F. Crystal Structures of Chiral Drugs

A crystalline solid is a solid composed of atoms or molecules arranged in a periodic pattern in three dimensions. The physical properties of chiral drugs discussed above are controlled by the packing mode and molecular interactions, which can be studied by analyzing their crystal structures. For example, different PXRD patterns, IR spectra, and solid state NMR spectra of the racemic compound and enantiomers are all the result of the different molecular packing and subsequently different molecular interactions in the crystals. On the other hand, the solubility and thermal properties reflect the lattice energy of the crystal. Therefore, it is necessary to analyze the crystal structure of a chiral drug to gain a fundamental understanding of its physical properties.

The nonsuperimposibility of the mirror image of a chiral molecule, which is the primary implication of its chirality, is reflected in its crystal structure. Although chirality does not imply the total lack of symmetry (asymmetry), the existence of chirality in a crystal does require the absence of improper rotation elements (a proper rotation combined with either an inversion center or a reflection plane). Consequently, the overall crystal structure of an enantiomer must be disymmetric (Brittain 1990). Geometrically, there are 230 unique ways of arranging objects repetitively in a three-dimensional pattern. These 230 space groups are classified into 32 crystal classes according to their symmetry. Enantiomers may then crystallize in only 11 enantiomorphous crystal classes of the 32 classes, which comprise 66 space groups (Jacques et al. 1981). However, racemic compounds are not subject to this restriction and potentially may crystallize in any space group, including a chiral one, as shown by achiral molecules that crystallize in chiral space groups (Jacques et al. 1981). A recent example is the achiral drug, cromolyn sodium hydrate, which crystallizes in the chiral space group $P1$ (Chen et al. 1999). Therefore a powder sample of this drug is essentially a racemic conglomerate of crystals with opposite chiralities. However, this designation is not particularly useful, because the molecules themselves are achiral.

Although enantiomers may, in principle, crystallize in any one of the 66 enantiomorphous space groups, a survey of 430 cases of enantiomers by Bel'skii and Zorkii (1971) revealed that enantiomers crystallize preferentially in certain selected space groups: 67% in the $P2_12_12_1$ space group, 27% in the $P2_1$ space group, 1% in the $C2$ space group, and 5% in the other 63 space groups. Racemates also frequently crystallize in certain space groups. A similar study of 792 cases of racemates revealed that 56% crystallized in the $P2_1/c$ space group, 15% in the $C2/c$ space group, and 13% in the $P\bar{1}$ space group, while 16% crystallized in the other 161 achiral space groups (Bel'skii and Zorkii 1971).

G. Comparison of the Crystal Structures of the Racemic Compound and Enantiomer

An interesting theoretical question relevant to chiral compounds is, which factors govern the formation of the different types of racemate? This question has practical importance, too, because conglomerates can be readily resolved by preferential crystallization, which is the most cost-effective method for enantiomeric separation. The answer lies in the different crystal stabilities of the racemic compound and enantiomer, which indicate differences between homochiral and heterochiral interactions introduced in Sect. B., "Nature of Racemates". Three principal factors are related to crystal stability, namely compactness, symmetry, and intermolecular forces, including hydrogen bonding, van der Waals forces and electrostatic interactions.

I. Compactness and Symmetry

After comparing the density of eight pairs of chiral compounds at room temperature, Wallach (1895) formulated a rule that crystals of the racemic compound tend to be denser than homochiral crystals. This rule implies that molecules in racemic crystals tend to be more tightly packed than in the corresponding homochiral crystals. At a low temperature, it is generally true that the more tightly packed form is more stable. However, BROCK et al. (1991) found that Wallach's rule is only valid for resolvable enantiomers, because this rule is derived from a biased group which excludes those racemic compounds that are markedly less stable. Based on this argument, it is impossible to compare the properties of racemic compounds and conglomerates without statistical bias. In spite of the statistical bias, the racemic compounds are, on average, more stable than the homochiral mixtures, i.e., conglomerates, because about 90% of chiral compounds form racemic compounds, while only 10% form conglomerates (COLLET et al. 1995). The greater compactness of the crystal structure of a racemic compound may, nevertheless, contribute to its greater stability than that of the conglomerate.

As discussed in Sect. F, enantiomers can crystallize only into asymmetric space groups which are devoid of inverse symmetry elements, while almost all racemates crystallize in those space groups that possess elements of inverse symmetry. This symmetry may contribute to close-packing, leading to greater stability of the racemic compound (JACQUES et al. 1981). Different packing arrangements also lead to the differences in van der Waals interactions and in the geometry of hydrogen bonding, which will be discussed in the next section.

II. Intermolecular Interactions in the Crystal

Recent developments in crystallographic and computational chemistry make it possible to estimate the strength of the intermolecular interactions in the crystal either qualitatively, i.e., by comparing the molecular packing mode and the geometry of the hydrogen bond network, or quantitatively, i.e., by comparing the crystal lattice energy (KRIEGER 1995). Both qualitative and quantitative analyses enable us to interpret the physical properties at the molecular level.

Qualitative structure analysis has been employed to study the structural basis of the formation of racemic compounds and conglomerates, using a series of salts formed by a chiral base and a series of structurally related acids as model systems. The rationale for preparing salts is that the probability of a chiral salt being a conglomerate is two to three times that for covalent compounds (JACQUES et al. 1981). On the other hand, salt forms of drugs are sometimes marketed because of their superior properties, such as dissolution rate and processing characteristics, over their covalent acidic or basic counterparts (BERGE et al. 1977). KINBARA et al. (1996a) compared the crystal structures of a series of salts of chiral primary amines with achiral carboxylic acids (Fig. 9). They found that both the formation and assembly of a characteristic columnar hydrogen-bond network, in which the ammonium cations and the carboxylate anions are aligned around a twofold screw axis, constituting a 2_1 column, are essential in the formation of conglomerates from these salts (Fig. 9). In another study of acidic salts of α-phenylethylamine with achiral dicarboxylic acids, BÖCSKEI et al. (1996) found that a conglomerate formed

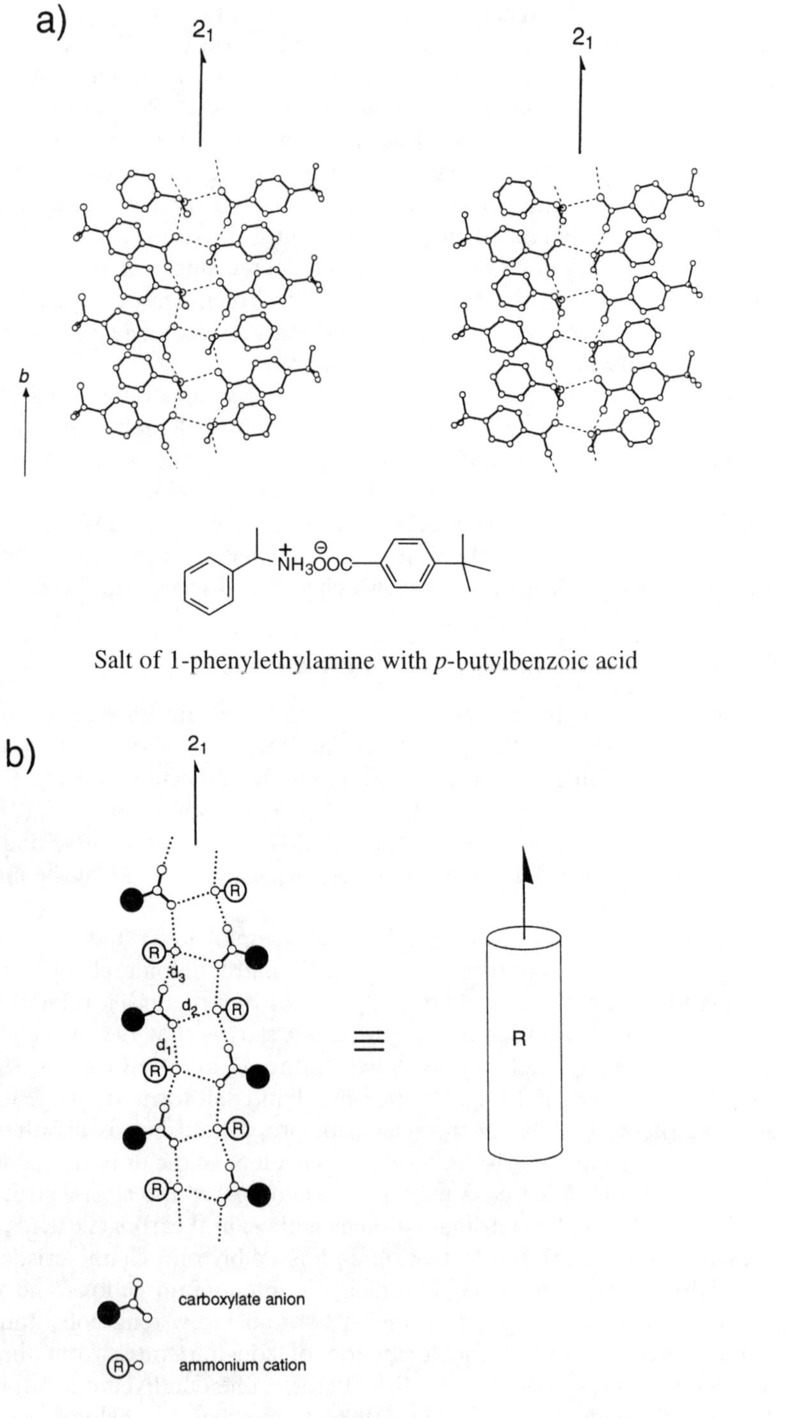
a)
2_1
2_1
b
Salt of 1-phenylethylamine with p-butylbenzoic acid
b)
2_1
d_3
d_2
d_1
≡
R
carboxylate anion
ammonium cation

only when there is a $-CH_2-CH_2-$ group between the two carboxylic groups and when the protonated and deprotonated carboxylic groups form hydrogen bonded chains rather than forming cyclic intramolecular hydrogen bonds.

These pioneering studies began a new chapter for understanding the structural basis of conglomerate formation using qualitative analysis of crystal structure. However, this approach is limited to the qualification of short-range interactions, particularly hydrogen bonds. If the difference in lattice energy between a conglomerate and a racemic compound arises from the long-range electrostatic or van der Waals interactions, this method is doomed to fail (LEUSEN 1993). To analyze quantitatively the lattice energy differences between a conglomerate and a racemic compound, molecular modeling (MM) tools may be applied to calculate the lattice energy. LI et al. (2001) found that van der Waals forces significantly contribute to the difference both in melting behavior and in the enthalpy of fusion, which are responsible for chiral discrimination in the solid state. In salts, the contribution of electrostatic energy towards the total lattice energy increases, which offsets the contribution of the van der Waals forces. This result may explain why conglomerates are more common among chiral salts. Currently, the accuracy of MM is still questionable. However, advances in this field will ultimately give us the power to probe the molecular interactions in the crystals and to predict the physical properties from the crystal structure.

H. Crystal Structural Basis of Diastereomer Separation

Another aspect of crystal structure studies focuses on the structural comparison of diastereomers, which is intended to elucidate the correlation between crystal structure and efficiency of resolution. The resolution of enantiomers via the formation and subsequent separation of a diastereomeric pair is a common practice. However, the choice of appropriate reagent is still based on trial and error. Although the efficiency of resolution may be modified by varying the crystallization conditions, such as solvent, cooling rate, and temperature, the efficiency depends mainly on the magnitude of the difference in solubility between the diastereomers. The difference of solubility is most likely controlled by the stability of the crystal that can be deduced from the crystal structure. Based on this argument, KINBARA et al. (1996b) studied the depen-

Fig. 9. a Hydrogen-bond network in the crystal of salt of 1-phenylethylamine with *p*-butylbenzoic acid, represented by the *dotted lines*. Two ammonium cation and carboxylate anion pairs form a unit through the hydrogen bonds between the ammonium hydrogens and the carboxylate oxygens. This unit forms an infinite columnar structure around a 2-fold screw axis along the *b axis* (2_1-column). **b** Schematic representation of a 2_1-column formed in conglomerate salts of chiral primary monoamines with achiral monocarboxylic acids. (KINBARA et al. 1996a, reproduced by permission of American Chemical Society)

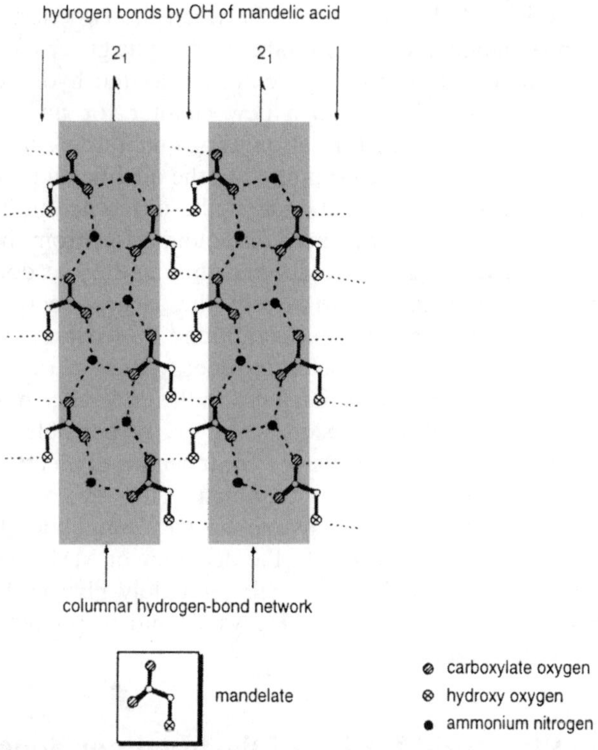

Fig. 10. Schematic representation of the hydrogen bond network of the less soluble salts of optically pure mandelic acid with 1-arylethylamines. Columnar supramolecular assemblies are formed by the hydrogen bonds between the carboxylate oxygens and the ammonium hydrogens, which are linked by OH···O hydrogen bonds to construct a tightly hydrogen-bonded supramolecular sheet. (Kinbara et al. 1998, reproduced by permission of Royal Society of Chemistry)

dence of resolution efficiency on the differential stability of a series of diastereomeric pairs. High resolution efficiency was achieved with those pairs, for which one crystal is stabilized by both hydrogen-bonding and van der Waals interactions. In the less soluble salt, a tightly columnar hydrogen bonded supramolecular sheet was formed (Fig. 10). Similarly, in another study, Kinbara et al. (1998) found that high resolution efficiency is achieved when a stable hydrogen bond column is formed in the less soluble salts. Low resolution efficiency is observed in those salts whose hydroxyl groups participate moderately in the formation of hydrogen bonds in each diastereomer.

Besides the hydrogen bond pattern, the conformation of molecules in the crystal is also an important factor, which will influence the geometry of the hydrogen bond network, i.e., bond lengths and angles, and its strength (Diego 1995). In addition to the strength of hydrogen bonding, CH· π interactions are also found to play a role in diastereomeric discrimination (Kinbara et al. 2000). These studies demonstrate that it is possible to design a resolving reagent by considering the number and orientation of interacting groups which may determine the type and strength of the molecular interactions thus formed in the diastereomeric salts.

I. Concluding Remarks

Chiral drugs comprise an important subgroup of pharmaceuticals. On account of their chirality, special attention should be paid to the physical characterization of chiral drugs. The various techniques that are employed in the characterization of polymorphs may be applied to chiral drugs. The nature of the racemate is the key property to be determined for a chiral drug, because it influences all the physical properties as well as the separation of the enantiomers. Classical characterization tools, such as thermodynamics and spectroscopy, are now routinely used. Meanwhile, crystal structure analysis is more extensively employed for physical characterization because the internal structure governs the solid-state properties. We are now in the era of understanding the physical properties from the point view of molecular packing and interactions in the solid-state, thanks to developments in X-ray crystallography and computational chemistry. The ultimate goal in the physical characterization of a drug is to predict its properties from the molecular structure. We may eventually achieve this goal by a combination of improved characterization tools and advancements in crystal structure analysis, in molecular-modeling of the solid state, in intermolecular force fields and energetics, and in molecular dynamics.

References

Bel'skii, Zorkii (1971) Distribution of molecular crystals by structural classes. Soviet Phys Crystallogr (English translation) 15:607

Berge SN, Bighley LD, Monkhouse DC (1977) Pharmaceutical salts. J Pharm Sci 66:1–19

Böcskei Z, Kassai C, Simon K, Fogassy E, Kozma D (1996) Racemic compound formation-conglomerate formation. Part 3. Investigation of the acidic salts of α-phenylethylamine by achiral dicarboxylic acids. Optical resolution by preferential crystallization and a structural study of (*R*)-α-phenylethylammonium hydrogen itaconate. J Chem Soc Perkin 2:1511–1515

Bonafede SJ, Ward MD (1995) Selective nucleation and growth of an organic polymorph by ledge-directed epitaxy on a molecular crystal substrate. J Am Chem Soc 117:7853–7861

Brienne MJ, Jacques J (1970) Une γ-lactone d'un type rare: L'acide trans π-camphanique. Tetrahedron 26:5087–5100

Brittain HG (1990) Crystallographic consequences of molecular disymmetry. Pharm Res 7:683–690

Brittain HG (1999) Methods for the characterization of polymorphs and solvates. In: Brittain HG (ed.) Polymorphism in Pharmaceutical Solids. Marcel Dekker. New York, NY, pp 227–278

Brock C, Schweizer WB, Dunitz JD (1991) On the validity of Wallach's rule: On the density and stability of racemic crystals compared with their chiral counterparts. J Am Chem Soc 113:9811–9820
Brocks D, Jamali F (1995) Stereochemical aspects of pharmacotherapy. Pharmacotherapy 15:551–564
Bundgaard H (1974) Influence of an acetylsalicylic anhydride impurity on the rate of dissolution of acetylsalicylic acid. J Pharm Pharmacol 26:535–540
Bugay DE (1995) Magnetic resonance spectrometry. In: Brittain HG (ed.) Physical Characterization of Pharmaceutical Solids. Marcel Dekker, New York, NY, pp 93–126
Bugay DE, Williams AC (1995) Vibration spectrometry. In: Brittain HG (ed.) Physical Characterization of Pharmaceutical Solids. Marcel Dekker, New York, NY, pp 59–92
Burger A, Rollinger JM, Brüggeller P (1997) Binary system of (*R*)- and (*S*)-nitrendipine-polymorphism and structure. J Pharm Sci 86:674–679
Burger A, Ramberger R (1979) On the polymorphism of pharmaceuticals and other molecular crystals. I Theory of thermodynamic rules, Mikrochim Acta [Wien] 2:259–271; II Applicability of thermodynamic rules, Mikrochim Acta [Wien] 2:273–316
Burt HM, Mitchell AG (1981) Crystal defects and dissolution. Int J Pharm 9:137–152
Byrn SR, Pfeiffer RR, Stowell JG (1999) Solid-State Chemistry of Drugs. 2nd edn. SSCI, Inc., West Lafayette, Indiana, IN
Chen LR, Young VG Jr., Lechuga-Ballesteros D, Grant DJW (1999) Solid state behavior of cromolyn sodium hydrates. J Pharm Sci 88:1191–1200
Collet A, Brienne MJ, Jacques J (1972) Dédoublements spontanes et conglomérats d'énantiomères. Bull Soc Chim Fr 1972:127–142
Collet A, Ziminski L, Garcia C, Vigne-Maeder F (1995) Chiral discrimination in crystalline enantiomer systems: facts, interpretations, and speculations. In: Siegel J (ed.) NATO ASI Series-Supramolecular Stereochemistry. Kluwer Academic, Netherland, pp 91–110
Diego HLD (1995) The crystal structure of (*S*-PEA)3(*S*-MA), Acta Cryst. C51: 253–256
Drayer D (1986) Pharmacodynamic and pharmacokinetic differences between drug enantiomers in humans: an overview. Clin Pharmacol Ther 40:125–133
Duddu S, Fung FKY, Grant DJW (1993) Effect of the opposite enantiomer on the physicochemical properties of (–)-ephedrinium 2-naphthalenesulfonate crystals. Int J Pharm 94:171–179
Duddu S, Fung FKY, Grant DJW (1996) Effects of crystallization in the presence of the opposite enantiomers on the crystal properties of (SS)-(+)-pseudoephedrinium salicylate. Int J Pharm 127:53–63
Duddu SP, Mehvar R, Grant DJW (1991) Liquid chromatographic analysis of the enantiomeric impurities in various (+)-pseudoephedrine samples. Pharm Res 8: 1430–1433
Eliel EL, Wilen SH, Mander LN (1994) Stereochemistry of Organic Compounds. John Wiley and Sons, New York, NY, pp 153–295
Elsabee M, Prankerd RJ (1992a) Solid-state properties of drugs. III. Differential scanning calorimetry of chiral drug mixtures existing as racemic solid solutions, racemic mixtures and racemic compounds. Int J Pharm 86:221–230
Elsabee M, Prankerd RJ (1992b) Solid-state properties of drugs. II. Peak shape analysis an deconvolution of overlapping endotherms in differential calorimetry of chiral mixtures. Int J Pharm 86:211–219
Gilman JJ, Johnston WG, Sears GW (1958) Dislocation etch pit formation in lithium fluoride. J Appl Phys 29:747–754
Giordano F, Bettini R, Donini C, Gazzaniga A, Caira MR, Zhang GGZ, Grant DJW (1999) Physical properties of parabens and their mixtures: Solubility in water, thermal behavior, and crystal structures. J Pharm Sci 88:1210–1216
Grant DJW (1999) Theory and origin of polymorphism. In: Brittain HG (ed.) Polymorphism in Pharmaceutical Solids. Marcel Dekker, New York, NY, pp 1–33

Grant DJW, Mehdizadeh M, Chow AHL, Fairbrother JE (1984) Nonlinear van't Hoff solubility-temperature plots and their pharmaceutical interpretation. Int J Pharm 18:25–38

Grant DJW, Higuchi T (1990) Solubility Behavior of Organic Compounds. John Wiley and Sons, New York, NY

Grunenberg A, Keil B, Henck J-O (1995) Polymorphism in binary mixtures, as exemplified by nimodipine. Int J Pharm 118:11–21

Gu C, Grant DJW (2000) Estimating thermodynamic stability relationships of polymorphs and solvates from heats of solution and either solubilities or dissolution rates. J Pharm Sci 90:1277–1287, 2001

Gu C, Grant DJW (2000a) Effects of crystallization in the presence of the diasteomer on the crystal properties of (*SS*)-(+)-pseudoephedrine hydrochloride. Enantiomer, 5:271–280

Gu C, Grant DJW (2002) Relationship between particle size and impurity incorporation during crystallization of (+)-pseudoephedrine hydrochloride, acetaminophen, and adipic acid from aqueous solution. Pharm Res 19:1068–1070

Guillory JK (1999) Generation of polymorphs, hydrates, solvates, and amorphous solids. In: Brittain HG (ed) Polymorphism of Pharmaceutics Solids. Marcel Dekker, New York, NY, pp 183–226

Islam MM, Bowen ID (1997) Pharmacological importance of stereochemical resolution of enantiomeric drugs. Drug Safety 17:149–165

Jacques J, Gabard J (1972) Étude des mélanges d'antipodes optiques. III. Diagrammes de solubilité pour les divers types de racémiques. Bull Soc Chim Fr, 342–350

Jacques J, Collet A, Wilen SH (1981) Enantiomers, Racemates, and Resolutions. John Wiley and Sons, New York, NY

Kagan HB (1995) Is there a preferred expression for the composition of a mixture of enantiomers? Recl Trav Chim Pays-Bas 114:203–205

Kinbara K, Hashimoto Y, Sukegawa M, Nohira H, Saigo K (1996a) Crystal structures of the salts of chiral primary amines with achiral carboxylic acids: Recognition of the common-occurring supramolecular assemblies of hydrogen-bond networks and their role in the formation of conglomerate. J Am Chem Soc 118:3441–3449

Kinbara K, Saiko K, Hashimoto Y, Nohira H, Saigo K (1996b) Chiral discrimination upon crystallisation of the diastereomeric salts of 1-aryethylamines with mandelic acid or *p*-methoxymandelic acid: interpretation of the resolution efficiencies on the basis of the crystal structures. J Chem Soc Perkin trans 2:2615–2622

Kinbara K, Kobyashi Y, Saigo K (1998) Systematic study of chiral discrimination upon crystallization. Part 2. Chiral discrimination of 2-arylalkanoic acids by (*1R,2S*)-2-amino-1,2,-diphenylethanol. J Chem Soc Perkin Trans 2:1767–1775

Kinbara K, Kobayashi Y, Saigo K (2000) Chiral discrimination of 2-arylalkanoic acids by (1S, 2R)-1-aminoindan-2-ol through the formation of a consistent columnar supramolecular hydrogen-bond network. J Chem Soc Perkin Tran 2:111–119

Kommuru TR, Khan MA, Reddy I (1998) Racemate and enantiomers of ketoprofen: Phase diagram, thermodynamic studies, skin permeability, and use of chiral permeation enhancers. J Pharm Sci 87:833–840

Krieger J (1995) New software expands role of molecular modeling technology. Chem Eng News Sept 4:30–40

Kuhnert–Brandstätter M, Ulmer R (1974) Beitrag zur thermischen Analyse optischer Antipoden: Mandelsäure, Mikrochim Acta [Wien], 927–935

Langhammer L (1975) Binary systems of enantiomeric nicotine derivatives. Arch Pharm 308:933–939

Law D (1994) Influence of composition of the crystallization medium on the physical properties and mechanical behavior of adipic acid crystals. Ph. D. Thesis, University of Minnesota, Minneapolis, MN, USA, pp 153–177

Leusen FJJ (1993) Rationalization of Racemate Resolution: a Molecular Modeling Study. Ph. D. Thesis, University of Nijmegen, Netherlands

Li ZJ, Grant DJW (1996) Effects of excess enantiomer on the crystal properties of a racemic compound: ephedrinium 2-naphthalenesulfonate. Int J Pharm 137:21–31
Li ZJ, Grant DJW (1997) Relationship between physical properties and crystal structures of chiral drugs. J Pharm Sci 86:1073–1078
Li ZJ, Zell MT, Munson EJ, Grant DJW (1999) Characterization of racemic species of chiral drugs using thermal analysis, thermodynamic calculation, and structural studies. J Pharm Sci 88:337–346
Li ZJ, Ojala WH, Grant DJW (2001) Molecular modeling study of chiral drug crystals: Lattice energy calculations. J Pharm Sci 90:1523–1539
Liu ST, Hurwitz A (1978) Effect of enantiomeric purity on solubility determination of dexclamol hydrochloride. J Pharm Sci 67:636–638
Midha KK, Mckay G, Rawson MJ, Hubbard JW (1998) The impact of stereoisomerism in bioequivalence studies. J Pharm Sci 87:797–802
Millership J, Fitzpatrick A (1993) Commonly used chiral drugs: A survey. Chirality 5:573–576
Mullin JW (1993) Crystallization. 3rd edn. Butterworth–Heinemann, London, UK, pp 202–260
Phadnis N, Suryanarayanan R (1997) Simultaneous quantification of an enantiomer and the racemic compound of ibuprofen by x–ray powder diffractometry. Pharm Res 14:1176–1180
Pifferi G, Perucca E (1995) The cost benefit ratio of enantiomeric drugs. Eur J Drug Metabolism and Pharmacokinetics 20:15–25
Piyarom S, Yonemochi E, Oguchi T, Yamamota K (1997) Effects of grinding and humidification on the transformation of conglomerate to racemic compound in optically active drugs. J Pharm Pharmacol 49:384–389
Repta AJ, Baltezor MJ, Bansal PC (1976) Utilization of an enantiomer as a solution to a pharmaceutical problem: Application to solubilization of 1,2-di(4-piperazine-2,6-dione)propane. J Pharm Sci 65:238–242
Shefter E, Higuchi T (1963) Dissolution behavior of crystalline solvated and nonsolvated forms of some pharmaceuticals. J Pharm Sci 52:781–791
Stewart M, Arnett E (1982) Chiral monolayers at the air -water interface. Top Stereochem 13:195
Stinson S (1998) Counting on chiral drugs. Chem Eng News Sept 21:83–103
Testa B, Trager WF (1990) Racemates versus enantiomers in drug development: dogmatism or pragmatism?. Chirality 2:129–133
Wainer I (1993) Drug stereochemistry: Analytical methods and pharmacology. Marcel Dekker, New York, NY
Wallach O (1895) Zur Kenntniss der Terpene und der ätherischen Oele. Liebigs Ann Chem 286:90–143
Wearley L, Antonacci B, Cacciapuoti A, Assenza S, Chaudry I, Eckhart C, Levine N, Loebenberg D, Norris C, Parmegiani R, Sequeira J, Yarosh-Tomaine T (1993) Relationship among physicochemical properties, skin permeability, and topical activity of the racemic compound and pure enantiomers of a new antifungal. Pharm Res 10:136–140
Weissbuch I, Zbaida D, Addadi L, Lahav M, Leiserowitz L (1987) Design of polymeric inhibitors for the control of crystal polymorphism. Induced enantiomeric resolution of racemic histidine by crystallization at 25°C. J Am Chem Soc 109:1869–1871
Weissbuch I, Porovitz-Biro R, Lahav M, Leiserowitz L (1995) Understanding and control of nucleation, growth, habit, dissolution and structure of two- and three-dimensional crystals using 'tailor-made' auxiliaries. Acta Cryst B51:115–148
York P (1983) Solid state properties of powders in the formulation and processing of solid dosage forms. Int J Pharm 14:1–28
Yoshioka R, Okamura K, Yamada S, Aoe K, Date T (1998) The role of water-solvation in the optical resolution of DL-leucine with (*S*)-(–)-2-phenylethanesulfonic acid-characterization and X-ray crystal structures of their diastereomeric salts. Bull Chem Soc Jpn 71:1109–1116

Yu L (1995) Inferring thermodynamic stability relationship of polymorphs from melting data. J Pharm Sci 84(8):966–974

Zhang G (1998) Influence of Solvents on Properties, Structures, and Crystallization of Pharmaceutical Solids. Ph. D. Thesis, University of Minnesota, Minneapolis, MN, USA, pp 70–122

Zhu H, Yuen C, Grant DJW (1996) Influence of water activity in organic solvent + water mixtures on the nature of the crystallizing drug phase, 1. Theophylline. Int J Pharm 135:151–160

Zografi G (1988) States of water associated with solids. Drug Dev Ind Pharm 14:1905–1926

Section II
Experimental Pharmacology

CHAPTER 6

Chiral Recognition in Biochemical Pharmacology: An Overview

B. Testa and J.M. Mayer

Dr. Joachim M. Mayer died in October 2001.

A. Introduction

The epoch-making work of Pasteur, van't Hoff and LeBel was published in 1860 and 1874, respectively, and marks the creation of stereochemistry as we understand it (reviewed by Holmstedt 1990). The observation by Piutti (1886) that the two enantiomers of asparagine have different tastes is perhaps the first literature report of enantioselectivity in a biological effect. Organoleptic differences between enantiomers have been discovered ever since (Gardner 1982). Around the end of the nineteenth century and the beginning of the twentieth century, a number of studies were carried out on differences in the biological effects of enantiomers, but it seems that the first clear proofs were provided by Cushny (see below).

Not seldom, the artist's intuition anticipates scientific discoveries, and enantioselectivity is no exception. Indeed, Charles Lutwidge Dodgson, better known by his nom de plume Lewis Carroll, wrote the following in *Through the Looking Glass and What Alice Found There* published in 1871:

Perhaps Looking-glass milk isn't good to drink ..." [Said Alice to her kitten].

Or madly squeeze a right-hand foot
Into a left-hand shoe ..." [Sang the White Knight to Alice].

The first quotation applies to enantioselectivity at a macroscopic biological level, as discussed in Sect. B., whereas the second quotation encapsulates chiral recognition (Sect. C.). Such is the sequence of arguments in this chapter, which serves as a general introduction to most of the following chapters in this volume.

B. Enantioselectivity at Macroscopic Biological Levels

The therapeutic activity of a drug can be broken down into two components, namely, what the drug does to the body (pharmacodynamics, PD), and what the body does to the drug (pharmacokinetics, PK). This is classical if not trivial knowledge, but it is not always realized just how closely PD and PK are intertwined and can influence each other (TESTA 1987). Thus, pharmacodynamic actions may influence disposition, for example, by modifying blood flow, whereas absorption, distribution, metabolism, and excretion will affect the intensity, duration, and localization of pharmacodynamic effects. Perhaps the best known PK-PD interactions are the production of active or toxic metabolites, and the induction or inhibition of its own metabolism by a drug (autoinduction or autoinhibition). Yet despite such interactions, an exposition of stereoselectivity cannot consider all pharmacological events simultaneously, hence the sequential presentation of the following subsections.

I. Stereoselectivity in Drug Action and Pharmacodynamics

1. Pfeiffer's Rule and Eudismic Analysis

The influence of chirality on pharmacological activities is illustrated in innumerable publications, an epoch-making work being that of CUSHNY (reviewed in 1926). Other important reviews are those of BECKETT (1959), PORTOGHESE (1970) and SIMONYI (1984).

A landmark generalization was made by PFEIFFER (1956), who first showed that some pharmacological data can be rationalized so that the greater the activity of the racemate, the larger the ratio of the enantiomers' activity. Indeed, for 14 drugs, PFEIFFER found a linear relationship between the logarithm of the activity ratio of enantiomers (dependent variable) and the logarithm of the average human dose of the racemate (independent variable), the slope being –0.35, the intercept 1.2 and the correlation coefficient $r > 0.9$. Despite its rightful impact, this study suffers from a number of limitations such as:

1. The possibility that a number of other observations were not included in the generalization.
2. The fact that the human dose is a highly hybrid parameter which depends on many biological parameters besides intrinsic activity, and particularly on bioavailability and clearance. Indeed, BARLOW (1990) has found numerous exceptions to Pfeiffer's rule.
3. The degree of optical purity plays a major role (see Sect. I.2.).

Nevertheless, ARIËNS and SIMONIS (1967) compiled some examples that follow Pfeiffer's rule, adding the important observation that the influence of configuration is large or small depending on whether the asymmetric center is located in a critical or noncritical moiety of the drug, i.e., a moiety playing an essential or accessory role in the binding to the receptor.

Analysis of chirality-activity relationships found a useful mathematical basis when LEHMANN and colleagues (LEHMANN 1986, 1987; LEHMANN et al. 1976) developed what they termed *eudismic analysis*. Briefly, they introduced the following definitions:

1. The more active and less active isomer are termed *eutomer* and *distomer*, respectively. A series of pairs will then consist of an eutomeric series and a distomeric series.
2. For any given pair the ratio of their activities (potencies, affinities, efficiencies as substrates, etc.) is called the *eudismic ratio* and its logarithm, the *eudismic index* (EI).

It is generally found that a plot of EI versus eutomer affinity (e.g., pIC_{50}, pK_i, pK_D) or potency (e.g., pED_{50}, pEC_{50}, pD_2, pA'_2) gives straight lines with positive slopes between 0.5 and 1.0. In the case of affinity, this slope, which has been termed the *eudismic affinity quotient* (EAQ), represents the increase in chiral discrimination per unit increase in affinity, and as such can be taken as a quantitative measure of the binding enantioselectivity displayed by a given receptor towards a series of stereoisomeric ligands. Numerous examples can be found in an extensive chapter by LEHMANN et al. (1976).

2. The Problem of Optical Purity

Eudismic analyses, like all other types of structure-activity relationships, can never be more reliable than the experimental data on which they are based. Yet one experimental factor is often overlooked when comparing the activities of enantiomers, namely their *degree of optical resolution*. Indications such as "the optical purity was better than 98%" are commonly found, but the limitations of the usual methods (optical rotation, NMR with chiral shift reagents, chiral chromatography) are ignored (CALDWELL and TESTA 1987; TESTA 1979, 1986a). That the degree of optical purity markedly and even dramatically influences the eudismic ratio was proven beyond doubt by BARLOW and colleagues (1972). Their simulations of stereospecific indices (i.e., eudismic ratios) versus degree of resolution should be a cause of worry for many workers, yet the warning has been all but ignored. It took almost two decades for a strong experimental confirmation to be published, when TROFAST and colleagues (1991) reported that marked differences in eudismic ratios exist between stereoisomers having good (ca. 98%) or excellent (ca. 99.7%) degrees of optical purity. The fear is thus real that many eudismic ratios in the literature are misleadingly underestimated.

II. Stereoselectivity in Some Pharmacokinetic Responses

Whereas the differences in pharmacological effects between enantiomers can be observed directly, stereoselectivity in pharmacokinetic processes may be more difficult to assess and has received belated attention compared to pharmacodynamic effects. The realization that enantiomers may show vastly different pharmacokinetic behavior and toxic properties, as well as spectacular advances in analytical methodologies, have led to an accelerating accumulation of data. In the following subsections, we summarize the role of stereoselectivity in the pharmacokinetic processes of absorption, distribution, excretion, and metabolism, as discussed at greater length in other chapters of this volume.

1. Oral Absorption

There are some pharmaceutical effects which may lead to the predominant appearance of one of the enantiomers in the plasma. Thus, liberation of a drug from its formulation matrix takes place in the gastrointestinal tract, and a stereoselective interaction with chiral excipients such as cellulose derivatives may slow down dissolution stereoselectively.

The absorption of many drugs occurs by passive diffusion through biological membranes. Since diffusion depends less on steric factors and more on physicochemical properties such as distribution coefficients (themselves a function of ionization and lipophilicity), absorption is generally not a stereoselective process (TESTA et al. 2000). However, the absorption of amino acids, sugars, vitamins, and other essential compounds involves active absorption by transporters. Because these must first recognize and bind their substrates (a condition for stereoselectivity, see Sect. C.), active transport usually shows some degree of stereoselectivity. An increasing number of drugs are now found to be absorbed or effluxed stereoselectively by transporters. Thus, the major antiparkinsonian drug L-dopa passes across the intestinal wall more rapidly than its inactive D-isomer which is not transported by the amino acid pump (WADE et al. 1973). Similarly, some β-lactam antibiotics are carried by the intestinal dipeptide transporter. In the rat intestine, L-cephalexin was shown to have a higher affinity for the carrier site than its D-isomer, but only the D-form could be detected in plasma since the L-isomer is also more susceptible to hydrolysis by peptidases in the intestinal wall (TAMAI et al. 1988). Overall, however, absorption plays only a minor role in stereoselective disposition.

2. Distribution

At equilibrium, the volume of distribution of drugs is a function of plasma and tissue binding. This can be demonstrated by a simple mass balance model:

$$V = V_P + V_T \cdot fu/fu_T$$

where V is the apparent volume of distribution, V_P the plasma volume, and V_T the tissue volume into which the drug distributes. The terms fu and fu_T are the fractions of drug unbound in plasma and in tissue, respectively. The volume of plasma is about 3 l for all drugs. Diffusion into tissues involves membrane crossing which, as discussed, depends mostly on physicochemical properties. Consequently, stereoselectivity in the total volume of distribution will be mostly be determined by the stereoselectivity in binding. The tissue binding capacity exceeds in most cases the binding capacity of plasma. However, fu_T is not very sensitive to stereoselective binding in a particular tissue since it is influenced by binding to a large number of sites, most of which are rather unspecific.

The major reason for differences in the volume of distribution of enantiomers is their binding to plasma proteins. The most important binding sites are on albumin for acids and on α_1-acid glycoprotein for bases. A large number of examples document differences in the binding of enantiomers to plasma proteins and in their tissue distribution (MÜLLER 1988). Nevertheless, the contribution of enantioselective distribution to overall stereoselectivity in drug disposition is usually modest.

3. Urinary Excretion

Excretion is the physical elimination of unchanged drug. Renal excretion is the net result of three processes, namely filtration, secretion, and reabsorption, which should be considered separately to understand stereoselective renal excretion.

Filtration does not discriminate between enantiomers. However, the rate of filtration depends on the fraction unbound in plasma, which may differ between enantiomers. Differences in the plasma binding of enantiomers will result in different rates of filtration. Tubular secretion is generally an active process and involves saturable binding to a carrier. This may result in stereoselectivity and even in enantiomeric interactions. Active secretion is believed to be responsible for the stereoselective renal clearance of, for example, pindolol, chloroquine, and disopyramine (HSYU and GIACOMINI 1985; GIACOMINI et al. 1986).

Drug reabsorption from the tubules and return to circulating plasma can be an active or passive process. L-Amino acids, for example, are reabsorbed stereoselectively from the proximal tubule. The importance of stereoselective renal reabsorption for a globally stereoselective drug disposition will be significant only if the fraction of drug excreted unchanged is greater than 25%.

4. Metabolism

Biotransformation by drug-metabolizing enzymes is the pharmacokinetic process exhibiting the greatest degree of stereoselectivity, although we can expect the world-wide ongoing research on transporters to uncover comparable level of chiral recognition. But metabolism is different from all other pharmacological processes because it displays not one but two types of stereoselectivity.

In all pharmacodynamic processes as well as in all pharmacokinetic processes except biotransformation, stereoselectivity manifests itself by the differential activities or disposition of stereoisomers. This is also seen in metabolic reactions, where differences between stereoisomers in terms of rates of biotransformation and/or nature of metabolites are manifestations of *substrate stereoselectivity*. In other words, substrate stereoselectivity is seen when stereoisomers are metabolized by the same biological system and under identical conditions (a) at different rates and/or, (b) to different products. Substrate stereoselectivity is a well-known and abundantly documented phenomenon under in vivo and in vitro conditions.

In addition to substrate stereoselectivity, metabolic reactions also display *product stereoselectivity*, which is defined as (a) the differential formation (in quantitative and/or qualitative terms) of two or more stereoisomeric metabolites, (b) from a single substrate having a suitable prochiral center or face (JENNER and TESTA 1973; MAYER and TESTA 1994; TESTA 1986b, 1988, 1989, 1990, 1995; TESTA and JENNER 1976, 1980; TESTA and MAYER 1988). Examples of metabolic pathways producing new centers of chirality in substrate molecules are: ketone reduction, reduction of carbon–carbon double bonds, hydroxylation of prochiral methylenes, oxygenation of tertiary amines to *N*-oxides, and oxygenation of sulfides to sulfoxides.

It is unfortunate to note that the fundamental difference between substrate and product stereoselectivity is not always recognized explicitly, resulting in some confusing presentations and interpretations of data.

There are some cases in which stereoselective metabolism is of toxicological relevance (MAYER and TESTA 1994). This situation can be illustrated with disopyramide and mianserin which undergo substrate stereoselective oxidation. *Disopyramide* (Fig. 1), a chiral antiarrhythmic agent, is marketed as the racemate. Although disopyramide is generally well tolerated, several cases of hepatic toxicity have been described. In vitro studies using rat hepatocytes revealed a considerably higher cytotoxicity of the (*S*)-enantiomer, as assessed by leakage of lactate dehydrogenase and morphological changes. The bio-

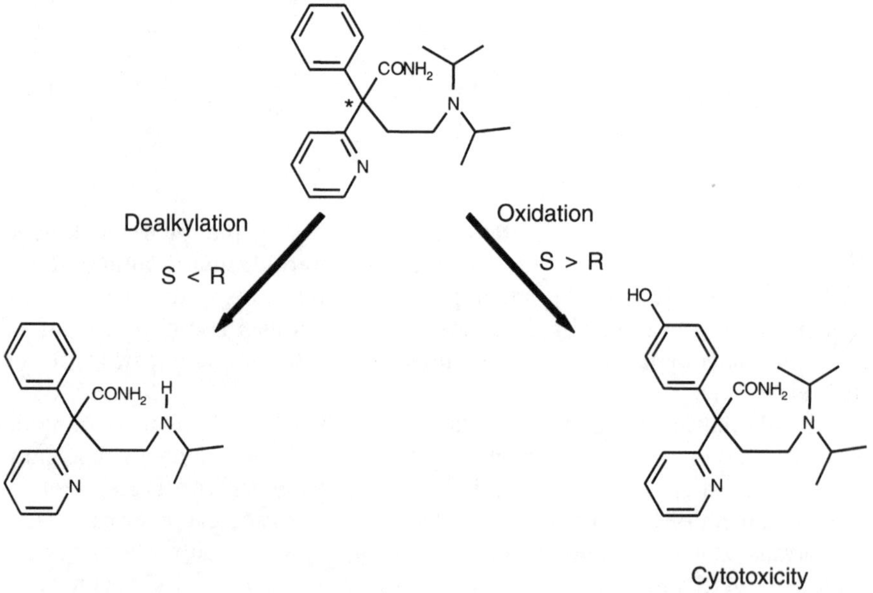

Fig. 1. Substrate stereoselective metabolism of disopyramide; *, chiral center. (LE CORRE et al. 1988)

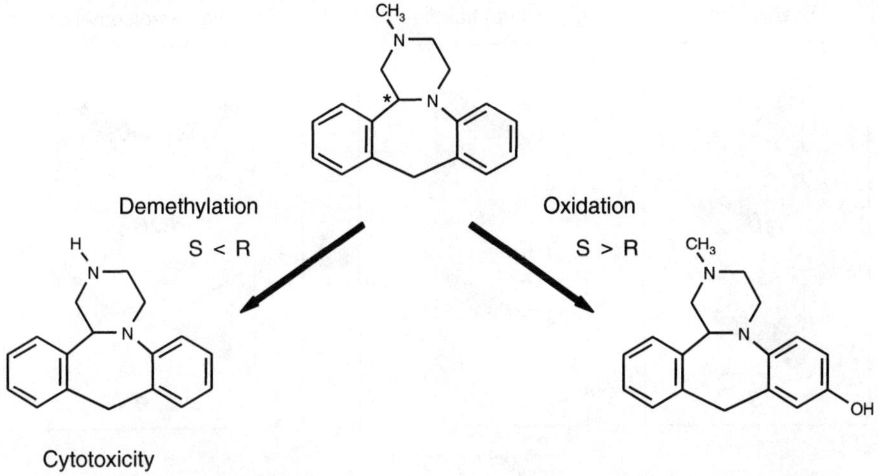

Fig. 2. Stereoselective metabolism of mianserin; *, chiral center. (RILEY et al. 1989)

transformation of disopyramide involves mono-*N*-dealkylation (which is stereoselective for (*S*)-enantiomer) and aromatic oxidation (which is stereoselective for the (*R*)-enantiomer). The low cytotoxicity of the *N*-dealkylated metabolites in rat hepatocytes led to the hypothesis that toxicity is mainly due to aromatic oxidation (LE CORRE et al. 1988).

The metabolism of the antidepressant drug *mianserin* (Fig. 2) revealed analogies and differences with disopyramide. Here, aromatic oxidation in human liver microsomes occurred more readily for the (*S*)-enantiomer, while *N*-demethylation was the major route for the (*R*)-enantiomer. At low drug concentrations, cytotoxicity towards human mononuclear leucocytes was due to (*R*)-mianserin more than to (*S*)-mianserin, and showed a significant correlation with *N*-demethylation (RILEY et al. 1989). Thus the toxicity of mianserin seemed associated with *N*-demethylation rather than with aromatic oxidation, in contrast to disopyramide. The chemical nature of the toxic intermediates was not established, but a comparison between disopyramide and mianserin emphasises that no a priori expectation should influence toxico-metabolic studies.

The mechanisms of substrate and product stereoselectivity are discussed in the next section.

C. Mechanisms of Chiral Recognition in Biochemical Pharmacology

I. Physicochemical Principles

Since enantiomers have identical chemical and physical properties, their discrimination or physical separation necessitates a "chiral handle." In stereo-

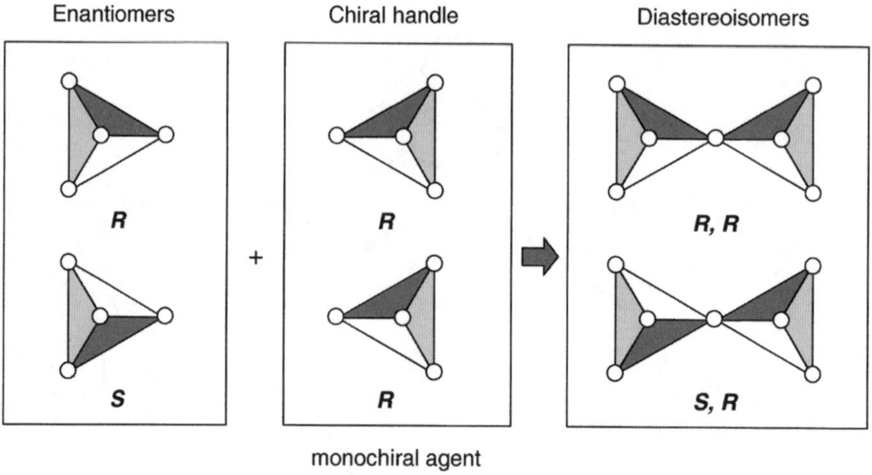

Fig. 3. Principle of chiral recognition: the reaction of a mixture of two enantiomers with an enantiomerically pure reagent (i.e., a "chiral handle") leads to two diastereoisomeric complexes

selective chromatography, chiral handles may be optically active solvents, derivatization agents or chiral columns (TESTA 1986a; Chap. 2, this volume). The two complexes resulting from such interactions are *diastereoisomeric* (Fig. 3) and as such have different physical properties (TESTA 1979). The discrimination between, or physical separation of, two enantiomers is made possible by these diastereomeric differences.

Living organisms, like the chiral handles used in synthetic chemistry, chromatography, and spectroscopy are monochiral at the molecular level. Indeed, only L-amino acids are encoded in proteins, whereas most sugars have the D-configuration. In addition, helicity in proteins is preferably right-handed. Further chirality results from secondary, supersecondary and tertiary structures. In consequence, hormones, enzymes, receptors, transporters, immunoglobulins, etc, are all chiral. When interacting with biological systems, two enantiomers will therefore form two diastereomeric complexes. This is the basis of chiral recognition in biochemical pharmacology.

II. The Model of Easson and Stedman

Our current understanding of biological enantioselectivity owes much to the three-point attachment model of EASSON and STEDMAN (1933). In comparing the activity of enantiomers containing a single center of asymmetry, they proposed a *three-point attachment model* to account for the observed selectivities. This model postulates three binding sites in the receptor (X, Y and Z) to which correspond three complementary (pharmacophoric) groups in the drug molecule (X′, Y′ and Z′) (Fig. 4). But only in the more active enantiomer (the

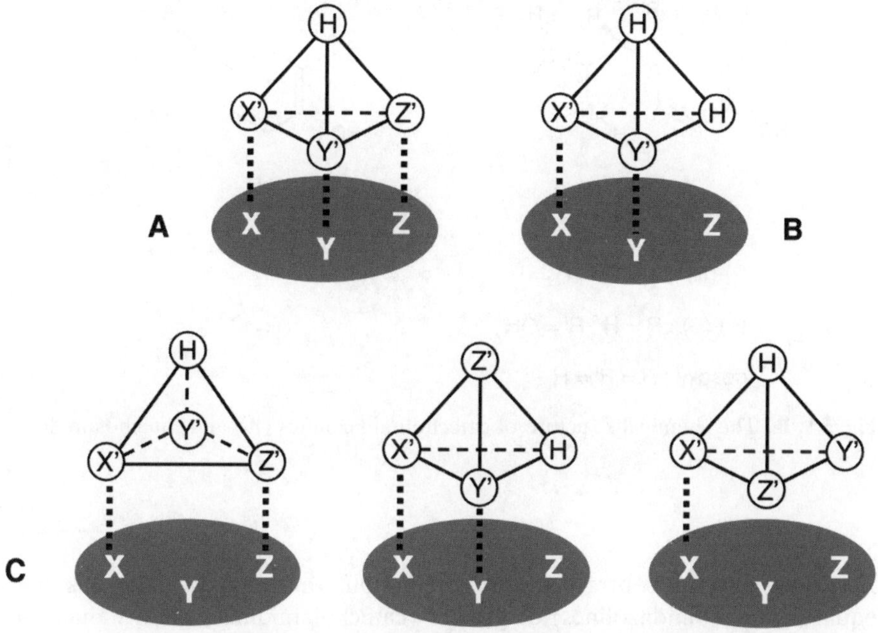

Fig. 4A–C. The three-point attachment model of Easson and Stedman schematizing chiral recognition at receptors and other biological binding sites. **A** Binding of an eutomer; **B** binding of a prochiral analog; **C** three out of six possible binding modes of a distomer. (Modified from EASSON and STEDMAN 1933; PORTOGHESE 1970)

eutomer) can the three groups X′, Y′ and Z′ be positioned so as to simultaneously interact with the three receptor sites X, Y and Z (Fig. 4A); the less active enantiomer (the distomer) can bind only via one or two out of its three complementary groups (Fig. 4C), hence its weaker affinity. The binding of a prochiral analog is shown in Fig. 4B.

An obvious feature of the current model is that it admits only attractive interactions between receptors and eutomers. But is current knowledge compatible with chiral recognition being based solely on attractive interactions? As clearly demonstrated by FERSHT (1987), unfavorable (repulsive) interactions are an important determinant of specificity. Such repulsive interactions may be steric or electrostatic in nature, implying that one site on the receptor may for example be a zone of steric hindrance. MEYER and RAIS (1989) have published vivid pictorial descriptions of chiral three-point recognition, including the case of one repulsive interaction.

The Easson-Stedman model was, for example, challenged with analogues of adrenaline, namely catecholimidazolines and catecholamidines (Fig. 5A and B, respectively) (RICE et al. 1987). In each series the hydroxylated enantiomers followed the order of affinity to the α-adrenoceptor predicted by the model.

A B

(-)-(R) : R = OH ; R' = H

(+)-(S) : R = H ; R' = OH

desoxy : R = R' = H

Fig. 5A, B. The chemical structure of catecholimidazolines (**A**) and catecholamidines (**B**)

However, and in contrast to the prediction, the desoxy analogues had equal (catecholimidazolines) or greater (catecholamidines) affinity than the respective eutomers. These finding were interpreted to mean that the Easson-Stedman model cannot be extended to catecholimidazolines and catecholamidines. However, such an interpretation may result from a static and too literal understanding of the model. Indeed, one of the NH/NH_2 groups in the protonated imidazolinyl and amidinyl functions fulfils the role of the NH_2 group in protonated adrenaline, whereas the other mimics the benzylic OH group as an H-bond donor to the receptor. The model thus retains its validity if *bioisosterism* is taken into account.

III. The Four-Location Model

Recent crystallographic data of broad significance have revealed one major ambiguity in the model of Easson and Stedman. As portrayed in Fig. 4, the model implies that the binding sites in the receptor are on a surface. However, the 3D model of the catalytic site of isocitrate dehydrogenase showed that three of its binding sites (X, Y and Z in Fig. 6, left) are located on the internal face of a cavity. As a result, these three sites alone would allow both enantiomers of the substrate to bind with similar affinity, the fourth group in the substrate pointing in either direction. Enantioselectivity is achieved by a fourth site, which proved to be Arg119 (W_1 in Fig. 6, right) in the metal-free enzyme, and Mg (W_2 in Fig. 6) in the magnesium-containing enzyme (MESECAR and KOSHLAND 2000). Figure 6 schematizes two enantiomers with groups X′, Y′ and Z′ bound to enzymatic sites X, Y and Z, respectively, chiral recognition being due to the group W′ in one enantiomer interacting with site W_1, and with site W_2 in the other enantiomer.

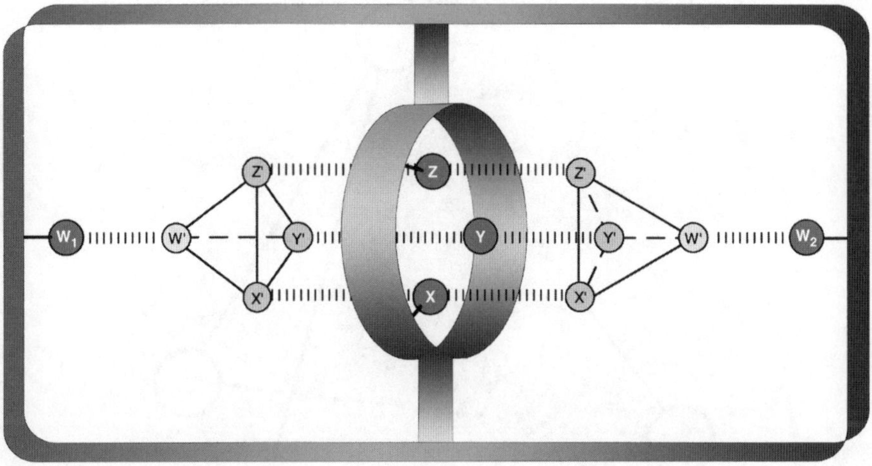

Fig. 6. A schematic representation of the four-location model of MESECAR and KOSHLAND (2000)

This significant result have led MESECAR and KOSHLAND (2000) to propose what they rightly labeled a *four-location model*, as schematized in Fig. 6, to explain a protein's ability to exhibit enantiospecificity. The four locations can be four attachment sites, or three attachment sites and one direction. *The latter case is in fact the model of Easson and Stedman, which must therefore be considered as a particular case of the more general four-location model.*

It is instructive to compare the four-location model with a four-point 3D model used in in silico *pharmacophore screening* of chemical libraries (LABAUDINIERE 1998). In this method, a pharmacophore (defined as the minimum necessary features for interacting with a biological target), is represented by chiral tetrahedron (Fig. 7) whose vertices can be of six types (H-bond donor or acceptor, aromatic ring, hydrophobic group, acid or base) separated by a number of predefined distances. The four-location model indeed offers a mechanistic rationale for the four-points 3D pharmacophoric model.

IV. Binding Versus Receptor Activation

Easson and Stedman in their epoch-making paper started from the implicit assumption that the enantiospecificity of receptors and enzymes is due to differential *affinity* (i.e., binding), and they explicitly and repeatedly stated that their model is one of *attachment*. Whether differential binding is the sole mechanism of chiral recognition in biochemical pharmacology is now challenged here. From a general viewpoint, the interaction of a xenobiotic, and more generally of any chemical compound, with a molecular machine (receptor, enzyme, transporter, etc.) can be broken down into two steps (TESTA 1989):

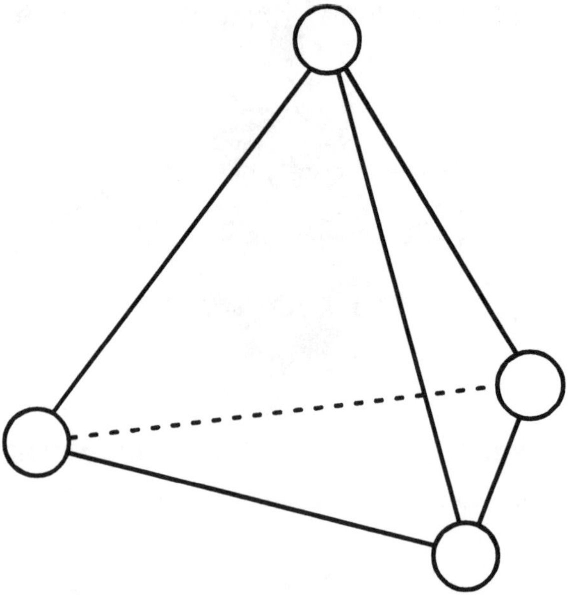

Fig. 7. Four-point 3D pharmacophore model used in in silico screening of chemical libraries. (Labaudiniere 1998)

1. *A binding step* (*recognition*) measured as affinity and involving the binding (complex formation) to the molecular machine.
2. *An activation step* (measured by potency, velocity, etc.) involving the functional response of the molecular machine, e.g., activation or blockade of a receptor, enzymatic catalysis leading to metabolite formation.

These two steps are conceptual ones, and as such have the utility and limitation of schematic perceptions. It is known, however, that both the binding and the activation step, either alone or together, can contribute to chiral recognition in pharmacodynamic and pharmacokinetic processes. A few examples are summarized below to demonstrate that chiral recognition does not need to be restricted to the binding step, but can also occur at the activation step.

Such a mechanism of enantioselectivity was uncovered in the activation of muscarinic presynaptic receptors by the enantiomers of *methacholine*, both of which acted as full agonists (Fuder et al. 1985). While the affinity ratio was ca. 180, the potency ratio was ca. 650. Thus, not only did (+)-methacholine show a higher affinity, it also had a higher efficacy, meaning that it had to occupy fewer receptors than its distomer to induce a given response. In other words, the high eudismic ratio in potency resulted from synergistic chiral recognition at both the binding and activation steps.

The effect of *dobutamine* enantiomers on α-adrenoceptors in rat aorta was due to a simpler mechanism of chiral recognition (RUFFOLO et al. 1981). Both enantiomers were partial agonists, the intrinsic activity of (–)- and (+)-dobutamine being 0.60 and 0.03, respectively. The (–) enantiomer was also 6 times more potent than the distomer. Both enantiomers however displayed identical affinity in two functional test models. In this example, chiral recognition is thus limited to the activation step, in complete contrast to the catecholimidazolines and catecholamidines discussed above (Fig. 5).

V. Binding Versus Enzyme Catalysis in Drug Metabolism

A similar story can be told for drug metabolism, if one is careful enough to distinguish between substrate and product enantioselectivity (see Sect. B.II.4.).

Metabolic transformations involve functionalization (oxidation, reduction. hydrolysis) and conjugation reactions. Accumulated evidence indicates that *substrate enantioselectivity* can originate from chiral recognition either at the binding step or at the catalytic step or at both. Here, *metabolic interactions* between enantiomers may be quite informative. Thus, the reaction of nicotine with guinea-pig lung azaheterocycle *N*-methyltransferase is enantiospecific in that only (+)-(*R*)-nicotine is *N*-methylated. The natural (–)-(*S*)-nicotine enantiomer is unreactive but acts as a strong competitive inhibitor of (*R*)-nicotine *N*-methylation (CUNDY et al. 1985). The complete lack of reactivity of (*S*)-nicotine despite its high affinity is of catalytic origin, presumably arising from an unproductive mode of binding which positions the target pyridyl nitrogen out of reach of *S*-adenosylmethionine. Less extreme examples were found in the microsomal oxidation of amphetamines (e.g., *para*-chloroamphetamine; AMES and FRANK 1982), where the (*S*)-enantiomer is a poorer substrate but a distinctly better ligand/inhibitor than the (*R*)-enantiomer. Such cases demonstrate that binding is a necessary but not sufficient condition for catalysis to occur; a proper positioning and alignment of the catalytic and target groups in the enzyme-substrate complex is also required (MENGER 1985).

Despite its limitations, Michaelis-Menten analysis offers another informative approach for assessing the binding and catalysis components of substrate enantioselectivity. Here, the Michaelis constant K_M represents affinity, whereas the catalytic rate is expressed by the maximal velocity V_{max} or the turnover number k_{cat}. A large compilation of literature data has been published (TESTA and MAYER 1988), uncovering cases of substrate enantioselective metabolism where the two enantiomers have similar K_M but different V_{max} values (catalysis-mediated recognition), other cases with different K_M but similar V_{max} values (binding-mediated recognition), and also cases where the enantiomers display differences in both K_M and V_{max} values.

Whereas substrate enantioselectivity involves the metabolic recognition of enantiomeric molecules, *product stereoselectivity* arises from the metabolic discrimination of enantiotopic or diastereotopic target groups in a single substrate compound. The phenomenon may be due to the action of distinct isoenzymes, or it may result from two distinct *binding modes* of a prochiral substrate to the active site of a single isoenzyme. In such a case, the two resulting enzyme-substrate complexes are diastereomeric and hence of different energy, and depending on the nature of both partners they interconvert rapidly or slowly relative to the rate of the enzymatic reaction. Thus, 2-phenylpropane ω-hydroxylation was found to involve two binding modes with cytochrome P450, each of which places one enantiotopic methyl group in the vicinity of the reactive site. Interconvertion of the two states was fast and its equilibrium constant was reflected in the ratio of the enantiomeric products (SUGIYAMA and TRAGER 1986). In other words, the observed product enantioselectivity was conditioned primarily by the binding step. Stated differently, the ratio of products depends on the relative probability of the two binding modes.

In contrast, the ratio of regioisomeric and stereoisomeric products resulting from the cytochrome P450LM2-mediated oxidation of camphor, adamantane, and adamandanone was found to parallel their chemical reactivity (WHITE et al. 1984). This suggested considerable movement of the substrate in the substrate-enzyme complex, and hence a catalytic control of the regio- and stereoselectivity.

VI. Current Limitations of the Three-Point and Four-Location Binding Models

To summarize the above, a general model of chiral recognition in biochemical pharmacology does not need to be limited to attractive interactions, but should also allow *one repulsive interaction* such as steric hindrance or electrostatic repulsion (Sect. C.II.). This is not a real limitation of the attachment models, provided it is kept in mind and made explicit.

A more serious limitation is the fact that within current knowledge, the attachment models remain *qualitative* in that they cannot rank the binding enantioselectivity of series of enantiomeric pairs. The development of a quantitative or even semi-quantitative model would be a major step forward. Progress in this direction should take advantage of intrinsic free energies of binding (positive and negative) of functional groups in drug-receptor interactions.

Another limitation is that the three-point and four-location models depict only affinity/binding. However, we know from Sects. C.IV and C.V that in drug-receptor interactions as well as in xenobiotic metabolism, enantioselectivity results from chiral recognition at the binding and/or at the activation step. This can be understood in molecular terms since the stereoelectronic conditions for high affinity and high efficiency or velocity may not be identical. Binding requires adequate functional groups and geometric features in the ligand molecule, as depicted in the three-point and four-location binding models. In contrast, the efficiency of the activation step depends on a *trigger* in the receptor or enzyme being activated. However, much remains to be understood regarding:

1. The structural and dynamic nature of the trigger
2. The involvement of binding sites in the activation step
3. The conformational dynamics that accompany and follow trigger activation (Wilson 2000), for example:
   a. Realignment of functional sites to form the catalytic triad of hydrolases
   b. Low-to-high spin transition in cytochrome P450
   c. Allosteric changes in receptors allowing coupling to other functional macromolecules such as G-proteins

In this molecular perspective, the difference between receptors, enzymes and transporters loses some of its significance, and all three are viewed as *molecular machines*. Whether the compound (endogenous or exogenous) acts on the effector or is acted upon or both, the basic phenomenon is that of a biological interaction which should be understood in its systemic globality.

At this stage of our functional understanding of molecular machines, it appears premature to transform the three-point and four-location attachment models into functional ones.

D. Conclusion

In this chapter, we have considered pharmacological processes of chiral recognition at two levels, the macroscopic and the molecular one. The two views are complementary and equally important. In vivo enantioselective processes allow clarification of the therapeutic and clinical significance of chiral recognition. At this stage, however, only information is gained, and little rationalization and understanding can be obtained.

Only at the molecular/mechanistic level is knowledge obtained, allowing understanding, prediction (if only qualitative), and a search for coherence with other phenomena. This is the path we have followed, hoping to provide the reader with a conceptual background useful to understand fully the following chapters in this volume.

References

Ames MM, Frank SK (1982) Stereochemical aspects of para-chloroamphetamine metabolism. Rabbit liver microsomal metabolism of R*S*-, R(–)- and S(+)-para-chloroamphetamine. Biochem Pharmacol 31:5–9

Ariëns EJ, Simonis AM (1967) Cholinergic and anticholinergic drugs, do they act on common receptors? Ann N Y Acad Sci 144:842–868

Barlow R (1990) Enantiomers: how valid is Pfeiffer's rule ? Trends Pharmacol Sci 11: 148–150

Barlow RB, Franks FM, Pearson JDM (1972) The relation between biological activity and the degree of optical resolution of optical isomers. J Pharm Pharmacol 24: 753–761

Beckett AH (1959) Stereochemical factors in biological activity. In: Jucker E (ed.) Progress in Drug Research, Volume 1. Birkhäuser, Basel, pp 455–530
Caldwell J, Testa B (1987) Criteria for the acceptability of experimental evidence for the enantiomeric composition of xenobiotics and their metabolites. Drug Metab Disposit 15:587–588
Cundy KC, Crooks PA, Godin CS (1985) Remarkable substrate-inhibitor properties of nicotine enantiomers towards a guinea pig lung aromatic azaheterocycle *N*-methyltransferase. Biochem Biophys Res Comm 128:312–316
Cushny AR (1926) Biological Relations of Optically Isomeric Substances. Williams and Wilkins, Baltimore
Easson LH, Stedman E (1933) Studies on the relationship between chemical constitution and physiological action. V. Molecular dissymmetry and physiological activity. Biochem J 27:1257–1266
Fersht AR (1987) The hydrogen bond in molecular recognition. Trends Biochem Sci 12:301–304
Fuder H, Jung B (1985) Affinity and efficacy of racemic, (+)-, and (−)-methacholine in muscarinic inhibition of [^{3}H]noradrenaline release. Br J Pharmacol 84:477–487
Gardner M (1982) The Ambidextrous Universe. Penguin Books, New York
Giacomini KM, Nelson WL, Pershe RA, Valdiviesco L, Turmer-Tamiyasu K, Blaschke TF (1986) In vivo interaction of the enantiomers of disopyramide in human subjects. J Pharmacokin Biopharm 14:335–356
Holmstedt B (1990) The use of enantiomers in biological studies: An historical review. In: Holmstedt B, Frank H, Testa B (eds) Chirality and Biological Activity. Liss, New York, pp 1–14
Hsyu PH, Giacomini KM (1985) Stereoselective renal clearance of pindolol in humans. J Clin Invest 76:1720–1726
Jenner P, Testa B (1973) The influence of stereochemical factors on drug disposition. Drug Metab Rev 2:117–184
Labaudiniere RF (1998) RPR's approach to high-speed parallel synthesis for lead generation. Drug Discovery Today 3:511–515
Le Corre P, Ratanasavanh D, Gibassier D, Barthel AM, Sado P, Le Verge R, Guillouzo A (1988) Rat hepatocyte cultures and stereoselective biotransformation of disopyramide enantiomers. In: Guillouzo A (ed.) Liver Cells and Drugs. Inserm, Paris, pp 321–324
Lehmann FPA (1986) Stereoisomerism and drug action. Trends Pharmacol Sci 7: 281–285
Lehmann FPA (1987) A quantitative stereostructure activity relationship analysis of the binding of promiscuous chiral ligands to different receptors. Quant Struct Act Relat 6:57–65
Lehmann FPA, Rodrigues de Miranda JF, Ariëns EJ (1976) Stereoselectivity and affinity in molecular pharmacology. In: Jucker E (ed.) Progress in Drug Research, Volume 20. Birkhäuser, Basel, pp 101–142
Mayer J, Testa B (1994) Stereoselectivity in metabolic reactions of toxication and detoxication. In: Welling PG, Balant LP (eds) Pharmacokinetics of Drugs, Springer Verlag, Berlin, pp 209–231
Menger FM (1985) On the source of intramolecular and enzymatic reactivity. Acc Chem Res 18:128–134
Mesecar AD, Koshland DE Jr (2000) A new model for protein stereospecificity. Nature 403:614–615
Meyer VR, Rais M (1989) A vivid model of chiral recognition. Chirality 1:167–169
Müller WE (1988) Stereoselective plasma protein binding of drugs. In: Wainer IW, Drayer DE (eds) Drug Stereochemistry, Analytical Methods and Pharmacology. Dekker, New York, pp 227–244
Pfeiffer CC (1956) Optical isomerism and pharmacological action, a generalization. Science 124:29–31

Piutti MA (1886) Sur une nouvelle espèce d'asparagine. Comp Rend Hebd Séances Acad Sci 134–138
Portoghese PS (1970) Relationships between stereostructure and pharmacological activities. Ann Rev Pharmacol 10:51–76
Rice PJ, Hamada A, Miller DD, Patil PN (1987) Asymmetric catecholimidazolines and catecholamidines: affinity and efficacy relationships at the alpha adrenoceptor in rat aorta. J Pharmacol Exp Therap 242:121–130
Riley RJ, Lambert C, Kitteringham NR, Park BK (1989) A stereochemical investigation of the cytotoxicity of mianserin metabolites in vitro. Br J Clin Pharmacol 27:823–830
Ruffolo RR Jr, Spradlin TA, Pollock GD, Waddell JE, Murphy PJ (1981) Alpha and beta adrenergic effects of the stereoisomers of dobutamine. J Pharmacol Exp Ther 219:447–452
Simonyi M (1984) On chiral drug action. Med Res Rev 4:359–413
Sugiyama K, Trager WF (1986) Prochiral selectivity and intramolecular isotope effects in the cytochrome P-450 catalyzed ω-hydroxylation of cumene. Biochemistry 25: 336–7343
Tamai I, Ling HY, Timbul SM, Nishikido J, Tsuji A (1988). Stereospecific absorption and degradation of cephalexin. J Pharm Pharmacol 40:320–324
Testa B (1979) Principles of Organic Stereochemistry. Dekker, New York
Testa B (1986a) The chromatographic analysis of enantiomers in drug metabolism studies. Xenobiotica 16:265–279
Testa B (1986b) Chiral aspects of drug metabolism. Trends Pharmacol Sci 7:60–64
Testa B (1987) Pharmacokinetic and pharmacodynamic events: Can they always be distinguished? Trends Pharmacol Sci 8:381–383
Testa B (1988) Substrate and product stereoselectivity in monooxygenase-mediated drug activation and inactivation. Biochem Pharmacol 37:85–92
Testa B (1989) Mechanisms of chiral recognition in xenobiotic metabolism and drug-receptor interactions. Chirality 1:7–9
Testa B (1990) Definitions and concepts in biochirality. In: Holmstedt B, Frank H, Testa B (eds) Chirality and Biological Activity. Liss, New York, pp 15–32
Testa B (1995) The Metabolism of Drugs and other Xenobiotics – Biochemistry of Redox Reactions. Academic Press, London
Testa B, Jenner P (1976) Drug Metabolism: Chemical and Biochemical Aspects. Dekker, New York
Testa B, Jenner P (1980) A structural approach to selectivity in drug metabolism and disposition. In: Jenner P, Testa B (eds) Concepts in Drug Metabolism, Part A. Dekker, New York, pp 53–176
Testa B, Mayer JM (1988) Stereoselective drug metabolism and its significance in drug research. In: Jucker E (ed.) Progress in Drug Research, Volume 32, Birkhäuser, Basel, pp 249–303
Testa B, Crivori P, Reist M, Carrupt PA (2000) The influence of lipophilicity on the pharmacokinetic behavior of drugs: Concepts and examples. Perspect Drug Discov Design 19:179–211
Trofast J, Osterberg K, Källström BL, Waldeck B (1991) Steric aspects of agonism and antagonism at β-adrenoceptors: synthesis of and pharmacological experiments with the enantiomers of formoterol and their diastereomers. Chirality 3:443–450
Wade DN, Mearrick PT, Morris JL (1973) Active transport of L-dopa in the intestine. Nature 242:463–465
Wilson EK (2000) Enzyme dynamics. Reactions require motions, according ot experiment and theory. Chem Eng News 78(29):42–45
White RE, McCarthy MB, Egeberg KD, Sligar SG (1984) Regioselectivity in the cytochromes P-450: Control by protein constraints and by chemical reactivities. Arch Biochem Biophys 228:493–502

CHAPTER 7

Enantioselectivity in Drug-Receptor Interactions

W. Soudijn, I. van Wijngaarden, and A.P. IJzerman

A. Introduction

G protein-coupled receptors (GPCRs) constitute one of the largest superfamilies of proteins in the human genome. It is estimated that there are at least 600 and probably over 1000 receptor species. The known ones are the target for approximately 60% of today's medicines for such diverse disease states as high blood pressure (β-adrenoceptor and angiotensin receptor antagonists), asthma (β-adrenoceptor agonists), and acid overproduction in the stomach (histamine H_2 receptor antagonists). Each member shares structural and/or sequence motifs and operates by common transduction mechanisms to mediate the transmission of extracellular signals into biochemical or electrophysiological responses within a cell. The signalling species is a specific endogenous molecule (e.g., neurotransmitter or hormone) which binds to the receptor. This results in activation of receptor and intracellularly located G proteins and propagation of the signal to effector molecules such as adenylyl cyclase. This enzyme converts ATP into cyclic AMP, the classical "second" messenger.

Many neurotransmitters and hormones are chiral compounds. Among them are proteins and peptides, such as FSH (follicle stimulating hormone) and the endorphins, respectively. They are composed of natural amino acids, which, except for glycine, are all chiral. The translation machinery for peptide and protein synthesis has evolved to utilize only one of the chiral forms of amino acids, the L-form. The reason for this is not clear. Amino acids themselves may also act as endogenous signalling molecules for GPCRs. Examples are GABA (γ-aminobutyric acid) acting on $GABA_B$ receptors, and metabotropic glutamate receptors for which L-glutamate and L-aspartate serve as signalling molecules. Other small endogenous molecules may serve a similar purpose. Among them are the classical biogenic amines, such as (nor)adrenaline, histamine, and serotonin, and numerous other examples, such as leukotrienes, adenosine, and ATP. This group hosts both nonchiral and chiral substances. Dopamine, serotonin, acetylcholine and histamine do not have a chiral center in their structure, whereas others possess one (adrenaline and noradrenaline) or more (adenosine and ATP). Interestingly, for a large number of cloned ("orphan") receptors, the endogenous ligand is not known (yet). An important effort to keep track of the latest information on receptors is the G protein-coupled receptor database which is maintained on the Internet (www.gpcr.org/7tm/).

The earliest evidence for stereoselective receptor recognition was reported by Piutti (1886) who pointed out that D-asparagine has a sweet taste, while the natural L-isomer is insipid. Interestingly, Pasteur (1886) ascribed this finding to the presence of an optically active substance in the nervous mechanism of taste. Given our current understanding of olfactory receptors as a special branch of the large superfamily of G protein-coupled receptors, Pasteur's statement was well ahead of its time. Cushny (1926) was the first to review and pay particular attention to the biological role of optically active substances. Even before the receptor concept had been introduced in an early form by Langley and Ehrlich (see Holmstedt and Liljestrand 1963) Cushny provided pharmacological proof that the two enantiomers of the muscarinic antagonist atropine showed large differences in potency, followed by similar observations, for example, adrenaline. Such findings led Easson and Stedman (1933) to hypothesize a three-point attachment for an asymmetric molecule to its receptor. For the first time then a rationale for stereoselective ligand recognition by GPCRs was proposed, which has become the dogma ever since. Interestingly, this concept was recently expanded by Mesecar and Koshland Jr (2000), who suggested a four-location rather than a three-point model. Their view, based on crystal structures of the enzyme isocitrate dehydrogenase complexed with either L- or D-isocitrate, is that such a four-location model is the minimal requirement for a protein to discriminate between L- and D-isomers.

In this chapter we will first review some typical, recent examples of chiral ligands interacting with their receptors. Such an overview is by definition arbitrary, since it is only self-evident from the receptor concept that different drugs, e.g., two enantiomers, may have different affinities, selectivities and intrinsic activities. In the latter part of this contribution we will include the macromolecular targets in our analysis and focus on the ligand-receptor interaction. Emphasis will be on receptor mutations that seem to have a direct impact on stereoselective recognition.

B. Receptor Ligands and Enantioselectivity

I. Ligands of Biogenic Amine Receptors

Serotonin (5-hydroxytryptamine, 5-HT, itself nonchiral) is a very ancient neurotransmitter (Peroutka 1994) that interacts with a variety of 5-HT receptor subtypes. It is involved in numerous physiological and pathophysiological processes. For instance, 5-HT_{1A} receptors are implicated in feeding, sexual activity, sleep, anxiety, and depression.

Table 1. Stereoselectivity ratios of aminotetralin derivatives

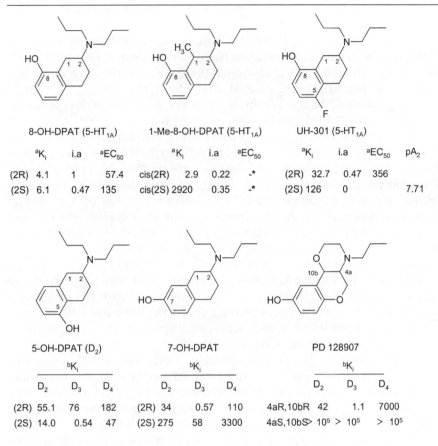

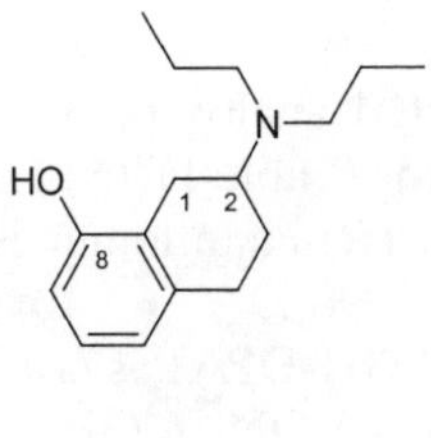

8-OH-DPAT (5-HT_{1A})

	^aK_i	i.a	$^aEC_{50}$
(2R)	4.1	1	57.4
(2S)	6.1	0.47	135

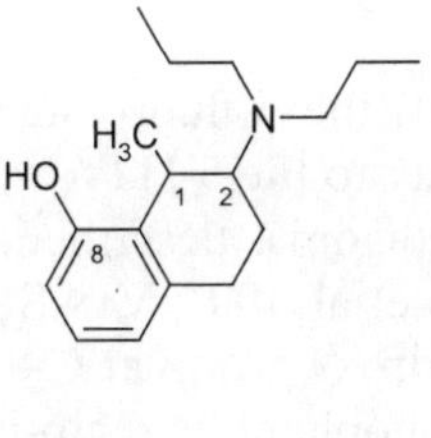

1-Me-8-OH-DPAT (5-HT_{1A})

	^aK_i	i.a	$^aEC_{50}$
cis(2R)	2.9	0.22	-*
cis(2S)	2920	0.35	-*

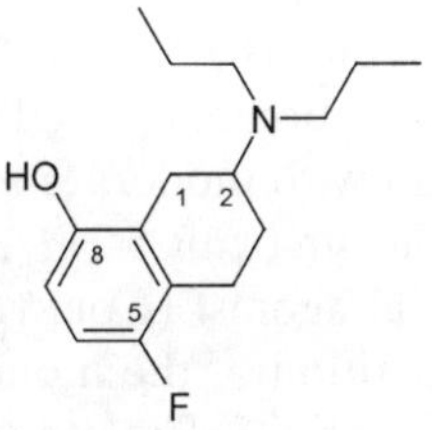

UH-301 (5-HT_{1A})

	^aK_i	i.a	$^aEC_{50}$	pA_2
(2R)	32.7	0.47	356	
(2S)	126	0		7.71

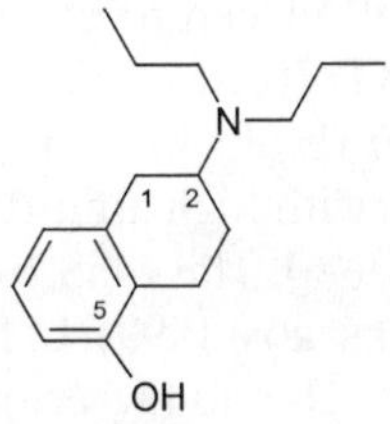

5-OH-DPAT (D_2)

	^bK_i		
	D_2	D_3	D_4
(2R)	55.1	76	182
(2S)	14.0	0.54	47

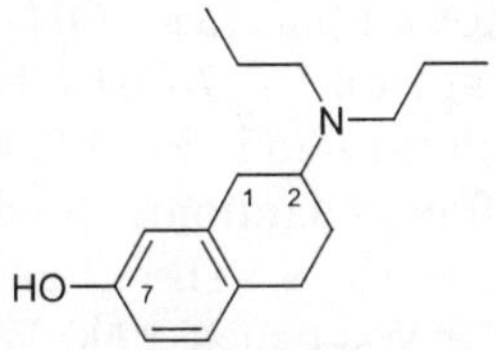

7-OH-DPAT

	^bK_i		
	D_2	D_3	D_4
(2R)	34	0.57	110
(2S)	275	58	3300

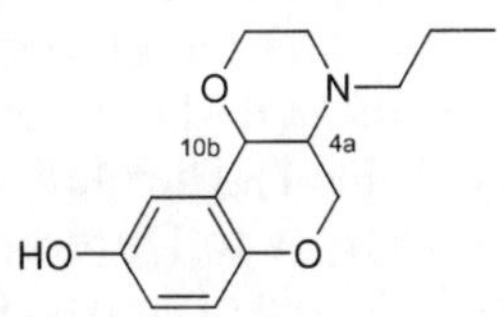

PD 128907

	^bK_i		
	D_2	D_3	D_4
4aR,10bR	42	1.1	7000
4aS,10bS	> 10^5	> 10^5	> 10^5

Affinity: K_i,nM.a:[^{3}H]-8-OH-DPAT, rat cortex membranes.b:[^{3}H]-N-0437, hD_{2L} / D_3 / $D_{4.2}$ Rs(CHO-K_i cells).
Intrinsic activity: i.a. Potency: EC_{50},nM;pA_2,a: FSK-stimulated cAMP, rat hippocampal membranes.
* Accurate EC_{50} values could not be calculated.

The racemate 8-OH-DPAT (Table 1) is the prototype of 5-HT_{1A} receptor ligands. In competition binding studies both enantiomers of 8-OH-DPAT are virtually equipotent. In functional tests, however, (*R*)-8-OH-DPAT is a full agonist, whereas (*S*)-8-OH-DPAT is a partial agonist (Table 1) (Cornfield et al. 1991). In human brain, however, both enantiomers act as full agonists (Palego et al. 2000). Introduction of a methyl group at C-1 of the nonaromatic ring of 8-OH-DPAT results in high stereoselectivity. The *cis*-(2*R*)-enantiomer has a high affinity for the 5-HT_{1A} receptor whereas the affinity of the *cis*-(2*S*)-enantiomer is 1000-fold lower. Both enantiomers are weak partial agonists (Table 1) (Cornfield et al. 1991). Modeling of 8-O*H*-DPAT and analogs shows that the methyl substituent at C-1 in the *cis*-(2*R*)-configuration does not interfere with the binding of the *N*-propyl chains or the 8-OH group to the 5-HT_{1A} receptor (Kuipers 1997a).

The enantiomers of UH-301, the 5-fluoro-substituted analog of 8-OH-DPAT, bind with moderate affinity to the 5-HT_{1A} receptor (Table 1). (*S*)-UH-301 was the first full 5-HT_{1A} antagonist described. The (*R*)-enantiomer is a weak partial agonist (Cornfield et al. 1991; Van Steen 1996).

The position of the monohydroxy group at C-8 of 8-OH-DPAT is crucial for 5-HT_{1A} activity. The other monophenolic regio-isomers 5-OH-, 6-OH- and 7-OH-DPAT are pure dopamine agonists (Seiler and Markstein 1984). Later studies provided evidence that 5-OH-DPAT and 7-OH-DPAT have preference for the novel dopamine D_3 receptor subtype (Van Vliet et al. 1996). The affinity resides mainly in the (*S*)-enantiomer of 5-OH-DPAT and the (*R*)-enantiomer of 7-OH-DPAT (Table 1). A similar reversal of the stereoselectivity was reported for the dopamine D_1 and D_2 receptors (Seiler and Markstein 1984). The authors showed that (*S*)-5-OH-DPAT can be superimposed on (*R*)-7-OH-DPAT but not on (*S*)-7-OH-DPAT. Restriction of the conformation of 7-OH-DPAT as illustrated in PD 128907 results in high stereoselectivity (Table 1). The (4aR,10bR)-enantiomer binds with high affinity and moderate selectivity to D_3 receptors (Van Vliet et al. 1996). The (4aS,10bS)-enantiomer is devoid of activity (De Wald et al. 1990; Wikström 1999). In functional tests (4aR,10bR)-PD 128907 acts as a full agonist at D_2 and D_3 receptors (De Wald et al. 1990; Pugsley et al. 1995).

Mixed dopamine D_2, D_3 and serotonin 5-HT_{1A} receptor binding properties have been demonstrated with 5-OMe-BPAT (Table 2). This compound is a derivative of 5-OMe-DPAT, the methoxy analog of 5-OH-DPAT. The enantiomers of 5-OMe-BPAT differ in their stereoselectivities for D_2, D_3, and 5-HT_{1A} receptors. (*R*)-5-OMe-BPAT prefers the dopamine receptors and (*S*)-5-OMe-BPAT the 5-HT_{1A} receptors. Both enantiomers act as full agonists at 5-HT_{1A} receptors. However the enantiomers differ in their intrinsic activity at the D_2 receptors. In vivo (*R*)-5-OMe-BPAT behaves as a D_2 receptor antagonist whereas (*S*)-5-OMe-BPAT acts as a D_2 receptor agonist (Homan et al. 1999a). The difference in intrinsic activity has been analyzed by molecular modeling, suggesting that (*R*)-and (*S*)-5-OMe-BPAT bind at the D_2 receptor in a different way. The interaction with the 5-HT_{1A} receptor is, however, very similar: both enantiomers are full agonists (Homan et al. 1999b). Compounds, such as (*R*)-5-OMe-BPAT, combining D_2 antagonism and 5-HT_{1A} agonism, may be useful antipsychotics.

Another 2-aminotetralin-containing compound is apomorphine (Table 2). The (6aR)-enantiomer binds with high affinity to dopamine D_2 receptors, whereas the (6aS)-enantiomer displays a 23-fold weaker affinity (Kula et al. 1985). In functional tests, however, both enantiomers have opposite effects: (6aR)-apomorphine acts as a potent full agonist and (6aS)-apomorphine acts as a moderately potent full antagonist (Goldman and Kebabian 1984). Substitution of the methyl group of apomorphine for *n*-propyl results in *N*-*n*-

Table 2. Stereoselectivity ratios of aminotetralin derivatives including apomorphine

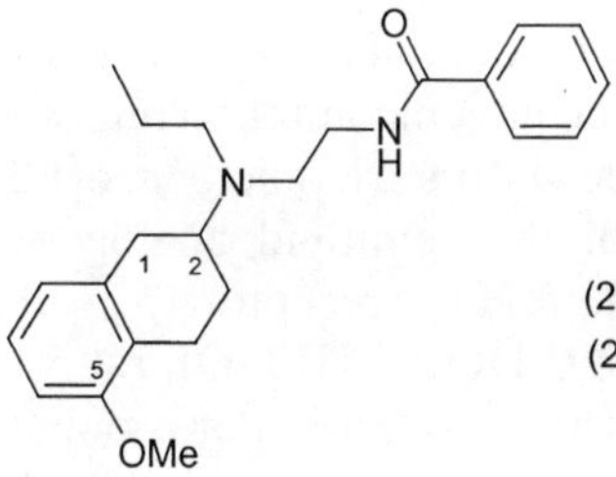

5-OMe-BPAT

	$^{a}K_i$ D_{2A}	D_3	$5\text{-}HT_{1A}$	i.a $5\text{-}HT_{1A}$	$^{a}IC_{50}$ $5\text{-}HT_{1A}$
(2R)	0.77	0.14	3.8	1	100
(2S)	3.4	0.42	0.20	1	< 100

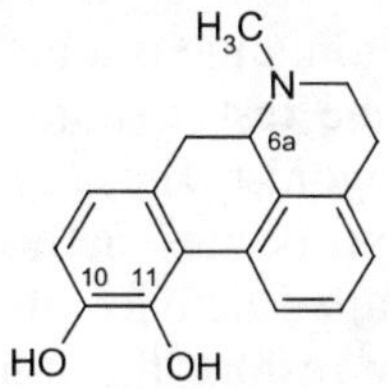

APO (D_2)

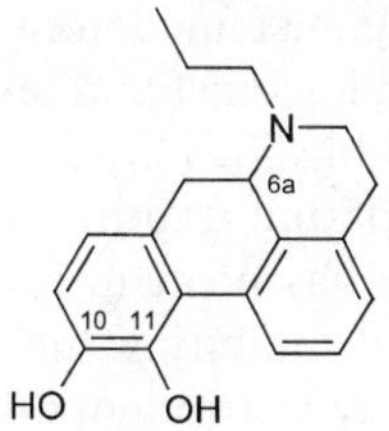

NPA (D_2)

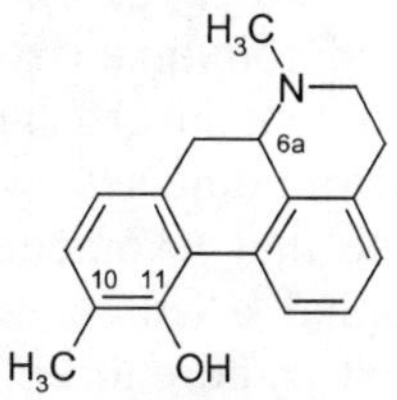

MHA ($5\text{-}HT_{1A}$)

	$^{b}IC_{50}$	i.a	$^{b}EC_{50}$	K_i
(2R)	4.3	1	14	
(2S)	98.4	0		495

	$^{b}IC_{50}$	i.a	$^{b}EC_{50}$
(6aR)	3	1	0.3
(6aS)	492	1	16.7

	$^{c}K_i$	i.a	$^{c}EC_{50}$	IC_{50}
(6aR)	3.1	1	50	
(6aS)	39.0	0		30

Affinity: K_i / IC_{50},nM.a:[^{3}H] - raclopride, hD_{2A} / D_3Rs (Ltk $^-$,CHO cells). b:[^{3}H] - ADTN, rat striatal membranes. c: [^{3}H]-8-OH-DPAT, rat cortex membranes. Intrinsic activity: i.a. Potency: EC_{50} / IC_{50} / K_i, nM. a: VIP-stimulated cAMP, GH_4ZD_{10} cells. b: 1R-αMSH release, rat pituitary cells. c: single shock-contractions, guinea pig ileum.

propyl norapomorphine (NPA) (Table 2). In contrast to the apomorphine enantiomers both enantiomers of NPA are full D_2 receptor agonists (Goldman and Kebabian 1984). In binding studies (6aR)-NPA is more potent than (6aS)-NPA (Kula et al. 1985).

Opposite pharmacological effects have also been observed in 10-methyl-11-hydroxy-aporphine (MHA) a close structural analog of apomorphine (Table 2). Unexpectedly, MHA interacts selectively with $5\text{-}HT_{1A}$ receptors. The (6aR)-enantiomer of MHA is a full agonist and the (6aS)-enantiomer is a full antagonist (Cannon et al. 1991). The $5\text{-}HT_{1A}$ receptor selectivity of MHA is due to the presence of the C-10 methyl substituent suggested to be capable of interacting with a lipophilic cavity unique for the 5-HT_{1A} receptor (Hedberg et al. 1995).

Potent 5-$HT_{2A/2C}$ receptor agonists were obtained by incorporation of the 2,5-dimethoxy substituents of the hallucinogenic amphetamines such as DOB (4-bromo-2,5-dimethoxyamphetamine) into a tricyclic benzo[1,2-b; 4,5-b′]difuran nucleus. The (*R*)-enantiomer of the semirigid analog of DOB, for instance, displays a high affinity for the 5-HT_{2A} receptor (K_i = 0.31 nM) and the 5-HT_{2C} receptor (K_i = 0.11 nM) ([^{3}H] DOB, [^{3}H] DOI, rat 5-HT_{2A}/5-HT_{2C} receptors, NIH-3T3 fibroblast cells). The (*S*)-enantiomer is only twofold less potent. In a functional test (phosphoinositide hydrolysis) the (*R*)-enantiomer acts as a full agonist at 5-HT_{2A} receptors (EC_{50} = 2.7 nM) and the (*S*)-enantiomer acts as a partial 5-HT_{2A} receptor agonist (EC_{50} = 19 nM, i.a. = 0.79) (Chambers et al. 2001).

Agonism versus inverse agonism has been reported for the enantiomers of the α_2-adrenoceptor agonist medetomidine (Jansson et al. 1998). Dexmedetomidine is a potent full agonist and levomedetomidine is a moderately potent inverse agonist (Table 3). However, in some test systems, levomedetomidine acts as a weak partial α_2-adrenoceptor agonist. These results indicate that levomedetomidine has protean agonistic properties: activating uncoupled α_2-adrenoceptors and inhibiting the constitutive activity of precoupled α_2-adrenoceptors. This type of compound may restore the normal function of the receptor. In binding studies dexmedetomidine is about 30 times more potent than levomedetomidine at α_{2A}-adrenoceptors (Table 3) (Jansson et al. 1994). Structurally related to medetomidine is RX821002 (Table 3). In contrast to dex- and levomedetomidine, both enantiomers of RX821002 are full antagonists at α_2-adrenoceptors. The affinity and activity reside mainly in the (2*S*)-enantiomer (Welbourn et al. 1986).

High stereoselectivity has also been observed in various compounds interacting with β-adrenoceptors. For instance the (*S*)-enantiomers of the well-known β-adrenergic blockers of the propranolol prototype are significantly more potent than the corresponding (*R*)-enantiomers (Table 3). This difference in affinity and activity has been the basis of combined receptor mutation and molecular modeling studies (vide infra). A number of β-blockers display affinity for the 5-HT_{1A} receptor. The enantioselectivity of the compounds for both receptors is quite similar (Table 3). A relatively novel β-adrenoceptor antagonist is nebivolol (Table 3). The compound is a pseudosymmetrical molecule with 4 centers of asymmetry. Nebivolol is a racemate consisting of the SRRR and RSSS enantiomers. The affinity and activity for the β_1- and β_2-adrenoceptors resides mainly in the (SRRR)-enantiomer. This enantiomer displays a moderate (binding) to high (cAMP-accumulation) selectivity for β_1-adrenoceptors (Table 3). The hemodynamic profile of nebivolol differs from that of the classical β-blockers. In contrast to the latter nebivolol improves left ventricular function, reduces systemic vascular resistance, and increases cardiac output. Both enantiomers contribute to this particular profile (Pauwels et al. 1991).

Table 3. Stereoselectivity ratios of α- and β-adrenoceptor ligands

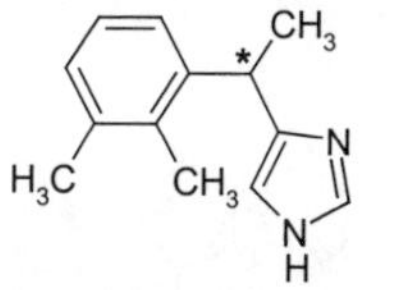

MEDETOMIDINE (α_{2A})

	^aK_i	i.a	$^aEC_{50}$	IC_{50}
(+)	0.78	1	17	
(-)	21.7	- 1		390

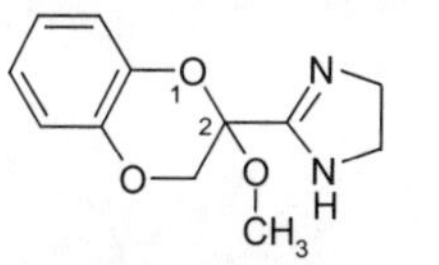

RX821002 (α_2)

	^bK_i	i.a	$^bpA_2$
(2R)	256	0	7.07
(2S)	0.36	0	9.69

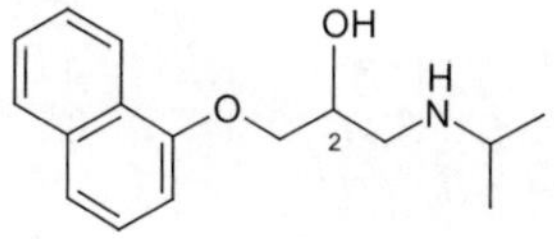

PROPRANOLOL ($\beta_{1,2}$,5-HT_{1A})

	^cK_i		i.a		^cK_i	
	$\beta_{1,2}$	5-HT_{1A}	$\beta_{1,2}$	5-HT_{1A}	$\beta_{1,2}$	5-HT_{1A}
(R)	100	3100	0	0	40	89000
(S)	0.6	68	0	0	0.5	50

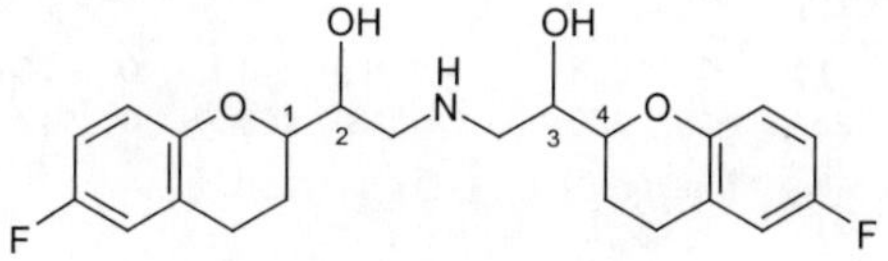

NEBIVOLOL (S,R,R,R) + (R,S,S,S)

	^dK_i		i.a		$^dIC_{50}$	
(1,2,3,4)	β_1	β_2	$\beta_{1,2}$	5-HT_{1A}	β_1	β_2
(S,R,R,R)	0.72	5.6	0	nt	0.41	36.7
(R,S,S,S)	200.8	139.4	0	nt	600	750

Affinity: K_i,nM. a:[^{3}H] - rauwolscine, hα_{2A}R (S115 cells). b:[^{3}H] - idazoxan, rat cortex membranes.
c:[^{125}I] IHYP, S49 mouse lymphoma cells; [^{3}H]-8-OH-DPAT, h5-HT_{1A}R (COS-7 cells).
d: [^{3}H] CGP-12177, hβ_1 / β_2Rs (CHO cells). Intrinsic activity: i.a. Potency: EC_{50} / IC_{50} / K_i, nM; pA_2.
a: FSK - stimulated cAMP, HEL92.1.7 cells. b: clonidine antagonism, rat vas deferens. c: isoproterenol - stimulated cAMP, S49 mouse lymphoma cells; FSK - stimulated cAMP, rat hippocampal membranes.
d: isoproterenol - induced cAMP, hβ_1 / β_2(CHO cells).

II. Ligands of Adenosine, Cannabinoid and Melatonin Receptors

Various compounds interacting with adenosine receptors display moderate to high stereoselectivity. For instance (*R*)-PIA, a reference agonist for adenosine A_1 receptors, binds 78 times better than (*S*)-PIA (Table 4). In functional tests a similar stereoselective ratio has been observed. Both enantiomers are full agonists (DALY 1982). Recently close analogs of (*R*)-PIA, e.g. 2-chloro-N^6-(1-phenoxy-2-propyl) adenosine, have been developed as potential antiischemic agents. These novel neuroprotective compounds display less cardiovascular side effects than (*R*)-PIA (KNUTSEN et al. 1999).

The enantiomers of the deazaadenine derivative ADPEP act as full antagonists at A_1 receptors (Table 4). (*R*)-ADPEP is approximately 30–35 times more potent than (*S*)-ADPEP (MÜLLER et al. 1990). A novel A_1 receptor

Table 4. Stereoselectivity ratios of adenosine A_1 receptor ligands

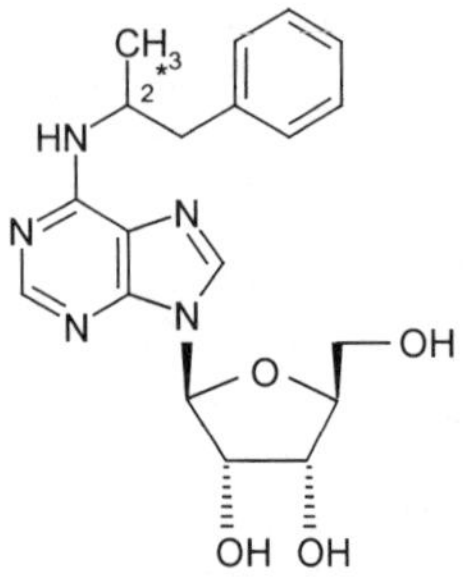

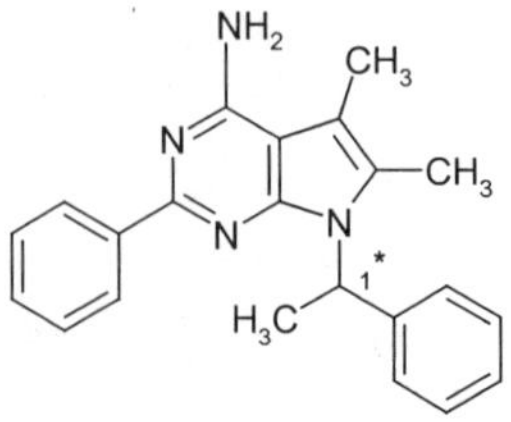

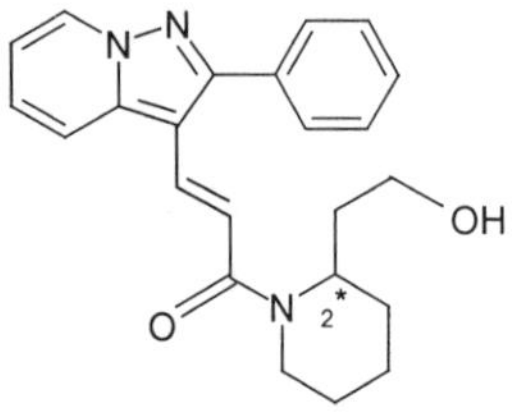

PIA (A_1)	^aK_i	i.a	^aK_i	ADPEP (A_1)	^bK_i	i.a	^bK_B	FK453 (2*R) (A_1)	$^cIC_{50}$	i.a	$^cIC_{50}$
(2*R)	1.6	1	3	(1*R)	4.7	0	3.8	(2*R)	17.5	0	0.56
(2*S)	125	1	200	(1*S)	165	0	112	(2*S)	10100	0	1180

Affinity: K_i / IC_{50}, nM. a:[^{3}H] - CHA, rat brain membranes. b:[^{3}H] - PIA, rat cortex membranes. c:[^{3}H] - CHA, rat cortex membranes. Intrinsic activity: i.a. Potency: K_i / K_B / IC_{50}, nM.
a: catecholamine-stimulated cAMP, rat adipocytes. b: antagonism inhibition cAMP by (R) - PIA, rat fat cells. c: inhibition adenosine-induced neg. inotropic act., guinea pig atria.

antagonist is FK453 (Table 4). The compound, synthesized in a program on diuretics, is structurally different from the well-known A_1 antagonists. The (*S*)-enantiomer is hardly active. These results indicate that the diuretic action of FK453 might be due to its selective A_1 antagonism (AKAHANE et al. 1996).

HU210 ((–)*R*,*R*-11-hydroxy-Δ^8-3-(1^1-dimethyl)heptyltetrahydrocannabinol) is a compound with high affinity for the cannabinoid receptor and a potent inhibitor of forskolin-stimulated cAMP production (Table 5; FELDER et al. 1992). Both affinity and potency of HU210 are much higher than those of its (+)*S*,*S*-enantiomer HU211 (Table 5) and also of the natural product (–)*R*,*R*-Δ^9-tetrahydrocannabinol. Compared to the natural product, the affinity and potency of HU210 are 2,750- and 880-fold higher (FELDER et al. 1992).

Melatonin, 5-methoxy-*N*-acetylserotonin, modulates circadian rhythm and the sleep–wake cycle in vertebrates by interaction with melatonin receptors. The affinities for the melatonin receptor and the functional activities of a series of chiral rigid melatonin analogs has been published by DAVIES et al. (1998). (*S*)-enantiomers have the highest affinity (Table 6). In a functional test both enantiomers of *N*-acetyl-4-aminomethyl-6-methoxy-9-methyl-1,2,3,4-tetrahydrocarbazole (AMMTC) act as full agonists and so does the (*S*)-enantiomer of its desmethoxy analog AMTC (DAVIES et al. 1998). However, the (*R*)-enantiomer of AMTC acts as a full antagonist. There is no statistical difference in the affinities of the racemate of *N*-(4-methoxy-2,3-dihydro-1*H*-phenalen-2-yl)acetamide and its enantiomers. In functional tests the com-

Table 5. Stereoselectivity ratios of ligands for cannabinoid, nociceptin, and cholecystokinin receptors

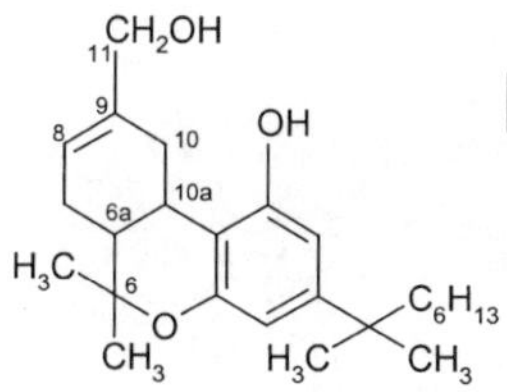

HU210 (6aR,10aR)($CB_{1,2}$)
HU211 (6aS,10aS)($CB_{1,2}$)

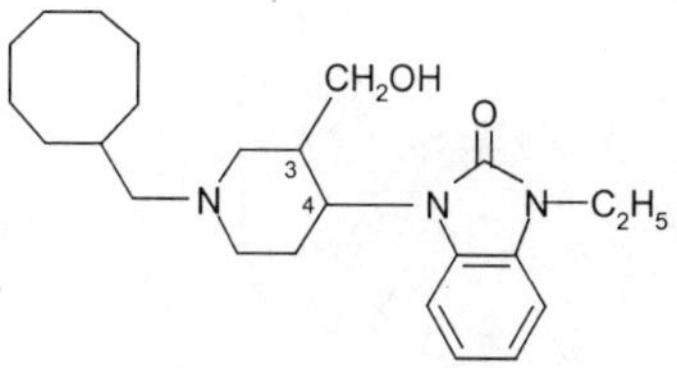

J-113397 (3R,4R)(ORL1)
J-112444 (3S,4S)(ORL1)

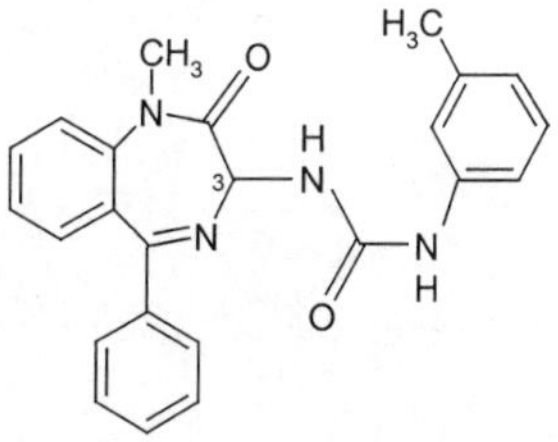

L-365,260 (3R)(CCK2)
L-365,346 (3S)(CCK1)

	$^{a}K_i$	i.a	$^{a}IC_{50}$
(6aR,10aR)	0.06	1	0.02
(6aS,10aS)	1364	1	191

	$^{b}IC_{50}$	i.a	$^{b}IC_{50}$	EC_{50}
(3R,4R)	2.3	0	5.6	$>10^4$
(3S,4S)	820	0	$>10^4$	$>10^4$

	$^{c}IC_{50}$		i.a	$^{d}IC_{50}$
	CCK1	CCK2	CCK2	CCK2
(3R)	740	8.5	0	39
(3S)	4.7	280		

Affinity: K_i / IC_{50}, nM. a:[^{3}H] CP55,940, hCBR L all membranes. b:[^{125}I] Tyr^{14}-nociceptin, hORL1R (CHO cells). c:[^{125}I] CCK-8, rat pancreatic (CCK1R); guinea pig brain (CCK2R). Intrinsic activity: i.a. Potency: EC_{50} / IC_{50}, nM. a: FSH-stimulated cAMP, rCBR (CHO cells). b: nociceptin-produced [^{35}S] GTPγS binding, hORL1R (CHO cells). d: inhibition CCK-8-induced IP, rCCK2R (CHO cells).

pounds lack agonist activity whereas the racemate acts as an antagonist (Table 6; JELLIMANN et al. 1999). A phenalene derivative possessing a partially constrained ethylamido chain is *N*-[(4-methoxy-2, 3-dihydrophenalen-1-yl)methyl] butanamide. The (–)-enantiomer (K_i = 0.8 nM) is 27-fold more potent than the (+)-enantiomer (K_i = 21.7 nM) (2-[^{125}I]-melatonin, chicken brain). In a functional test (melanophore contraction in *X. laevis* tadpoles) the racemate acts as a full agonist (EC_{50} = 1.61 nM) (JELLIMANN et al. 2000).

III. Ligands of Peptide Receptors

ORL-1, a recently cloned opioid receptor-like GPCR with significant homology to the classical μ-, δ- and κ-opioid receptors has a low affinity for endogenous and synthetic μ-, δ- and κ-ligands. The endogenous ligand of the ORL-1 receptor is nociceptin, a peptide consisting of 17 amino acids with a low, if any, affinity for the classical opioid receptors (HENDERSON and MCKNIGHT (1997). The first highly potent and selective nonpeptide antagonist 1-[(3R,4R)-1-cyclooctylmethyl-3-hydroxymethyl-4-piperidyl]-3-ethyl-1,3-dihydro-2*H*-benzimidazol-2-one (J-113397) was developed by KAWAMOTO et al. (1999). J-113397 binds with high affinity to the ORL-1 receptor but has a low affinity for the μ-, δ- and κ-receptors. The compound is a potent inhibitor of the nociceptin-stimulated GTPγS binding to G proteins (Table 5). The (3*S*,4*S*)-enantiomer J-112444 on the other hand has a 400-fold lower affinity for the

Table 6. Stereoselectivity ratios of melatonin receptor ligands

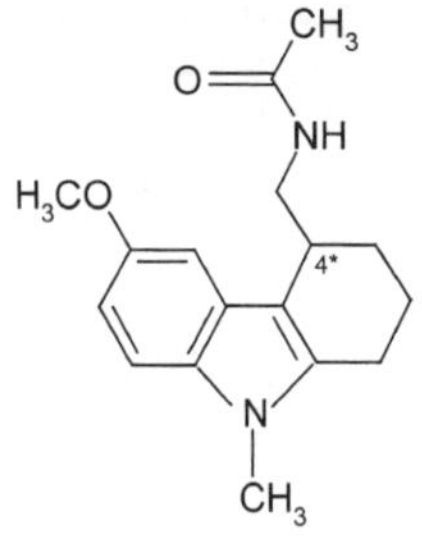

AMMTC (MT)

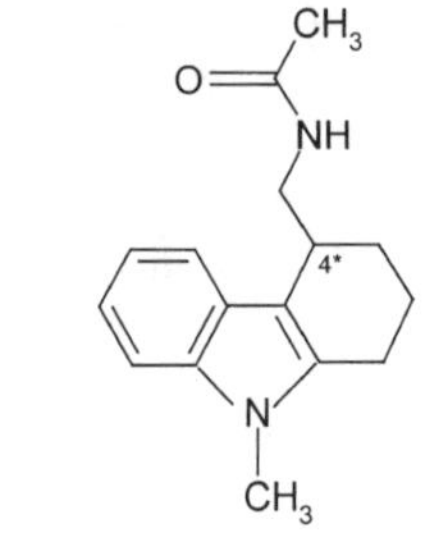

AMTC (MT)

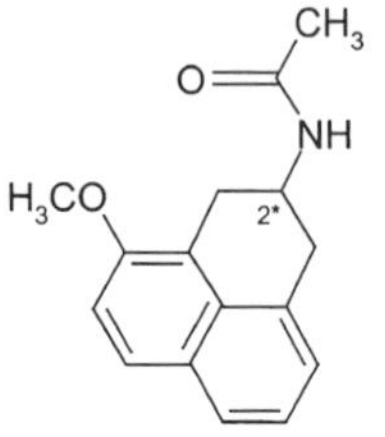

(MT)

AMMTC (MT)	K_i	i.a	$^{a}EC_{50}$	AMTC (MT)	K_i	i.a		(MT)	K_i	i.a	$^{b}IC_{50}$
(4*R)	48.4	1	52.3	(4*R)	708	0	$^{a}IC_{50}$=6400nM	(R)	38.3	0	(RS) 487
(4*S)	0.372	1	0.23	(4*S)	40	1	$^{b}EC_{50}$= 189nM	(S)	28.7	0	

Affinity: K_i,nM. [2-^{125}I] - melatonin, chicken brain membranes. Intrinsic activity: i.a
Potency: EC_{50} / IC_{50}, nM. a: X.laevis dermal melanophore cell line. b: X.laevis tadpoles.

ORL-1 receptor and is virtually inactive as nociceptin antagonist (Table 5). The inhibiting effect of nociceptin on forskolin-stimulated cAMP accumulation was dose-dependently inhibited by J-113397 with an IC_{50} of 26 ± 3 nM (OZAKI et al. 2000). The selectivity of action of J-113397 on ORL-1 receptor was confirmed in ORL-1 receptor-deficient mice (ICHIKAWA et al. 2001).

Cholecystokinins (CCKs) are endogenous neuropeptides interacting with G protein-coupled central and peripheral CCK-receptors of which there are at present two subtypes known, i.e., $CCK_1 = CCK_A$, mainly situated in the gastrointestinal tract, and $CCK_2 = CCK_B$, predominantly present in the brain. L-365 260, 3R(+)-*N*-(2,3-dihydro-1-methyl-2-oxo-5-phenyl-1*H*-1,4-benzodiazepin-3-yl)-N^1-(3-methylphenyl) urea is a potent CCK_2 selective antagonist, whereas the selectivity is reversed in the (*S*)-enantiomer L-365 346, a potent CCK_1 antagonist (Table 5) as shown by BOCK et al. (1993). It was reported by JAGERSCHMIDT et al. (1996) that histidine His381 of the rat CCK_2 receptor determines its selectivity for antagonists. The search for CCK_2 antagonists with improved aqueous solubility and brain penetration compared to L-365 260 resulted in L-740 093 a close analog of L-365 260 in which the C-5 phenyl group was replaced by a 3-azabicyclo[3,2,2]nonan-3-yl moiety (PATEL 1994). Both the affinity of L-740 093 for the CCK_2 receptor (IC_{50} 0.1 nM) as well as its selectivity versus the CCK_1 receptor (16 000-fold) are extremely high. The penetration in mouse brain of L-740 093 or L-365 260 after intravenous injection was studied by the ex vivo binding of ^{125}I-B*H*-CCK-8S to brain tissue homogenates. The ID_{50} values for L-740 093 and L-365 260 were 0.2 and 13 mg/kg, respectively, indicating that for studying central effects of

CCK_2 receptor inhibition in patients L-740093 is probably the better tool (PATEL et al. 1994).

The endogenous undecapeptide substance P (SP) stimulates central and peripheral neurokinin (NK) receptors. A large number of stereoisomeric antagonists has been synthesized and tested for affinity and potency. The four stereoisomers of 2-(diphenylmethyl)-3-((3,5-bis(trifluoromethyl)benzyl)-oxy)-1-azabicyclo[2.2.2]octane (Table 7) were prepared by SWAIN et al. (1995) and tested for affinity on human NK_1 receptors expressed in CHO cells and for potency in inhibiting substance P-induced extravasation in the skin of guinea pigs. It was shown that optimal affinity and potency was highly dependent on the absolute configuration of the substituent at the C3-atom of the quinuclidinyl moiety and much less on the *cis*- or *trans*- configurations of the substituents. This was confirmed by testing the affinity for the NK_1 receptor of the four stereoisomers of 2-((3,5-bis(trifluoromethyl)benzyl)-oxy)-3-phenylmorpholine (Table 7) as reported by HALE et al. (1996). Note that the numbers of the asymmetric C-atoms in the quinuclidinyl and morpholine structures are reversed. The enantiomers L-733060 (2*S*,3*S*) and L-733061 (2*R*,3*R*) are piperidine analogs of the morpholine compound (Table 7) and again the (2*R*,3*R*) enantiomer displays the lowest affinity for the human NK_1 receptor, IC_{50} 350nM versus 1nM for the (2*S*,3*S*) isomer (HARRISON et al. 1994). Introduction of additional substituents to the morpholine derivative shown in Table 7 resulted in MK869 = L-754030, a highly potent NK_1 receptor antagonist in the (1*R*,2*R*,3*S*) conformation (HALE et al. 1998). Deletion of either the (1R)-CH_3 group or the para F-atom of MK869 does not affect the affinity for the NK_1 receptor. However when both the para F-atom is deleted and the configuration of the 1-CH_3 group is changed from R to S a tenfold loss in affinity results. The ID_{50}s of MK869 and the structurally modified drugs in functional in vivo tests were comparable at the time of peak effect but the modified drugs were shorter-acting. The "(1*S*,2*R*,3*S*) minus F" derivative was not tested (HALE et al. 1998). The search for potential therapeutic applications of NK antagonists has covered a wide range of indications. So far, antiemetic, antidepressant, and antianxiety properties appeared to be clinically significant. Intracerebroventricular infusion of NK agonists induces locomotor activation and vocalization in guinea pigs. The effects can be abolished by pretreatment with brain permeating NK antagonists such as L-733060, but not by its less active enantiomer L-733061. Structures of both compounds and their affinities for the NK_1 receptor are shown in Table 7. Stress-induced vocalizations by guinea pig pups during transient maternal separation were also inhibited by antagonists such as MK869, L-760735 (a MK869 analog), and L-733060. The enantiomers of the latter two antagonists (L-770765 and L-733061) with a respective 873- and 350-fold lower affinity for the NK_1 receptor did not inhibit this type of vocalizations (KRAMER et al. 1998). A novel NK_1 antagonist is a nonbasic tri-substituted propane derivative: 2-[*N*-(2-(4-hydroxymethylphenoxy) acetyl) amino]-3-(1-*H*-indol-3-yl)-1-[*N*-(2-methoxybenzyl)-*N*-acetylamino] propane. The activity resides mainly in the (*R*)-enantiomer

Table 7. Stereoselectivity ratios of NK_1 receptor ligands

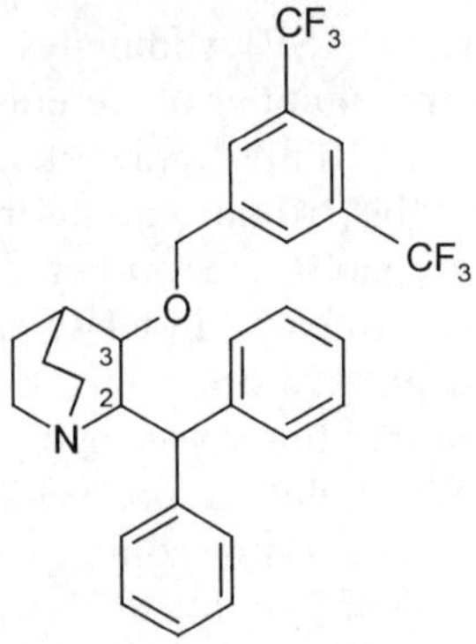

L-709,210 (2S,3S)

	IC_{50}	i.a	[a]Inhib.
(2R,3R)	300	0	0%
(2S,3S)	0.2	0	76%
(2R,3S)	0.2	0	70%
(2S,3R)	125	0	nt

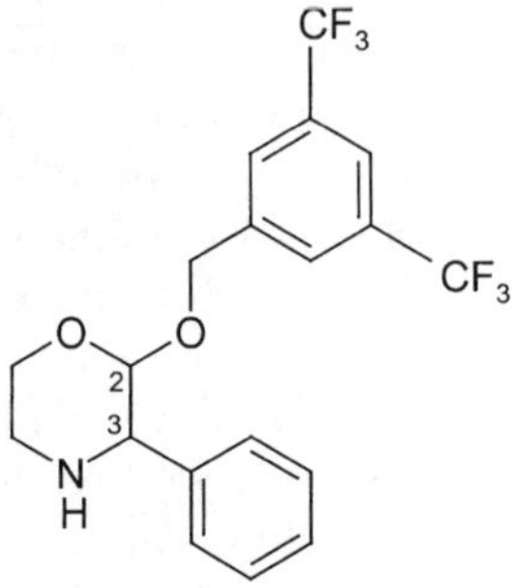

	IC_{50}
(2R,3R)	287
(2S,3S)	2.4
(2R,3S)	376
(2S,3R)	59

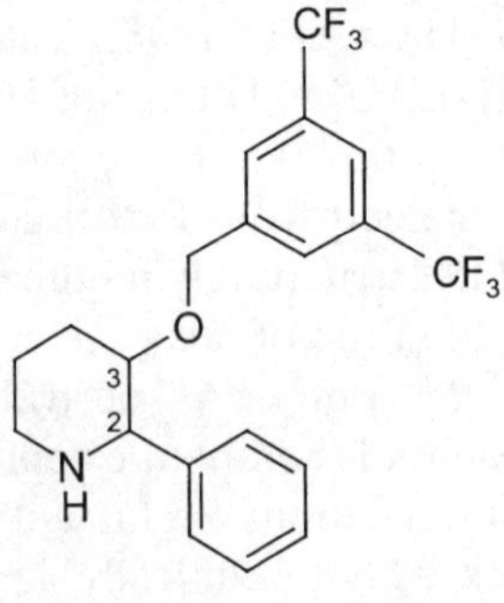

L-733060 (2S,3S)
L-733061 (2R,3R)

	IC_{50}	i.a
(2R,3R)	350	0
(2S,3S)	1.0	0

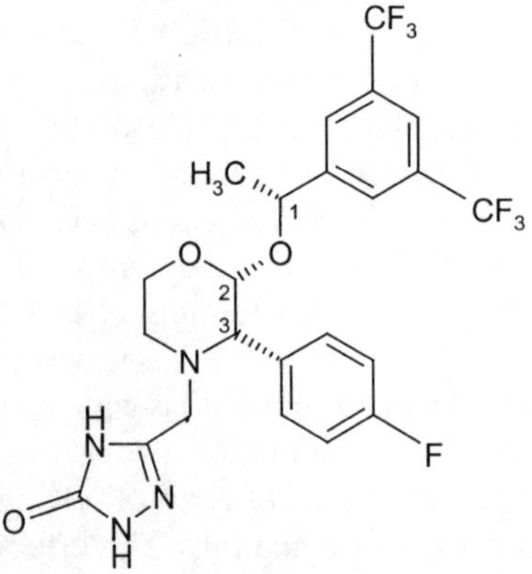

L-754,030 = MK869 (1R,2R,3S)

	IC_{50}	i.a	[b]ID_{50} 4 hr	[b]ID_{50} 24 hr
(1R,2R,3S)	0.09	0	0.04	0.33

Affinity: IC_{50},nM. [^{125}I] SP, hNK_1R (CHO cells). Intrinsic activity: i.a Potency:
a: inhibition SP-induced extravasation skin guinea pigs, dose 10 mg / kg po.
b: ID_{50}, mg / kg i.v. inhibition GR73632-induced foot tapping, gerbils.

($IC_{50} = 0.4\,nM$). The (*S*)-enantiomer is more than 100-fold less potent ([^{125}I]SP binding, h NK_1R(IM-9 cells). Functional data have not been published yet (Fritz et al. 2001).

C. Receptors and Enantioselectivity

I. Introduction

GPCRs are membrane-bound proteins. Multiple sequence alignments suggest there are seven quite hydrophobic stretches of 20–25 amino acids in virtually all receptors. This observation has led to the hypothesis that these domains are membrane-spanning, arranged in a barrel-like fashion and α-helical in nature, such that the more hydrophobic amino acids are facing the lipid environment. As a consequence other, on average more hydrophilic, residues are facing the central core of the protein, which may be involved in the recognition of endogenous and synthetic ligands among other functions. The membrane-embedded regions are connected via intra- and extracellular domains, with the *N*-terminus at the outside of the cell and the C-terminus located in the cytosol. These structural features find strong support in the atomic architecture of bacteriorhodopsin which also has this seven-helices-template. Bacteriorhodopsin is one of the few membrane-bound proteins for which electron-diffraction and crystallographic data are available (Henderson et al. 1990; Luecke et al. 1999) yielding a high resolution 3D structure, but it is a bacterial proton pump rather than a GPCR. Nevertheless, we have used bacteriorhodopsin as a template to develop a backbone of the transmembrane domains for, among others, the adenosine A_1 receptor (IJzerman et al. 1992), quite similar to the seminal models developed by Hibert and coworkers (Hibert et al. 1991; Trump-Kallmeyer et al. 1992). The visual pigment rhodopsin, which is a GPCR, has also been subjected to biophysical experimentation. At present resolution data are of significantly lower quality, suggesting, though, that the helical organization is similar to that in bacteriorhodopsin, but certainly not identical. As a consequence a slightly different receptor model emerged, when we used the α-carbon atoms of the transmembrane domains of rhodopsin (Baldwin et al. 1997) as a template for the construction of a model for the human adenosine A_1 receptor interacting with prototypic agonists (Rivkees et al. 1999). Obviously, all such models can only be rough approximations of biochemical reality. In our opinion they are best used as a visualization of known structure-activity/affinity relationships and as filters to, for example, select amino acid residues for mutation. This latter feature has proven successful, also in the delineation of the molecular mechanisms involved in stereoselective recognition.

In this section we will illustrate this with examples of α- and β-adrenoceptors, 5-HT_{1A} and melanocortin receptors. We will address only those studies in which point mutations in the receptor substantiate and corroborate the reported findings and derived models with respect to stereoselective interactions.

II. Adrenoceptors

β_2-Adrenoceptors serve as prototypic GPCRs. They were one of the first to be cloned and subjected to site-directed mutagenesis. In these studies an aspartate residue in helix III (Asp113 in the human receptor) and two serines in helix V (Ser204 and 207) were identified as anchor points for the protonated amine function and the catechol group of the endogenous agonist, adrenaline, respectively (Strader et al. 1988, 1989). Interestingly, an amino acid responsible for stereoselective recognition could not be identified, despite the strong preference of the receptor for the (*R*)-enantiomer of adrenaline. We identified, using receptor models based on the bacteriorhodopsin template (Fig. 1), two potential amino acids for interaction with the β-OH group in the ligand, which defines the chiral center (Wieland et al. 1996). Either Ser165 (helix IV) or Asn293 (helix VI) were the most likely candidates for direct interaction. Mutation of Ser165 to Ala, however, did not change the binding of the two stereoisomers of isoprenaline (a close and more stable analog of adrenaline). In contrast, an Asn293Leu mutation showed a substantial loss in the stereospecificity of catecholamine binding (transient receptor expression in HEK293 cells) and receptor activation after stable expression in CHO cells. The maximal stimulation of adenylyl cyclase by (*R*)-isoprenaline was similar in CHO cell membranes expressing either wild-type or mutant receptor, indicating that the mutant receptor was fully capable of activating the G protein pool available (Fig. 2). Concentration-response curves of the two isoprenaline

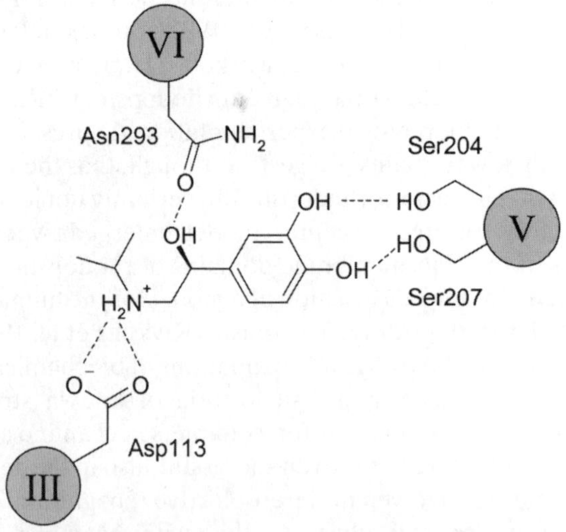

Fig. 1. Schematic representation of the binding of (*R*)-isoprenaline to the human β_2-adrenoceptor. The receptor α-helical backbone is viewed from the extracellular side. (*R*)-Isoprenaline and the receptor residues thought to be involved in ligand binding – Asp113 (helix III), Ser204 and Ser207 (helix V) and Asn293 (helix VI) – are shown. *Dashed lines*, potential hydrogen bonds

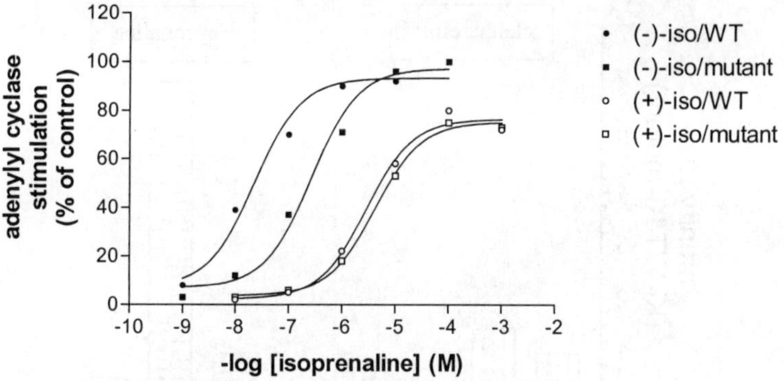

Fig. 2. Stimulation of adenylyl cyclase in membranes from CHO cells expressing wild-type (*circles*) or Ans293Leu mutant (*squares*) β_2-adrenoceptors by (–)-isoprenaline (*solid symbols*) or (+)-isoprenaline (*open symbols*). Data are means of four independent experiments, performed in duplicate

isomers showed a substantial stereoselectivity (over 130-fold) on the wild-type receptor. This was reduced to less than 13-fold on the mutant receptor, all due to the reduced potency of the more active R-isomer. Similar data were obtained for adrenaline and noradrenaline. From this we reasoned that Asn293 is critical in, although probably not solely responsible for, the stereoselective recognition of catecholamines.

Partial agonists such as terbutaline and clenbuterol were less affected, if at all. All these compounds share the ethanolamine side chain with the chiral carbon atom, and differ in the aromatic moiety only. Hence, we considered it worthwhile to further characterize the atomic details of this stereoselective interaction. We synthesized "mutated" derivatives of both isoprenaline and clenbuterol, and analyzed their effects on both the wild-type and mutant receptor (Zuurmond et al. 1999). Such an approach, we hypothesized, might lead to a "gain of function" in terms of affinity, indeed providing further physical evidence for a direct interaction between a specific receptor residue and a functional group in the ligand. The β-hydroxy group in the parent molecules isoprenaline and clenbuterol was replaced by either hydrogen, methyl or methoxy substituents. The six new derivatives were also tested on the wild-type receptor and the Asn293Leu mutant. It appeared that one analog of isoprenaline, but not of clenbuterol, showed a sixfold gain in affinity on the mutant receptor (Fig. 3). This derivative had a methyl substituent instead of the hydroxy group, fully compatible with the more lipophilic nature of the leucine side chain versus the hydrophilic asparagine residue. On the wild-type receptor a 100-fold affinity difference was observed between isoprenaline and this "methyl" derivative, which was fully canceled on the mutant receptor. Interestingly, the methyl derivative did not prove superior to isoprenaline on the mutant receptor, suggesting that the receptor environment around the chiral center remains largely hydrophilic, despite the presence of the leucine

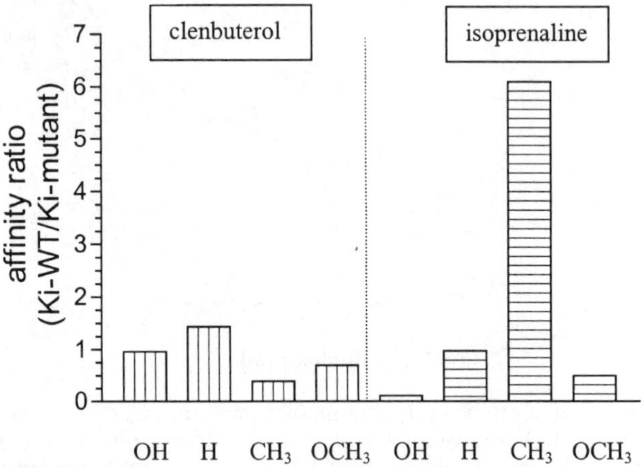

Fig. 3. Affinity ratios ($K_{i,\ wild\text{-}type}/K_{i,\ Asn293Leu}$) for clenbuterol, isoprenaline, and derivatives. Data are means of three independent experiments, performed in duplicate

side chain. All these findings also provide evidence that there is not just one single agonist binding site on the receptor. Hence, the critical involvement of Ser204 and Ser207 in agonist binding may be limited to catecholamines only. These two amino acids are supposed to hydrogen bond to the two catechol OH-groups, and as a consequence (Fig. 1), isoprenaline is directed to position its β-OH group in the vicinity of Asn293. In binding clenbuterol and derivatives Asn293 does not seem to play any particular role, and, thus, Ser204 and Ser207 are thought to be of less importance too. Preliminary experiments with a "double mutant" β_2-adrenoceptor (Ser204Ala, Ser207Ala) strongly corroborated this idea. Isoprenaline was over 100-fold more potent on the wild-type receptor, whereas clenbuterol's affinity was only 3-fold diminished on the double mutant receptor.

Ruffolo and coworkers studied the stereoselective recognition by α_{2A}-adrenoceptors (HEHR et al. 1997; HIEBLE et al. 1996). They too found that a serine in helix IV (also on position 165) did not play a role, whereas Ser90 (in helix II at the cytosolic end) and Ser419 (helix VII) seemed to serve a comparable role as Asn293 on the β_2-adrenoceptor. Thus, the affinities of the (*S*)-(+)-enantiomers of noradrenaline and adrenaline were only marginally affected, whereas a dramatic drop in affinity (40–70-fold) was observed for the active (*R*)-(−)-enantiomers. Ser90 is conserved in only the α-adrenoceptor subtypes, whereas the serine on position 419 is also conserved in the β-adrenoceptors. Molecular modeling showed that Ser90, however, is rather remote from the other amino acids thought to be important for binding, i.e., the equivalents of Asp113, Ser204 and Ser207 in the β_2-adrenoceptor. All in all the outcome of these studies seems somewhat less unequivocal than for the β_2-adrenoceptor.

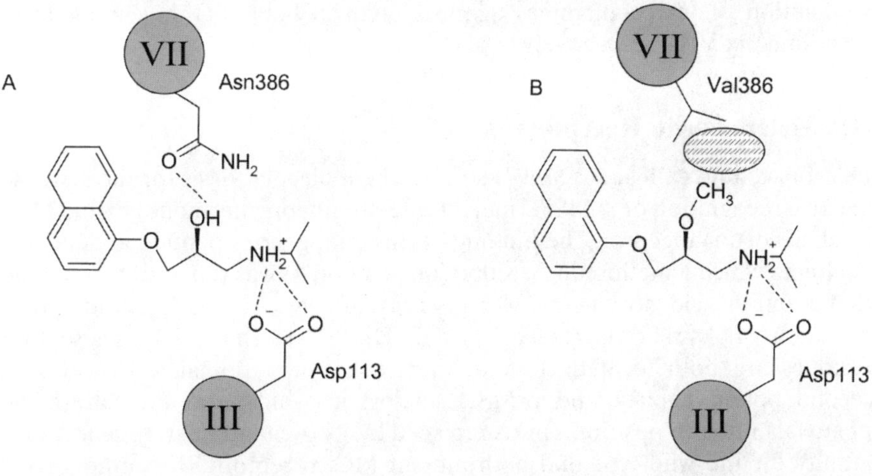

Fig. 4. Schematic representation of putative interactions of *S*-propranolol with the wild-type 5-HT_{1A} receptor (**A**), and its methylated analog with the Asn386Val mutant receptor (**B**), respectively. In **B** the *shaded area* suggests a favorable lipophilic contact of Val386 with the methyl moiety of the ligand's β-methoxy group. *Dashed lines* (**A**, **B**), potential hydrogen bonds

III. 5-HT_{1A} Receptors

KUIPERS et al. (1997b) analyzed the stereoselective interaction between aryloxypropanolamine antagonists and the human 5-HT_{1A} receptor. These ligands, all classical β-adrenoceptor blockers, display significant affinity for this serotonin receptor subtype as well. Thus, the (*R*)- and (*S*)-enantiomers of propranolol, penbutolol, and alprenolol were studied with respect to their ability to bind to the wild-type and a mutant 5-HT_{1A} receptor. This Asn386Val (helix VII) mutant appeared to act as a chiral discriminator. In all cases both enantiomers displayed lower affinity for the mutant receptor, but the more active (*S*)-enantiomer was always more affected. The mutation did not have any influence on other, structurally unrelated, 5-HT receptor ligands. This prompted us to synthesize a series of propranolol analogs with structural variation in the oxypropanolamine moiety (the "side chain"), and determine their affinity for both the wild-type and the Asn386Val mutant receptor. In particular, methylation of the β-OH group increased the affinity of the (*S*)-enantiomer for the mutant receptor to a value identical to the wild-type receptor (K_i = 160 nM in both cases). These and other findings led to the model depicted in Fig. 4, in which the lipophilic valine residue accommodates the methoxy group of the propranolol derivative, leading to a substantial gain of affinity. Interestingly, an asparagine residue is also present on the same location in helix VII of the β-adrenoceptor, suggesting that the antagonist binding model developed for the 5-HT_{1A} receptor also holds for the β-adrenoceptor. In that case two asparagine residues are responsible for the stereoselective recognition of (catecholamine) agonists (Asn in helix VI) and antagonists (Asn in helix VII), respectively.

III. Melanocortin Receptors

Frändberg and colleagues (1994) studied the molecular basis for the stereoselective recognition of α-MSH (melanocyte-stimulating hormone) by the MC_1 (melanocortin) receptor. The binding of this endogenous peptide and a close analog in which a methionine residue (on position 4) was replaced by the synthetic amino acid norleucine, was severely affected (up to 250-fold) upon mutation of Asp117Ala (helix III) and His260Ala (helix VI). These two residues are both located close to the extracellular domains. However, a second potent agonist and α-MSH analog, in which also the natural L-phenylalanine (on position 7) was replaced by its D-enantiomer, retained high affinity for the wild-type and both mutant MC_1 receptors. The authors concluded that different binding epitopes on GPCRs appear to exist for peptide ligand stereoisomers.

D. Concluding Remarks

From a simplified perspective, stereoisomers can be regarded as a rather special pair-wise class of different drugs. In such a view it is only logical that two enantiomers may have different affinities, selectivities, and intrinsic activities. This notion, however, somehow denies the continued interest in the development of enantiomers as potential new drugs, and the special emphasis on chirality in drug design. It is anticipated that the future elucidation of the 3D structure of receptors will fuel our understanding of the molecular and even atomic mechanisms behind stereoselective recognition. Only then will the old concepts of, for example, Easson and Stedman have their true structural correlate.

References

Akahane A, Katayama H, Mitsunaga T, Kita Y, Kusunoki T, Terai T, Yoshida K, Shiokawa Y (1996) Discovery of FK453, a novel nonxanthine adenosine A_1 receptor antagonist. Bioorg Med Chem Lett 6:2059–2062

Baldwin JM, Schertler GFX, Unger VM (1997) An alpha-carbon template for the transmembrane helices in the rhodopsin family of G protein-coupled receptors. J Mol Biol 272:144–164

Bellier B, McCort-Tranchepain I, Ducos B, Danaseimento S, Meudal H, Noble F, Garbay C, Roques BP (1997) Synthesis and biological properties of new constrained CCK-B antagonists: discrimination of two affinity states of the CCK-B receptor on transfected CHO cells. J Med Chem 40:3947–3956

Bock MG, Dipardo RM, Evans BE, Rittle KE, Whitter WL, Garsky VM, Gilbert KF, Leighton JL, Carson KL, Mellin EC, Veber DF, Chang RSL, Lotti VJ, Freedman SB, Smith AJ, Patel S, Anderson PS, Freidinger RM (1993) Development of 1,4-benzodiazepine cholecystokinin type B antagonists. J Med Chem 36:4276–4292

Cannon JG, Moe ST, Long JP (1991) Enantiomers of 11-hydroxy-10-methyl-aporphine having opposing pharmacological effects at 5-HT_{1A} receptors. Chirality 3:19–23

Chambers JJ, Kurrasch-Orbaugh DM, Parker MA, Nichols DE (2001) Enantiospecific synthesis and pharmacological evaluation of a series of super-potent, conformationally restricted 5-$HT_{2A/2C}$ receptor agonists. J Med Chem 44:1003–1010

Cornfield LJ, Lambert G, Arvidsson L-E, Mellin C, Vallgårda J, Hacksell U, Nelson DJ (1991) Intrinsic activity of enantiomers of 8-hydroxy-2-(di-*n*-propylamino)tetralin and its analogs at 5-hydroxytryptamine$_{1A}$ receptors that are negatively coupled to adenylate cyclase. Mol Pharmacol 39:780–787

Cushny AR (1926) Biological relations of optically isomeric substances. William & Wilkins, Baltimore

Daly JW (1982) Adenosine receptors: targets for future drugs. J Med Chem 25:197–207

Davies DJ, Garratt PJ, Tocher DA, Vonhoff S, Davies J, Teh M-T, Sugden D (1998) Mapping the melatonin receptor 5. Melatonin agonists and antagonists derived from tetrahydro-cyclopent [b] indoles, tetrahydrocarbazoles and hexahydrocyclohept [b] indoles. J Med Chem 41:451–457

De Wald HA, Heffner TG, Jaen JC, Lustgarten DM, Mc Phail AT, Meltzer LT, Pugsley TA, Wise LD (1990) Synthesis and dopamine agonist properties of (±)-*trans*-3,4,4a,-10b-tetrahydro-4-propyl-2H,5*H*-[1] benzopyrano [4,3-b]-1,4-oxazin-9-ol and its enantiomers. J Med Chem 33:445–450

Easson LH, Stedman E (1933) Studies on the relationship between chemical constitution and physiological action. V. Molecular dissemetry and physiological activity. Biochem J 27:1257–1266

Felder CC, Veluz JS, Williams HL, Briley EM, Matsuda LA (1992) Cannabinoid agonists stimulate both receptor- and non-receptor-mediated signal transduction pathways in cells transfected with and expressing cannabinoid receptor clones. Mol Pharmacol 42:838–845

Frändberg PA, Muceniece R, Prusis P, Wikberg J, Chhajlani V (1994) Evidence for alternate points of attachment for a-MSH and its stereoisomer [Nle^4,D-Phe^7]-α-MSH at the melanocortin-1 receptor. Biochem Biophys Res Commun 202:1266–1271

Fritz JE, Hipskind PA, Lobb KL, Nixon JA, Threlkeld PG, Gitter BD, McMillian CL, Kaldor SW (2001) Expedited discovery of second generation NK-1 antagonists: identification of a nonbasic aryloxy substituent. Bioorg Med Chem Lett 11: 1643–1646

Goldman ME, Kebabian JW (1984) Aporphine enantiomers. Interaction with D-1 and D-2 dopamine receptors. Mol Pharmacol 25:18–23

Hale JJ, Mills SG, MacCoss M, Finke PE, Cascieri MA, Sadowski S, Ber E, Chicchi GG, Kurtz M, Metzger J, Eiermann G, Tsou NN, Tattersall D, Rupniak NMJ, Williams AR, Rycroft W, Hargreaves R, MacIntyre DE (1998) Structural optimization affording 2-(*R*)-(1-(*R*)-3,5-bis(trifluoromethyl)phenylethoxy)-3-(*S*)-(4-fluoro)phenyl-4-(3-oxo-1,2,4-triazol-5-yl) methylmorpholine a potent, orally active, long-acting morpholine acetal human NK-1 receptor antagonist. J Med Chem 41:4607–4614

Hale JJ, Mills SG, MacCoss M, Shah SK, Qi H, Mathre DJ, Cascieri MA, Sadowski S, Strader CD, MacIntyre DE, Metzger JM (1996) 2(*S*)-((3,5-bis(trifluoromethyl) benzyl)oxy)-3(*S*)-phenyl-4-((3-oxo-1,2,4-triazol-5-yl)methyl) morpholine. J Med Chem 39:1760–1762

Harrison T, Williams BJ, Swain CJ, Ball RG (1994) Piperidine-ether based hNK_1 antagonists: determination of the relative and absolute stereochemical requirements. Bioorganic Med Chem Lett 4:2545–2550

Hedberg MH, Johansson AM, Nordvall G, Yliniemelä A, Li HB, Martin AR, Hjörth S, Unelius L, Sundell S, Hacksell U (1995) (*R*)-11-hydroxy- and (*R*)-11-hydroxy-10-methyl-aporphine: synthesis, pharmacology and modeling of D_{2A} and 5-HT_{1A} receptor interactions. J Med Chem 38:647–658

Hehr A, Hieble JP, Li YO, Bergsma DJ, Swift AM, Ganguly S, Ruffolo Jr RR (1997) Ser^{165} of transmembrane helix IV is not involved in the interaction of catecholamines with the α_{2a}-adrenoceptor. Pharmacology 55:18–24

Henderson G, McKnight AT (1997) The orphan opioid receptor and its endogenous ligand nociceptin/orphanin FQ. TiPS 18:293–300
Henderson R, Baldwin JM, Ceska TA, Zemlin F, Beckmann E, Downing KH (1990) J Mol Biol 213:899–929
Hibert MF, Trumpp-Kallmeyer S, Bruinvels A, Hoflack J (1991) Three-dimensional models of neurotransmitter G-binding protein-coupled receptors. Mol Pharmacol 40:8–15
Hieble JP, Hehr A, Li YO, Naselsky DP, Ruffolo Jr RR (1996) α_2-Adrenergic receptors: structure, function and therapeutic applications. In: Lanier SM, Limbird LE (eds) Harwood, pp 43–51
Holmstedt B, Liljestrand G (1963) Readings in Pharmacology. Pergamon, Oxford
Homan EJ, Copinga S, Unelius L, Jackson DM, Wikström HV, Grol CJ (1999a) Synthesis and pharmacology of the enantiomers of the potential antipsychotic agents 5-OMe-BPAT and 5-OMe-(2,6-di-OMe)-BPAT. Bioorg Med Chem 7: 1263–1271
Homan EJ, Wikström HV, Grol CJ (1999b) Molecular modeling of the dopamine D_2 and serotonin 5-HT_{1A} receptor binding modes of the enantiomers of 5-OMe-BPAT. Bioorg Med Chem 7:1805–1820
Ichikawa D, Ozaki S, Azuma T, Nambu H, Kawamoto H, Iwasawa Y, Takeshima H, Ohta H (2001) In vitro inhibitory effects of J-113397 on nociceptin/orphinan FQ-stimulated [^{35}S]GTPγS binding to mouse brain. Neuroreport 12:1757–1761
IJzerman AP, Van Galen PJM, Jacobson KA (1992) Molecular modeling of adenosine receptors. I. The ligand binding site on the A_1 receptor. Drug Des Disc 9:49–67
Insel PA, Maguire ME, Gilman AG, Bourne HR, Coffino P, Melmon KL (1976) Beta adrenergic receptors and adenylate cyclase: products of separate genes? Mol Pharmacol 12:1062–1069
Jagerschmidt A, Guillaume-Rousselet N, Vikland M-L, Goudreau N, Maigret B, Roques BP (1996) His 381 of the rat CCK_B receptor is essential for CCK_B versus CCK_A receptor antagonist selectivity. Eur J Pharmacol 296:97–106
Jansson CC, Marjamäki A, Luomala K, Savola JM, Scheinin M, Äkerman KEO (1994) Coupling of human a_2-adrenoceptor subtypes to regulation of cAMP production in transfected S115 cells. Eur J Pharmacol 266:165–174
Jansson CC, Kukkonen JP, Näsmann J, Huifang G, Wurster S, Virtanen R, Savola JM, Cockfort V, Äkerman KEO (1998) Protean agonism at a_{2A}-adrenoceptors. Mol Pharmacol 53:963–968
Jellimann C, Mathé-Allainmat M, Andrieux J, Renard P, Delagrange P, Langlois M (1999) Melatonergic properties of the (+)- and (−)-enantiomers of *N*-(4-methoxy-2,3-dihydro-1*H*-phenalen-2-yl) amide derivatives. J Med Chem 42:1100–1105
Jellimann C, Mathé-Allainmat M, Andrieux J, Kloubert S, Boutin JA, Nicolas J-P, Bennejean C, Delagrange P, Langlois M (2000) Synthesis of phenalene and acenaphthene derivatives as new conformationally restricted ligands for melatonin receptors. J Med Chem 43:4051–4062
Kawamoto H, Ozaki S, Itoh Y, Miyaji M, Arai S, Nakashima H, Kato T, Ohta H, Iwasawa Y (1999) Discovery of the first potent and selective small molecule opioid receptor- like (ORL1) antagonist:1-[(3R,4R)-1-cyclooctylmethyl-3-hydroxymethyl-4-piperidyl]-3-ethyl-1,3-dihy-dro-2*H*-benzimidazol-2-one(J-113397). J Med Chem 42:5061–5063
Knutsen LJS, Lau J, Petersen H, Thomsen C, Weis JU, Shalmi M, Judge ME, Hansen AJ, Sheardown MJ (1999) *N*-Substituted adenosines as novel neuroprotective A_1 agonists with diminished hypotensive effects. J Med Chem 42:3463–3477
Kramer MS, Cutler N, Feighner J, Shrivastava R, Carman J, Sramek JJ, Reines SA, Liu G, Snavely D, Wyatt-Knowles E, Hale JJ, Mills SG, MacCoss M, Swain CJ, Harrison T, Hill RG, Hefti F, Scolnick EM Cascieri MA, Chicchi GG, Sadowski S, Williams AR, Hewson L, Smith D, Carlson EJ, Hargreaves RJ, Rupniak NMJ (1998) Distinct mechanism for antidepressant activity by blockade of central substance P receptors. Science 281:1640–1645

Kuipers W (1997a) Structural characteristics of 5-HT_{1A} receptors and their ligands. In: Olivier B, van Wijngaarden I, Soudijn W (eds) Serotonin receptors and their ligands. Elsevier, Amsterdam, p 45

Kuipers W, Link R, Standaar PJ, Stoit AR, van Wijngaarden I, Leurs R, IJzerman AP (1997b) Study of the interaction between aryloxypropranolamines and Asn386 in Helix VII of the human 5-hydroxytryptamine$_{1A}$ receptor. Mol Pharmacol 51:889–896

Kula NS, Baldessarini RJ, Bromley S, Neumeyer JL (1985) Effects of isomers of apomorphines on dopamine receptors in striatal and limbic tissue of rat brain. Life Sci 37:1051–1057

Luecke H, Schobert B, Richter HT, Cartailler JP, Lanyi JK (1999) Structure of bacteriorhodopsin at 1.55 Å resolution. J Mol Biol 291:899–911

McLean S, Ganong A, Seymour PA, Snider RM, Desai MC, Rosen T, Bryce DK, Longo KP, Reynolds LS, Robinson G, Schmidt AW, Siok C, Heym J (1993) Pharmacology of CP-99,994: a nonpeptide antagonist of the tachykinin neurokinin-I receptor. J Pharmacol and Exp Therapeutics 267:472–479

Mesecar AD, Koshland Jr DE (2000) A new model for protein stereospecificity. Nature 403:614–615

Müller CE, Hide I, Daly JW, Rothenhäusler K, Eger K (1990) 7-Deaza-2-phenyl-adenines:structure-activity relationships of potent A_1 selective adenosine receptor antagonists. J Med Chem 33:2822–2828

Oksenberg D, Peroutka SJ (1988) Antagonism of 5-hydroxytryptamine (5-HT_{1A}) receptor-mediated modulation of adenylate cyclase activity by pindolol and propranolol isomers. Biochem Pharmacol 37:3429–3433

Ozaki S, Kawamoto H, Itoh Y, Miyaji M, Azuma Y, Ichikawa D, Nambu H, Iguchi T, Iwasawa Y, Ohta H (2000) In vitro and in vivo pharmacological characterization of J-113397, a potent and selective non-peptidyl ORL1 receptor antagonist. Eur J Pharmacol 402:45–53

Palego L, Giromella A, Marazziti D, Giannaccini G, Borsini F, Bigazzi F, Naccarato AG, Lucacchini A, Cassano GB, Mazzoni MR (2000) Lack of stereoselectivity of 8-hydroxy-2(di-*N*-propylamino)-tetralin-mediated inhibition of forskolin-stimulated adenylyl cyclase activity in human pre- and post-synaptic brain regions. Neurochem Int 36:225–232

Pasteur L (1886) note on Piutti's contribution (*vide infra*) Comptes Rend 103:138

Patel S, Smith AJ, Chapman KL, Fletcher AE, Kemp JA, Marshall GR, Hargreaves RJ, Ryecroft W, Iversen LL, Iversen SD, Baker R, Showell GA, Bourrain S, Neduvelil JG, Matassa VG, Freedman SB (1994) Biological properties of the benzodiazepine amidine derivative L-740093 a cholecystokinin-B/gastrin receptor antagonist with high affinity in vitro and high potency in vivo. Mol Pharmacol 46:943–948

Pauwels PJ, van Gompel P, Leysen JE (1991) Human b_1- and b_2-adrenergic receptor binding and mediated accumulation of cAMP in transfected chinese hamster ovary cells. Biochem Pharmacol 42:1683–1689

Peroutka SJ (1994) Molecular biology of serotonin (5-HT) receptors. Synapse 18: 241–260

Piutti A (1886) Sur une nouvelle espèce d'asparagine. Comptes Rend 103:134–137

Pugsley TA, Davis MD, Akunne HC, Mackenzy RG, Shih YH, Damsma G, Wikström H, Whetzel SZ, Georgic LM, Cooke LW, Demattos SB, Corbin AE, Glase SA, Wise LD, Dijkstra D, Heffner TG (1995) Neurochemical and functional characterization of the preferentially selective D_3 agonist PD128907. J Pharmacol Expt Therap 275:1355–1366

Schertler GFX, Villa C, Henderson R (1993 Projection structure of rhodopsin. Nature 362:770–772

Seiler MP, Markstein R (1984) Further characterization of structural requirements for agonists at the striatal dopamine D_2 receptor and a comparison with those at the striatal dopamine D_1 receptor. Mol Pharmacol 26:452–457

Snider RM, Constantine JW, Lowe III JA, Longo KP, Lebel WS, Woody HA, Drozda SE, Desai MJ, Vinick FJ, Spencer RW, Hess HJ (1991) A potent nonpeptide antagonist of the substance P (NK_1) receptor. Science 251:435–437

Strader CD, Sigal IS, Candelore MR, Rands E, Hill WS, Dixon RAF (1988) Conserved aspartic acid residues 79 and 113 of the β-adrenergic receptor have different roles in receptor function. J Biol Chem 263:10267–10271

Strader CD, Candelore MR, Hill WS, Sigal IS, Dixon RAF (1989) Identification of two serine residues involved in agonist activation of the β-adrenergic receptor. J Biol Chem 264:13572–13578

Swain CJ, Seward EM, Cascieri MA, Fong TM, Herbert R, MacIntyre OE, Merchant KJ, Owen SN, Owens AP, Sabin V, Teall M, Van Niel MB, Williams BJ, Sadowski S, Strader C, Ball RG, Baker R (1995) Identification of a series of 3-(benzyloxy)-1-azabicyclo [2.2.2] octane human NK_1 antagonists. J Med Chem 38:4793–4805

Trumpp-Kallmeyer S, Hoflack J, Bruinvels A, Hibert M (1992) Modeling of G protein-coupled receptors: application to dopamine, adrenaline, serotonin, acetylcholine, and mammalian opsin receptors. J Med Chem 35:3448–3462

Van Steen BJ (1996) Structure-affinity relationship studies on 5-HT_{1A} receptor ligands. Ph D thesis, Leiden University

Van Vliet LA, Tepper PG, Dijkstra D, Damsma G, Wickström H, Pugsley TA, Akunne HC, Heffner TG, Glase SA, Wise LD (1996) Affinity for dopamine D_2, D_3 and D_4 receptors of 2-aminotetralins. Relevance of D_2 agonist binding for determination of receptor subtype selectivity. J Med Chem 39:4233–4237

Welbourn AP, Chapleo CB, Lane AC, Myers PL, Roach AG, Smith CFC, Stillings MR, Tulloch IF (1986) a_2-adrenoreceptor antagonists. 4. Resolution of some potent selective prejunctional α_2-adrenoceptor antagonists. J Med Chem 29:2000–2003

Wieland K, Zuurmond HM, Krasel C, IJzerman AP, Lohse MJ (1996) Involvement of Asn-293 in stereospecific agonist recognition and in activation of the β_2-adrenergic receptor. Proc Natl Acad Sci USA 93:9276–9281

Wikstrøm H (1999) personal communication

Zuurmond HM, Hessling J, Bluml K, Lohse M, IJzerman AP (1999) Study of interaction between agonists and Asn293 in helix VI of human β_2-adrenergic receptor. Mol Pharmacol 56:909–916

CHAPTER 8

Mechanisms of Stereoselective Binding to Functional Proteins

G. KLEBE

A. Introduction

The three-dimensional structure of a drug molecule determines and evolves its biological properties at a given receptor. The spatial configuration of a molecule results from the connectivity of the atoms composing the molecule under consideration. If this spatial configuration encodes for asymmetry, the molecule gives rise to optical activity and two enantiomers exist. Without breaking bonds such a pair of chiral molecules cannot be transformed from one to the other form (neglecting special situations of slowly converting atropisomers; see also Chap. 4, this volume.). As long as such enantiomers are exposed to an achiral environment they possess identical properties. However, once presented to a handed surrounding, image and mirror-image will be recognized differently and will produce different effects. Receptors, for example enzymes, often being the target of drug molecules, create a chiral environment. They are constructed from chiral building blocks, such as L-amino acids. Other targets, e.g., DNA, RNA, or ribozymes, are composed by D-ribose or D-desoxyribose as chiral building blocks. Furthermore, as a whole, these molecules adopt conformations that correspond to handed objects. Accordingly, a different biological response can be expected once the two mirror-symmetrical molecules bind to such a chiral receptor. For example, the flavor of the two enantiomeric forms of carvone and limone (1 and 2 in Fig. 1) create quite different smelling characteristics once binding to the corresponding receptors (FRIEDMAN and MILLER 1971). These belong to the class of G-protein coupled receptors, present in our nose (Fig. 1). Many examples are known that demonstrate the deviating biological properties of chiral molecules, producing distinct profiles either in the intensity or quality of their biological response (ARIENS 1984, 1993). For example, the drug labetalol (3; Fig. 2) has two stereogenic centers giving rise to two pairs of diastereomers. Labetalol is claimed to be a mixed antagonist and it acts at the α_1-, β_1-, and β_2-adrenergic receptors (ARIENS et al. 1988) with deviating profiles depending on its stereochemistry.

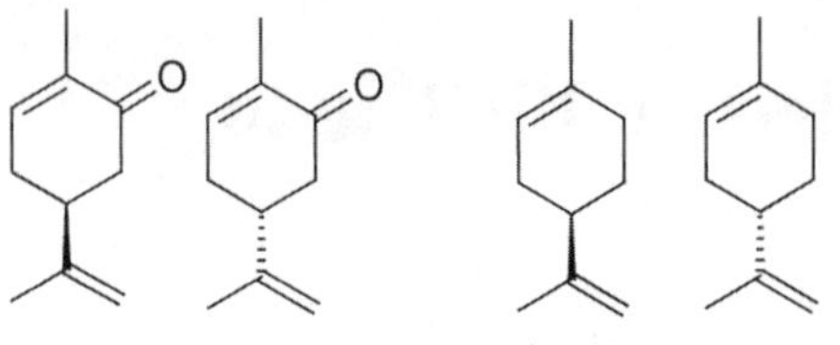

1 2

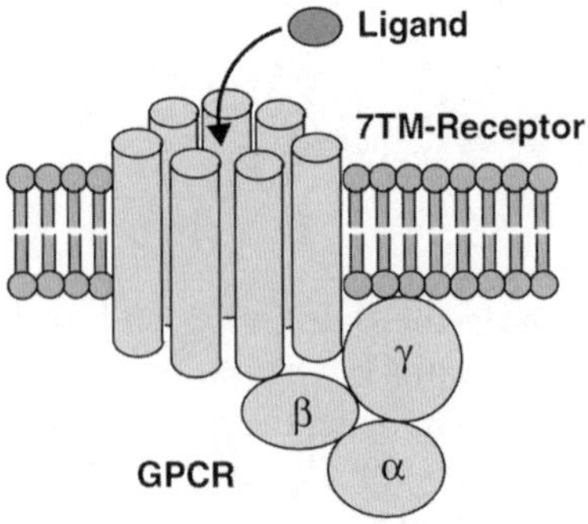

Fig. 1. The enantiomers of carvone (1) and limonene (2) create quite different smelling characteristics via binding to their receptor. This difference results from distinct chiral recognition of the enantiomers at a G-protein coupled receptor constructed from seven transmembrane helices

Labetalol

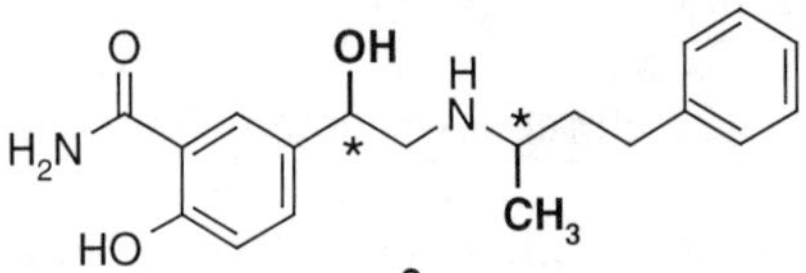

3

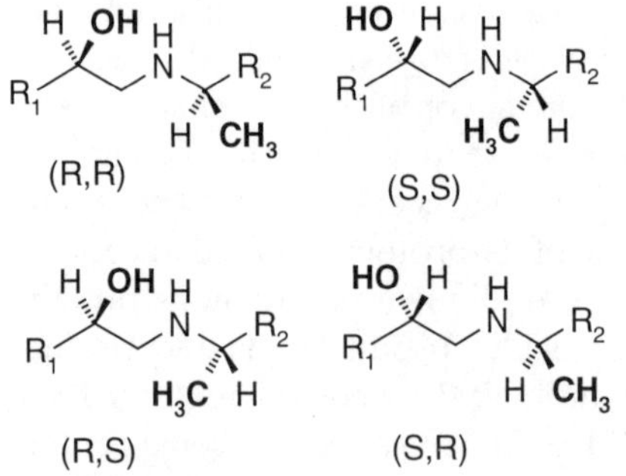

Fig. 2. Labetalol (3) has two asymmetric centers giving rise to four diastereomers which produce different pharmacological responses at distinct G-protein coupled receptors. The antagonistic effects decrease for the α_1-receptor as S,[1]R [2]≫ S,S ~ R,R > R,S; for the β_1-receptor R,R ≫ R,S > S,S ~ S,R; and for the β_2-receptor R,R [3]≫ R,S ≫ S,S ~ S,R

B. Chiral Molecules in a Crystal Environment

Crystallography is the most powerful tool to elucidate the three-dimensional structure of molecules. Exploiting the effects of anomalous scattering of particularly the heavier atoms, this technique allows determination of the absolute configuration of molecules.

The crystallization of a racemate can either result in the formation of racemic crystals showing both enantiomers side-by-side in the same unit cell, or in a spontaneous resolution leading to an agglomerate of morphologically enantiomorphic crystals. These crystals are composed either by the image or the mirror-image. The latter crystals do not possess any symmetry operations corresponding to a mirror, an inversion center or a glide plane. Thus, out of the 230 possible space groups, only 65 can accommodate chiral molecules. In a crystal, molecules have to pack densely together. The obtained self-recognition of a racemate or an enantiomeric pure compound produce quite different packing patterns that could parallel with different conformations adopted in the solid state. For example, Fig. 3 shows the packing of the amino acid histidine in a racemic and enantiomeric-pure crystal (EDINGTON and HARDING 1974; MADDEN et al. 1972). Proteins being chiral objects and found in nature as single enantiomers can only crystallize in one of the 65 enantiomorphic space groups. As a curiosity the polypeptide chain of Rubredoxin has been synthesized from D-amino acids (ZAWADZKE and BERG 1993). A 1:1

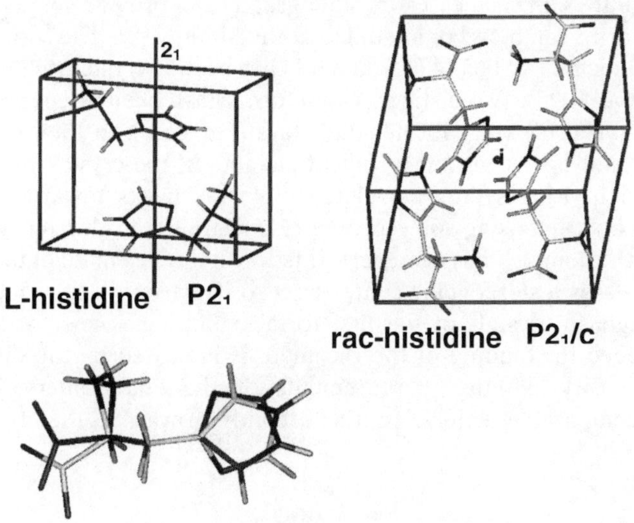

Fig. 3. The enantiopure L-histidine crystallizes in the acentric space group $P2_1$ with two molecules in the unit cell (*left*). They are related via a twofold screw axis. The racemic histidine crystallizes in the centrosymmetrical space group $P2_1/c$ with four molecules in the unit cell (*right*). In the *center* of the cell (*i*) a center of symmetry is indicated. In both crystal structures, the local packing interactions are quite different and imply in the present case different conformations of the amino acid (superposition below)

mixture with the natural L-protein resulted in a racemate forming racemic crystals in a centrosymmetrical space group.

C. Recognition of Chiral Ligands at Protein Binding Sites

At a protein binding site, enantiomers are recognized as different species. Accordingly, distinct affinities are the rule, similar affinities the exception. HIV protease is a key enzyme in the replication of the human immunodeficiency virus (WEST and FAIRLIE 1995). It cleaves the produced polypeptide chain into functional proteins. This protease is a homodimer of two 99-amino-acid peptide chains. Also this enzyme has been synthesized purely from D-amino acids (MILTON et al. 1992; JUNG 1992). Interestingly enough, the L-protein can only cleave the natural L-substrate, whereas the D-protein operates only on the all D-substrate. The enantiomers of chiral inhibitors discriminate between both handed forms of the enzyme, whereas an achiral inhibitor reveals the same inhibitory potency toward both forms.

Binding modes of ligands can be studied by determining their crystal structures when complexed with a protein. Since usually the affinity of two enantiomers toward a given protein differs by some orders of magnitude, only the binding geometry with one of the mirror images can be studied successfully. Therefore, of particular interest are cases where both species reveal comparable affinities. In such cases, crystallography can provide some insight into how both forms can be accommodated at the binding site. The serine protease inhibitor Daiichi DX 9065 (4 in Scheme 1) has been cocrystallized with trypsin (STUBBS et al. 1995). Two distinct crystal forms have been discovered showing deviating crystal packings in the solid state (Fig. 4a,b). In the first form, the binding pocket opens toward a solvent channel in the crystal. In the second form, a neighboring symmetry-related molecule packs toward the binding site of the first, imposing some geometry constraints in this region. Crystallization of the complex has been performed with the racemate of the inhibitor, which possesses a stereogenic center next to its carboxylate group. Interestingly enough, in crystals of the first form exhibiting access to the solvent, clear evidence for binding of the racemate is indicated by the difference in electron density. Affinity measurements of the enantiomers in solution revealed comparable values. In the second form with the geometrically

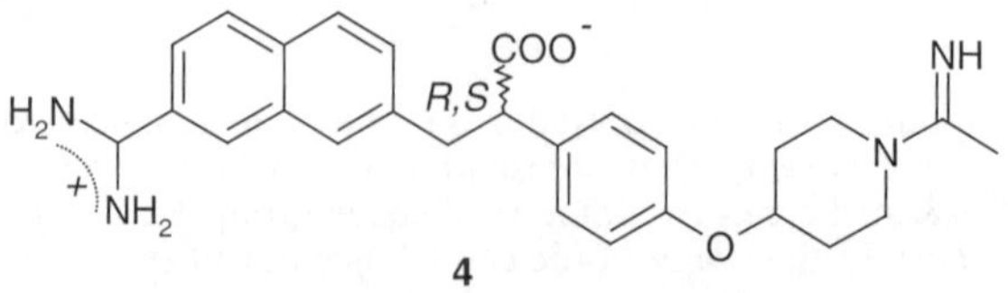

4

Scheme 1.

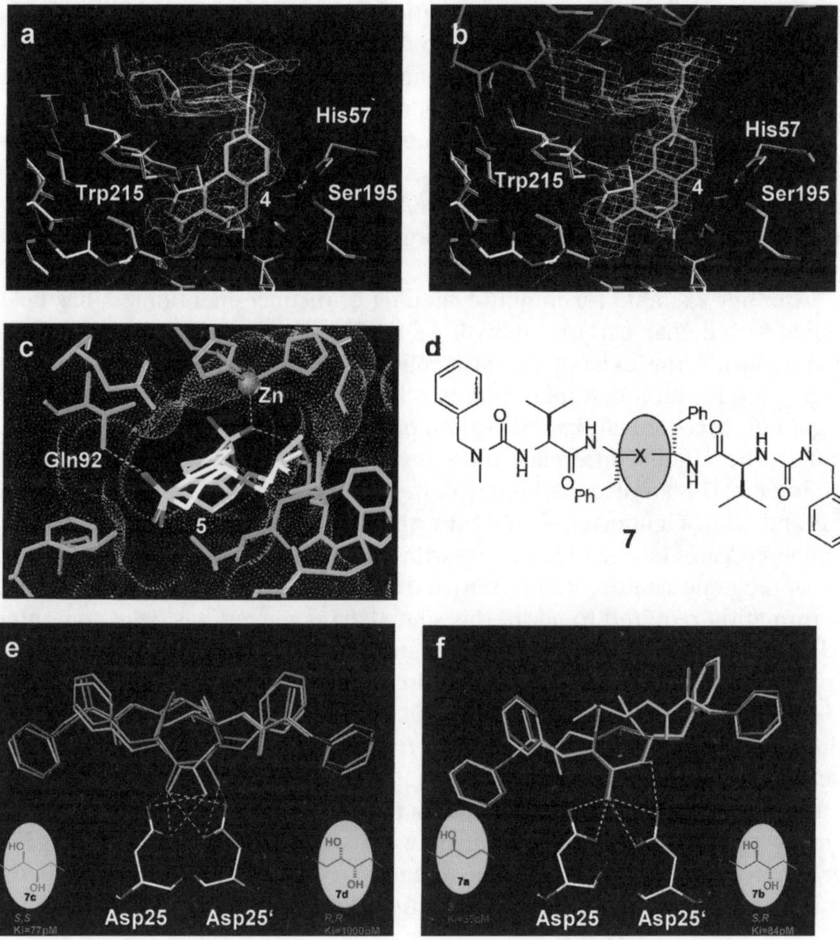

Fig. 4. a, b Binding geometry of the serine protease inhibitor DX9065a (4, Scheme 1) to two different crystal forms of trypsin. On the *left*, the structure of a crystal form with the binding pocket opening to a solvent channel is shown. This form accommodates both enantiomers by giving a different orientation to the carboxylate group. A second crystal form, displayed on the *right*, closes up the binding site due to contacts with a neighboring molecule (*magenta*) in the densely packed crystal. The residual electron density indicates that this crystal form is only occupied by one enantiomer. **c** The two enantiomers of the sulfonamide 5 (Scheme 2) bind with comparable orientation to carbonic anhydrase. They coordinate to zinc through their sulfonamido groups. The hydrophobic isobutylamino groups, attached to the stereogenic center, orient similarly into a hydrophobic pocket. Energetically distinct ligand conformations are adopted to achieve this comparable binding mode, but the difference in affinity represents two orders of magnitude. **d–f** HIV protease is a homodimer with C2 symmetry. Inhibitors such as 7 have been synthesized and studied crystallographically, reflecting this C2 symmetry. The two enantiomers *S,S* (*green*) and *R,R* (*magenta*) (**e**) differ by a factor of 13 in affinity. This difference can be explained by a less efficient hydrogen bonding pattern formed by the central *R,R* diol unit compared to the *S,S* unit. The pseudo-symmetrical *S* (*red*) and *S,R* inhibitors (*blue*) bind with comparable affinity and very similar binding geometry (**f**). Obviously, the additional *H*-bond forming OH group in the *S,R*-diol does not improve binding affinity. The gain of one hydrogen-bond is compensated by less favorable solvation/desolvation properties

constrained binding site, only one enantiomer can be detected. Obviously, the contacts to the neighboring molecule in the crystal packing generate a binding site that favors binding of one of the enantiomers. Apparently, the conditions experienced in the first form, accommodating both enantiomers, better resembles the situation in solution and, most likely, under in vivo conditions. However, this example demonstrates that chiral recognition requires pronounced and well-discriminating molecular interactions in the close proximity of the stereo center thus creating local molecular asymmetry for the considered ligands.

Another example for inhibitor binding of distinct enantiomers has been studied with human carbonic anhydrase II (Fig. 4c) (GREER et al. 1994). Both enantiomeric forms of a potent sulfonamide (5 in Scheme 2) show quite similar binding modes, although their affinity differs by approximately two orders of magnitude. Since solvation/desolvation properties are identical for both enantiomers, the affinity difference must result from distinct interactions at the binding site. Both molecules bind with their sulfonamide group to the catalytic zinc and the neighboring Thr199. Furthermore, the SO_2 group in the six-membered rings interact identically with Gln92. The iso-butylamino group at the stereogenic center orients into the same hydrophobic pocket, but the conformation required to adopt this spatial arrangement has been computed as less favorable for the less active enantiomer. This difference provides an explanation for the observed affinity discrepancy.

A high resolution structure determination has been reported on the enantiomeric discrimination of two agonists (6 in Scheme 3) binding to the human nuclear retinoic acid receptor (KLAHOLZ et al. 2000). Both enantiomers bind to the structurally conserved binding pocket of the receptor. With their benzoic acid moiety, both form a network of hydrogen bonds to Arg278, Ser289, a water, and Leu233. The orientation of the hydroxy group in the central bridge gives rise to the two enantiomeric forms. In both, this OH group

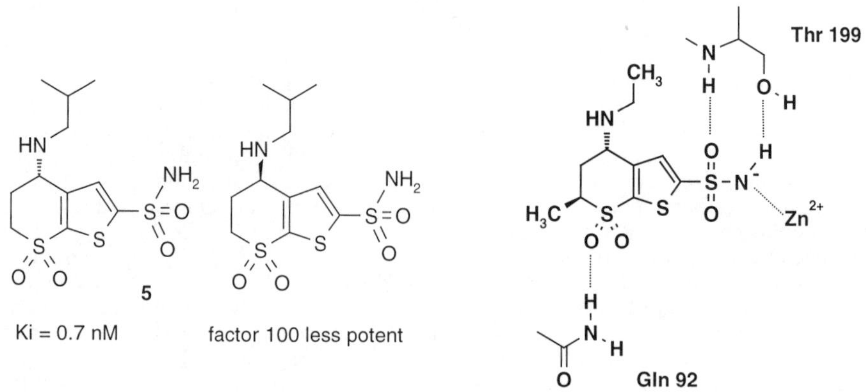

Scheme 2.

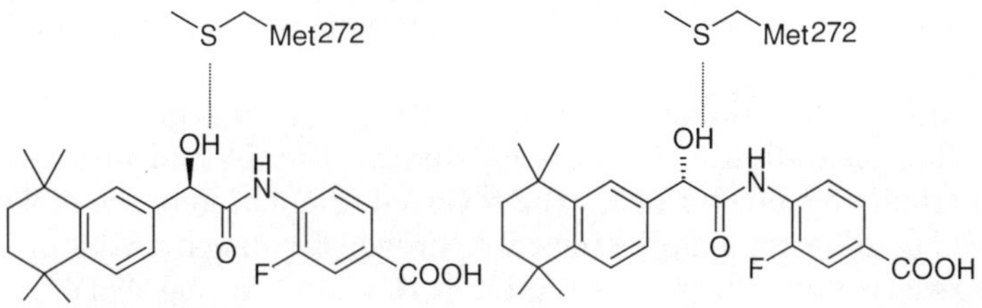

R-**6** *S*-**6**

Scheme 3.

forms an H-bond with good geometry to the sulfur of Met272. Interestingly enough, the tetrahydronaphthalene moiety adopts in both cases a different conformation with respect to the bond next to the hydroxymethylene bridge. Thus, the rotation of the tetrahydronaphthalene moiety together with the rearrangement of the amido group occurs to maintain the H-bond to Met272, simultaneously accommodating the naphthalene moiety. A comparison of the bound conformations of both enantiomers with the energetically most favorable arrangement in the free state suggests that the largely inactive *S*-enantiomer binds with an energetically less favorable conformation compared to the active *R*-enantiomer. It is reported that the affinity of both compounds toward the receptor differs by three orders of magnitude. Accordingly, and in analogy with the example of carbonic anhydrase binding, the affinity difference and thus chiral discrimination result from an adaptation of the less active enantiomer in a rather strained unfavorable conformation to fit into the structurally unchanged binding pocket.

D. Recognition of Chiral Building Blocks in Stereoisomers at the Protein Binding Site

The above-mentioned HIV protease belongs to the class of aspartyl proteinases. Its catalytic center is composed by two aspartates, each residing in one of the two C2-symmetrical homodimers. A set of four inhibitors (7 in Fig. 4) has been studied with this enzyme (KEMPF et al. 1990; HOSUR et al. 1994). The genuine skeleton of these inhibitors has been designed to resemble the C2-symmetry of the protein (Fig. 4d–f). To mimic possible transition state analogs, OH groups with varying stereochemistry were attached, resulting in four different compounds. The inhibitors 7a with X = *S*-CH(OH)CH_2 and 7b with X = *S*-CH(OH)-*R*-CH(OH) were found to adopt nearly identical binding modes and to have very similar affinities. In both inhibitors the *S*-CH(OH) groups are recognized by Asp25′ and Asp25. The additional OH group, present in the second inhibitor 7b, only interacts with Asp25. Obviously, this additional hydrogen bond does not contribute to binding affinity and is even compensated by solvation effects resulting in a slightly reduced binding affinity (by ca. 2.4).

Two other stereoisomers 7c and 7d, being mirror-symmetrical to each other in their central diol moiety, were studied. The *S,S* and *R,R* derivatives differ in affinity by a factor of 12. The reversed orientation of the diol portion is believed to allow for a more efficient hydrogen bonding network in the case of the *S,S*-derivative 7c compared to the *R,R*-analog 7d (Fig. 4e,f).

Inversion of configuration at C1 in *N*-acetylglucosamide (8 in Scheme 4) results in different diastereomers. This local change strongly modifies the binding mode of the ligand (IMOTO et al. 1972) (Fig. 5a). Although still recognized by Asp59 and Ala107 through their amido groups, the glucose moieties of both isomers orient in opposite directions of the binding pocket, thus interacting with different binding-site residues. Changes in the molecular skeleton of a ligand can also require changes in the local stereochemistry of a ligand. The two thrombin inhibitors Napap (9) and Argatroban (10) (Scheme 5) bind with their amidino or guanidino group, respectively into the specificity pocket of this serine protease (SHAFER and GOULD 1993). The altered topology of the basic side-chain along with the modified length of the sulfonamido linker requires inverted stereochemistry at the central C_α atom for potent enzyme inhibition (Fig. 5b).

Without knowing the binding mode of a particular substrate, the stereochemical prerequisites for a competitive inhibitor are difficult to estimate. The

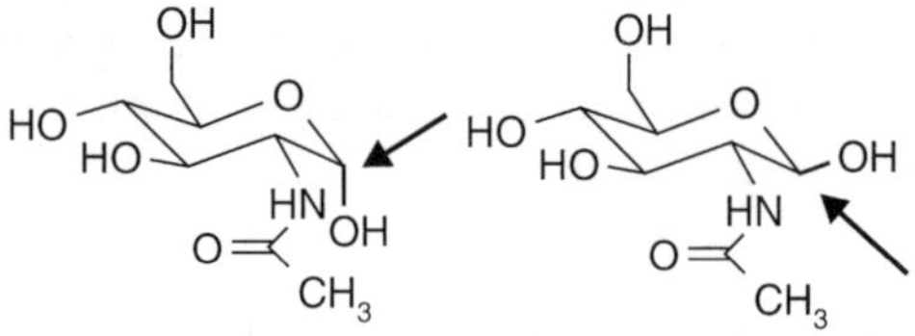

8

Scheme 4.

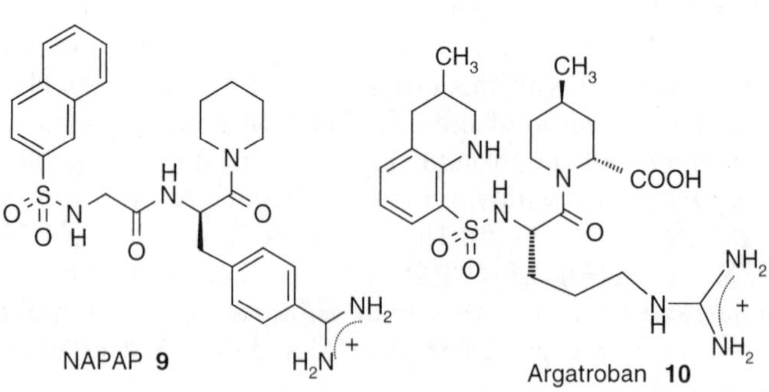

NAPAP **9**

Argatroban **10**

Scheme 5.

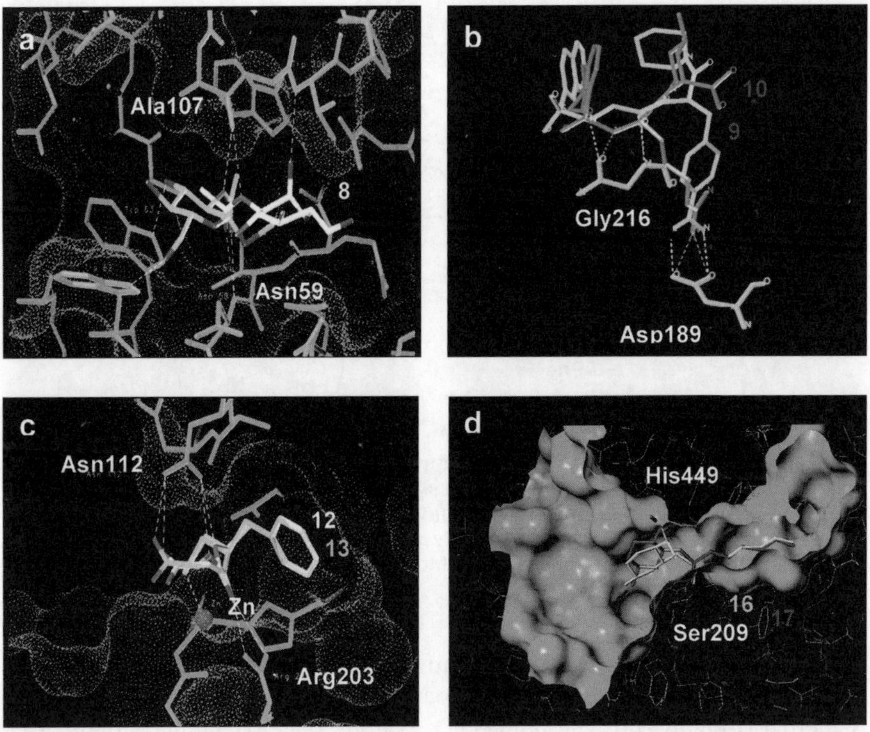

Fig. 5. **a** The two sugar moieties of 8 (Scheme 4) exhibit different orientation of the OH group at C1 giving rise to different diastereomers. Due to this difference the recognition properties are modified substantially and quite different binding modes are observed. The amido groups are commonly recognized by Asn59 and Ala107. The remaining parts of the molecules, including the distinct C1 OH group, are placed into different binding pockets. **b** Napap 9 (*green*) and Argatroban 10 (*red*) (Scheme 5) are recognized with their basic amidino- or guanidino group by Asp189 in the specificity pocket of thrombin, the piperidine and naphthyl or tetrahydroquinoline moiety pack together and occupying similarly the proximal and distal hydrophobic S2/S3 pockets in thrombin. Due to the recognition differences of the basic groups, the rather similar placements of the hydrophobic molecular portions are only achieved with inverted stereochemistry at C_α carbon. **c** Thiorphen (12, *white*) and retro-thiorphen (13, *yellow*) (see also Fig. 7) differ by their reverse sequence of the amino- and carbonyl group in the central peptide bond. However, the side-chains and the reversed peptide bond are similarly recognized by the residues exposed to the binding site of thermolysin. **d** The two enantiomers of menthyl phosphonate transition state analogs (15, 16) have been cocrystallized with *Candida rugosa* lipase. Both inhibitors adopt slightly different orientations at the catalytic center. The isopropyl groups interfere with the neighboring residue. In the 1*S*-derivative in particular this group pushes the catalytically important His449 out of place. This reorientation has been linked to the lower reactivity of the ester corresponding to the related translation state analog

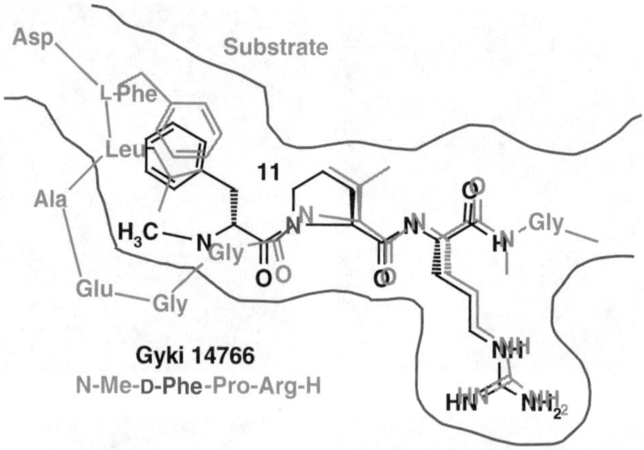

Fig. 6. Gyki (11, black), like the natural fibrinopeptide (gray), occupies the specificity pockets of thrombin. Both molecules possess an arginine residue with identical binding orientation. The proline moiety mimics valine in the substrate and the terminal phenylalanine resembles a similar residue in the peptide. However, D-Phe configuration is required in Gyki to position its phenyl moiety in the same volume area as the substrate

natural substrate of thrombin, fibrinogen, has a L-phenylalanine at its *N*-terminal part, just next to the cleaved peptide bond (Fig. 6). The inhibitor Gyki *N*-methyl-D-phenyl-prolylarginine (11), exhibits a phenylalanine with inverted stereochemistry. The role of this inversion becomes obvious when examining the crystal structures of the enzyme with the substrate and Gyki *N*-methyl-D-phenyl-prolylarginine, indeed, the benzyl moiety of D-phenylalanine mimics a hydrophobic surface of the substrate formed by a L-phenylalanine and L-leucine residue (BAJUSZ 1993; MARTIN et al. 1992) (Fig. 6).

E. Chemical Reactions in Proteins Using Stereoisomeric Substrates

Nature has provided several classes of enzymes to cleave polypeptide chains. Usually these catalysts operate efficiently only on peptides composed by L-amino acids. Nevertheless, some naturally occurring peptide antibiotics contain D-amino acids, resulting in metabolically more stable compounds. For the same reasons, D-amino acids have been introduced in some synthetic drugs of peptidic origin. These often exhibit higher activity and better metabolic stability. A special case are synthetic peptides with retro-inverse configuration, where the sequence of amino and carboxy groups is interchanged (Fig. 7). In order to retain the same relative configuration, L-amino acids are replaced by their D-analogs. Apparently, this exchange can mislead some enzymes or receptors and recognition, similar to the natural peptide, is observed. As an advantage the retro-inverse peptides are usually more resistant to metabolic

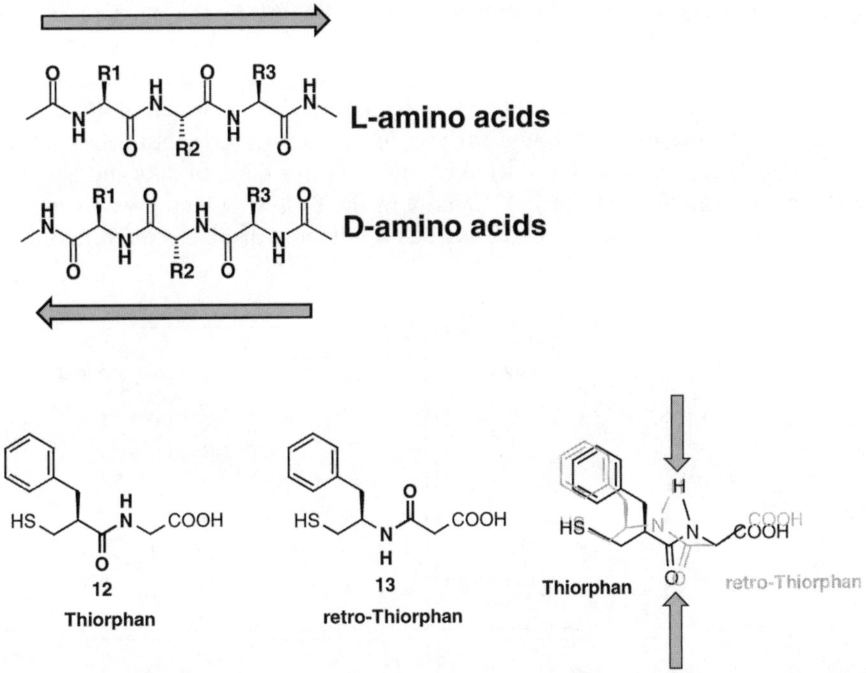

Fig. 7. Peptides composed by D-amino acids often possess better metabolic stability. To retain the same relative configuration, L-amino acids in the original peptide have to be replaced by their D-analogs while reversing the order of the peptide bond to retro-inverse configuration. Cases are known where this exchange can mislead enzymes or receptors resulting in a recognition of the reversal peptide similar to the natural one. The experimentally observed binding modes (see also Fig. 5c) of thiorphan (12) and retro-thiorphan (13) provide some explanation how this comparable recognition might be achieved on a molecular level

degradation. The experimentally observed binding modes of thiorphen (12) and retro-thiorphen (13) at the binding site of thermolysin (RODERICK et al. 1989) possibly provide a first glimpse of how the similar recognition of peptides with reversed sequence could be achieved (Figs. 5c, 7). Both inhibitors are equally potent. At the binding site, they adopt a geometry that places their peptide bonds in similar orientation even though they are being reversed in the two cases. The backbone amido and carbonyl groups of both ligands can form equally stable H-bonds with adjacent residues Asn112 and Arg203 (Figs. 5c, 7).

Catalytic pathways in enzymes are stereospecific. This property qualifies them as valuable stereoselective catalysts for organic synthesis. Organic reactions are often performed in nonaqueous solvents, but most biomolecules degrade under such conditions. Due to their biological function to operate at hydrophobic interfaces, lipases usually remain intact in such a milieu (SCHMID and VERGER 1998). Accordingly, they have frequently been used as stereospecific catalysts in organic synthesis. Lipases belong to the class of serine hydrolases which cleave ester bonds in triglycerides. Their catalytic center is composed by a triad formed by a serine, histidine, and aspartate or glutamate residue. The cleavage reaction proceeds via a nucleophilic attack of the serine OH group on the ester carbonyl carbon, resulting in a tetrahedral transition state (Fig. 8). Subsequently, the weakened C–O ester bond breaks and a covalent acyl intermediate is formed. Usually in the following step a water molecule attacks this acyl intermediate as nucleophile and displaces the remaining

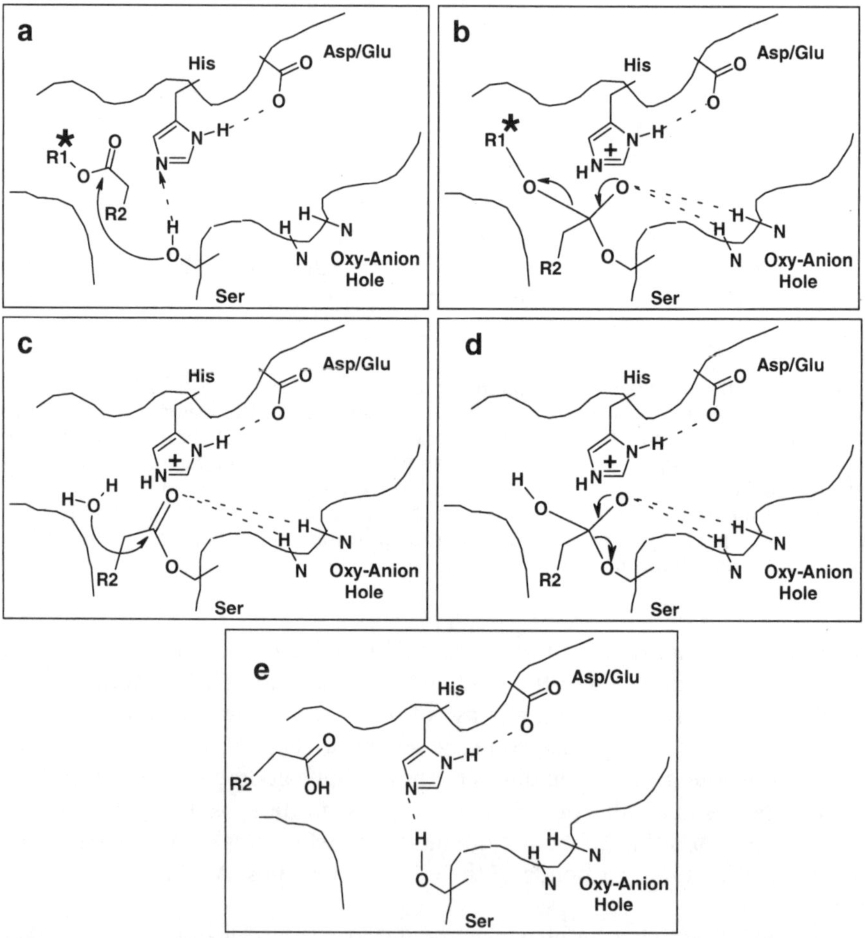

Fig. 8a–e. General reaction path of ester hydrolysis in a serine hydrolase. The reaction proceeds via two intermediate tetrahedral transition states (**b**, **d**) and a trigonal covalently bound acyl intermediate (**c**). This latter form decomposes via a nucleophilic attack by a water molecule, or by an alcohol or amide under nonaqueous conditions. If for example the substituent R1 bears a stereogenic center (**a**, **b**), diastereomeric transition states are produced whose energies will differ

part of the cleaved substrate from the catalyst. Under nonaqueous conditions, other nucleophiles can also be used to react with the acyl intermediate, e.g., alcohols or amines. These reagents will result in the formation of a new ester or amide bond. It has been shown that different nucleophiles react with deviating efficacy. Accordingly, many examples exist of lipases discriminating specifically between different nucleophiles. Since enantiomers are recognized as distinct species by a protein catalyst, lipases can discriminate between the chiral forms of a nucleophile attacking the intermediately formed acyl complex (CHEN and SIH 1989; GUTMAN and SHAPIRA 1995).

F. Structural Basis for Chiral Resolution in Lipases

Various molecular properties are responsible for chiral discrimination in the decomposition step of the acyl intermediate. The reaction is performed on a particular time scale and differences in the reaction velocity will result in chiral discrimination. Since the reaction at a binding site of a protein involves several consecutive steps, it is difficult to estimate which of the various steps is the rate-determining one and whether or how this provokes chiral discrimination. The same step will not necessarily be the rate-determining one for different reagents.

Kinetic, energetic, and structural considerations are involved. As a first insight, structural evidence has been associated with kinetic findings. As described, the reaction proceeds via a tetrahedral transition state. Accordingly, chemically stable transition-state-analogs have been synthesized. Usually a phosphorous atom is selected to mimic the intermediately formed tetravalent carbon.

For example, the enantio-preference of *Candida rugosa* lipase to hydrolyze both forms of menthyl pentanoate (14, 15 in Fig. 9) has been studied by CYCLER et al. (1994). Kinetic evidence has been collected which indicates that the *R*-enantiomer (16) is favored. To mimic the tetrahedral transition state of the hydrolysis step, the phosphonate analog of menthyl heptanoate (16, 17) has been synthesized and covalently linked to Ser209 in the active site. The corresponding 1*R*-isomer mimics the transition state for the fast-reacting enantiomer while the 1*S*-isomer resembles the transition state for the slow-reacting enantiomer. Crystals were obtained for the inactivated enzyme containing either inhibitor 16 or 17 (Fig. 9). Subsequently, the structures with both isomers have been determined (Fig. 5d). Both phosphonate groups bind similarly with a covalent link to the terminal oxygen of Ser209, and show *S*-configuration at their phosphorous. The terminal phosphonyl oxygens of both inhibitors occupy the oxyanion hole with hydrogen bonds to the amido NH groups of Ala210 and Gly124. The hexyl chain orients along a hydrophobic channel that extends toward the center of the enzyme. In both inhibitors the menthyl portion is positioned toward the entrance of the binding pocket where it opens to the solvent. The orientation of the six-membered rings in the two isomers exposes the iso-propyl groups of both isomers in a similar spatial area

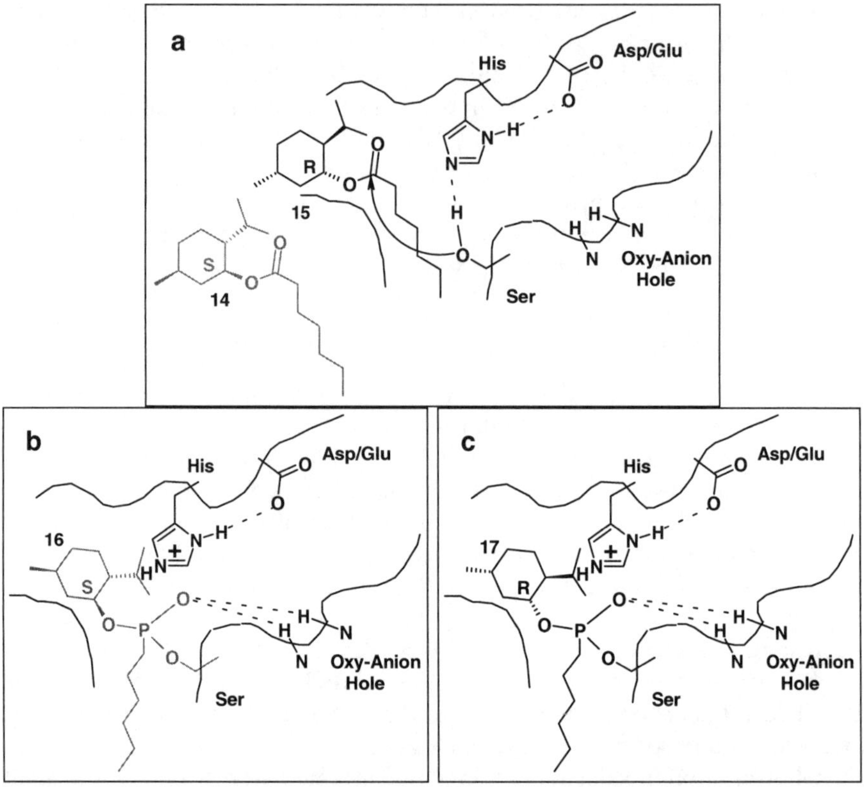

Fig. 9a–c. The ester hydrolysis of racemic menthyl pentanoate (14, 15) was studied in detail (**a**). Deviating reaction kinetics were observed. To study the geometry of the proceeded transition states, phosphonate analogs (16, 17) were synthesized and cocrystallized with the lipase *Candida rugosa* (**b**, **c**). Inhibitor complexes with deviating binding geometry were observed (see Fig. 5d)

next to the catalytic histidine. However, in the 1*S*-derivative this group is closely directed toward Phe344 and displaces the rings of Phe345 and His449. Compared to the apoenzyme and the structure of the 1*R*-isomer this rotation amounts to 60°. This reorientation of the His449 ring has been linked to the differences in reactivity for the two esters toward hydrolysis.

In the complex with the 1*R*-isomer, which mimics the transition state of the fast-reacting enantiomer, the imidazole ring of His449 adopts the same orientation as in the uncomplexed state and Nε2 forms a bifurcated hydrogen bond to Oγ of the catalytic Ser209 and to O1 of the mentyl moiety. This arrangement is in agreement with the usual assumptions about the catalytic mechanism in serine-type hydrolases. For the 1*S*-derivative, the transition-state-analog pushes His449 out of place and thus perturbs the hydrogen bond between Nε2 and O1 of the menthyl moiety. Simultaneously, conformational rearrangement modifies the torsion angles around phosphorous and increases the latter distance by 0.3 Å compared to the 1*R* derivative. The interaction geometries formed by the tetrahedral intermediates along the reaction path are less favorable for the 1*S*-isomer. These differences are assumed to be responsible for the experimentally observed differences in reactivity along the two stereoisomeric enzyme reactions.

G. Conclusion

Biological macromolecules, which are frequently receptors of small molecule ligands (e.g., drugs), create a chiral molecular environment. They are constructed from chiral building blocks and arrange as a whole in a handed conformation. Accordingly, enantiomers will be recognized as different species at the binding sites of such receptors. Some rare cases have been reported on the crystallographically determined binding mode of both enantiomers together with the protein.

Usually quite distinct binding affinities are observed for both forms. However, since solvation/desolvation properties are identical in these cases the energy difference has been attributed to conformational differences. A more strained and thus less favorable binding configuration is detected for the less active enantiomer. This distinct immobilization of stereoisomers in the handed environment of a binding pocket can also be exploited in enzyme kinetics, e.g., to achieve more favorable metabolic stability of retro/inverse peptides or to succeed in chiral resolution of amines and alcohols with lipases. Further crystallographic work with carefully selected and optimized crystallization conditions will hopefully reveal a more detailed insight into the spatial discrimination of stereoisomers at the binding site of proteins.

Acknowledgements. The author is grateful to Prof. D. Moras (Strasbourg, France) for making an example of his work available. The color images Figs. 4c, 5a–c were reproduced from the textbook of Böhm, Klebe and Kubinyi, "Wirkstoffdesign," Spektrum Akad. Verlag, Heidelberg, with the kind permission of the publisher.

References

Ariens EJ (1984) Stereochemistry, a Basis for Sophisticated Nonsense in Pharmacokinetics and Clinical Pharmacology, Eur J Clin Pharmacol 26:663–668

Ariens EJ (1993) Nonchiral, Homochiral, and Composite Chiral Drugs, Trends Pharmacol Sci 14:68–75

Ariens EJ, Wuis EW, Veringa EJ (1988) Stereoselectivity of Bioactive Xenobiotics. A Pre-Pasteur Attitude in Medicinal Chemistry Pharmacokinetics and Clinical Pharmacology, Biochemical Pharmacology 37:9–18

Bajusz S (1993) Chemistry and Biology of the Peptide Anticoagulant D-MePhe-Pro-Arg-H (GYKI-14766), Adv Exp Med Biol 340:91–108

Chen CS, Sih CJ (1989) General Aspects and Optimization of Enantioselective Biocatalysis in Organic Solvents. The Use of Lipases, Angew Chem Int Ed Engl 28:695–708

Cygler M, Grochulski P, Kazlauskas RJ, Schrag JD, Bouthillier F, Rubin B, Serreqi AN, Gupta AK (1994) A Structural Basis for the Chiral Preferences of Lipases, J Am Chem Soc 116:3180–3186

Edington P, Harding MM (1974) The Crystal Structure of DL-Histidine, Acta Crystallogr Sect B 30:204–206

Friedman L, Miller JG (1971) Odor Incongruity and Chirality, Science 172:1044–1046

Greer J, Erickson JW, Baldwin JJ, Varney MD (1994) Application of the Three-Dimensional Structures of Protein Target Molecules in Structure-Based Drug Design, J Med Chem 37:1035–1054

Gutman AL, Shapira M (1995) Synthetic Applications of Enzymatik Reactions in Organic Solvents, Adv Biochem Eng Biotechnol; Vol. 52. In: Fiechter A (ed.) Springer Heidelberg 87–128

Hosur MV, Bhat TN Kempf DJ, Baldwin ET, Liu B, Gulnik S, Wideburg NE, Norbeck DW, Appelt K, Erickson JW (1994) Influence of Stereochemistry on Activity and Binding Modes for C_2 Symmetry-Based Diol Inhibitors of HIF-1- Protease, Am Chem So 116:847–855

Imoto T, Johnson LN, North ACT, Phillips JA, Rupley JA (1972) Vertebrate Lysozyme in "The Enzymes", Vol. VII. In: Boyer PD (ed.), 3rd ed., Academic Press, New York, pp 665–868

Jung G (1992) Proteine aus der D-chiralen Welt, Angew. Chemie 104:1484–1486 (int. edition)

Kempf DJ, Norbeck DW, Codacovi L, Wang XC, Kohlbrenner WE, Wideburg NE, Paul DA, Knigge MF, Vasavanonda S, Craig-Kennard AC, Saldivar A, Rosenbrook W Jr, Plattner JJ, Erickson (1990) Structure-based, C2 Symmetric Inhibitors of HIV Protease. J Med Chem 33:2687–2689

Klaholz BP, Mitschler A, Belema M, Zusi C, Moras D (2000) Enantiomer Discrimination Illustrated by high Resolution Structures of the Human Nuclear Receptor hRARγ, Proc Nat Acad Sci 97:6322–6372

Madden JJ, McGandy EL, Seemann NC, Harding MM, Hoy A (1972)The Crystal Structure of the Monoclinic Form of L-Histidine, Acta Crystallogr Sect B 28:2382–2389

Madden JJ, McGandy EL, Seemann NC (1972) The Crystal Structure of the Orthorhombic Form of L-(+)-Histidine, Acta Crystallogr Sect B 28:2377–2383

Martin PD, Robertson W, Turk D, Huber R, Bode W, Edwards BFP (1992) The Structure of Residues 7–16 of the Alpha Chain of Human Fibrinogen Bond to Bovine Thrombin at 2.3 Angstroms Resolution, J Biol Chem 267:7911–7920

Milton RC, Milton SCF, Kent SBH (1992) Total Chemical Synthesis of a D-Enzyme: the Enantiomers of HIV-1 Protease show Reciprocal Chiral Substrate Specificity, Science 256:1445–1448

Roderick SL, Fournie-Zaluski MC, Roques BP, Matthew BW (1989) Thiorphan and *retro*-Thiorphan Display Equivalent Interactions When Bound to Crystalline Thermolysin, Biochemistry 28:1493–1497

Schmid RD, Verger R (1998) Lipases: Interfacial Enzymes with Attractive Applications, Angew Chem. 110:1694–1720 (int. Edition)

Shafer JA, Gould RJ Hrsg. (1993) Design of Antithrombotic Agents, Persp. Drug, Discov. Design 1:419–548

Stubbs MT, Huber R, Bode W (1995) Crystal Structures of Factor Xa specific Inhibitors in Complex with Trypsin: Structural Grounds for Inhibition of Factor Xa and Selectivity against Thrombin, FEBS Letts 375:103–107

West ML, Fairlie DP (1995) Targeting HIV-1 Protease: A Test for Drug Design Methodologies, Trends Pharm Sci 16:67–74

Zawadzke LE, Berg JM (1993) The Structure of a Centrosymmetric Protein Crystal, Proteins 16:301–305

CHAPTER 9
Stereoselective Drug-Channel Interactions

C. Valenzuela

All artificial bodies and all minerals have superposable images. Opposed to these are nearly all organic substances which play an important role in plant and animal life. These are asymmetric, and indeed have the kind of asymmetry in which the image is not superposable with the object.
Louis Pasteur, 1860

A. Ion Channels as Drug Targets

Ion channels are integral components of cellular signaling pathways in almost all living cells. In fact, our ability to do exercise, to perceive colors and sounds, to process language, and, more generally, to initiate muscle contraction depend on electrical messages produced as ion channels in cell membranes open and close (Armstrong and Hille 1998). Various ion channels mediate sensory transduction, electrical propagation over long distances, and synaptic transmission. In this chapter, I will focus on voltage-gated ion channels, the family of channels that includes Na^+, K^+ and Ca^{2+} channels responsible for nerve and muscle excitability. Our basic understanding of these proteins maintains the framework and rigor established 50 years ago by Hodgkin and Huxley (1952), enriched by much new molecular information (Catterall 1988) and by insights gained from patch-clamp methods (Hamill et al. 1981). Ionic channels are membrane proteins that are blocked by many different types of drugs, such as local anesthetics, antiarrhythmics, anticonvulsants, antihypertensive drugs, etc. Therefore, they represent the binding sites for these therapeutic agents and, in fact, ion channels are considered their drug targets, whose topology, at the ion channel level, has been analyzed by studying the interactions of specific ion channel blockers and site-directed mutant ion channels (Ragsdale et al. 1994; Hockerman et al. 1997a; Yeola et al. 1996; Franqueza et al. 1997; Nau et al. 1999a).

In the best case, drug targets would have several characteristics: known biological function; robust assay systems for in vitro characterization and high-throughput screening; and finally, they have to be specifically modified by and accessible to small molecular weight compounds in vivo. Ion channels have many of these attributes and, thus, they can be viewed as suitable targets for small-molecule drugs. However, target-based drug discovery programs require the identification, characterization and validation of molecular targets. An approach to these considerations is to associate the gene that encodes the expression of such ion channels with a specific cellular process, disease indication, or mammalian model of disease. In fact, involvement of a gene in an inherited disease provides direct evidence that a given protein is functionally important. During the last decade, enormous progress has been made identifying mutations in ion channel genes that cause human and animal diseases that result from defects in ion channel function (*channelopathies*).

Stereoselective interactions are especially interesting because they can reveal three-dimensional relationships between drug and channel with otherwise identical biophysical and physicochemical properties (ARIËNS 1993; FRANKS and LIEB 1994). Furthermore, stereoselectivity suggests direct and specific receptor-mediated action, and identification of such stereospecific interactions may have important clinical consequences (ARIËNS 1993; ÅBERG 1972). The fact that drug targets would be able to discriminate between the enantiomers present in a racemic drug is the consequence of the ordered asymmetric macromolecular units that form living cells. In fact, it is this ordered asymmetry which gives the macromolecules the necessary information to discriminate between the optical isomers of monomeric substrates such as amino acids and sugars. However, almost 25% of drugs used in the clinical practice are racemic mixtures, the individual enantiomers of which frequently differ in both their pharmacodynamic and pharmacokinetic profile. Moreover, different effects induced by one of the enantiomers of a racemic drug may contribute to undesired effects, which can be similar to or different from the pharmacological effect of the drug. In other cases, the pharmacological effects induced by the two enantiomers on the molecular target are opposite, as exemplified by the dihydropyridines at L-type Ca^{2+} channels.

Ion channels can be differentiated into those that open or close in response to changes in transmembrane voltage, and those in which opening and closing mechanisms are modulated by binding of extracellular or intracellular ligands by activation of a cation-specific conductance across the plasma membrane (CATTERALL 1994). Ion channels from the first category are named *voltage-gated ion channels* and mediate rapid, voltage-gated changes in ion permeability during action potentials in excitable cells and also modulate membrane potential and ion permeability in many unexcitable cells (HILLE 1992). In contrast, those ion channels which open and close in response to the concentration of extracellular or intracellular ligands are called *ligand-gated ion channels*. These include the nicotinic acetylcholine, γ-aminobutyric acid (GABA), and glycine receptors, and they mediate local increases in ion conductance at chemical synapses and thereby depolarize or hyperpolarize the subsynaptic area of the cell (CATTERALL 1988). Both types of ion channels belong to the superfamily of plasma-membrane cation channels.

B. Voltage-Dependent Ion Channels

I. Na^+ Channels

1. Structure

The first ion channel gene to be cloned was that encoding the sodium channel of the *eel electroplax* (NODA et al. 1984) (Fig. 1). The activation of sodium channels after depolarization of the cell membrane leads to their brief opening (for <1 ms) after a short latency, and afterwards to their inactivation (a closed nonconducting state) which persists until the membrane is hyperpolarized (Fig. 2). Voltage-gated sodium channels cloned from a variety of other species and tissues share a striking degree of homology to the original cDNA (KALLEN et al. 1993). In humans, at least six separate genes encode homologous sodium channel isoforms expressed in brain, heart, and skeletal muscle (RODEN and GEORGE 1997). Sodium channels, as well as calcium and potassium channels, are hetero-oligomeric complexes of an α subunit that forms the ion pore and accessory subunits (β subunits) which regulate the function of the channel. The α subunit of Na^+ channels is composed of around 2000 residues. The sequence contains four homologous domains (I to IV). Hydropathy analyses suggest that each of these domains contains six transmembrane segments (called S1 to S6) and that the residues that link the four repeats, like the N and the C termini, lie on the cytoplasmic side of the membrane.

2. Activation

Voltage-gated Na^+ channels are extremely sensitive to small changes in membrane potential. HODGKIN and HUXLEY (1952) proposed that Na^+ channel activation must result from movement of charges within the membrane, the so called "gating currents," a small charge movement generated by the voltage-driven conformational changes that open these channels (ARMSTRONG and HILLE 1998). Since the cloning of the first Na^+ channel, the amphipatic S4 α helices in each of these four domains were immediately proposed as the voltage-sensor of the channel. These segments contain between 5 and 8 charged residues each, spaced at three amino acid intervals. Thus, it was proposed that these α helices would move in response to changes in membrane potential, changing the conformation of the protein to open or close the ion pore. Site-directed mutagenesis in which some of the positive charged amino acids of S4 were replaced by neutral residues (STÜHMER et al. 1989) supported this hypothesis. More recently, mutation of selected amino acids at the S4 segment to cysteines, followed by application of cysteine-modifying agents to the internal or the external surface of cells expressing recombinant Na^+ channels, have revealed that certain residues are accessible to the inside of the membrane at rest and are accessible to the exterior of the membrane during depolarization (YANG and HORN 1995; YANG et al. 1996). These results not only indicate that this segment moves in response to membrane depolarization, but

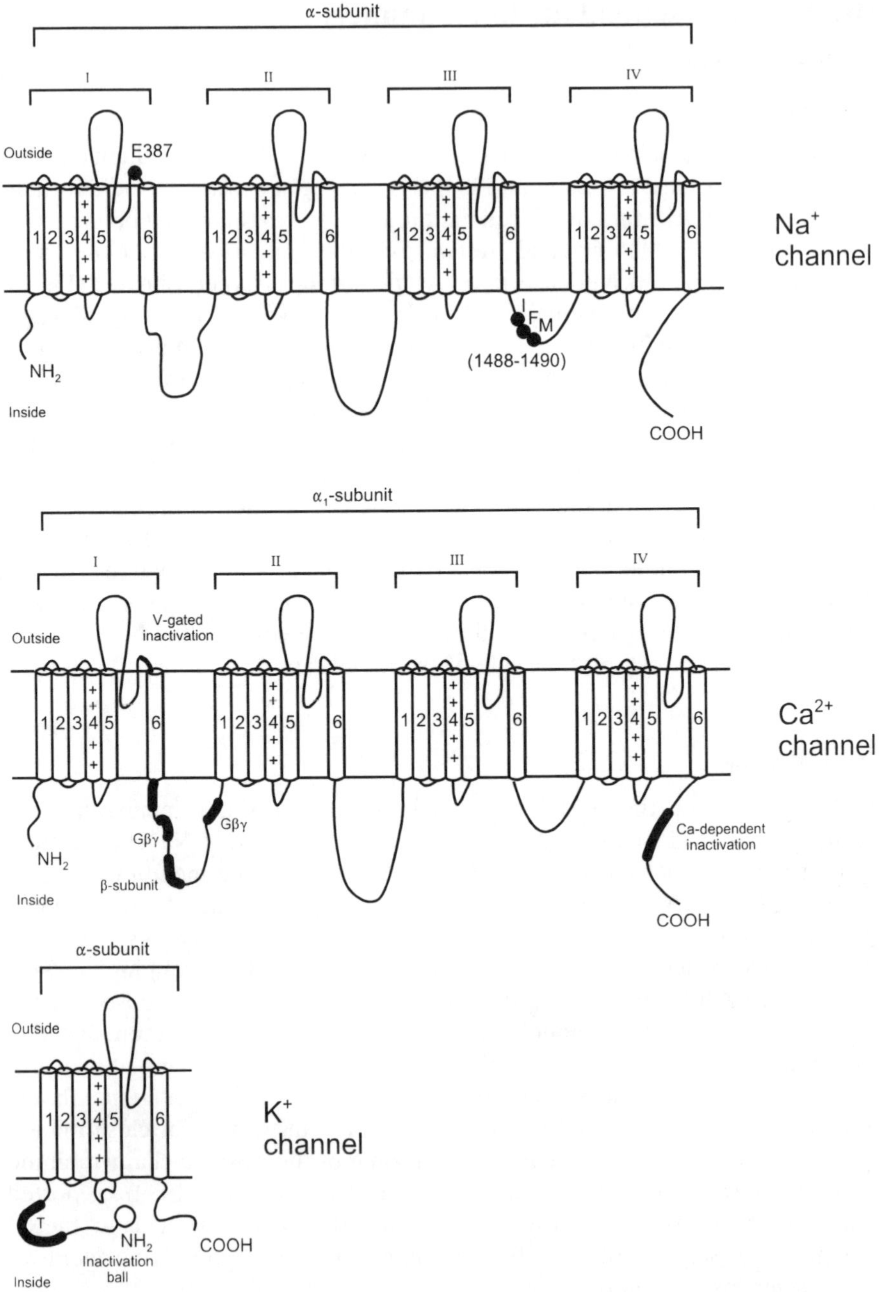
α-subunit
I
II
III
IV
Outside
E387
Na⁺ channel
I F M
(1488-1490)
NH2
Inside
COOH
α1-subunit
V-gated inactivation
Ca2+ channel
Gβγ
β-subunit
Ca-dependent inactivation
K⁺ channel
T
Inactivation ball

Fig. 1. Topology of the α subunits of Na^+, Ca^{2+}, and K^+ channels deduced from hydropathy, mutagenesis, and antibody studies. In the Na^+ channel, the amino acids involved in TTX binding (*E387*) and in the inactivation (*IFM*) are marked in *black*. In the Ca^{2+} channel, the portions of the channel which interact with the β-subunit and with the $G_{\beta\gamma}$ subunits at the linker binding domains I and II are indicated. The C terminus participates in the Ca^{2+}-dependent Ca^{2+} inactivation, while the S6 segment of the domain I is involved in voltage-dependent inactivation. The scheme of the K^+ channel shows the inactivation ball at the N-end, the "ball" receptor at the S4-S5 linker. Functional K^+ channels are tetramers, that can be homo- or heterotetramers of these α subunits. Tetramerization with other α subunits requires the tetramerization domain of the α subunit (*T*). The ion pore of the channel lies on the S5-S6 loop. Heteromultimeric functional K^+ channels only appear with members of the same subfamily (Kv1.1 with Kv1 members, but not with Kv2, Kv3 or Kv4 members)

also suggests that the S4 voltage sensor moves in its own relatively narrow "channel" during gating.

3. Inactivation

Sodium channels open in response to depolarization, leading Na^+ ions to the intracellular media and, thus, driving the membrane voltage positive during the upstroke or phase 0 of action potentials. Then they inactivate spontaneously (stop conducting) and the repolarization of the membrane potential begins. This process (fast inactivation) acquires most of its voltage dependence from coupling to voltage-dependent activation, and it can be specifically prevented by treatment of the intracellular surface of the Na^+ channel with proteolytic enzymes (Armstrong et al. 1974; Valenzuela and Bennett 1994), which suggests the presence of an inactivation particle in the channel that prevents reclosing, a foot-in-the-door mechanism. Cutting off the protein foot (the inactivation particle) with proteolytic enzymes removed this interference, making it easy to close the activation door at any time. These observations led Bezanilla and Armstrong (1977) to propose the ball-and-chain model, in

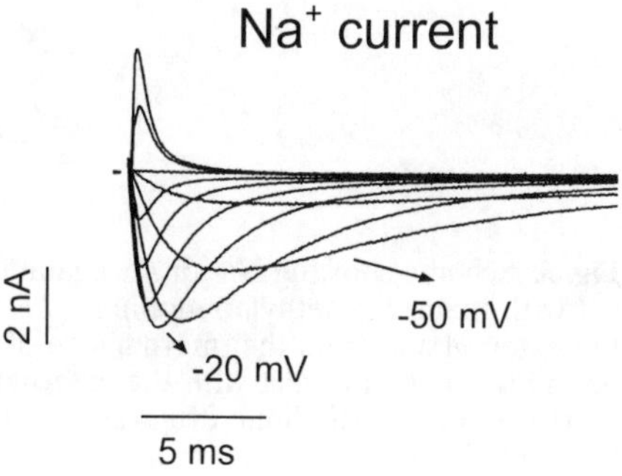

Fig. 2. Original records of Na^+ current from a guinea pig ventricular myocyte. The holding potential was maintained at -120 mV and depolarizing pulses between -80 and $+50$ mV were applied every 5 ms

which an inactivation particle tethered on the intracellular surface of the channel binds to a receptor site in the intracellular mouth of the pore, and blocks it. Studies using antibodies antipeptide as well as mutagenesis analysis identified the short intracellular segment that connects domains III and IV as the inactivation gate of the Na^+ channel (VASSILEV et al. 1988; STÜHMER et al. 1989). In particular, mutation of hydrophobic amino acid residues in this segment IFM to QQQ results in sodium channels that activate normally but inactivate much slower and less completely than wild-type channels (STÜHMER et al. 1989; PATTON et al. 1992; HARTMANN et al. 1994; BENNETT et al. 1995).

4. Ion Pore and Selectivity

A combination of site-directed mutagenesis and toxin-binding studies have determined that the extracellular parts of the loop between the fifth and sixth transmembrane domains dip down into the membrane and form the outer part of the Na^+ channel pore. Tetrodotoxin (TTX) and local anesthetics have been used to map the outer and the inner mouth of the Na^+ channel pore, respectively (Fig. 3). The study of the effects of TTX on different site-directed mutant Na^+ channels has revealed that the glutamic at position 387 (E387) is a molecular determinant of TTX block (TERLAU et al. 1991). This residue lies within the region of repeat I that links S5 and S6, which is called SS1-SS2. Mutation of each of the negatively charged residues in the equivalent position in the other three repeats also prevented TTX binding (TERLAU et al. 1991). This indicated that the SS1-SS2 region of all four repeats contribute to the outer pore

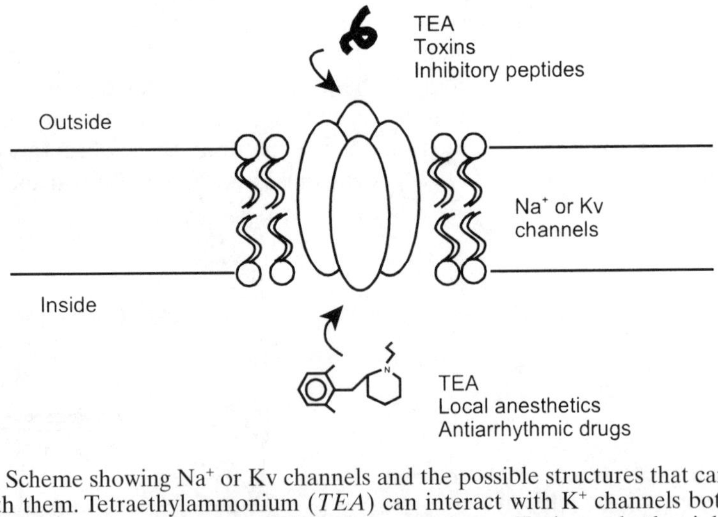

Fig. 3. Scheme showing Na^+ or Kv channels and the possible structures that can interact with them. Tetraethylammonium (*TEA*) can interact with K^+ channels both from the external and from the internal side of the membrane. Toxins and other inhibitory peptides usually interact with the external mouth of the ion pore. Finally, local anesthetics and antiarrhythmic drugs act on the inner mouth of the ion pore, both at the Na^+ and K^+ channels

of the Na^+ channel and define the TTX binding site as a ring of negative charges surrounding the external opening of the ion pore. The linkers between S5 and S6 of the four domains of Na^+ and Ca^{2+} channels have similar sequences. Both channels have two rings of charge encircling the pore, each ring containing four residues, one from each homologous domain. The outer ring is entirely negative and is thought to be relatively distant from the pore axis. The inner ring, two to three residues deeper into the pore, is composed of DEKA (Asp Glu Lys Ala) in the Na^+ channel and EEEE (Glu Glu Glu Glu) in the Ca^{2+} channel. These differences in the amino acid sequence (an increased number of negative charges) seem to be the expected ones for a channel conducting doubly charged Ca^{2+}. Mutations in the SS1-SS2 region (K1422E, A1712E) induce dramatic modifications in ion selectivity (HEINEMANN et al. 1992), converting a Na^+ channel into a Ca^{2+} preferring channel. This demonstrates that K1422 and A1712 constitute part of the selectivity filter of the Na^+ channel and that these two amino acids lie within the permeation pathway of the ion pore.

II. Ca^{2+} Channels

1. Structure

Voltage-gated Ca^{2+} channels come in a surprising number of different types. Ca^{2+} channels are classified as T, L, N, P, Q, and R and are distinguished by their sensitivity to pharmacological blockers, single channel conductance, kinetics, and voltage dependence. L-type Ca^{2+} channels have a wide distribution, being found in skeletal, cardiac and smooth muscle, endocrine cells, and other tissues. In contrast, P, N, and Q channels appear to be confined to the nervous system and some types of endocrine cells. Finally, T-type Ca^{2+} channels are found in cardiac muscle and some types of nerve cells. The first calcium channel cloned was the L-type Ca^{2+} channel of skeletal muscle (CURTIS and CATTERALL 1984; TANABE et al. 1987). Since most pharmacological studies have been performed in L-type Ca^{2+} channels, and since they represent the more widely distributed Ca^{2+} channels in different living cells, the chapter will focus on this type of Ca^{2+} channels. The L-type calcium channel is a heteroligomeric complex of five subunits, α_1, α_2, β, γ and δ, arranged in a 1:1:1:1:1 stoichiometry (CATTERALL 1988), with the ion channel pore formed by the α_1 subunit. The L-type α_1 subunit family includes α_{1S}, which is expressed in skeletal muscle (TANABE et al. 1987), α_{1C}, which is expressed in cardiac and smooth muscle, neurons, and many other cell types (TSIEN et al. 1995), and α_{1D}, which is expressed in endocrine and neuronal cells (SEINO et al. 1992). The α_1 cDNA encodes a protein of 1873 amino acids with a molecular weight of 212 kDa. Ca^{2+} channels, as well as Na^+ channels, activate and inactivate in response to changes in membrane potential (Fig. 4). The magnitude of the current through Ca^{2+} channels is increased upon application of brief depolarizing steps through a phenomenon called "facilitation" (ZYGMUNT and MAYLIE 1990). Both ways

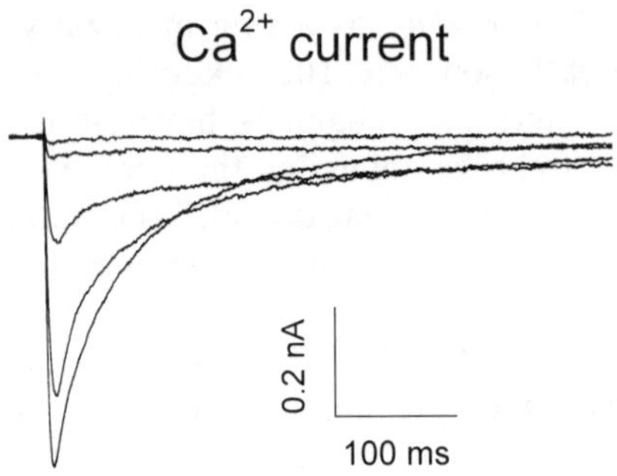

Fig. 4. Original records of Ca^{2+} current from a guinea pig ventricular myocyte. The holding potential was maintained at –40 mV and depolarizing pulses between –40 and +40 mV were applied every 30 s

of autoregulation, inactivation and facilitation of the current, modify the amount of Ca^{2+} ions that enter the cell during repetitive activity. This increase in the magnitude of the Ca^{2+} current is Ca-dependent, since it is not observed in the presence of Ba^{2+} as the charge carrier, and is modulated by the binding of calmodulin to a motif of the α_1 subunit located at the C terminus (ZÜLKE et al. 1999).

2. Activation

Similarly to Na^+ channels, the α_1 subunit of the Ca^{2+} channels contains four internal repeats (I to IV) with homologous sequences, each having six putative α-helical membrane spanning segments (S1 to S6) with the S4 segments exhibiting positively charged residues every third or fourth amino acid (TANABE et al. 1987). Transmembrane S4 segments possess a number of highly conserved positive charges and, by analogy with Na^+ channels, is thought to serve as the voltage sensor for Ca^{2+} channel opening (Fig. 1). Similar to what happens in Na^+ channels, the loop linking S5 and S6 dips back down into the membrane and contributes to the pore.

3. Inactivation

Inactivation of Na^+ channels is caused by the cytosolic loop linking repeats III and IV, which acts as a hinged lid and occludes the inner mouth of the pore. Thus, it was expected that a similar mechanism might occur in Ca^{2+} channels. Surprisingly, this was not the case and mutations or deletions at this loop do not modify the inactivation process of Ca^{2+} channels (ZHANG et al. 1994). Moreover, Ca^{2+} channels exhibit two different types of inactivation: voltage-dependent and Ca^{2+}-dependent inactivation (DELEON et al. 1995; ASHCROFT 1999). Voltage-dependent inactivation is modified after site-directed mutagenesis of residues within and immediately adjacent to the sixth transmembrane domain I (Fig. 1). An additional residue located at the loop connecting domains I and II (within the $G_{\beta\gamma}$-binding site of the G_s protein) has also been involved in voltage-dependent inactivation of Ca^{2+} channels (HERLITZE et al. 1997). It has been demonstrated that calmodulin tethers to Ca^{2+} channels and acts as the Ca^{2+} sensor for inactivation in L-type Ca^{2+} channels (PETERSON et al. 1999). The calcium-dependent inactivation is likely to occur via Ca^{2+}-dependent interaction of tethered calmodulin with the calmodulin-binding isoleucine-glutamine ("IQ") motif in the carboxy terminus of the α_{1C} channel subunit (ZÜLKE et al. 1998; PETERSON et al. 1999). Moreover, it has been demonstrated that calmodulin is a critical Ca^{2+} sensor for both inactivation and facilitation, and that the nature of the modulatory effect depends on residues in the "IQ" motif important for calmodulin binding (ZÜLKE et al. 1999).

4. Ion Pore and Selectivity

The ion pore of Ca^{2+} channels is formed by the loops linking the S5 and S6 segments of the four domains, similarly to that previously described for Na^+ channels. A ring of four negatively charged glutamate residues, one in each loop, forms a high-affinity binding site for Ca^{2+} and provides a structural basis for the high Ca^{2+} selectivity of the channel (YANG et al. 1993). The importance of these residues was first pointed out in previous works on Na^+ channels in which a single point mutation equivalent to glutamate conferred Ca^{2+} selectivity to the Na^+ channel (HEINEMANN et al. 1992). Thus, studies in which these glutamates were mutated to the equivalent amino acids present in Na^+ channels made the Ca^{2+} channel permeable to monovalent ions such as Na^+ and Li^+ (YANG et al. 1993). Moreover, in contrast to what happens in wild-type channels, the current through these mutant Ca^{2+} channels was not affected by external Ca^{2+}, thus confirming that the glutamate residues are responsible for Ca^{2+} selectivity, although they do not contribute equally to the Ca^{2+}-binding site, with that in domain III having the most crucial effect.

III. K^+ Channels

The first K^+ channel to be identified came from the cloning of the *Shaker* gene of *Drosophila* which causes flies to shake when exposed to ether (PAPAZIAN et al. 1987). However, voltage-gated potassium channels are found in virtually all cells and can be divided into two main structural families depending on whether they possess six or only two transmembrane domains (CATTERALL 1988; RODEN and GEORGE 1997; SNYDERS 1999). The latter include the inwardly rectifying K^+ channels. Finally, another family of K^+ channels has been cloned consisting of four transmembrane domains (for review see SNYDERS 1999). In the present review, I will focus on those K^+ channels that exhibit six transmembrane spanning domains (6Tm-1P). Initially, four *Drosophila* subfamilies were described (*Shaker*, *Shab*, *Shal* and *Shaw*), and the first cloned mammalian K^+ channels were related to these subfamilies (DOLLY and PERCEJ 1996). In the mammalian KvX.Y nomenclature Kv reflects K^+ channel, voltage-gated, X

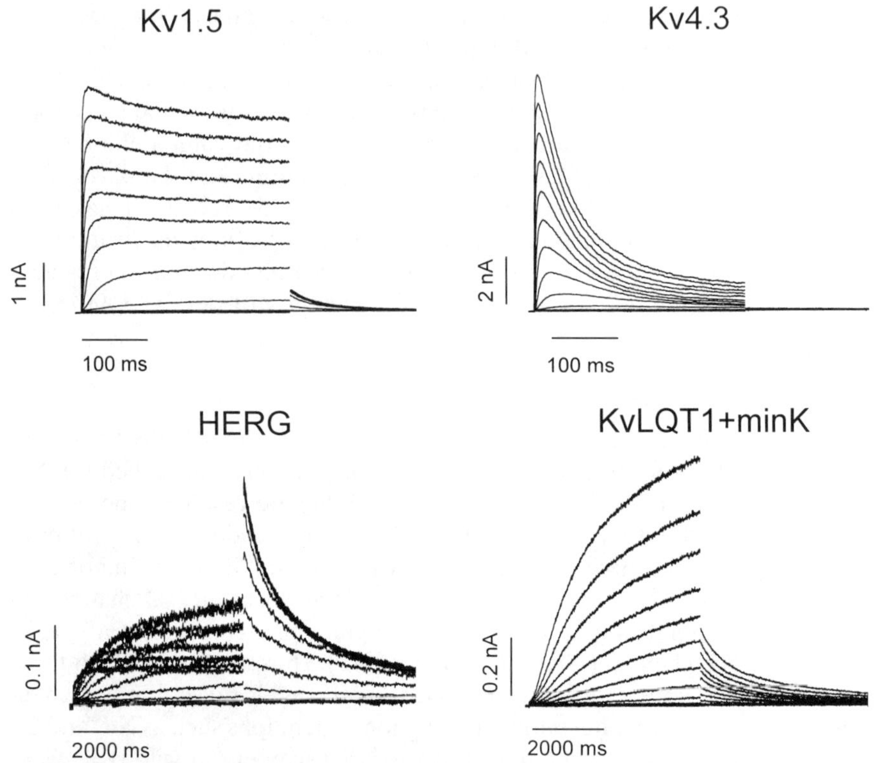

Fig. 5. Original records obtained after transient transfection of cDNA encoding Kv1.5, Kv4.3, HERG, and KvLQT1+minK in Ltk^{-} cells (for Kv1.5 channels) and in CHO cells (for the other three types of channels). Holding potential was maintained in all cases at –80 mV. Cells transfected with cDNA encoding the expression of Kv1.5 or Kv4.3 channels were pulsed every 10 s from –80 and +60 mV. Cells transfected with cDNA encoding the expression of HERG and KvLQT1+minK were pulsed every 30 s from –80 and +60 mV

represents the subfamily, and Y the number of the gene within the subfamily. Assembly of four α subunits into a tetrameric structure is needed to create a functional K^+ channel (MACKINNON 1991; DOYLE et al. 1998). Other K^+ channel subunits with similar molecular structure have been cloned in part on the basis of their involvement in congenital arrhythmias: KvLQT1 and HERG. The activation of these channels after applying depolarizing pulses leads to the genesis of different K^+ currents (Fig. 5). Unfortunately they have not been assigned formal numbers in the KvX.Y convention, thus resulting in the present hybrid nomenclature. All these channels are closed at the resting membrane potential of the cell and open upon depolarization. Therefore, they are involved in the repolarization of the action potential, and thus in the electrical excitability of nerve and muscle fibers, including cardiac muscle. Mutation of genes encoding members of these K^+ channel subfamilies lead to a number of human diseases, such as episodic ataxia, long QT syndrome and epilepsy.

1. Structure

6Tm-1P K^+ channels are made up of pore-forming α subunits that may associate with one of a number of different types of β subunits. To date, nine different subfamilies of Kv channel α subunits have been described (Kv1–9) in addition to KvLQT1 and HERG channels (Deal et al. 1996). The α subunit of the Kv channels corresponds to a single domain of the Na^+ or Ca^{2+} channels and four such subunits form the functional K^+ channel pore. Each α subunit consists of six transmembrane spanning segments (S1 to S6), which are highly conserved, and intracellular N and C termini of variable length (Fig. 1). The S4 segment shows a highly homology with the voltage-gated Na^+ channel: it is amphipatic, having a positive charged amino acid at every third position. Like in Na^+ channels, the S4 segment is involved in voltage-dependent activation of K^+ channels and the amino terminus is involved in the voltage-dependent inactivation. The linker between S5 and S6 segments dips back into the membrane and participates in formation of the channel pore. It is equivalent to the SS1-SS2 of the Na^+ channels, and is called H5 region or P loop in K^+ channels. A huge variety of voltage-dependent K^+ channels have been described, far more than was originally anticipated. Part of this diversity is the consequence of the existence of different genes which encode different Kv channels. However, other factors such as alternative splicing, the formation of heteroligomeric channels and their association with different types of β subunits, also contribute to this high diversity.

2. Activation

Experiments similar to those performed in Na^+ channels have revealed that the S4 segment represents the voltage sensor of K^+ channels (Papazian et al. 1991). Thus, this segment represents the major component of the voltage sensor for gating, although negative charges in S2 and S3 also contribute to this process. Depolarization of the membrane causes a physical (outward) movement of S4 which then induces further conformational changes that open the channel and permit selective K^+ permeation. This movement has been monitored electrically as the gating current (Bezanilla and Stefani 1998), or by means of fluorescence (Mannuzzu et al. 1996).

3. Inactivation

Upon depolarization, voltage-gated K^+ channels activate more or less rapidly, rising to a maximum peak amplitude before the current declines in a slow (C-type) or a fast process (*N*-type). The *N*-type inactivation involves the *N*-terminus of the α subunit. Thus, upon opening of the channel, the *N*-terminal domain moves into the internal mouth and occludes the pathway of K^+ ions through the pore (HOSHI et al. 1990). After removal of this region, inactivation can be restored by the corresponding synthetic peptide. Since functional K^+ channels are composed by the assembly of four α subunits, each K^+ channel has four inactivation balls, which in *Shaker* channels correspond to the first 20 amino acids of the N terminus. Only one of these balls binds to the receptor in the mouth of the ion pore, but the four are needed for normal inactivation rate of the channel (GOMEZ-LAGUNAS and ARMSTRONG 1995). This inactivation ball binds to a receptor located at the S4-S5 linker, which involves a threonine (ISACOFF et al. 1991). In contrast, C-type inactivation is modified by mutations in the S6 and P regions, and is usually slow and incomplete (HOSHI et al. 1991). This kind of inactivation is present in almost all K^+ channels and it may reflect a constriction of the pore involving residues on the extracellular surface of the ion pore.

4. Ion Pore and Selectivity

The ion conduction pathway consists of a sequence of approximately 20 amino acids (P loop) between the S5 and S6 segments with contributions of the S6 and the S4-S5 linker. The four subunits are oriented such that the S5-P-S6 sections face each other, creating the central pore. Within the P region, the amino acid sequence motif (TxTTx)GYG is now considered the K^+-selectivity signature motif (HEGINBOTHAM et al. 1992). Mutagenesis studies have revealed that the selectivity filter forms an essential part of the permeation pathway. Recently, MACKINNON and his colleagues (DOYLE et al. 1998) obtained the first crystal structure of a K^+ channel, the bacterial K^+ channel (KcsA) cloned from *Streptomyces lividans* with a 3.2-Å resolution. The X-ray analysis confirms that the channel contains four identical subunits, each with two transmembrane α-helices and a P domain. The subunits create an inverted cone or teepee, in which the α-helices resemble the poles of the teepee. The selectivity filter fills the wider base on the extracellular face. The pore has a length of 45 Å; it starts from the inside with a tunnel (18 Å long, ~6 Å wide) which opens into a 10-Å-wide cavity followed by the narrow selectivity filter (12 Å). Very recently, it has been proposed that the structure of Kv channels differs from that of the nonvoltage-gated bacterial potassium channel (KcsA) by the introduction of a sharp-bend in the inner (S6) helices. This bend would occur at a Pro-X-Pro sequence that is highly conserved in the Kv channels (DEL CAMINO et al. 2000). Part of the P region of each subunit is α-helical, with their electronegative carboxyl end directed toward the central cavity, which helps to stabilize the potassium ions. The selectivity filter is lined by the carbonyl oxygen atoms of the GYG signature sequence. The tyrosine (Y) points away from the pore and interacts with other conserved aromatic residues of the pore helices. Thus, a fairly rigid donut of 12 interlocking aromatic amino acids is formed around the selectivity filter that holds the GYG backbone at the optimal distance from the center to achieve potassium selectivity: the four backbones form a pore with the carbonyl oxygens at the proper distance to compensate the cost of dehydration of a K^+ ion (~3 Å). Interestingly, the inner tunnel is largely coated with hydrophobic side chains, explaining why binding of open channel blockers such as tetraethylammonium (TEA) derivatives, antiarrhythmic drugs like quinidine, or local anesthetics such as bupivacaine are stabilized by hydrophobic interactions (CHOI et al. 1993; SNYDERS and YEOLA 1995; YEOLA et al. 1996; FRANQUEZA et al. 1997).

C. Stereoselective Interactions Between Local Anesthetic-Like Drugs and Na^+ Channels

The sodium channel is the site of action of local anesthetics, certain anticonvulsants, and antiarrhythmic drugs used to decrease sodium channel activity in syndromes of hyperexcitability such as cardiac arrhythmias and epileptic seizures (HILLE 1992; HONDEGHEM and KATZUNG 1977; VAUGHAN WILLIAMS 1984; ROGAWSKI and PORTER 1990). All these drugs induce voltage- and/or use-dependent blockade of Na^+ and other ion channels. Most of these therapeutic agents are weak bases and several lines of evidence indicate that most of their blocking effects are due to the binding of their charged form to a receptor site located in the internal mouth of the ion pore (HILLE 1992; STRICHARTZ 1987). In fact, permanently charged lidocaine analogs (QX-314, QX-222) do not block Na^+ channels when applied extracellularly, whereas they inhibit Na^+ current when applied intracellularly (HILLE 1992; STRICHARTZ 1987). The effects of individual drugs belonging to these three pharmacological groups have been generally interpreted under the framework of the modulated receptor hypothesis proposed by HILLE (1977) and HONDEGHEM and KATZUNG (1977). This hypothesis states that local anesthetic-like drugs, including antiarrhythmics and some anticonvulsants, bind with different affinities to the different conformational states of the channel. In particular, drug affinity for depolarized conformations of the channel (open and inactivated) is much higher than for hyperpolarized conformations (rested). High-affinity local anesthetic binding induces additional nonconducting states stabilizing the channel in its inactivated state (VEDANTHAM and CANNON 1999). Local anesthetics and related drugs inhibit Na^+ current by blocking Na^+ channels, whereas toxins such as veratridine, aconitine, grayanotoxin, and batrachotoxin increase the influx of Na^+ ions into the cell by causing persistent activation of sodium channels as a consequence of their specific binding to the so-called neurotoxin receptor site 2 (CATTERALL 1980). Batrachotoxin (BTX) is a highly hydrophobic steroidal alkaloid obtained from the skin secretions of South American frogs of the genus *Phyllobates terribilis* (DALY et al. 1980). Binding of BTX to sodium channels shifts the voltage dependence of sodium channel activation in the negative direction, eliminates almost completely fast and slow inactivation, and alters ion selectivity (WANG and WANG 1994), thereby inhibiting the inactivation process. Binding of BTX is strongly dependent on the state of the Na^+ channel with a strong preference for the open state, and it is allosterically inhibited by binding of tetrodotoxin (BROWN 1986) or local anesthetic-like drugs (LINFORD et al. 1998).

The cloning and expression of the Na^+ channel have enabled more detailed structure-function studies of the molecular determinants of Na^+ channel blockade by local anesthetic drugs. In fact, several studies using alanine-scanning mutagenesis have identified the amino acids F1764 and Y1771 in segment S6 at domain IV (D4-S6) in the α subunit of the rat brain IIA Na^+ channel as important molecular determinants of binding of local anesthetics (RAGSDALE et al. 1994). Moreover, these amino acids are also involved in binding of class I antiarrhythmic and anticonvulsant drugs, which strongly suggest that they share a common receptor site (RAGSDALE et al. 1994, 1996). Since these residues are oriented on the same face of the α helix, it was suggested that they face the channel pore and interact with local anesthetics molecules through hydrophobic or cation-π electron interactions (RAGSDALE et al. 1994). It has also been reported that BTX binds specifically to the S6 segment at domain I of the α subunit of the Na^+ channel, involving an isoleucine, an asparagine and a leucine at positions I433, N434 and L437, respectively, in the skeletal muscle μ1-Na^+ channel (WANG and WANG 1998). Indeed, their replacement by lysine renders voltage-gated Na^+ channels resistant to BTX (WANG and WANG 1998; LINFORD et al. 1998). The potency of bupivacaine enantiomers was modified in mutant μ1-Na^+ channels at the position N434, an amino acid involved in BTX binding. In fact, different amino acid substitutions at this position led to an increase or to a decrease in bupivacaine potency that depended on the physicochemical characteristics of the amino acid introduced. Thus, the potency of bupivacaine enantiomers was increased in N434 mutant μ1-Na^+ channels in which the native asparagine had been substituted by amino acids containing an aromatic group (phenylalanine, tryptophan, or tyrosine), a polar group (cysteine) or a negative charge (aspartic), and decreased in a mutation containing a positive charge (lysine) (WRIGHT et al. 1998; NAU et al. 1999a). All these results suggest that segments D1-S6 and D4-S6 align adjacently along the pore of the Na^+ channel, forming a domain-interface site for binding of BTX and of local anesthetics in close proximity. Finally, it has been shown that the selectivity filter of the channel is involved in interactions with the hydrophilic part of the drugs, and that it normally limits extracellular access to and escape from their binding site just within the selectivity filter. Indeed, mutations at the selectivity filter of Na^+ channel allowed the access of quaternary lidocaine derivatives to the receptor site at the internal mouth of the ion pore when applied extracellularly (SUNAMI et al. 1997; QU et al. 1995).

Most antiarrhythmic drugs and local anesthetics are racemic mixtures. Although there have been reported several cases of stereoselective interactions between anticonvulsant or antiarrhythmic drugs with Na^+ channels (TRICARICO et al. 1991; NORRIS and KING 1997; SMALLWOOD et al. 1989; KOHLHARDT and FICHTNER 1988; GÖDICKE et al. 1992), stereoselective blockade of both cardiac and nerve Na^+ channels have been better characterized with

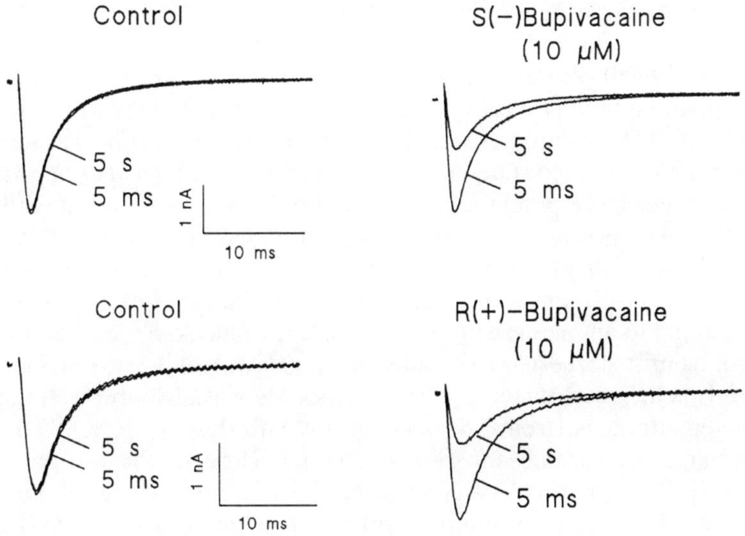

Fig. 6. Differential effects of 10 μM (*S*)-(–)-bupivacaine and (*R*)-(+)-bupivacaine on cardiac Na^+ currents. Note that the block induced by (*R*)-(+)-bupivacaine is higher than that observed in the presence of (*S*)-(–)-bupivacaine. In all panels, two current traces elicited after depolarizing the membrane potential from –120 to –20 mV are shown: one elicited after the application of a 5-ms conditioning pulse to 0 mV and a second one elicited after applying a 5-s conditioning pulse to 0 mV. Under control conditions there were not any noticeable changes in the magnitude of the current measured after a conditioning depolarizing pulse of 5 ms or 5 s. In the presence of either bupivacaine enantiomer, there was a higher decrease after long (5 s) than after short (5 ms) conditioning pulses (block of the inactivated state), and this effect was more evident in the presence of (*R*)-(+)-bupivacaine

local anesthetics such as RAC109, bupivacaine or cocaine (Clarkson 1989; Wang and Wang 1992; Valenzuela et al. 1995a; Nau et al. 1999a,b). The blockade of the rested or open states of unmodified Na^+ channels induced by these therapeutic agents has been reported to be nonstereoselective, whereas a weakly stereoselective blockade of the inactivated state of these ion channels has been observed for the local anesthetics bupivacaine, tocainide, and RAC109. In general, one enantiomer of these drugs was about 1.5 times more efficient in blocking Na^+ channels than the other (Valenzuela et al. 1995a; Clarkson 1989) (Fig. 6). In contrast to unmodified Na^+ channels, it has been demonstrated that the affinity of batrachotoxin-modified Na^+ channels for (*S*)-(–)-bupivacaine is 30 times higher than for (*R*)-(+)-bupivacaine. In other words, modified Na^+ channels exhibit a greater and opposite stereoselective blockade compared to the unmodified ones (Wang and Wang 1992; Valenzuela et al. 1995a; Nau et al. 1999a,b); this suggests that in BTX-modified channels, the chiral part of bupivacaine may be oriented toward the BTX binding site or toward sites that are allosterically changed in the presence of BTX (Nau et al. 1999a).

There exists only one study in which the molecular determinants of stereoselective interactions of local anesthetics and Na^+ channels have been studied (NAU et al. 1999a). The effects of bupivacaine enantiomers were analyzed on several point mutations at N434 in the S6 at the domain I of the μ1 Na^+ channel (involved in BTX binding). A weakly stereoselective interaction was observed between the inactivated state of wild type channels, with (*R*)-(+)-bupivacaine being 1.5 times more potent than (*S*)-(–)-bupivacaine, as reported earlier for other Na^+ channel isoforms (VALENZUELA et al. 1995a; STRICHARTZ 1987). In μ1-Na^+ channels in which N434 was replaced by an arginine (N434R), the potency of (*S*)-(–)-bupivacaine decreased more than that of (*R*)-(+)-bupivacaine, leading to an increase in the stereoselective blockade induced by bupivacaine, from a stereoselective ratio ($K_{D,S(-)}/K_{D,R(+)}$) of 1.5 to 3 (NAU et al. 1999a). This increase in stereoselective blockade was also observed with the cocaine enantiomers (from 2.4 to 4) but not with those of RAC109. In both cocaine and bupivacaine, the chiral carbon is close to the tertiary amine, whereas in RAC109 it is located close to the aromatic part of the molecule, the moiety that might not interact with the residue at position N434. Additionally, the bulkier structure of RAC109 may cause steric hindrance in binding to resting and inactivated μ1-Na^+ channels. These results demonstrate that N434 in D1-S6 interacts directly with the positively charged moiety of local anesthetics and also that this amino acid determines the degree of stereoselective blockade induced by local anesthetics in Na^+ channels (NAU et al. 1999a).

In addition to the weakly stereoselective effects observed in the blocking properties of local anesthetics, there are racemic drugs whose enantiomers induce opposite effects on Na^+ channels, the most characteristic being the piperazinylindole derivative DPI 201–106, with cardiotonic and antiarrhythmic properties (KOHLHARDT et al. 1986). The *R*-(+)- enantiomer of this drug prevents depolarization-induced activation of the Na^+ channel, while the *S*-(–)-enantiomer inhibits Na^+ channel inactivation (ROMEY et al. 1987; WANG et al. 1990). These agonist and antagonist effects of the enantiomers of DPI 201–106 on Na^+ channels resemble the effects of the enantiomers of 1,4-dihydropyridines on L-type Ca^{2+} channels (HOCKERMAN et al. 1997b). The agonist and antagonist effects induced by *S*-(–)-DPI 201–106 and *R*-(+)-DPI 206–106, respectively, are the consequences of a slowing of the Na^+ channel fast inactivation and of the binding to both rest and open channels, respectively (WANG et al. 1990). Radio-labeled binding experiments have demonstrated that DPI 201–106 allosterically controls binding of BTX (ROMEY et al. 1987).

As pointed out at above, stereoselective interactions can have important clinical consequences (ÅBERG 1972; ARIËNS 1993). Indeed, weakly stereoselective interactions like those described between local anesthetics and Na^+ channels can lead to dramatic differences in cardiac and central nervous toxicity (ÅBERG 1972; LUDUENA et al. 1972). A good representative example is bupivacaine, with (*R*)-(+)-bupivacaine being only 1.6 times more potent in blocking cardiac Na^+ channels than (*S*)-(–)-bupivacaine, but with significant differences in cardiotoxicity, the *S*-(–)- enantiomer being significantly safer than the *R*-(+)- one (GRISTWOOD et al. 1994). Moreover, (*S*)-(–)-bupivacaine is not only less cardiotoxic than the racemic mixture, but its local anesthetic action is longer than that of the racemate. These lines of evidence together with other findings on other cardiac ion channels (see Sect. E.) have led recently to the marketing of (*S*)-(–)-bupivacaine as a less cardiotoxic alternative to racemic bupivacaine and even to ropivacaine (the *S*-(–)- enantiomer of *N*-propyl-pipecoloxilidide) (GRISTWOOD et al. 1994; VALENZUELA et al. 1995b, 1997).

D. Stereoselective Interactions Between Calcium Channel Antagonists and L-Type Ca^{2+} Channels

At least five different types of voltage gated Ca^{2+} channels exist in electrically excitable mammalian cells. Only one type, L-type Ca^{2+} channels, contains high-affinity binding domains within their α_1-subunits for different chemical classes of drugs (Ca^{2+} channel antagonists, exemplified by verapamil, diltiazem, and nifedipine). Their stereoselective, high-affinity binding induces blockade of L-type Ca^{2+} channels in heart and smooth muscles, resulting in antihypertensive, cardiodepressive, and antiarrhythmic effects. Amino acids involved in drug binding have been identified recently using photoaffinity labeling, chimeric α_1 subunits and site-directed mutagenesis (STRIESSNIG et al. 1998). The α_1 subunits confer the characteristic pharmacological and functional properties of Ca^{2+} channels, although their function is modulated by the association with auxiliary subunits (CATTERALL 1994). All L-type Ca^{2+} channels share a common pharmacological profile, with a high sensitivity to several classes of drugs that have been used as selective L-type Ca^{2+} channel blockers. The three main classes of L-type-selective Ca^{2+} channel blockers currently available for clinical use are phenylalkylamines, benzothiazepines, and dihydropyridines.

Verapamil is the prototype phenylalkylamine and is the only drug of this class currently available for clinical use. It is used as an antihypertensive and antiarrhythmic drug (TRIGGLE 1991). Phenylalkylamines have a chiral center, their levorotatory enantiomers being more potent in blocking L-type Ca^{2+} channels than their dextrorotatory enantiomers (HOCKERMAN et al. 1997a). Analysis of the effects of phenylalkylamines on chimeras and site-directed mutant L-type Ca^{2+} channels have demonstrated that three amino acids at the S6 segment of domain IV (Y1463, A1467 and I1470) (YAI motif), and four residues at IIIS6 (Y1152, F1164, V1165 and I1153), are molecular determinants of phenylalkylamines binding at L-type Ca^{2+} channels (HOCKERMAN et al. 1995, 1997c). These results indicate that the receptor site for phenylalkylamines at L-type Ca^{2+} channels is located in both domains III and IV, i.e., at the interface between these domains (HOCKERMAN et al. 1997a; STRIESSNIG et al. 1998). Although L-type Ca^{2+} channels are more sensitive to phenylalkylamines, other types of Ca^{2+} channels (DIOCHOT et al. 1995), as well as Na^+ and K^+ channels are also sensitive to these drugs (Galper and Catterall 1979; Rauer and Grissmer 1999; Zhang et al. 1999). More important, mutagenesis studies have suggested that M395 in the P-loop and G421 at S6 in Kv1.3 channels are involved in verapamil binding to Kv1.3 channels, and represent the residues homologous to those involved in verapamil binding at L-type Ca^{2+}channels, and also to those involved in the binding of local anesthetics in Na^{+} channels (Rauer and Grissner 1999).

Diltiazem is the prototype benzothiazepine and is the only clinically used drug of this class (Triggle 1991). Benzothiazepines contain two adjacent chiral centers, and thus, exist as four stereoisomers, the (+)-cis form one being the most potent (Floyd et al. 1992). In contrast to phenylalkylamines, benzothiazepines bind to an external domain of the L-type Ca^{2+} channel. In fact, the extracellular application of their quaternary derivatives induces more similar blocking effects than the application of uncharged compounds. However, the intracellular application of either agent (quaternary or tertiary ammonium derivative) did not result in a significant reduction of the L-type Ca^{2+} current (Hering et al. 1993). The receptor site for benzothiazepines in L-type Ca^{2+} channels seems to overlap with that reported for phenylalkylamines. The importance of the IVS6 amino acid residues Y1463, A1467, and I1470 for benzothiazepine blockade has been demonstrated, since the transfer of these three amino acids to a P-type or a Q-type channel (less sensitive to diltiazem than L-type Ca^{2+} channels) rendered P- or Q-type channels with the same sensitivity to diltiazem than L-type Ca^{2+} channels (Hering et al. 1993).

Dihydropyridines (DHP) are allosteric modulators that may act on L-type Ca^{2+} channels as agonists or antagonists (McDonald et al. 1994). Mutations of residues that differ between L-type and non-L-type Ca^{2+} channels have revealed multiple amino acids in IIIS5, IIIS6, and IVS6 transmembrane segments involved in the high-affinity binding of DHP agonists and antagonists to L-type Ca^{2+} channels (Mitterdorfer et al. 1996; Peterson et al. 1997). Thus, it is proposed that DHP bind to a single site at the interface of domains III and IV (Hockerman et al. 1997b). Mutagenesis studies have concluded that all amino acids responsible for the differences in DHP sensitivity between L-type and non-L-type Ca^{2+} channels are located within domains III and IV including IIIS5, IIIS6, and IVS6 as well as at the linker IVS5-IVS6 (Hockerman et al. 1997a).

Binding of DHPs induces a conformational change of the Ca^{2+} channel that stabilizes either its open or inactivated state. In fact, the DHP receptor site can exist in at least two different affinity states, and the high-affinity state for antagonists is stabilized by Ca^{2+} binding (Striessnig et al. 1998). A recent study in which a total of nine amino acids from L-type α_1 subunits were substituted into the S5 and S6 transmembrane segments of domain III and the S6 transmembrane segment of domain IV of the P/Q-type α_{1A} subunit rendered a chimeric α subunit that exhibits a high-affinity receptor site for DHP that responds appropriately to both DHP agonists and antagonists and is stereoselective like the native DHP receptor of L-type Ca^{2+} channels (Hockerman et al. 1997b). Moreover, in these chimeras, DHP agonists inhibit high-affinity binding of DHP antagonists, consistent with the notion that both agonist and antagonist effects are mediated by the binding of the drugs to the same receptor site that displays stereoselectivity (HOCKERMAN et al. 1997b). Therefore, molecular differences in the interactions of these two classes of drugs with the same receptor site must be responsible for the dramatic differences in their pharmacological effects.

Phenylalkylamines, benzothiazepines, and DHP interact with receptor sites located at the IIIS6 and IVS6 transmembrane segments and are thus well positioned to alter protein-protein interactions at the interface between the allosterically coupled domains III and IV (HOCKERMAN et al. 1997a). Unfortunately, no study has examined the molecular determinants of stereoselective blockade induced by these three classes of calcium antagonists. The study of the structural basis for the opposite, stereoselective effects of DHP agonists and antagonists on L-type Ca^{2+} channels will provide further information on the mechanism of action of these drugs and on the interactions between these Ca^{2+} channel domains in channel gating.

E. Stereoselective Interactions Between Local Anesthetics and K^+ Channels

Unlike Na^+ channels, whose activation induces a transient inward current with similar biophysical properties among the different isoforms, voltage-gated K^+ channels constitute the more diverse family of ion channels. In fact, in cardiac tissues only, more than 12 K^+ currents have been described (DEAL et al. 1996; SNYDERS 1999). K^+ currents can be classified in transient outward, delayed rectifiers, and inward rectifiers (Fig. 5). This chapter will focus on transient and delayed rectifiers, whose cloned counterparts belong to those with six transmembrane segments (6Tm-1P). Voltage-gated K^+ channels (including Kv, HERG, and KvLQT1) represent the molecular site of action of class III antiarrhythmic drugs, which act by prolonging the cardiac action potential and, as a consequence, the refractory period (VAUGHAN WILLIAMS 1984). This lengthening of the refractoriness of cardiac tissue can, theoretically, lead to the cessation of reentrant arrhythmias (RODEN 1993). However, most class I and class IV antiarrhythmic drugs (such as quinidine, verapamil, and nifedipine) (SNYDERS et al. 1992; RAMPE et al. 1993; ZHANG et al. 1997) also exhibit class III properties, and, thus, most of them also bind to one or more K^+ channels (SNYDERS et al. 1992; DELPÓN et al. 1995, 1996; VALENZUELA et al. 1996; FRANQUEZA et al. 1998; CARMELIET and MUGABWA 1998). K^+ channels are widely expressed in all living cells. Depending on their localization, they play very different physiological roles, from excitability to secretion functions. The involvement of Kv1.3 channels in the immune response has even been described following the activation of T-cells (CHANDY et al. 1993). These channels may thus became molecular targets for the development of candidate immunosuppressant drugs (RAUER and GRISSMER 1999).

Like Na^+ channels which represent the molecular site of action of local anesthetic-like drugs but are also sensitive to very different compounds, K^+ channels are effectively blocked by a highly heterogeneous range of agents which do not belong to a unique pharmacological family. As a direct consequence of the differences between K^+ channels (and currents) (Fig. 5), each type of K^+ channel exhibits a higher affinity for a given drug structure. Since there are more pharmacological studies and more structure-function data for Kv1.5, KvLQT1, and HERG, this chapter will focus on the pharmacology of these three types of K^+ channels. It is worthwhile noting that these are involved in hereditary diseases, such as the long QT syndrome, and in acquired diseases such as atrial fibrillation (Ashcroft 1999).

The slowly activating delayed rectifier K^+ current (I_{Ks}) is generated by the activation of KvLQT1 channels modulated by the β-subunit minK (Sanguinetti et al. 1996), and is involved in the repolarization of the cardiac action potential. Until now, only one specific blocker of this current is available, chromanol 293B, which induces a use-dependent blockade of I_{Ks} in human ventricular myocytes (Bosch et al. 1998). The blocking effects induced by chromanol 293B on I_{Ks} are due to its actions on KvLQT1 and not on the minK regulatory subunit, since potency of chromanol 293B in blocking KvLQT1 and KvLQT1+minK is similar (Loussouarn et al. 1997). Although it is suggested that this drug binds to an internal site in the channel, the molecular determinants of its binding are unknown. Only one report exists to show stereoselective interactions between a drug and KvLQT1 channels. A new benzodiazepine derivative [L-364,373[(3-*R*)-1,3-dihydro-5-(2-fluorophenyl)-3-(1*H*-indol-3-ylmethyl)-1-methyl-2*H*-1,4-benzodiazepin-2-one] interacts with KvLQT1 channels in a stereoselective manner, such that the (*R*)- enantiomer activates the current, whereas the (*S*)- enantiomer blocks KvLQT1 channels (Salata et al. 1998). As in the case of chromanol 293B, the effects of this new agent are only due to its interactions with KvLQT1 channels, but, unfortunately, the molecular determinants of such stereoselective effects are still unknown.

The rapidly activating delayed rectifier K^+ current (I_{Kr}) results from the activation of HERG channels modulated by the MirP regulatory subunit (Sanguinetti and Jurkiewicz 1990; Curran et al. 1995; Abbott et al. 1999). This current is characterized by its inward rectification, its sensitivity to micromolar concentrations of La^{3+}, and by its specific blockade caused by methanesulfonanilide drugs like E4031 and dofetilide (Sanguinetti and Jurkiewicz 1990). These drugs bind to HERG channels near the internal mouth of the channel pore; a serine at the S5-S6 linker (S620) as well as intact C-type inactivation are crucial for binding dofetilide-like drugs at HERG channels (Wang et al. 1997; Ficker et al. 1998). In fact, the mutation S620T, which abolishes high-affinity dofetilide blockade of HERG channels, has been correlated with a loss of C-type inactivation. Unfortunately, no study exists to show the effects of enantiomers on HERG channels.

The ultrarapid delayed rectifier K^+ current (I_{Kur}) results from the activation of Kv1.5 channels that are abundantly expressed in human atria (Wang et al. 1993). These K^+ channels are sensitive to most class I antiarrhythmic drugs and local anesthetics (SNYDERS et al. 1992; DELPÓN et al. 1996; VALENZUELA et al. 1996; FRANQUEZA et al. 1998; CARMELIET and MUGABWA 1998). More importantly, the concentrations needed to inhibit these channels are in the same range as those required to block Na^+ channels (SNYDERS et al. 1990, 1992; VALENZUELA et al. 1995a,b). All these compounds, as well as internal TEA, are open channel blockers that bind to a common internal receptor site at Kv1.5 channels. Mutagenesis studies have revealed that quinidine binds to a receptor site formed by amino acids located in the S6 segment, in particular by T505 and V512 (YEOLA et al. 1996), in such a way that binding of quinidine to Kv1.5 channels requires one polar interaction at position 505 (T505) and a hydrophobic one at position 512 (V512).

Stereoselective interactions between drugs and Kv1.5 channels have been extensively studied with bupivacaine-like enantiomers (VALENZUELA et al. 1995b, 1997; FRANQUEZA et al. 1997, 1999; LONGOBARDO et al. 1998; GONZÁLEZ et al. 2001). Bupivacaine blockade of Kv1.5 channels is stereoselective with (*R*)-(+)-bupivacaine being sevenfold more potent than its enantiomer (VALENZUELA et al. 1995b) (Fig. 7). Bupivacaine blocked the open state of Kv1.5 channels in a state-, time- and voltage-dependent manner similar to quinidine. Mutagenesis studies have identified the amino acids T505, L508, and V512 in the S6 segment as molecular determinants of stereoselective blockade induced by bupivacaine and Kv1.5 channels (FRANQUEZA et al. 1997). In fact, stereoselective interactions between bupivacaine and Kv1.5 channels require a polar interaction at position 505 and two hydrophobic interactions at positions 508 and 512. When the threonine present in wild-type channels at position 505 was substituted by hydrophobic amino acids (alanine, valine, or isoleucine), stereoselective blockade was completely abolished, whereas a substitution that maintained the polar character of the native threonine (T505S),

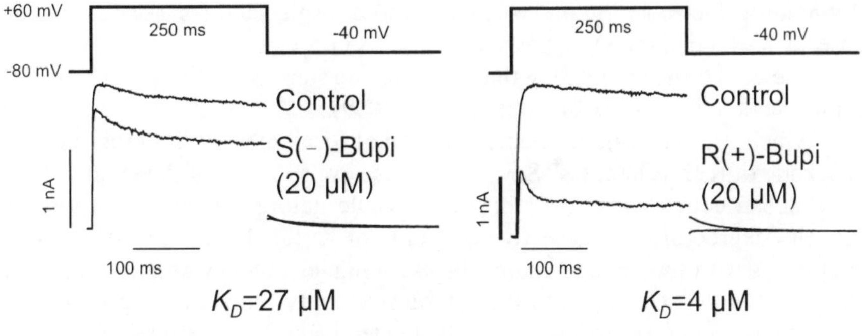

Fig. 7. Original Kv1.5 current records obtained in the absence and in the presence of either (*S*)-(–)-bupivacaine or (*R*)-(+)-bupivacaine at 20 μM. Both enantiomers induced a fast initial decline of the current that indicates an open channel block mechanism, the steady-state block being higher in the presence of (*R*)-(+)-bupivacaine

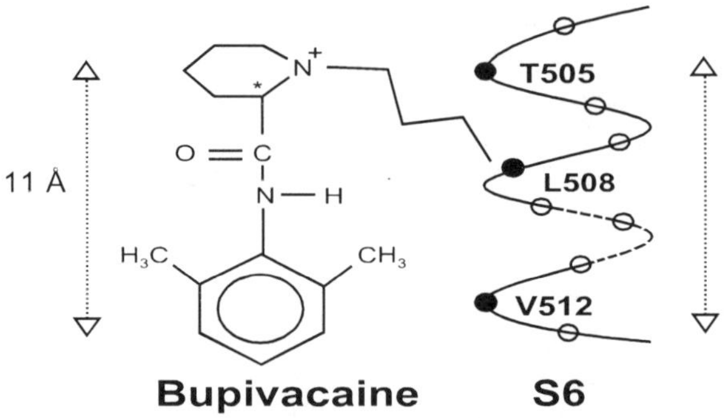

Fig. 8. Proposed model of the interaction between bupivacaine and the Kv1.5 channel. The amino acids in *black* are molecular determinants of stereoselective bupivacaine blockade. Note the similarities with the molecular determinants of local anesthetic binding site in Na^+ channels, and those of DHP in L-type Ca^{2+} channels (see Sect. F.)

altered the potency of both enantiomers but preserved stereoselectivity (FRANQUEZA et al. 1997), thus indicating that stereoselective blockade of wild-type Kv1.5 channels induced by bupivacaine involves a specific interaction between the –OH moiety of the residue at position 505 and the tertiary amino group of the drug. Replacement of L508 or V512 by a more polar methionine abolished stereoselectivity. In contrast, replacement of V512 by an alanine that preserved the hydrophobic nature of the residue, maintained stereoselectivity (FRANQUEZA et al. 1997). T505 and V512 are ~11 Å apart and bupivacaine is ~11 Å in length with the tertiary amine positively charged and hydrophobic moieties at either end (SINGH and COURTNEY 1990). A model was therefore proposed (Fig. 8) in which the positively charged end could interact with T505, and the aromatic ring could establish a hydrophobic interaction with V512. Interestingly, blockade of Kv2.1 and Kv4.3 channels that carry a valine at the T505 equivalent position and a leucine and an isoleucine (or a valine, in the case of Kv4.3 channels) at L508 and V512 Kv1.5 equivalent positions is not stereoselective, thus indicating that bupivacaine stereoselectivity displays subfamily selectivity (FRANQUEZA et al. 1997, 1999).

Which are the structural determinants of local anesthetic stereoselective blockade of Kv1.5 channels? Several studies have shown that the length of the alkyl substituent at position 1 in the molecule determines both the potency and the degree of stereoselective blockade of Kv1.5 channels, in such a way that drugs with longer side-chains display a higher potency and also a higher degree of stereoselectivity. In fact, stereoselectivity requires the presence of more than one $-CH_2$ group in the alkyl side-chain (VALENZUELA et al. 1997; LONGOBARDO et al. 1998; GONZÁLEZ et al. 2001). This hydrophobic side-chain could interact with the L508 of Kv1.5 channels, although further experiments with site-directed mutant channels at this position are needed.

F. Drug Receptor Sites at Na^+, Ca^{2+} and K^+ Channels

Local anesthetics and related antiarrhythmic and anticonvulsant drugs are state-dependent pore blockers of Na^+ channels. Analysis of their receptor binding site by site-directed mutagenesis has identified three amino acid residues in transmembrane segment IVS6 and another one in segment IS6, required for high affinity binding and effective frequency-dependent blockade of the channel (RAGSDALE et al. 1994; NAU et al. 1999a). Residues at positions I1764 and Y1771 (separated by two helical turns) would correspond to G502 and L508 in Kv1.5 channels, in which T505 and V512 represent the major molecular determinants of quinidine binding (YEOLA et al. 1996). Moreover, stereoselective bupivacaine blockade of Kv1.5 channels requires a polar interaction at position 505 and at least two hydrophobic interactions at positions 508 and 512 (FRANQUEZA et al. 1997). Thus, the local anesthetic binding site in Kv1.5 channels would be shifted one turn on the S6 helix, consistent with the differences in the values of fractional electrical distance (~0.5–0.7 in Na^+ channels versus ~0.2 in Kv1.5 channels).

The amino acid residues identified in Na^+ channels as molecular determinants of local anesthetic binding are in analogous positions to the YAI motif in the IVS6 segment of L-type Ca^{2+} channels, which is required for high affinity binding of phenylalkylamines and diltiazem (HOCKERMAN et al. 1995). Therefore, all major drugs that block Na^+, K^+, and Ca^{2+} channels in a state-dependent manner interact with the IVS6 segments of these channels. This is a remarkable confluence of binding sites for drugs with a wide range of structures. Further elucidation of the detailed mechanisms of interaction of these drugs with their receptor sites may reveal the common mechanisms through which they affect Na^+, K^+, and Ca^{2+} channel function.

G. Future Directions

During the last decade, enormous progress has been made in identifying binding receptor sites for toxins, local anesthetics, class III antiarrhythmic agents, and Ca^{2+} antagonists (RAGSDALE et al. 1994; FRANQUEZA et al. 1997; HOCKERMAN et al. 1997a; FICKER et al. 1998). These studies have led to a better understanding of the interactions between drugs and ion channels. However, there are only a few studies in the literature dealing with the molecular determinants of stereoselective drug-channel effects. Because stereoselective interactions reveal three dimensional relationships between drug and channel, a better knowledge of the three-dimensional structure of receptor sites will contribute to a more rational drug design. Also, identifying such stereospecific interactions may have important clinical consequences. In fact, since 1992, the US Food and Drug Administration (FDA) and the European Committee for Proprietary Medicinal Products have required manufacturers to research and characterize each enantiomer in all drugs proposed to be marketed as a mixture. Hopefully, the knowledge of the crystal structure of ion channels, which began with the first crystal structure of the KcsA K^+ channel (DOYLE et al. 1998; GULBIS et al. 1999), will help molecular modeling techniques to improve our understanding of ion channel-drug interactions.

List of Abbreviations

TEA tetraethylammonium
BTX batrachotoxin

Acknowledgements. The author thanks Teresa González and Drs. Eva Delpón, Mónica Longobardo, Angel Cogolludo, Ricardo Caballero and Juan Tamargo for their thoughtful comments on the manuscript. During the period over which this chapter was written, financial support was provided in form of research Grants (CICYT SAF98–0058 and U.S.-Spain Science & Technology Program 98131).

References

Abbott G, Sesti F, Splawski I, Buck M, Lehmann, Timothy K, Keating M, Goldstein S (1999) MirP1 forms I_{Kr} potassium channels with HERG and is associated with cardiac arrhythmia. Cell 97:175–187
Åberg G (1972) Toxicological and local anesthetic effects of optically active isomers of two local anesthetic compounds. Acta Pharmacol Toxicol 31:273–286
Ariëns E (1993) Nonchiral, homochiral and composite chiral drugs. Trends Pharmacol Sci 14:68–75
Armstrong C, Bezanilla F, Rojas E (1974) Destruction of sodium conductance inactivation in squid axons perfused with pronase. J Gen Physiol 62:375–391
Armstrong C, Hille B (1998) Voltage-gated ion channels and electrical excitability. Neuron 20:371–380
Ashcroft F (1999) Ion Channels and Disease. Academic Press, San Diego
Bennett P, Valenzuela C, Chen L-Q, Kallen R (1995) On the molecular nature of the lidocaine receptor of cardiac Na^+ channels: Modification of block by alterations in the α-subunit III-IV interdomain. Circ Res 77:584–592
Betz H (1990) Ligand-gated ion channels in the brain: The amino acid receptor superfamily. Neuron 5:383–392
Bezanilla F, Armstrong C (1977) Inactivation of the sodium channel. I. Sodium current experiments. J Gen Physiol 70:549–596
Bezanilla F, Stefani E (1998) Gating currents. Methods Enzymol 293:331–352
Bosch R, Gaspo R, Busch A, Lang H, Li G-R, Nattel S (1998) Effects of chromanol 293B, a selective blocker of the slow, component of the delayed rectifier K^+ current, on repolarization in human and guinea pig ventricular myocytes. Cardiovasc Res 38:441–450
Brown G (1986) 3*H*-batrachotoxinin-A benzoate binding to voltage-sensitive sodium channels: inhibition by the channel blockers tetrodotoxin and saxitoxin. J Neurosci 6:2065–2070
Carmeliet E, Mugbawa K (1998) Antiarrhythmic drugs and cardiac ion channels: mechanisms of action. Prog Biophys Mol Biol 70:1–72
Catterall W (1980) Neurotoxins that act on voltage-sensitive sodium channels in excitable membranes. Annu Rev Pharmacol Toxicol 20:15–43
Catterall W (1988) Structure and function of voltage-sensitive ion channels. Science 242:50–61
Catterall W (1994) Molecular properties of a superfamily of plasma-membrane cation channels. Curr Opin Cell Biol 6:607–615

Chandy K, Gutman G, Grissmer S (1993) Physiological role, molecular structure and evolutionary relationships of voltage-gated potassium channels in T lymphocytes. Sem Neurosci 5:125–134

Choi K, Mossman C, Aube J, Yellen G (1993) The internal quaternary ammonium receptor site of *Shaker* potassium channels. Neuron 10:533–541

Clarkson C (1989) Stereoselective block of cardiac sodium channels by RAC109 in single guinea-pig ventricular myocytes. Circ Res 65:1306–1323

Curran M, Splawski I, Timothy K, Vincent G, Green E, Keating M (1995) A molecular basis for cardiac arrhythmia: HERG mutations cause long QT syndrome. Cell 80:795–803

Curtis B, Catterall W (1984) Purification of the calcium antagonist receptor of the voltage-sensitive calcium channel from skeletal muscle transverse tubules. Biochemistry 23:2113–2118

Daly J, Myers C, Warnick J (1980) Levels of batrachotoxin and lack of sensitivity to its action in poison-dart frogs (*Phyllobates*). Science 208:1383–1385

Deal K, England S, Tamkun M (1996) Molecular physiology of cardiac potassium channels. Physiol Rev 76:49–76

DeLeon M, Wang Y, Jones L, Pérez-Reyes E, Wei X, Soong T, Snutch T, Yue D (1995) Essential Ca^{2+}-binding motif for Ca^{2+}-sensitive inactivation of L-type Ca^{2+} channels. Science 270:1502–1506

Del Camino D, Holmgren M, Liu Y, Yellen G (2000) Blocker protection in the pore of a voltage-gated K^+ channel and its structural implications. Nature 403:321–325

Delpón E, Valenzuela C, Pérez O, Casis O, Tamargo J (1995) Propafenone preferentially blocks the rapidly activating component of delayed rectifier K^+ current in guinea pig ventricular myocytes. Voltage-independent and time-dependent block of the slowly activating component. Circ Res 76:223–235

Delpón E, Valenzuela C, Pérez O, Franqueza L, Gay P, Snyders D, Tamargo J (1996) Mechanisms of block of a human cloned potassium channel by the enantiomers of a new bradycardic agent: *S*-16257–2 and *S*-16260–2. Br J Pharmacol 117:1293–1301

Diochot S, Richard S, Baldy-Moulinier M, Nargeot J, Valmier J (1995) Dihydropyridines, phenylalkylamines and benzothiazepines block *N*-, P/Q- and R-type calcium currents. Pflügers Arch 431:10–19

Dolly J, Parcej D (1996) Molecular properties of voltage-gated K^+ channels. J Bioenerg Biomembr 28:231–253

Doyle D, Cabral J, Pfuetzner R, Kuo A, Gulbis J, Cohen S, Chait B, MacKinnon R (1998) The structure of the potassium channel: molecular basis of K^+ conduction and selectivity. Science 280:69–77

Ficker E, Jarolimek W, Kiehn J, Baumann A, Brown A (1998) Molecular determinants of dofetilide block of HERG K^+ channels. Circ Res 82:386–395

Floyd D, Kimball S, Krapcho J, Das J, Turk C (1992) Benzazepinone calcium channel blockers. 2. Structure-activity and drug metabolism studies leading to potent antihypertensive agents. Comparison with benzothiazepinones. J Med Chem 35:756–772

Franks N, Lieb W (1994) Stereospecific effects of inhalation general anesthetic optical isomers on nerve ion channels. Science 254:427–430

Franqueza L, Longobardo M, Vicente J, Delpón E, Tamkun M, Tamargo J, Snyders, Valenzuela C (1997) Molecular determinants of stereoselective bupivacaine block of hKv1.5 channels. Circ Res 81:1053–1064

Franqueza L, Valenzuela C, Delpón E, Longobardo M, Caballero R, Tamargo J (1998) Effects of propafenone and 5-hydroxy-propafenone on hKv1.5 channels. Br J Pharmacol 125:969–978

Franqueza L, Valenzuela C, Eck J, Tamkun M, Tamargo J, Snyders D (1999) Functional expression of an inactivating potassium channel (Kv4.3) in a mammalian cell line. Cardiovasc Res 41:212–219

Galper J, Catterall W (1979) Inhibition of sodium channels by D600. Mol Pharmacol 15:174–178

Gödicke J, Herzig S, Mescheder A, Steinke F (1992) Enantioselectivity of asocainol studied at different conditions: a novel approach to check the feasibility of molecular models of antiarrhythmic action. Naunyn-Schmiedeberg's Arch Pharmacol 346:345–351

Gomez-Lagunas F, Armstrong C (1995) Inactivation in ShakerB K^+ channels: a test for the number of inactivating particles on each channel. Biophys J 68:89–95

González T, Longobardo M, Caballero R, Delpón E, Sinisterra JV, Tamargo J, Valenzuela C (2001) Stereoselective effects of the enantiomers of a new local anesthetic, IQB-9302, on a human cardiac potassium channel (Kv1.5). Br J Pharmacol 132:385–392

Gristwood R, Bardsley H, Baker H, Dickens J (1994) Reduced cardiotoxicity of levobupivacaine compared with racemic bupivacaine (Marcaine): new clinical evidence. Exp Opin Invest Drugs 3:1209–1212

Gulbis J, Mann S, MacKinnon R (1999) Structure of a voltage-dependent K^+ channel β subunit. Cell 97:943–952

Hamill O, Marty A, Neher E, Sakmann B, Sigworth F (1981) Improved patch-clamp techniques for high-resolution current recording from cells and cell-free membrane patches. Pflügers Archiv Eur J Physiol 391:85–100

Hartmann H, Tiedeman A, Chen S, Brown A, Kirsch G (1994) Effects of III-IV linker mutations on human heart Na^+ channel inactivation gating. Circ Res 75:114–122

Heinemann S, Terlau H, Stühmer W, Imoto K, Numa S (1992) Calcium channel characteristics conferred on the sodium channel by single mutations. Nature 356: 441–443

Heginbotham L, Abramson T, MacKinnon R (1992) A functional connection between the pores of distantly related ion channels as revealed by mutant K^+ channels. Science 258:1152–1155

Hering S, Savchenko A, Strubing C, Lakitsch M, Striessnig J (1993) Extracellular localization of the benzothiazepine binding domain of L-type Ca^{2+} channels. Mol Pharmacol 43:820–826

Herlitze S, Hockerman G, Scheuer T, Catterall W (1997) Molecular determinants of inactivation and G protein modulation in the intracellular loop connecting domains I and II of the calcium channel alpha 1A subunit. Proc Natl Acad Sci USA 94:1512–1516

Hille B (1977) Local anesthetics: Hydrophilic and hydrophobic pathways for the drug-receptor reaction. J Gen Physiol 69:497–515

Hille B (1992) Ionic Channels of Excitable Membranes. Second edition. Sinauer Associates Inc., Sunderland, MA, USA

Hockerman G, Johnson B, Scheuer T, Catterall W (1995) Molecular determinants of high affinity phenylalkylamine block of L-type calcium channels. J Biol Chem 270:22119–22122

Hockerman G, Peterson B, Johnson B, Catterall W (1997a). Molecular determinants of drug binding and action on L-type calcium channels. Annu Rev Pharmacol Toxicol 37:361–396

Hockerman G, Peterson B, Sharp E, Tanada T, Sheuer T, Catterall W (1997b) Construction of a high-affinity receptor site for dihydropyridine agonists and antagonists by single amino acid substitutions in a non-L-type Ca^{2+} channel. Proc Natl Acad Sci USA 94:14906–14911

Hockerman G, Johnson B, Abbott M, Scheuer T, Catterall W (1997c) Molecular determinants of high affinity phenylalkylamine block of L-type calcium channels in transmembrane segment IIIS6 and the pore region of the alpha1 subunit. J Biol Chem 272:18759–18765

Hodgkin A, Huxley A (1952) A quantitative description of membrane current and its application to conduction and excitation in nerve. J Physiol (Lond.) 117:500–544

Hondeghem L, Katzung B (1977) Time- and voltage-dependent interactions of antiarrhythmic drugs with cardiac sodium channels. Biochim Biophys Acta 472:373–398

Hoshi T, Zagotta W, Aldrich R (1990) Biophysical and molecular mechanisms of *Shaker* potassium channel inactivation. Science 250:533–538

Hoshi T, Zagotta W, Aldrich R (1991) Two types of inactivation in *Shaker* K^+ channels: effects of alterations in the carboxy-terminal region. Neuron 7:547–556
Isacoff E, Jan Y, Jan L (1991) Putative receptor for the cytoplasmic inactivation gate in the *Shaker* K^+ channel. Nature 353:86–90
Kallen R, Cohen S, Barchi R (1993) Structure, function and expression of voltage-dependent sodium channels. Mol Neurobiol 7:383–428
Kohlhardt M, Fröbe U, Herzig J (1986) Modification of single cardiac Na^+ channels by DPI 201–106. J Membr Biol 89:163–172
Kohlhardt M, Fichtner H (1988) Inhibitory effects of diprafenone stereoenantiomers on cardiac Na^+ channels. Eur J Pharmacol 156:55–62
Linford N, Cantrell A, Qu Y, Sheuer T, Catterall W (1998) Interaction of batrachotoxin with the local anesthetic receptor site in transmembrane segment IVS6 of the voltage-gated sodium channel. Proc Natl Acad Sci USA 95:13947–13952
Longobardo M, Delpón E, Caballero R, Tamargo J, Valenzuela C (1998) Structural determinants of potency and stereoselective block of hKv1.5 channels induced by local anesthetics. Mol Pharmacol 54:162–169
Loussouarn G, Charpentier F, Mohammad-Panah R, Kunzelmann K, Baró I, Escande D (1997) KvLQT1 potassium channel but not IsK is the molecular target for *trans*-6-cyano-4-(*N*-ethylsulfonyl-*N*-methylamino)-3-hydroxy-2,2-dimethyl-chromane. Mol Pharmacol 52:1131–1136
Luduena F, Bogado E, Tullar B (1972) Optical isomers of mepivacaine and bupivacaine. Archiv Int Pharmacodyn Ther 200:359–369
MacKinnon R (1991) Determination of the subunit stoichiometry of a voltage-activated K channel. Nature 350:232–235
Mannuzzu L, Moronne M, Isacoff E (1996) Direct physical measure of conformational rearrangement underlying potassium channel gating. Science 271:213–216
McDonald T, Peltzer S, Trautwein W, Peltzer D (1994) Regulation and modulation of calcium channels in cardiac, skeletal and smooth muscle cells. Physiol Rev 74: 365–507
Mitterdorfer J, Wang Z, Sinnegger M, Hering S, Striessnig J, Grabner M, Glossmann H (1996) Two amino acid residues in the IIIS5 segment of L-type calcium channels differentially contribute to 1,4-dihydropyridine sensitivity. J Biol Chem 271: 30330–30335
Nau C, Wang S-Y, Strichartz G, Wang G (1999a) Point mutations at N434 in D1–S6 of μ1 Na^+ channels modulate binding affinity and stereoselectivity of local anesthetic enantiomers. Mol Pharmacol 56:404–413
Nau C, Vogel W, Hempelmann G, Bräu M (1999b) Stereoselectivity of bupivacaine in local anesthetic-sensitive ion channels of peripheral nerve. Anesthesiology 91:786–795
Noda M, Shimuzi S, Tanabe T, Takai T, Kayano T, Ikeda T, Takahashi H, Nakayama H, Kanaoka Y, Minamino N, Kangawa K, Matsuo H, Raftery M, Hirose T, Inayama S, Hayashida H, Miyata T, Numa S (1984) Primary structure of *Electrophorus electricus* sodium channel deduced from cDNA sequence. Nature 312:121–127
Norris S, King A (1997) The stereo-isomers of the anticonvulsant ARL 12495AA limit sustained repetitive and modify action potential properties of rat hippocampal neurons in Vitro. J Pharmacol Exp Ther 281:1191–1198
Papazian D, Schwarz T, Tempel B, Jan Y, Jan L (1987) Cloning of genomic and complementary DNA from *Shaker*, a putative potassium channels gene from Drosophila. Science 237:749–753
Papazian D, Timpe L, Jan Y, Jan L (1991) Alteration of voltage-dependence of *Shaker* potassium channel by mutations in the S4 sequence. Nature 349:305–310
Patton D, West J, Catterall W, Goldin A (1992) Amino acid residues required for fast Na^+-channel inactivation: charge neutralization and deletions in the III-IV linker. Proc Natl Acad Sci USA 89:10905–10909
Peterson B, Johnson B, Hockerman G, Acheson M, Scheuer T, Catterall W (1997) Analysis of the dihydropyridine receptor site of L-type calcium channels by alanine-scanning mutagenesis. J Biol Chem 272:18752–18758

Peterson B, DeMaria C, Yue D (1999) Calmodulin is the Ca^{2+} sensor for Ca^{2+}-dependent inactivation of L-type calcium channels. Neuron 22:549–558
Qu Y, Rogers J, Tanada T, Scheuer T, Catterall W (1995) Molecular determinants of drug access to the receptor site for antiarrhythmic drugs in the cardiac Na^+ channel. Proc Natl Acad Sci USA 92:11839–11843
Ragsdale D, McPhee J, Scheuer T, Catterall W (1994) Molecular determinants of state-dependent block of Na^+ channels by local anesthetics. Science 265:1724–1728
Ragsdale D, McPhee J, Scheuer T, Catterall W (1996) Common molecular determinants of local anesthetic, antiarrhythmic, and anticonvulsant block of voltage-gated Na^+ channels. Proc Natl Acad Sci USA 93:9270–9275
Rampe D, Wible B, Fedida D, Dage R, Brown A (1993) Verapamil blocks a rapidly activating delayed rectifier K^+ channel cloned from human heart. Mol Pharmacol 44:642–648
Rauer H, Grissmer S (1999) The effect of deep pore mutations on the action of phenylalkylamines on the Kv1.3 channel. Br J Pharmacol 127:1065–1074
Roden D (1993) Current status of class III antiarrhythmic drug therapy. Am J Cardiol 72:44B–49B
Roden D, George A (1997) Structure and function of cardiac sodium and potassium channels. Am J Physiol 273:H511–H525
Rogawski M, Porter R (1990) Antiepileptic drugs: pharmacological mechanisms and clinical efficacy with consideration of promising developmental stage compounds. Pharmacol Rev 42:223–286
Romey G, Quast U, Pauron D, Frelin C, Renaud J, Lazdunski M (1987) Na^+ channels as sites of action of the cardioactive agent DPI 201–106 with agonist and antagonist enantiomers. Proc Natl Acad Sci USA 84:896–900
Salata J, Jurkiewicz N, Wang J, Evans B, Orme H, Sanguinetti M (1998) A novel benzodiazepine that activates cardiac slow delayed rectifier K^+ currents. Mol Pharmacol 54:220–230
Sanguinetti M, Jurkiewicz N (1990) Two components of cardiac delayed rectifier K^+ current. Differential sensitivity to block by class III antiarrhythmic agents. J Gen Physiol 96:195–215
Sanguinetti M, Curran M, Spector P, Zou A, Shen J, Atkinson D, Keating M (1996) Coassembly of KvLQT1 and minK (IsK) to form cardiac I_{Ks} potassium channel. Nature 384:80–83
Seino S, Chen L, Seino M, Blondel O, Takeda J, Johnson J, Bell G (1992) Cloning of the alpha 1 subunit of a voltage-dependent calcium channel expressed in pancreatic beta cells. Proc Natl Acad Sci USA 89:584–588
Singh B, Courtney K (1990) The classification of antiarrhythmic mechanisms of drug action: experimental and clinical consideration. In: Zipes D, Jalife J (eds) Cardiac Electrophysiology: From Cell to Bedside. WB Saunders, Philadelphia, pp 882–897
Smallwood J, Robertson D, Steinberg M (1989) Electrophysiological effects of flecainide enantiomers in canine Purkinje fibres. Naunyn-Schmiedeberg's Arch Pharmacol 339:625–629
Snyders D, Hondeghem L (1990) Effects of quinidine on the sodium current of ventricular guinea-pig myocytes: evidence for a drug-associated rested state with altered kinetics. Circ Res 66:565–579
Snyders D, Knoth K, Roberds S, Tamkun M (1992) Time-, voltage-, and state-dependent block by quinidine of a cloned human cardiac potassium channel. Mol Pharmacol 41:322–330
Snyders D, Yeola S (1995) Determinants of antiarrhythmic drug action. Electrostatic and hydrophobic components of block of the human cardiac hKv1.5 channel. Circ Res 77:575–583
Snyders D (1999) Structure and function of cardiac potassium channels. Cardiovasc Res 42:377–390
Strichartz G (1987) Local Anesthetics. Handbook of Experimental Pharmacology. Vol 81. Springer-Verlag, Berlin

Striessnig J, Grabner M, Mitterdorfer J, Hering S, Sinnegger M, Glossmann H (1998) Structural basis of drug binding to L Ca^{2+} channels. Trend Pharmacol Sci 19: 108–115

Stühmer W, Conti F, Suzuki H, Wang X, Noda M, Yahagi N, Kubo H, Numa S (1989) Structural parts involved in activation and inactivation of the sodium channel. Nature 339:597–603

Sunami A, Dudley S, Fozzard H (1997) Sodium channel selectivity filter regulates antiarrhythmic drug binding. Proc Natl Acad Sci USA 94:14126–14131

Tanabe T, Takeshima H, Mikami A, Flockerzi V, Takahashi H, Kangawa K, Kojima M, Matsuo H, Hirose T, Numa S (1987) Primary structure of the receptor for calcium channel blockers from skeletal muscle. Nature 328:313–318

Terlau H, Heinemann S, Stühmer W, Pusch M, Conti F, Imoto K, Numa S (1991) Mapping the site of block by tetrodotoxin and saxitoxin of sodium channel II. FEBS Lett 293:93–96

Tricarico D, Fakler B, Spittelmeister W, Ruppersberg J, Stützel R, Franchini C, Tortorella V, Conte-Camerino D, Rüdel R (1991) Stereoselective interaction of tocainide and its chiral analogs with sodium channels in human myoballs. Pfügers Arch 418:234–237

Triggle D (1991) Calcium-channel drugs: structure-function relationships and selectivity of action. J Cardiovasc Pharmacol 18:S1–S6

Tsien R, Lipscombe D, Madison D, Bley K, Fox A (1995) Reflections on Ca(2+)-channel diversity, 1988–1994. Trends Neurosci 18:52–54

Valenzuela C, Bennett P (1994) Gating of cardiac Na^+ channels in excised membrane patches after modification by α-chymotrypsin. Biophys J 67:161–171

Valenzuela C, Snyders D, Bennett P, Tamargo J, Hondeghem L (1995a) Stereoselective block of cardiac sodium channels by bupivacaine in guinea pig ventricular myocytes. Circulation 92:3014–3024

Valenzuela C, Delpón E, Tamkun M, Tamargo J, Snyders D (1995b) Stereoselective block of a human cardiac potassium channel (Kv1.5) by bupivacaine enantiomers. Biophys J 69:418–427

Valenzuela C, Delpón E, Franqueza L, Gay P, Pérez O, Tamargo J, Snyders D (1996) Class III antiarrhythmic effects of zatebradine. Time-, state-, use-, and voltage-dependent block of hKv1.5 channels. Circulation 94:562–570

Valenzuela C, Delpón E, Franqueza L, Gay P, Snyders D, Tamargo J (1997) Effects of ropivacaine on a potassium channel (hKv1.5) cloned from human ventricle. Anesthesiology 86:718–728

Vassilev P, Scheuer T, Catterall W (1988) Identification of an intracellular peptide segment involved in sodium channel inactivation. Science 241:1658–1661

Vaughan Williams E (1984) A classification of antiarrhythmic actions reassessed after a decade of new drugs. J Clin Pharmacol 24:129–147

Vedantham V, Cannon S (1999) The position of the fast inactivation gate during lidocaine block of voltage-gated Na^+ channels. J Gen Physiol 113:7–16

Wang G, Dugas M, Ben-Armah I, Honerjager P (1990) Interaction between DPI 201–106 enantiomers at the cardiac sodium channel. Mol Pharmacol 37:17–24

Wang G, Wang S (1992) Altered stereoselectivity of cocaine and bupivacaine isomers in normal and batrachotoxin-modified Na^+ channels. J Gen Physiol 100:1003–1020

Wang G, Wang, S (1994) Modification of cloned brain Na^+ channels by batrachotoxin. Pfügers Archiv 427:309–316

Wang G, Wang S (1998) Point mutations in segment I-S6 render voltage-gated Na^+ channels resistant to batrachotoxin. Proc Natl Sci Acad USA 95:2653–2658

Wang S, Morales M, Liu S, Strauss H, Rasmusson R (1997) Modulation of HERG affinity for E-4031 by $[K^+]_o$ and C-type inactivation. FEBS Lett 417:43–47

Wang Z, Fermini B, Nattel S (1993) Sustained depolarization-induced outward current in human atrial myocytes. Evidence for a novel delayed rectifier K^+ current similar to Kv1.5 cloned channel currents. Circ Res 73:1061–1076

Wright S, Wang S-Y, Wang G (1998) Lysine point mutations in Na^+ channel D4–S6 reduce inactivated channel block by local anesthetics. Mol Pharmacol 54:733–739
Yang J, Ellinor P, Sather W, Zhang J-F, Tsien R (1993) Molecular determinants of Ca selectivity and ion permeation in L-type Ca channels. Nature 366:158–161
Yang N, Horn R (1995) Evidence for voltage-dependent S4 movement in sodium channels. Neuron 15:213–218
Yang N, George A, Horn R (1996) Molecular basis of charge movement in voltage-gated sodium channels. Neuron 16:113–122
Yeola S, Rich T, Uebele V, Tamkun M, Snyders D (1996) Molecular analysis of a binding site for quinidine in a human cardiac delayed rectifier K^+ channel: role of S6 in antiarrhythmic drug binding. Circ Res 78:1105–1114
Zhang J, Ellinor P, Aldrich R, Tsien R (1994) Molecular determinants of voltage-dependent inactivation in calcium channels. Nature 372:97–100
Zhang S, Zhou Z, Gong Q, Makielski C, January C (1999) Mechanism of block and identification of the verapamil binding domain to HERG potassium channels. Circ Res 84:989–998
Zhang X, Anderson J, Fedida D (1997) Characterization of nifedipine block of the human heart delayed rectifier, hKv1.5. J Pharmacol Exp Ther 281:1247–1256
Zülke R, Reuter H (1998) Ca^{2+}-sensitive inactivation of L-type Ca^{2+} channels depends on multiple cytoplasmic amino acid sequences of the α_{1C} subunit. Proc Natl Acad Sci USA 95:3287–3294
Zülke R, Pitt G, Deisseroth K, Tsien R, Reuter H (1999) Calmodulin supports both inactivation and facilitation of L-type calcium channels. Nature 399:159–162
Zygmunt A, Maylie J (1990) Stimulation-dependent facilitation of the high threshold calcium current in guinea-pig ventricular myocytes. J Physiol (Lond) 428:653–671

CHAPTER 10

Stereoselective Bioactivation and Bioinactivation – Toxicological Aspects

N.P.E. VERMEULEN

A. Introduction

Stereoisomerism manifests itself in various forms, such as those related to enantiomers, diastereoisomers, epimers, meso- and geometrical isomers. Generally speaking, the consequences of stereoisomerism for the biological action of xenobiotics in living organisms are still difficult to predict because they are complex and of multi-factorial origin. In recent years, however, significant progress has been made in rationalization and understanding of toxicity at the molecular level. In the development of a toxic effect, nowadays the following stages are usually distinguished: (a) toxicokinetics (i.e., comprising absorption, distribution, and elimination); (b) biotransformation, either resulting in bioactivation or bioinactivation; (c) reversible or irreversible interactions with cellular or tissue components; (d) protection or repair mechanisms; and (e) the nature and extent of the toxic effect for the organism. By taking these stages into consideration and by realizing the intrinsic asymmetry of receptors, enzymes, and other types of macromolecules, which leads to the phenomenon of "chiral recognition," rationalization of stereoselectivity in toxicity becomes more and more feasible.

In this chapter, first a number of examples will be presented to illustrate the relevance of stereoselectivity in the toxicity of drugs and other xenobiotics. Subsequently, a few examples will be discussed more extensively to delineate some of the underlying molecular mechanisms, usually involving bioactivation by biotransformation enzymes. Finally, some general conclusions will be drawn.

B. Toxins of Natural Origins: Complex Stereochemistry

The earliest written reports on natural toxins date from 1500 B.C. (Ebers papyrus) and a first classification, into origin from plants, animals, and minerals, was made by Dioscorides around 50 A.D.. Currently, a vast body of natural toxins has been identified and classified (GALLO et al. 1996). A majority of these toxins occur naturally as stereoisomers; some of them have extremely interesting, complex, and stereochemical structures and extremely strong

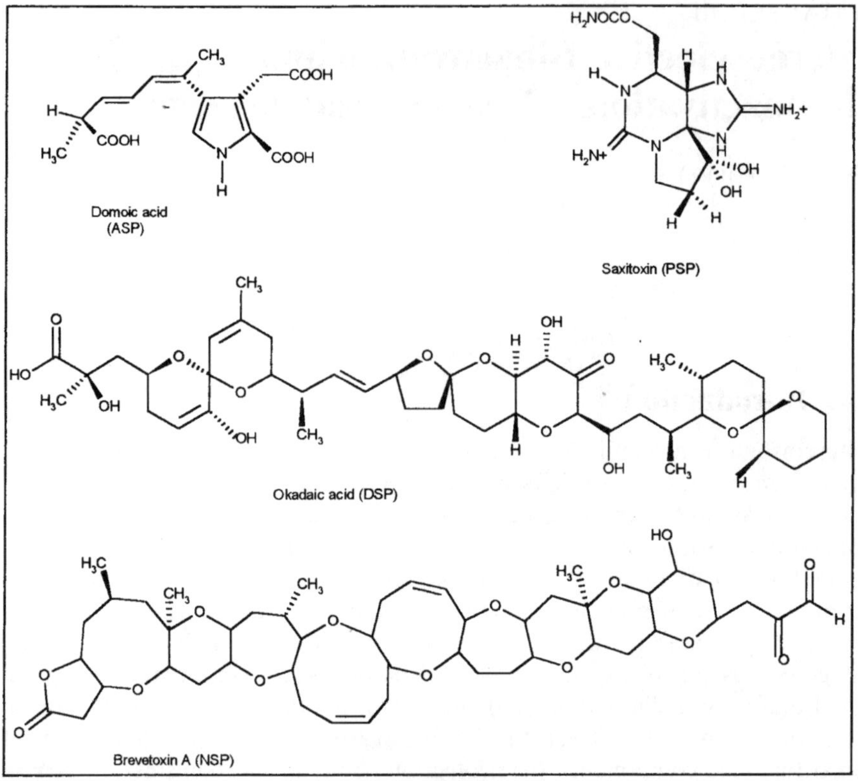

Fig. 1. Chemical structures of some shellfish-poisoning toxins illustrating the structural and stereochemical variability and complexity

and sometimes selective toxic effects (GOETZ and MEISEL 2000). In Fig. 1, structures of some paralytic shellfish-poisoning toxins are shown.

The prominent role of stereochemical factors in the toxicity of natural toxins is clearly shown in the bioactivation of pulegone, a fragrance and flavoring monoterpene constituent of pennyroyal oil. According to folklore, pennyroyal was also used as a herbal medicine to control menses and to terminate pregnancy (RIDDLE and ESTES 1992). The *S*-(−)-pulegone enantiomer was found to be significantly less hepatotoxic in mice than *R*-(+)-pulegone. The general scheme for the oxidative bioactivation via menthofuran, the major proximate-metabolite, concerns oxidative metabolism of the isopropylidene methyl group *syn*- to the ketone, via Cytochrome P450 into an *E*-allylic alcohol, followed by an intramolecular dehydration to the furan (Fig. 2). Interesting mechanistic studies with $^{18}O_2$ and D-labeling supported this mechanism (NELSON 1995). A second pathway of bioactivation was shown to involve stereoselective hydroxylation at the C-5 position to form 5-hydroxy pulegone, which gets transformed into piperitenone (Fig. 2). Recent studies of a 5-dimethyl analogue of *R*-(+)-pulegone suggest that stereoselective C5-hydroxylation also

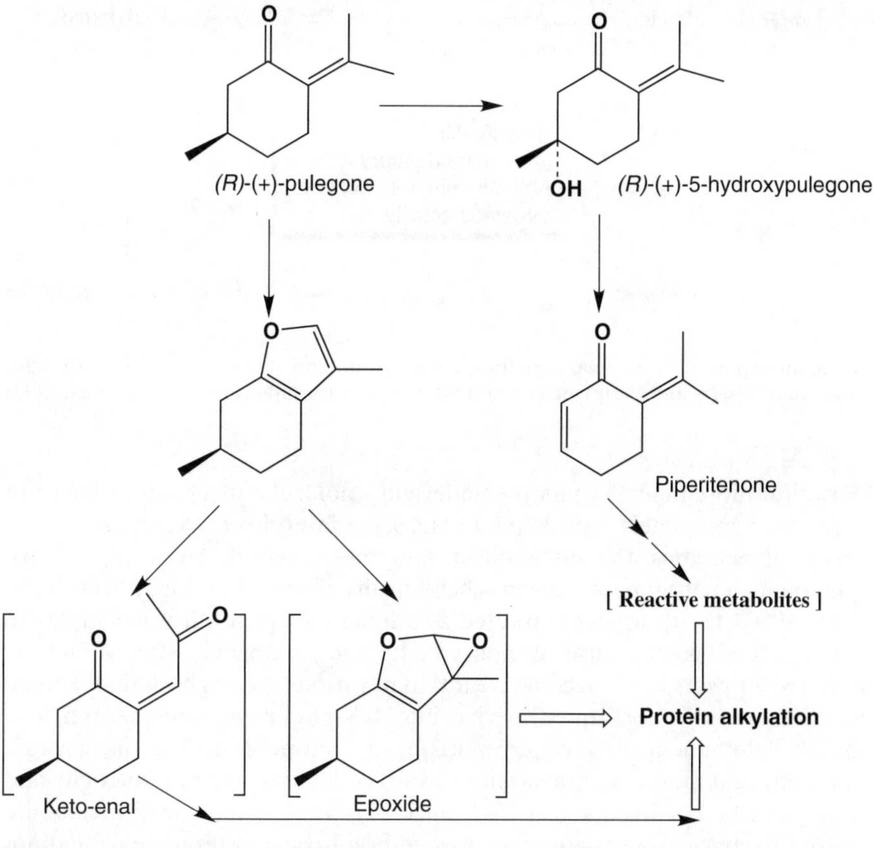

Fig. 2. General scheme for the stereoselective oxidative bioactivation of *R*-(+)-pulegone via menthofuran to a reactive γ-keto enal and/or epoxide that may alkylate proteins. Alternatively, stereoselective hydroxylation at the C-5 position ultimately leading to piperitenone occurs as a second route of bioactivation. (Adapted from NELSON 1995 and THULASIRAM et al. 2000)

may play a considerable role in *R*-(+)-pulegone-mediated hepatotoxicity (THULASIRAM et al. 2000). Based on comparative metabolic profiling and induction studies of *S*-(–)- and *R*-(+)-pulegone in the rat, differences in urinary metabolite profiles were related to the stereoselectivity in the observed hepatotoxicity of both pulegone enantiomers (MADYASTA and GAIKWAD 1998).

From the literature it becomes clear that natural compounds usually exert a very high selectivity, often due to stereoselectivity, both in terms of undesired toxic and desired biological activities.

C. Thalidomide: A Classic Example of Stereoselectivity

A classic example of toxicity-related stereoselectivity and its complexity is that of thalidomide. Despite the fact that the stereoselectivity involved has been

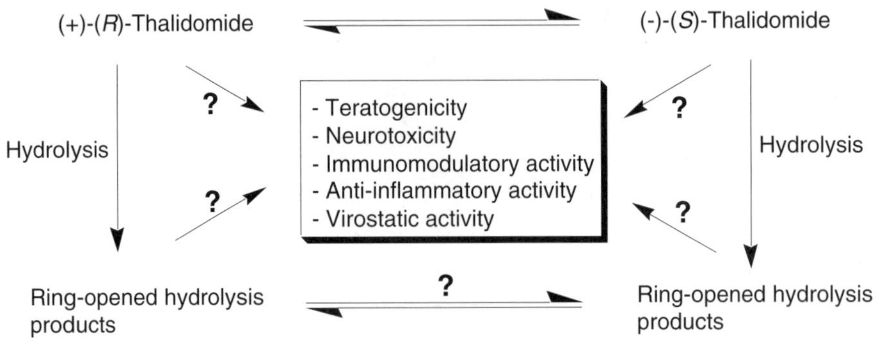

Fig. 3. Interplay of biological activities, metabolism, and stereoselectivity of thalidomide enantiomers and their ring-opened metabolites. (Adapted from REIST et al. 1998)

known for more than 25 years, the underlying molecular mechanisms have not as yet been completely elucidated. In Fig. 3, the interplay between the various biological activities, the metabolism, and the stereoselectivity of the two thalidomide enantiomers is given schematically (REIST et al. 1998). At the end of the 1950s, the drug was marketed as a hypnotic agent but was withdrawn in the early 1960s because of embryotoxic and teratogenic effects. Only in more recent years has it become clear that in various species both thalidomide enantiomers are transformed in vivo into 1-*N*-phthaloylglutamine and 1-*N*-phthaloylglutamic acid. Several reports have been published on the sensitivity of different mouse and rat strains towards thalidomide embryotoxicity and teratogenicity. For instance, in SWS mice only the glutamic acid metabolite derived from the *S*(−)-enantiomer was embryotoxic and teratogenic and not the one derived from the *R*(+)-isomer (OCKZENFELZS et al. 1976). In contrast, in rabbits both thalidomide enantiomers were found to be equally teratogenic. At the end of the 1980s, however, a rapid racemization of thalidomide enantiomers of α-C-substituted analogues (BLASCHKE and GRAUTE 1987) and of a phthalimide analogue of thalidomide was suggested to occur under physiological conditions. Later, chiral interinversion of *R*-(+) and *S*-(−)-thalidomide was shown in vivo in humans (ERIKSSON et al. 1995). In vitro, chiral inversion of thalidomide enantiomers is catalyzed by albumin, amino acids (Arg and Lys), phosphate, and hydroxyl ions. Hydrolysis of the thalidomide enantiomers into ring-opened products was also base-catalyzed (REIST et al. 1998).

However, the reaction mechanisms of the chiral inversion and the hydrolysis were suggested to be different: base-catalyzed electrophilic substitution vs. nucleophilic substitution. Still, enantioselectivity in the observed teratogenicity could result from fast hydrolysis to chirally stable teratogenic metabolites, rather than from intrinsic stereoselectivity in the parent drug. The suggestion that the use of *R*-thalidomide would have avoided the toxicity problems associated with the racemic drug is still premature at this time. Remarkably, despite the fact that thalidomide has been seen for a long time as a model toxin for embryotoxicity and teratogenicity, illustrating the pertinent role of stereoselectivity, thalidomide is currently being redeemed because of promising activities in a variety of therapeutic areas: leprosy, regulation of TNF-α activity, and probably also other cytokines, autoimmune diseases (e.g., aphthous ulcers in AIDS patients), and cancer (KLING 2000).

D. Chemotherapeutic Agents and Stereoselectivity

The prominent role of stereochemical factors in toxicity is also clearly seen with chemotherapeutic drugs. The desired toxicity of these compounds (i.e. cytostatic effect), as well as undesired toxicity towards nontumor tissue (i.e. cytotoxic effect), depends directly on the steric configuration of the parent drug. For instance, platinum drugs require a *cis*- and not *trans*-configuration to coordinate bifunctionally to DNA, thus inhibiting DNA replication and transcription (SINGH and KOROPATNICK 1998). The dose-limiting peripheral neurotoxicity of ormaplatin and oxaliplatin, observed in clinical trials in humans, was shown to be reflected in vivo in rats upon repeated dosing by enantioselective neurotoxicity, the *S,S*-isomers being less neurotoxic (SCRENCI et al. 1997). The folate antimetabolite, methotrexate, a potent inhibitor of dihydrofolate reductase, also demonstrates high stereochemical requirements for interaction with a biological template. In a number of other cytostatic drugs, stereochemistry indirectly plays a crucial role, since substrate- or product-stereoselective biotransformation is involved. For instance, gossypol, a polyhydroxylated 2,2′-bi-naphthalene originating from cotton seeds (Fig. 4), was found to possess cytotoxic properties in vitro and in vivo. The effect of (–)-gossypol was at least twofold greater than that of (+)-enantiomer (JOSEPH et al. 1986). Multidrug-resistant MCF/ADR cells appeared more resistant to (–)-gossypol than the parent cell line, which was tentatively attributed to the inactivation of (–)-gossypol by GSH-transferases present in the MCF/ADR cells. Interestingly, (–)-gossypol was a threefold more potent inhibitor of GSH-transferase α- and π-isoenzymes than the (+) enantiomer (BENZ et al. 1990).

Metabolism of the cytostatic drug cyclophosphamide (CP), containing a chiral phosphorous atom (CP; Fig. 4) involves hydroxylation at the C-4 position by Cytochrome P450. Subsequently, a number of detoxification reactions can occur (e.g., oxidation to the 4-keto-derivative, dechloroethylation and formation of a carboxylic acid). The phosphoramide mustard resulting from spontaneous decomposition of 4-hydroxy-CP is thought to be the cytotoxic chemotherapeutic species. The chiral nature of the phosphorus atom results in a twofold greater therapeutic index (LD_{50}/ED_{90}) for the *S*-(–)-enantiomer (against the ADJ/PC6 plasma cell tumor in mice) without differences in metabolism by rat liver microsomes (COX et al. 1976). Recently, in cancer patients differences were also found for the clearance of the CP-enantiomers and for the formation clearance of their dechlorinated metabolites (WILLIAMS

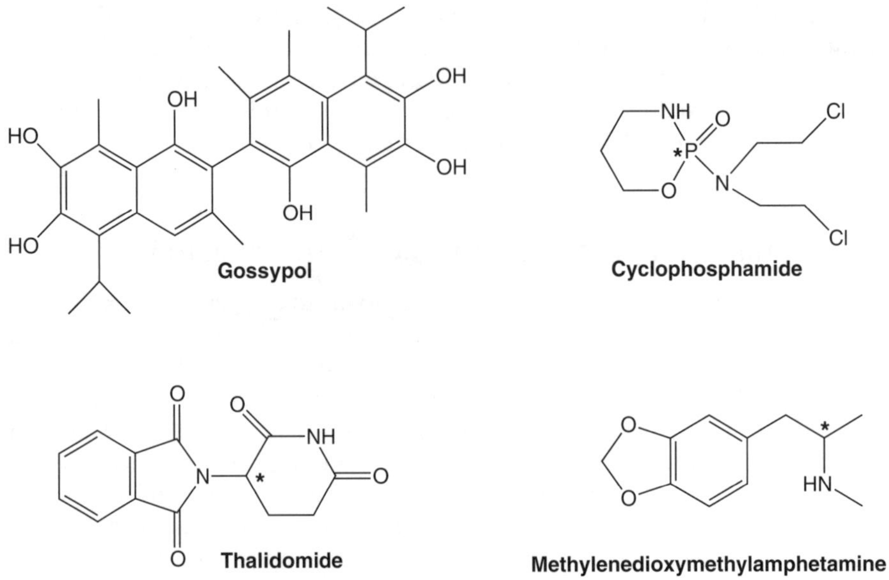

Fig. 4. Example of compounds that exert stereoselective toxicity. Mechanisms of toxicity and stereoselectivities involved are described in Sect. C and D.

et al. 1999a). Interestingly, in bone marrow patients of the dechlorinated metabolite from *S*-(−)-CP was induced by phenytoin coadministration, while having no effect on the dechlorination of the *R*-(+)-enantiomer. This suggests that in vivo, in humans, different cytochrome P450 enzymes are responsible for this metabolic reaction (WILLIAMS et al. 1990).

For ifosfamide it was shown that dechlorination of the *S*-enantiomer is primarily mediated by Cytochrome P450 2B6, and the *R*-enantiomer primarily by Cytochrome P450 3A4/5 isoenzymes (GRANVIL et al. 1999). When comparing ifosfamide and cyclophosphamide (CP), both alkylating antitumor agents, continuation of the use of racemic ifosfamide has been questioned for two reasons: first, stereochemistry plays a relatively minor role in the efficacy and the toxicity of CP, but it is a major factor in the neurotoxicity associated with ifosfamide, and second, (*R*)-ifosfamide contains the unique antitumor activity of this agent and the ifosfamide-associated neurotoxicity is primarily related to (*S*)-ifosfamide.

Modification of the ring structure of CP introduces a second chiral center. For instance, a methyl- or phenyl-substitution at the 4-position results in four stereoisomers (two *trans*-isomers, *RR* and *SS*, and two *cis*-isomers, *RS* and *SR*). 4-Methyl-CP isomers lack a difference in metabolism by rat liver microsomes and in chemotherapeutic efficiency against ADJ/PC6 plasma cell tumors in mice. In contrast, it was suggested that the *trans*-diastereoisomers of 4-phenyl-CP are metabolized to phosphoramide mustards, whereas the *cis*-diastereoisomers are not.

E. Stereoselective Biotransformation and Bioactivation by Cytochromes P450

One of the most important drug metabolizing enzyme systems in mammals and nonmammals is the so-called mixed-function oxidase (MFO) system, consisting of a large family of Cytochrome P450 isoenzymes with a broad range of different substrates (substrate selectivity), as well as products that can be formed from one substrate (product selectivity). For instance, the anticoagulant warfarin is metabolized to dehydrowarfarin, 4′- 6-, 7-, and 10-hydroxywarfarin by a number of Cytochrome P450 isoenzymes. Regio- and stereoselective hydroxylation reactions of *R*(+)- and *S*(−)-warfarin have been suggested to be useful as a probe for the functional characterization of the multiplicity of hepatic Cytochrome P450 isoenzymes (Kaminsky et al. 1984; Thijssen and Baars 1987).

Among the best documented early examples of the stereoselective action of Cytochrome P450 isoenzymes are those provided by studies of the stereoselective bioactivation of polycyclic aromatic hydrocarbons (PAHs) to carcinogens by Cytochrome P450 1A1 (Fig. 5) (Jerina et al. 1985). This isoenzyme, among others, was found to catalyze with a high regio- and stereoselectivity the epoxidation of benzo[a]pyrene to its (+)-*7R,8S*-epoxide, which induces over 14 times more lung tumors in newborn mice than its enantiomeric epoxide. Among the four stereoisomeric diol-epoxides, the (+)-*7R,8S*-

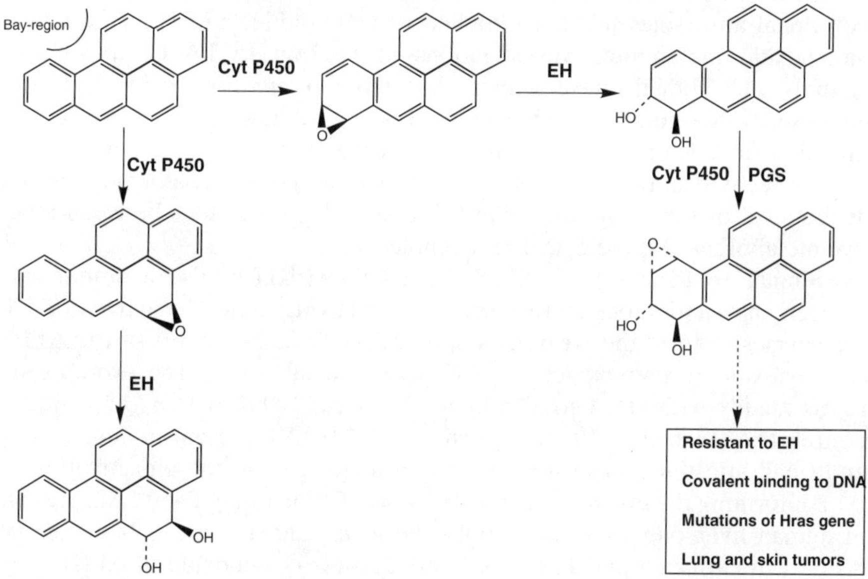

Fig. 5. Regio- and stereoselectivity of Cytochrome P450 1A1/2 and epoxide hydrolase (EH) in the formation of the bay-region diol-epoxide diastereoisomers, which is primarily responsible for the carcinogenicity of benzo[a]pyrene

dihydrodiol-*9S,10R*-epoxide was several-fold more mutagenic to Chinese hamster V79 cells than the corresponding (–)-enantiomer and, in addition, it was the only isomer with a high carcinogenic activity on mouse skin and in newborn mice. Thus, the (+)-*7R,8S*-diol-*9S,10R*-epoxide was established as the ultimate carcinogen of benzo[a]pyrene. At the DNA level, stereoselectivity in adduct-formation between the diol-epoxides and the exocyclic 2-amino group of guanine has been shown to be responsible for the observed stereoselectivity in this type of toxicity. From comparative investigations on the tumorigenicity of various PAHs (benzo[a]pyrene, benz[a]anthracene, chrysene), it has been derived that a bay-region diol-epoxide functionality is of crucial importance in the expression of tumors and, furthermore, that common stereochemical factors such as relative configuration (pseudo-diequatorial hydroxyl groups) and absolute configuration (*R*-configuration at benzylic carbon atom) appear to be far more important in this regard than chemical reactivity (JERINA et al. 1985). The enantioselectivity of GSTs is preferential for epoxides and diolepoxides with an (*R*-)configuration at the benzylic oxirane carbon (SEIDEL et al. 1998).

The carcinogenicity and mutagenicity of dibenz[c,h]acridine further support the idea that the highest biological activity is attributable to a bay-region diolepoxide that is superimposable on the highly active *R,S,S,R*-bay-region diol-epoxides of the PAHs mentioned above (WOOD et al. 1986). Interestingly, the isomeric dibenzo[a,e]pyrene, a very weak tumor initiator, was not metabolized to the corresponding proximate dihydrodiol (DEVANESAN et al. 1990). In recent years, immunochemical detection with monoclonal or polyclonal antibodies has been applied to DNA adducts of benzo[a]pyrene diolepoxide metabolites. Investigations in relation to PAH exposure of animals and humans have shown that the (+)-*anti*-diolepoxide-2 deoxyguanosine DNA adduct formed in vivo was more persistent in mouse epidermis than rat epidermis (ROJAS and ALEXANDROV 1986). Clearly, apart from stereoselectivity at the level of biotransformation, stereoselectivity also occurs at the level of toxicologically relevant molecular interactions between reactive metabolites and biological macromolecules.

Similar to benzo[a]pyrene, aflatoxin B1 (AFB1) is also a potent carcinogen that undergoes stereoselective bioactivation by Cytochrome P450 isoenzymes to form the AFB1 *exo*-epoxide plus small amounts of the AFB1 *endo*-epoxide (BAERTSCHI et al. 1989; RANEY et al. 1992). The *exo*-epoxide reacts readily with DNA to give high adduct yields, but the *endo*-epoxide is nonreactive (IYER et al. 1994). Cytochrome P450 3A4, present in human liver and small intestines, was shown to play a major role in the bioactivation of AFB1, forming the *exo*-8,9-epoxide. However, Cytochrome P450 1A2, present in human liver only, is forming both the *endo*- and the *exo*-8,9-epoxide of AFB1 (GUENGERICH and JOHNSON 1999). The *exo*-8,9-epoxide of AFB1 is by far the most mutagenic and likely also carcinogenic metabolite product of AFB1 (Fig. 6) (IYER et al. 1994). The fate of AFB1 epoxides and hence their ultimate biological activities are governed by their rates of reaction (both

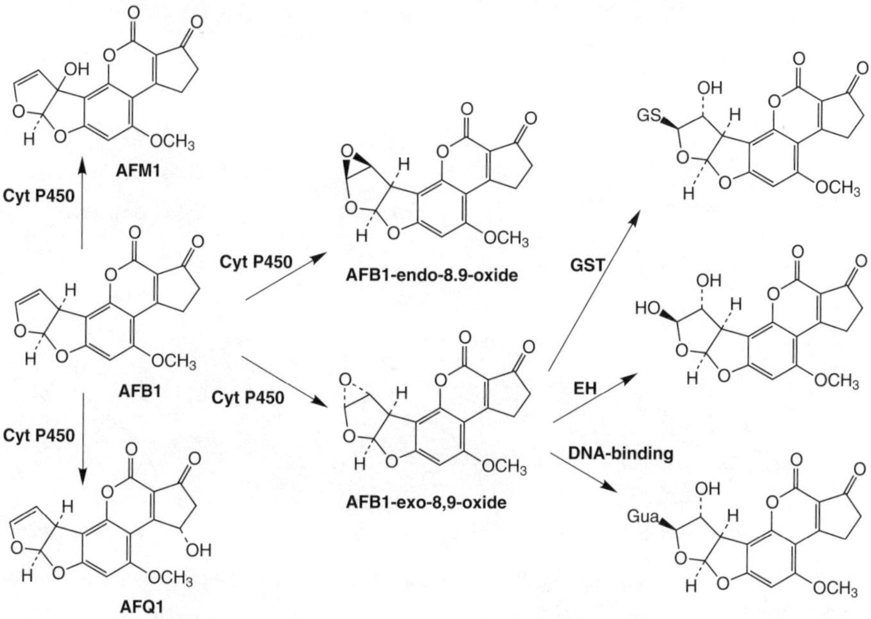

Fig. 6. Regio- and stereoselective metabolic activation and inactivation pathways of aflatoxin B1. Apart from stereoselective reactions by Cytochrome P450s, epoxide hydrolase (EH) and GSH-transferases (GSTs) stereoselective reactions with H_2O, DNA, and proteins also play a role

spontaneous and enzymatically catalyzed) with H_2O, GSH and DNA. From kinetic studies it has been concluded that the rate of hydrolysis is not further enhanced by epoxide hydrolase, but is competed with some GSH-transferase isoenzymes. Studies pointed at the polymorphic GSH-transferase M1-1 as most important isoenzyme and various GSTs were demonstrated to differ considerably in stereochemical selectivity. Another striking feature of AFB1 8,9-epoxide was the remarkable stereoselectivity difference in reaction of the isomers with DNA. The *exo*-isomer is about 10^3-fold more reactive than the *endo*, as judged from differences in both genotoxicity and DNA-binding (Iyer et al. 1994; Guengerich et al. 1999). The reaction with DNA involves intercalation and acid catalysis of both hydrolysis and conjugation.

F. Stereoselectivity and Genetic Polymorphisms in Drug Oxidations

Biotransformation enzymes constitute a major determinant of interindividual differences in the metabolism and the rate of elimination of numerous drugs and other xenobiotics and, consequently, in their pharmacological or toxicological response. For a number of enzymes (Cytochromes P450, GSH-

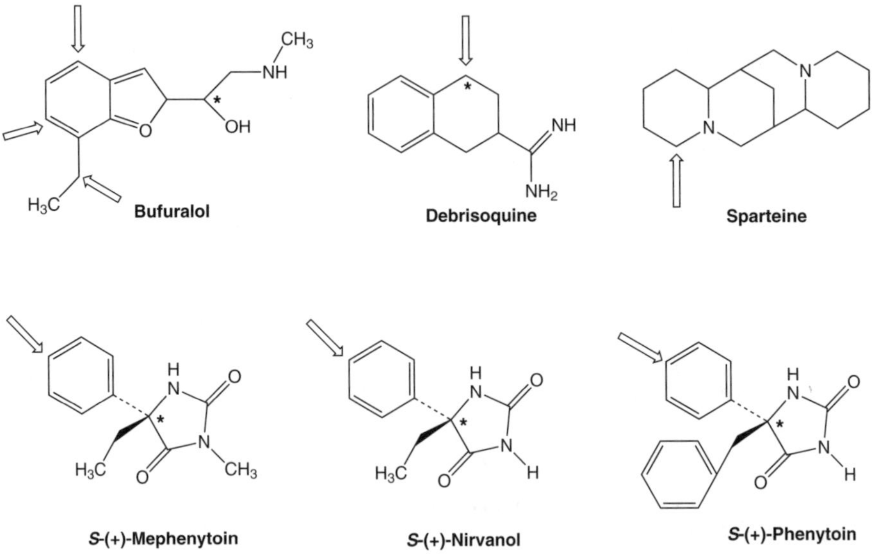

Fig. 7. Substrates of Cytochrome P450 2D6 undergoing stereoselective oxidative metabolic reactions

transferases, epoxide hydrolases, and *N*-acetyltransferases) the genetically determined absence of enzyme activity in a fraction of the population was reported. In addition to a lack of expression of protein, gene mutations have also been shown to lead to proteins with no activity, altered activity, or even multicopies of protein, thus leading to distinct phenotypes (for Cytochrome P450 2D6, for example, to poor, intermediate, extensive, and ultrarapid metabolizers) (for a review, see WORMHOUDT et al. 1999).

Among the many drugs whose oxidation has been reported to be impaired in so-called poor metabolizer phenotypes is the β-adrenoceptor antagonist bufuralol (Fig. 7). It is a chiral drug for which 1′-hydroxylation is under genetic control and selective for the (+)-isomer. Hydroxylation of bufuralol at the 4- and 6-positions is under the same genetic control and favors the (−)-isomer. The stereoselectivity for the aliphatic 1′- and aromatic 2,4-oxidations is virtually abolished in poor metabolizers phenotypes. Various early investigations (e.g., MEYER et al. 1986) have provided evidence for the hypothesis that 1′-hydroxylation of bufuralol is mediated by the polymorphic Cytochrome P450 2D6 isoenzyme, which also stereoselectively catalyzes the hydroxylation at the prochiral benzylic C_4-position of debrisoquine (Fig. 7) (MEESE et al. 1988), C-oxidation in sparteine (Fig. 7) (EBNER et al. 1995), *O*-demethylation of dextrometorphan, E10-hydroxylation of nortriptyline, *trans*-4′-hydroxylation of perhexiline, α-hydroxylation of (+)-metoprolol, and several other reactions in various substrates (e.g., BOOBIS et al. 1985; DISTLERATH et al. 1985).

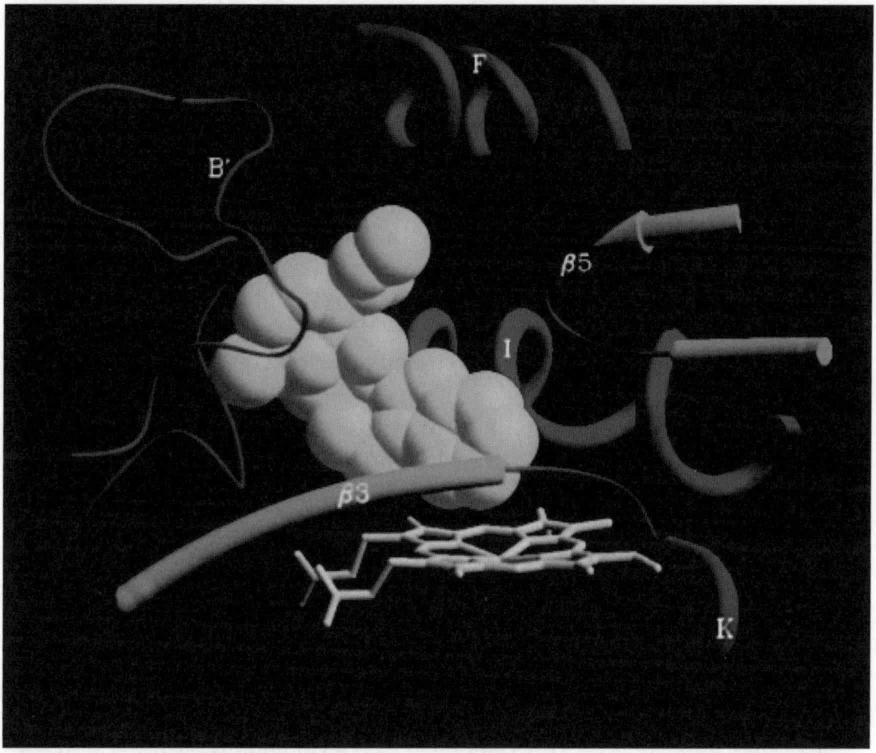

Fig. 8. In silico-generated model of the active site of Cytochrome P450 2D6, in this case with ajmalicine docked in it. Apart from ajmalicine, the I-helix, the heme porfyrine and some other details are also visible. (Adapted from De Groot et al. 1996)

Koymans et al. (1992) used 16 substrates, accounting for 23 regio- and stereochemically controlled metabolic reactions, and molecular modeling techniques to derive a predictive substrate model for Cytochrome P450 2D6. The predictive value of the model was evaluated by investigating the P450 2D6-mediated metabolism of 4 compounds (alfentanil, astemizole, resperidone, and nebivolol), comprising 14 oxidative metabolic routes, 13 of which proved to be correctly predicted as far as the involvement of P450 2D6 was concerned. More recently, several refined substrate models (De Groot et al. 1997) and, by homology modeling techniques, protein models of this Cytochrome P450 (De Groot et al. 1996, 1999) have been reported (Fig. 8).

With these and similar other in silico models it has been possible to rationalize the regio- and stereoselectivity of substrates of CYT P450 2D6 [e.g., of debrisoquine (Lightfoot et al. 2000)], as well as the stereoselectivity of this isoenzyme [e.g., of betaxalol and fluoxetine (De Groot et al. 1999)], and of MDMA, metoprolol, methoxyphenamine and mexilitine (De Groot et al. 1996, 1997). In fact, similar in silico models have been reported for most of the other important Cytochrome P450 isoenzymes as well (for a review, see Ter Laak and Vermeulen 2001). Similar findings on stereoselectivity have been made with other Cytochrome P450 isoenzymes and other classes of drugs, for example, 5-substituted-5-phenyl-hydantoins (Fig. 7) (Meyer et al. 1986). For instance, the 4′-hydroxylation of *S*(+)-mephenytoin (*N*-methyl-5-ethyl-5-phenyl-hydantoin) is subject to genetic polymorphism. In poor metabolizers, *S*(+)-mephenytoin is not significantly *para*-hydroxylated and, like the *R*-enantiomer, is available for *N*-demethylation. In extensive metabolizers, extensive 4′-hydroxylation of *S*(+)-mephenytoin occurs (at greater rates than *N*-demethylation). In contrast, *R*(–)-mephenytoin is metabolized by *N*-demethylation [affording pharmacologically active *R*(–)-5-ethyl-5-phenyl-hydantoin, also known as nirvanol, (Fig. 7)]. Nirvanol hydroxylation also favors the *S*-enantiomer and cosegregates with *S*(+)-mephenytoin hydroxylation in humans. Phenytoin (5,5-diphenyl-hydantoin), a prochiral drug, also undergoes polymorphic 4′-hydroxylation (Fig. 7). In humans, Cytochrome P450-mediated metabolism of phenytoin to 5-(4′-hydroxyphenyl)-5-phenylhydantoin results almost exclusively in the *S*(–)-enantiomer (95%). The product enantioselectivity in this reaction appears to be controlled by the relative affinities of the prochiral diphenylhydantoin to two different Cytochrome P450 isoenzymes, belonging to the P450 2C isoenzyme subfamily (Relling et al. 1990). More recently, evidence was presented that P450 2C9 and 2C19 are responsible for the stereoselective metabolism of phenytoin into its major metabolite (Bajpai et al. 1996).

Another important Cytochrome P450 isoenzyme family showing stereoselective metabolism is the P450-3A isoenzyme subfamily. P450 3A4 appears to play a major role in the oxidation of many drugs, including nifedipine oxidation, *R*(+)-warfarin 9,10-dehydrogenation, *S*(–)-warfarin 10-hydroxylation, quinidine *N*-hydroxylation, cortisol 6β-hydroxylation, testosteron 6β-hydroxylation, but also in the bioactivation of aflatoxin B1 and G1, 6-aminochrysene, benzo[a]pyrene, and pyrrolizidine alkaloids (Guengerich 1991 and 1999).

In fact, for all relevant Cytochrome P450 isoenzymes stereoselective metabolic reactions, some of which have significant toxicological consequences, have been reported (Vermeulen and Te Koppele 1993; Mason and Hutt 1997; Baillie and Schultz 1997; and Caldwell 1999). Obviously, information on stereoselectivity and structure-biotransformation relationships may be worthwhile to rationalise or predict this type of toxicity-related biotransformation.

Stereoselective neurotoxicity was reported for methylenedioxymethylamphetamine (MDMA) (Fig. 7). It causes degeneration of serotonergic and, to a lesser extent, catecholaminergic neurons. MDMA is a widely abused amphetamine derivative (Ecstasy), of which the *S*(+)-enantiomer has the highest pharmacological as well as toxicological potency. Following Cytochrome P450-mediated demethylation, 3,4-methylenedioxyamphetamine (MDA) is formed; it also has been shown to be a serotonergic neurotoxin. *S*(+)-MDMA undergoes this demethylation reaction to a far greater extent than the *R*(–)-enantiomer (Lin et al. 1997). The profile of neurotoxicological reactions is identical to that of *para*-chloroamphetamine (pCA). Altogether, stereoselective formation of as yet unknown metabolites of MDMA (and of MDA and pCA) may also be involved in the stereoselective toxicity of MDMA.

G. Glucuronyl Transferases and Stereoselective Bioactivation

UDP-D-glucuronic acid is an important optically active cofactor for microsomal UDP-glucuronyl-transferases (GTs), which catalyze the conjugation of substrates containing O-, N- or C-functional groups to this "glucuronic acid." In a number of cases, glucuronidation plays a significant role in the generation of toxic effects (MULDER et al. 1990). Apart from increasing the instability and/or reactivity of procarcinogens (e.g., *O*-glucuronide of *N*-hydroxy-2-acetaminofluorene), the glucuronides may serve as transport forms of procarcinogens to various tissues (e.g., the *N*-glucuronide of *N*-hydroxy-2-naphtylamine).

Apart from substrate-stereoselective glucuronidation of hydroxylated hexobarbital, considerable enantioselective differences have been shown in vitro for propranolol, fenoterol, carvedilol, dihydrodiols of polycyclic aromatic hydrocarbons, and oxazopam. Moreover, (–)-morphine has been shown to be glucuronidated exclusively at the 3-hydroxy position, whereas (+)-morphine is preferentially glucuronidated at the 6-hydroxy position. In vivo, enantioselectivity is often less pronounced than in vitro. 3-Methylcholanthrene, 8-naphtoflavone, inducers of the so-called GT1-isoenzyme, and ethanol have been shown to increase enantioselectivity in the formation of diastereomeric glucuronides from racemic oxazepam in hepatic microsomes of rabbits, and to produce a large increase in enantioselectivity, whereas inducers of the GT2-isoenzyme, such as phenobarbital, only increased oxazepam glucuronidation without affecting the enantioselectivity. Inducers of forms other than GT1 and GT2, such as *trans*-stilbene-oxide and clofibric acid, generally cause increases in activities, coupled with increases in *R/S*- ratios.

As yet, little is known about the active sites of these enzymes. The enzyme mechanism is probably consistent with a single displacement, SN2-type nucleophilic attack of the aglycone on UDP-glucuronic acid in a ternary complex. Further studies seem necessary for a better understanding of the stereochemical behavior and active-site geometry of UDP-glucuronyl transferases (ARMSTRONG et al. 1988; MULDER et al. 1990).

Formation of acylglucuronides, catalyzed by glucuronyl transferases, plays a significant role in many chiral carboxylic acid-containing drugs, such as ketoprofen, naproxen, carprofen, fenoprofen, and ximoprofen (AKIRA et al. 1998; HAYBALL 1995). In several of these drugs, irreversible covalent binding via reactive acylglucuronides has been reported. Two mechanisms have been proposed: transacylation and a rearrangement glycation mechanism. From these

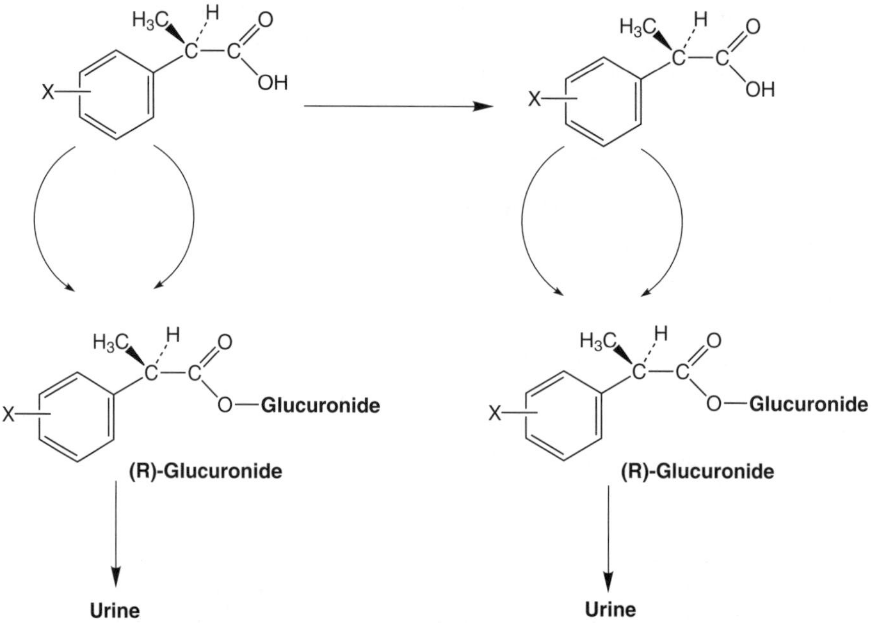

Fig. 9. Formation, hydrolysis, and urinary excretion of epimeric acyl glucuronides of NSAIDs which undergo chiral inversion. Not shown is the internal acyl migration of the *O*-acyl glucuronides, which may also be stereoselective. (Adapted from HAYBALL 1995)

NSAIDs and several other examples it becomes clear that the stereoselective formation of acylglucuronides and the stereoselective reactivity of these conjugates is crucial, since this may contribute to the overall enantioselectivity of carboxylic acid-containing drugs (HAYBALL 1995). In Fig. 9, the formation, hydrolysis, and renal clearance of epimeric acyl/glucuronides of NSAIDs undergoing chiral inversion are shown. Not shown is the internal acyl migration of the *O*-acyl glucuronides, which may also be stereoselective.

H. Sulfotransferases and Stereoselective Bioactivation

Sulfotransferases (STs) catalyze the transfer of a sulfate group from 3′-phosphoadenosine 5′-phosphosulfate (PAPS) to nucleophiles, such as alcohols, phenols, and amines. Several isoenzymes, grouped into several classes based on substrate selectivity towards phenols and alcohols, have been purified (MULDER et al. 1990).

Generally speaking, solid conclusions regarding stereoselective sulfation in vivo are difficult because of competition between (stereoselective) glucuronidation and sulfation for the same substrates. Stereoselective sulfation of 4′-hydroxypropranol enantiomers was demonstrated in vitro in hepatic tissue

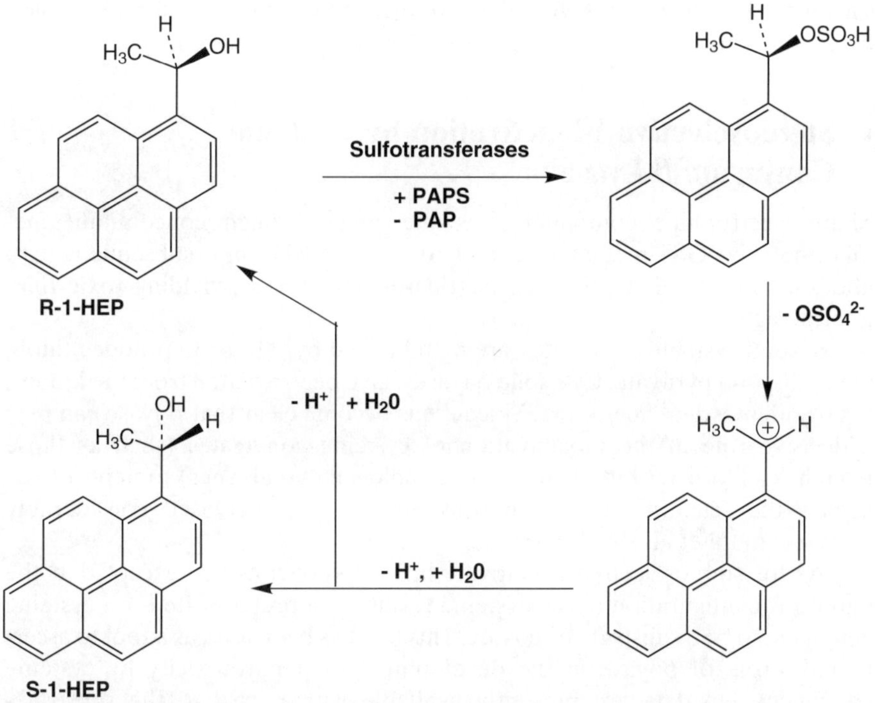

Fig. 10. Stereoselective mutagenicity of 1-hydroxyethylpyrene enantiomers through stereoselective bioactivation by human estrogen sulfotransferase (hEST). (Adapted from HAGEN et al. 1998)

of various species. Enantiomeric (–/+)-4′-hydroxypropranolol sulfate ratios varied from 1.07 in rat and 0.73 in dog to 0.62 in hamster. Terbutaline and prenalterol, a structural analogue of terbutaline, both showed preferential sulfation for the respective (+)-enantiomers in human liver cytosol (WALLE and WALLE 1990). At least three ST-isoenzymes have been isolated from rat liver. Few examples are yet known demonstrating a causal relationship between sulfation and stereoselective toxicity. Metabolic activation, however, has been shown to occur in the case of 7-hydroxymethyl-12-methylbenz[a]anthracene, the sulfation of which is followed by alkylation of DNA via the sulfate ester. Recently, various human sulfotransferases (hP-PST, hM-PST, hH-PST, and TA1538-hEST) and rat sulfotransferases were expressed in Ames *Salmonella* strains. The two 1-hydroxyethylpyrene enantiomers were sulfated to a mutagenic sulfate-conjugate by both hEST and hHST-isoenzymes, but they differed substantially in their enantioselectivity for this compound (Fig. 10) (HAGEN et al. 1998). A complicating factor, however, may occur in cases where sulfate conjugates are reactive to water. In this case sulfotransferases may act as a kind of invertase for hydroxyl groups at chiral carbon atoms. This has been shown for 1-hydroxyethylpyrene enantiomers which undergo chiral inversion by enantioselective sulfotransferases (Landsiedel et al. 1998).

I. Stereoselective Bioactivation by Cysteine Conjugate *β*-Lyase

Many electrophilic compounds become toxic through concomitant metabolism via GST-mediated conjugation to GSH and subsequent catabolism, among others, through peptidases and *β*-lyase, yielding toxic thiol intermediates.

Several cysteine conjugates, are metabolized by *β*-lyase to produce thiols, ammonia, and pyruvate. Cytosolic *β*-lyases have been isolated from the kidney, liver, and intestinal microflora. Since it has become clear that *β*-lyase can play a decisive role in the bioactivation of cysteine conjugates (such as those from hexachloro-1,3-butadiene or other halogenated alkenes) to nephrotoxic agents, the scientific interest in this enzyme has increased considerably (Commandeur et al. 1990, 1995).

As for stereoselectivity, relatively little is known as yet, except that the natural *R* configuration in the cysteine residue is a prerequisite for a cysteine conjugate to be a substrate of *β*-lyase. This fact has been used as a tool to ascertain the role of *β*-lyase in the development of nephrotoxicity by cysteine conjugates. No data are presently available with regard to the substrate-stereoselective effects of the noncysteine part of the thioether substrates of *β*-lyase (Commandeur and Vermeulen 1990). Such stereoselective effects may be anticipated however since, for example, the regioisomeric 1,2- and 2,2-dichlorovinyl-L-cysteine conjugates have also been shown to cause strongly different *β*-lyase-mediated mutagenic and nephrotoxic effects (Commandeur et al. 1991).

J. General Conclusions

This chapter has demonstrated that it is generally recognized at present that stereochemical factors play a significant role in the pharmacological or toxicological action of drugs and other xenobiotics. As for the toxicity of xenobiotics, stereoselectivity in biotransformation (i.e., bioactivation or bioinactivation) is of major importance. As a consequence, the biological actions of xenobiotics, in general, are difficult to predict because of the complex regulation of the processes involved.

In recent years, considerable progress has been made in the elucidation of stereochemical mechanisms of several important biotransformation enzymes. However, the multiplicity in terms of isoenzymes, the widely varying substrate selectivity of each isoenzyme, as well as secondary metabolism (which in itself can be stereoselective), considerably complicates predictions in metabolism as well as in toxic effects. In principle, the same factors do complicate the correlation of data on stereoselective effects obtained in vitro with those obtained in vivo, even when it concerns the same species. The prediction of stereoselective effects from one species to another or from one chemical to another is even more difficult.

As this chapter has also demonstrated, new insights into stereochemical mechanisms at a molecular level [i.e., at the level of binding of substrate molecules to active (or binding) sites of isolated and purified (iso)-enzymes (or other proteins)] in some cases provided more or less simplified models or working hypotheses that might be helpful in predicting stereoselective biotransformations and toxic effects. Progress in research along these lines will be most promising, from not only an academic but also a more practical point of view, namely, that concerned with predictability.

From a toxicological point of view, it is important to stress that knowledge of stereoselective effects at the different levels of the development of biologically active drugs or other chemicals should be obtained as early as possible. Without this knowledge, it is almost impossible to interpret the toxicodynamics and toxicokinetics of chiral compounds.

References

Armstrong RN, Andre JC, Bessems JGM (1988) Mechanistic and stereochemical investigations of UDP-glucuronosyltransferases. In: Siest G, Magdalou J, Burchell B (eds) Cellular and Molecular Aspects of Glucuronidation Colloque. Inserm. John Libbey Eurotext Ltd., London, 173:51–58

Akira K, Taira T, Hasegawa H, Sakuma C, Shinehara Y (1998) Studies on the stereoselective internal acyl migration of ketoprofen glucuronides. Drug Metab Disp 26:457–464

Baertschi SW, Raney KD, Shimada T, Harris TM, Guengerich FP (1989) Comparison of rates of enzymatic oxidation of aflatoxin B1, aflatoxin G1, and sterigmatocystin and activities of the epoxides in forming guanyl-N7 adducts and inducing different genetic responses. Chem Res Toxicol 2:114–122

Baillie TA, Schultz (1997) Stereochemistry in drug development; role of chirality as a determinant of drug action, metabolism and toxicity. In: Aboul-Enein HY, Wainer IW (eds) The impact of stereochemistry on drug development and use. John Whiley & Sons, New York, pp 21–42

Bajpai M, Roskos LK, Shen DD, Levy RH (1996) Roles of cytochrome P450 2C9 and 2C19 in the stereoselective metabolism of phenytoin to its major metabolite. Drug Metab Disp 24:1401–1403

Benz CC, Kenry MA, Ford JA, Townsend AJ, Cox FW, Palayoor S, Matlin SA, Hait WN, Cowan KH (1990) Biochemical correlates of the antitumor and antimitochondrial properties of gossypol enantiomers. Mol Pharmacol 37:840–848

Blaschke G, Graute WF (1987) Enantiomere des Konfigurationsstabilen C-Methyl-thalidomids. Liebigs Ann Chem 647–648

Boobis AR, Murray S, Hampden CE, Davies DS (1985) Genetic polymorphism in drug oxidation: in vitro studies of human debrisoquine 4-hydroxylase and bufuralol 1′-hydroxylase activities. Biochem Pharmacol 34:65–71

Caldwell (1999) Through the looking glass in chiral drug development. Modern Drug Discovery 51–60

Commandeur JNM, Vermeulen NPE (1990) Molecular and biochemical mechanisms of chemically-induced nephrotoxicity: a review. Chem Res Toxicol 3:171–194

Commandeur JNM, Stijntjes GJ, Wijngaard J, Vermeulen NPE (1991) Metabolism of L-cysteine *S*-conjugates and *N*-(trideuteroacetyl)-L-cysteine *S*-conjugates of four fluoroethylenes in the rat. Role of balance of deacetylation and acetylation in relation to the nephrotoxicity of mercapturic acids. Biochem. Pharmacol 42:31–38

Commandeur JNM, Stijntjes GJ, Vermeulen NPE (1995) Enzymes and transport systems involved in the formation and disposition of glutathione *S*-conjugates. Role in bioactivation and detoxication mechanisms of xenobiotics. Pharmacol Rev 47:271–330

Cox PJ, Farmer PB, Jarman M, Jones J, Stec WJ, Kinas R (1976) Observations on the differential metabolism and biological activity of the optical isomers of cyclophosphamide. Biochem Pharmacol 25:993–999

De Groot MJ, Vermeulen NPE, Kramer JD, Van Acker FAA, Donné-Op den Kelder GM (1996) A three-dimensional protein model for human cytochrome P450 2D6 based on the crystal structures of P450 101, P450 102 and P450 108. Chem Res Tox 9:1079–1091

De Groot MJ, Bijloo GJ, Martens BJ, Van Acker FAA, Vermeulen NPE (1997) A refined substrate model for human Cytochrome P450 2D6. Chem Res Toxicol 10:41–48

De Groot MJ, Ackland MJ, Horne VA, Alex AA, Jones BC (1999) A novel approach to predicting P450 mediated drug metabolism. CYP 2D6 calatyzed *N*-dealkylation reactions an qualitative metabolite predictions using a combined protein and pharmacophore model for CYP 2D6. J Med Chem 42:4062–4070

Devanesan PD, Cremonesi P, Nunnally JE, Rogan EG, Cavalieri EL (1990) Metabolism and mutagenicity of dibenzo[a,e]pyrene and the very potent environmental carcinogen dibenzo[a,e]pyrene. Chem Res Toxicol 3:580–586

Distlerath LM, Reilly PEB, Martin MV, Wilkinson GR, Guengerich FP (1985) Immunochemical characterization of the human liver cytochrome P450 involved in debrisoquine hydroxylation. In: Boobis AR, Caldwell J, De Matters F, Elcombe CR (eds) Microsomes and Drug Oxidations. Taylor & Francis, London, pp 380–389

Ebner T, Meese CO, Eichelbaum M (1995) Mechanism of Cytochrome P450 2D6-catalyzed sparteine metabolism. Molec. Pharmacol 48:1078–1086

Eriksson T, Björkman S, Roth B, Fyge A, Høglund P (1995) Stereospecific determination, chiral inversion in vitro and pharmacokinetics in humans of the enantiomers of thalidomide. Chirality 7:44–52

Gallo MA, Ill HCP, Dragan YP, Anthony DC, Montine TJ, Graham DG, Russel FE, Norton S (1996) In: Casarett and Doull's Toxicology – The Basic Science of Poisons. McGraw-Hill, New York, chapters 1, 8, 26 and 27

Goetz CG, Meisel E (2000) Biological Neurotoxins. Neurol Clin 18:719–740

Granvil CP, Madan A, Sharkawi M, Parkinson A, Wainer IW (1999) Role of CYP2B6 and CYP3A4 in the in vitro *N*-dechloroethylation of (*R*)- and (*S*)-ifosfamide in human liver microsomes. Drug Metab Dispos 27:533–541

Guengerich FP (1991) Oxidation of toxic and carcinogenic chemicals by human cytochrome P450 enzymes. Chem Res Toxicol 4:391–407

Guengerich FP, Johnson WW (1999) Kinetics of hydrolysis and reaction of aflatoxin B_1 *exo*–8,9-epoxide and relevance to toxicity and detoxication. Drug Metab Rev 31:141–158

Hagen M, Pabel U, Landsiedel R, Bartsch I, Falany CN, Glatt HR (1998) Expression of human estrogen sulfotransferase in *S*-typhimurium: differences between hHST and hEST in the enantioselective activation of 1-hydroxyethylpyrene to a mutagen. Chem Res Interact 109:249–253

Hayball PJ (1995) Formation and reactivity of acyl-glucuronides. The influence of chirality. Chirality 7:1–9

Iyer R, Coles B, Raney KD, Their R, Guengerich FP, Harris TM (1994) *Title follows* . J Amer Chem Soc 116:1603–1619

Jerina DM, Sayer JM, Yagi H, Van Bladeren PJ, Thakker DR, Levin W, Chang RL, Wood AW, Conney AH (1985) In: Boobis AR, Caldwell J, De Mattheis F,

Elcombe R (eds) Microsomes and Drug Oxidations. Taylor & Francis, London, pp 310–319
Joseph AEA, Matlin SA, Knox P (1986) Cytotoxicity of enantiomers of gossypol. Brit J Cancer 54:511–518
Kaminsky LS, Dunbar DA, Wang PP, Beaune P, Larrey D, Guengerich FP, Schnellmann RG, Sipes IG (1984) Human hepatic cytochrome P450 composition as probed by in vitro microsomal metabolism of warfarine. Drug Metab Dispos 12:470–476
Kling J (2000) Redeeming thalidomide: clinical trials for this notorious drug suggest uses against cancer and autoimmune diseases. Mod Drug Discov, p 35
Koymans LMH, Vermeulen NPE, van Acker SABE, te Koppele JM, Heykants JJP, Lavrijsen K, Meuldermans W, Donné-Opdenkelder GM (1992) A predictive model for substrates of cytochrome P450-debrisoquine (2D6) Chem Res Toxicol 5:211–219
Landsiedel R, Pabel U, Engst W, Ploske J, Seidel A, Glatt HR (1998) Chiral inversion of 1-hydroxyethylpyrene enantiomers mediated by enantioselective sulfo transferases. Biochem Biophys Res Comm 247:181–185
Lightfoot T, Ellis SW, Mahling J, Acklands MJ, Blaney FE, Bijloo GJ, de Groot MJ, Vermeulen NPE, Blackburn GM, Lennard MS, Tucker GT (2000) Regioselective hydroxylation of debrisoquine by cytochrome P450 2D6: implications for active site modelling. Xenobiotica 30:219–233
Lin LY, DiStefano EW, Schmitz DA, Hsu L, Ellis SW, Lennard MS, Tucker GT, Ch AK (1997) Oxidation of methamphetamine and methylenedioxymethamphetamine by Cyp 2D6. Drug Metab Disp 25:1059–1064
Madyastha KM, Gaikwad NW (1998) Metabolic fate of *S*-(–)-pulegone in rat. Xenobiotica 28:723–734
Mason JP and Hut AJ (1997) Stereochemical aspects of drug metabolism (1997) In: Aboul-Enein HY, Wainer IW (eds) The impact of stereochemistry on drug development and use. John Whiley & Sons, New York, pp 45–106
Meese CO, Fisher C, Eichelbaum M (1988) Stereoselectivity of the 4-hydroxylation of debrisoquine in man, detected by GC-MS. Biomed Environ Mass Spectrom 15:63–71
Meyer UA, Gut J, Kronback T, Scoda C, Meier UT, Catin T, Dayer P (1986) The molecular mechanism of two common polymorphisms of drug oxidation – evidence for functional changes in cytochrome P450 isoenzymes catalyzing bufuralol and mephenytoin oxidation. Xenobiotica 16:449–464
Mulder GJ, Coughtrie MWH, Burchell B (1990) Glucuronidation, In: Mulder GJ (ed.) Conjugation Reactions in Drug Metabolism: An Integrated Approach Taylor & Francis, London, pp 51–107
Nelson SD (1995) Mechanisms of the formation and disposition of reactive metabolites that can cause acute liver injury. Drug Metab Rev 27:147–177
Ockzenfelz H, Köhler E, Meise W (1976) Teratogene Wirkung und Stereospezifität eines Thalidomid-Metaboliten. Pharmazie 31:492–493
Raney KD, Meyer DJ, Ketterer B, Harris TM, Guengerich FP, (1992) Glutathione conjugation of aflatoxin B1 *exo*- and *endo*-epoxides by rat and human glutathione *S*-transferases. Chem Res Toxicol 5:470–478
Reist M, Carrupt P-A, Francotte E, Testa B (1998) Chiral inversion and hydrolysis of thalidomide: mechanisms and catalysis by bases and serum albumin, and chiral stability of teratogenic metabolites. Chem Res Toxicol 11:1521–1528
Relling MV, Ayoma T, Gonsales FJ, Meyer UA (1990) Tolbutamide and mephenytoin hydroxylation by human cytochrome P450 s in the CYP2Cu subfamily. J Pharmacol Exp Ther 252:442–447
Riddle JM and Estes JW (1992) Oral contraceptives in ancient and medieval times. Amer Sci 80:226–233
Rojas M, Alexandrov K, (1986) In vivo formation and persistence of DNA adducts in mouse and rat skin exposed to benzo[a]pyrene-dihydrodiols and -dihydrodiolepoxides. Carcinogenesis 7:1553–1556

Screnci D, Hambley TW, Galettis P, Brouwer W, McKeage MJ (1997) Stereoselective peripheral sensory neurotoxicity of diaminocyclohexane platinum enantiomers related to ormaplatin and oxaliplatin. Br J Cancer 76:502–510

Seidel A, Friedberg F, Löllmann B, Schwierzoek A, Funk M, Frank H, Holler Romy, Oesh F, Glatt HR (1998) Detoxication of optically active bay region polycyclic aromatic hydrocarbon dihydrodiol epoxides by human GST P1–1 expressed in Chinese hamster V79 cells. Carcinogenesis 19:1975–1981

Singh G, Koropatnick J (1988) Differential toxicity of *cis* and *trans* isomers of dichlorodiammineplatinum. J Biochem Toxicol 3:223–233

Ter Laak AM, Vermeulen NPE (2001) Molecular modeling approaches to predict metabolism and toxicity. In: Testa B, Waterbeemd H vd, Folkers G, Guy R (eds) The Second logP Symposium Lipophilicity in Drug Disposition, Verlag Helvetica, Chimica Acta, Zürich

Thulasiram HV, Gada AK, Madyastha MK (2000) Role of C-5 chiral center in *R*-(+)-pulegone-mediated hepatotoxicity: metabolic disposition and toxicity of 5,5-dimethyl-2-(1-methylethylidene)-cyclohexagone in rats. Drug Metab Disp 28: 833–844

Thijssen HHW and Baars LGM (1987) The biliary excretion of acenocoumarol in the rat: stereochemical aspects. J Pharm Pharmacol 39:655–657

Vermeulen NPE, Te Koppele JM (1993) Stereoselective biotransformation. Toxicological consequences and implications. In: Wainer IW (ed.) Drug Stereochemistry (2nd edition, revised and expanded), Marcel Dekker, Inc., New York/Basel/Hong Kong, pp 245–280

Walle T, Walle UK, (1990) Stereoselective sulphate conjugation of racemic terbutaline by human liver cytosol. Brit J Clin Pharmacol 30:127–133

Williams KM (1990) Enantiomers in arthritic disorders. Pharmacol. Ther 46:273–295

Williams ML, Wainer IW, Granvil CL, Bernstein ML, Ducharme MP (1999a) Pharmacokinetics of (*R*)- and (*S*)-cyclophosphamide and their dechloroethylated metabolites in cancer patients. Chirality 11:301–308

Williams ML, Wainer IW (1999c) Cyclophosphamide versus ifosfamide: to use ifosfamide or not to us, that is the three-dimensional question. Curr Pharm Des 5:665–672

Williams ML, Wainer IW, Embree L, Barnett M, Granvil CL, Ducharme MP (1999b) Enantioselective induction of cyclophosphamide metabolism by phenytoin. Chirality 11:569–574

Wood AW, Chang RL, Kumar S, Shirai N, Jerina DM, Lehr RE, Conney AH (1986) Bacterial and mammalian cell mutagenicity of four optically active bay-region 3,4-Diol-1,2-epoxides and other derivatives of the nitrogen heterocycle dibenz[c,h] acridine. Cancer Res 46:2760–2766

Wormhoudt LW, Commandeur JNM, Vermeulen NPE (1999) Genetic polymorphisms of human *N*-acetyltransferase, Cytochrome P450, Glutathione *S*-Transferase and epoxide hydrolase enzymes – Relevance to Xenobiotic Metabolism and Toxicity. Crit Rev Toxicol 29:59–124

Section III
Drug Absorption, Distribution, Metabolism, Elimination

CHAPTER 11

Intestinal Drug Transport: Stereochemical Aspects

H. SPAHN-LANGGUTH, C. DRESSLER, and C. LEISEN

Dedicated to our dear colleague and collaborator in the transporter field Professor Dr. Axel Büge, who unexpectedly died September 17, 2001.

A. General Aspects Regarding Intestinal Transport of Drugs: Mechanisms, Transporters, Techniques

I. The Gut Wall, its Physiological Functions, and Factors Affecting Drug Absorption from the Gastrointestinal Tract

The gut wall is a complex physiological system which, in addition to its function in homeostasis (electrolyte and water balance), allows absorption of essential nutrients, contributes to metabolism of dietary molecules, and possesses defence mechanisms against toxins.

Most important for absorptive processes is the inner layer of the gut tissue, the mucosa, which consists of the epithelial cell layer, the lamina propria, and a thin muscular layer. The blood vessels reached by most compounds that enter the systemic circulation are located in the submucosal area. The intestinal epithelium is composed of a number of different cell types, of which the absorptive cells (*enterocytes*) are most important. Such cells are highly polarized upon differentiation, and apical and basolateral membranes are separated by tight junctions. The apical membrane forms microvilli, and various transporters and enzymes are located in this membrane (brush-border membrane).

A drug passing through the intestinal lumen at variable transit-related velocities may be subject to various processes: pH influences, food- and mucus-interactions, intraluminal degradation, biotransformation via endogenous and bacterial enzymes, uptake into the enterocytes and subsequent entrance into the blood vessels by different inside-directed processes (e.g., passive diffusion, carrier-mediated uptake) and/or exsorption (intestinal secretion) back into the luminal space.

For compounds absorbed via passive absorption and those exhibiting carrier-mediated uptake or secretion, these processes may be highly variable so that permeability and transporter expression can vary from site to site, i.e., between gut segments, and even depend on diet or previous or concomitant drug therapy (HANAFY et al. 2001; WAGNER et al. 2001).

Since expression of transporters for absorption or secretion may vary between segments, significant intestinal secretion from the enterocyte to the lumen might be the predominant process in one intestinal segment, while later the secreted drug reenters the gut wall. As a consequence, discontinuous input profiles might result, as those detected in studies with the β-adrenoceptor antagonists talinolol and celiprolol (WETTERICH et al. 1996; HARTMANN et al. 1989), particularly with lower doses. For carrier-mediated transport, processes which are saturable, the drug concentration at the absorption site codetermines the relevance of the respective process. Preferentially, for most drugs, uptake into the enterocyte and the blood appears to occur via passive diffusion, the most prominent process in drug absorption.

Effective permeability represents the one parameter that serves to characterize the permeation of compounds through cell layers. This permeability may be different in the various gut segments (e.g., SINKO et al. 1991). It summarizes all passive and active transport and all metabolic processes in the intestine leading to a decreased luminal concentration.

II. Transporters

Transport proteins include inside- as well as outside-directed carrier systems (Fig. 1), most of which are located in the apical membrane of the enterocytes (TSUJI and TAMAI 1996). In general, carriers may be involved in 3 types of transport processes: facilitated diffusion, cotransport (symport) and countertransport (antiport). "Secondary active transport" is feasible for co- as well as countertransporters using the energy from the downhill transport of one transported compound to drive the uphill transport of another substrate. As opposed to carriers, pumps perform primary active transport by linkage of transport to an external source of energy (hydrolysis of a phosphate bond) (WAGNER et al. 2001).

Carrier-mediated absorptive or exsorptive processes are saturable and inhibitable (Fig. 1). Each transporter may be characterized with respect to K_m and V_{max}, i.e., the substrate concentration, at which half-maximal transport velocity is observed, and the maximum transport capacity.

Well-known examples for absorptive transport are amino acid and peptide transporters (TSUJI and TAMAI 1996). Small peptides are absorbed from the small intestine by a carrier-mediated uptake process. The uptake of small peptides is stimulated by an inwardly directed H^+-gradient, which is generated by the combined action of a Na^+/H^+-exchanger in the brush-border membrane and a Na^+/K^+-ATPase in the basolateral membrane. This transport system for oligopeptides is shared by orally active α-amino-β-lactam antibiotics (GANAPATHY and LEIBACH 1986; MURER et al. 1976).

Similar to metabolic processes in the intestine, intestinal exsorptive transport also became evident as soon as the new generation of highly specific, potent and low-dosed drugs were developed. Limited peroral bioavailability or lack of bioavailability of various newly developed compounds were indicative of bioavailability-limiting processes.

Secretory transporters have been extensively studied in cancer chemotherapy during the past years. Extrusion of drugs from the tumor cell may occur via multidrug transporters, is less structure-specific than inside-directed transport, and relates to a variety of different drugs, many of which are not associated with cancer chemotherapy. By actively extruding cytostatic agents out of the tumor cell, different transmembrane proteins confer the multidrug resistance (MDR) phenotype to various malignancies. Well-known members of these ATP-dependent transporters are 170-P-glycoprotein (P-gp, Fig. 2) and the multidrug resistance-related protein (MRP). By pharmacological inhibition of such proteins using so-called reversal agents (e.g., the cyclosporine A analogue PSC833), different in vitro and partially in vivo studies have shown promising results to overcome MDR. Since P-gp and MDR are also physiologically expressed in different blood-organ or organ-environment barriers (liver, kidney, blood-brain barrier, intestine), the general pharmacokinetic behavior of P-gp/MRP substrates may be affected when coadministered with reversal agents. Genetically modified cell lines as well as mdr/mrp-deficient mice may serve as experimental test models for mechanistic studies and as a basis for clinical investigations.

The biochemistry, pharmacology, and structure-activity relationships of P-gp and its substrates have been described in a number of publications (e.g., Wagner et al. 2001). Besides P-gp, secretory transport of several organic anions is mediated by the multidrug resistance-associated protein family (MRP), which contains six members (Borst et al. 1999) and has previously been shown to contribute to the primary active (ATP-dependent) transport of various glutathione, sulfate, and glucuronide conjugates and various anions. In liver, kidney, and gut mainly MRP2 (apical membrane of polarized cells like P-gp), MRP3 (basolateral membrane), and MRP6 appear (Borst et al. 1999). Unlike P-gp and MRP, which are ATP-dependent extrusion pumps, the organic cation transporter (OCT) family includes various OCT subtypes, which represent an electrogenic import system for organic cations. Furthermore, other efflux transporters may mediate intestinal secretion of anions in addition to P-gp and MRPs (e.g., Wagner et al. 2001).

Carrier-mediated transport processes for drugs may be further complicated by the fact that substrates exist, which show affinities to *multiple* carrier systems, e.g., the HMG-CoA reductase inhibitor atorvastatin, which is secreted from the enterocytes by P-gp and taken up by the H^+-monocarboxylic acid cotransporter (MCT) (Wu et al. 2000). An alternative example from the group of "bisubstrates" (Ullrich et al. 1993) is fexofenadine (Cvetkovic et al. 1999).

Another aspect is the overlapping substrate specificity detected for several compounds and/or their metabolites (Wacher et al. 1995) with respect to their interaction with cytochrome-P-450 isozymes and P-glycoprotein.

III. Bioavailability-Reducing Metabolic Processes in Addition to Countertransport and Liver First-Pass: Luminal Metabolism, Metabolism at/in Enterocytes

It is known today that – in addition to the presence of enzymes in the gut lumen, such as lipases – a wide variety of different enzymes are present in the enterocytes catalyzing various phase-I- and phase-II metabolic reactions. It may be difficult to differentiate between gut and liver (first-pass) metabolism when only blood data are available (Doherty and Pang 1997). It may further be difficult to differentiate between intestinal exsorption and metabolism, since both may reduce the amount entering the portal blood in a dose-dependent manner leading to low bioavailability at low doses and higher bioavailabilities at higher doses (Fig. 3). Differentiation using in vivo studies – on the clinical level – appears to be only possible by selective inhibition using process-specific inhibitors (Wu et al. 1995; Spahn-Langguth et al. 1997).

Fig. 1a,b. The amount of drug absorbed from the gut lumen per unit of time depends on two different processes, passive membrane passage and active transport. Passive diffusion is determined by the physicochemical properties of the drug, such as lipophilicity and pK_a, but also by the molecular weight. Active transport depends upon the respective transporter affinity as well as transport velocity and capacity, and it is saturable. **a** Two inside-directed transport processes: Simplified graph depicting the gut with its enterocyte layer and *active plus passive* processes responsible for the absorption of drugs (compounds included as examples: D-enantiomer of methotrexate (absorption via passive diffusion) and L-methotrexate (passive diffusion plus active process via the folic acid carrier) and folic acid, the physiological substrate of the folic acid transporter). Active processes are saturable and inhibitable, while passive processes are not. The insert depicts the relationship between (drug) concentration and *flux*, when active and passive processes are separated and combined, respectively, and when the active process has low (*upper graph*) or intermediate impact (*lower graph*) on the total absorptive flux. (Note: There may also be the situation where passive fluxes are negligible and the active process is almost solely responsible for absorption, and, hence, total absorption is saturable.) In terms of effective intestinal *permeabilities* (not shown), a straight line with a slope of zero in the flux graph means that in this concentration range the permeability would no longer be dependent upon the luminal concentration, while permeabilities would be decreasing with increasing concentration in the first segment of the curve. **b** One (passive) inside- and one outside-directed process: Graph similarly simplified as that in **a**, yet depicting the simultaneous occurrence of absorption via (inside-directed) passive diffusion and of intestinal secretion (exsorption) from the enterocyte (or blood) into the lumen (example: talinolol). Like inside-directed active transport, outside-directed active transport is also saturable and may be affected by competitive and noncompetitive inhibitors. Here, the insert depicts the concentration-dependence of total absorptive *flux* when the extent of exsorption is low (*upper graph*) or significant (*lower graph*). When considerable exsorption occurs, systemic availability may be low for a particular dose or concentration range. Upon saturation of the transport process, the effective intestinal *permeabilities* (not shown) would remain constant even though luminal concentrations are higher, while permeabilities would be increasing with increasing concentration in the first segment of the curve, i.e., when the transporter is not yet operating at maximum velocity

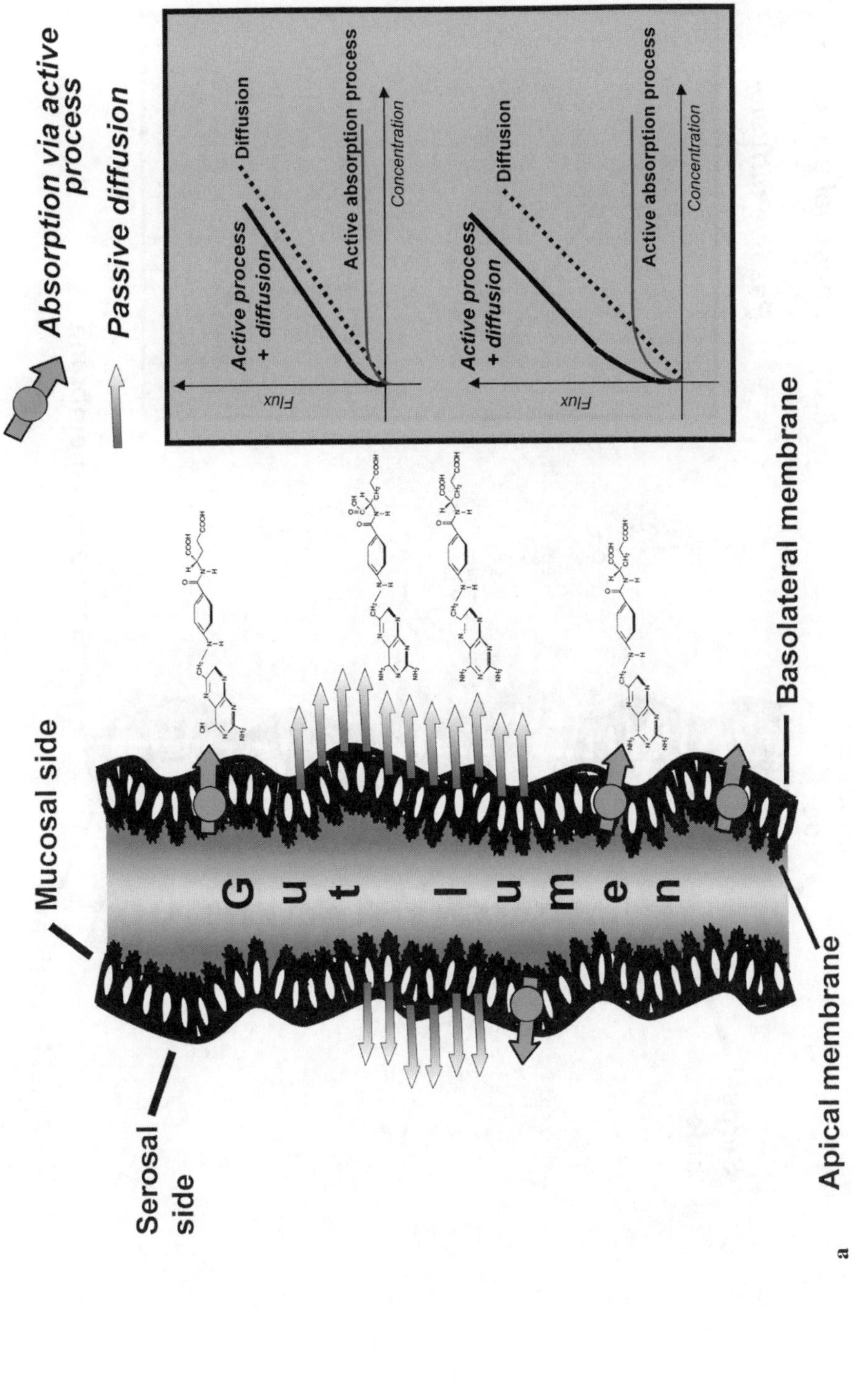
Absorption via active process
Passive diffusion
Active process + diffusion
Diffusion
Active absorption process
Flux
Concentration
Active process + diffusion
Diffusion
Active absorption process
Flux
Concentration
Mucosal side
Serosal side
Gut lumen
Basolateral membrane
Apical membrane

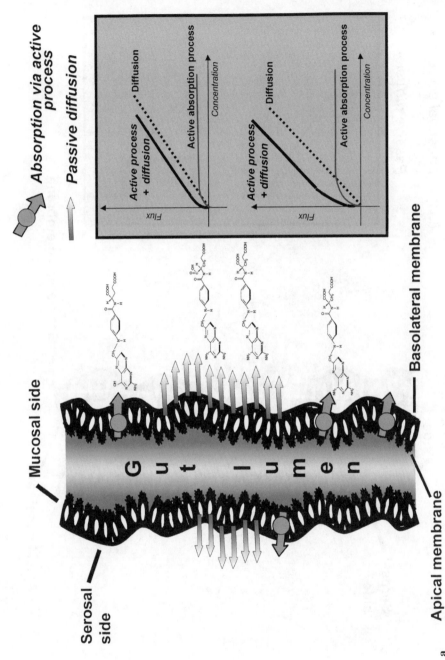

a

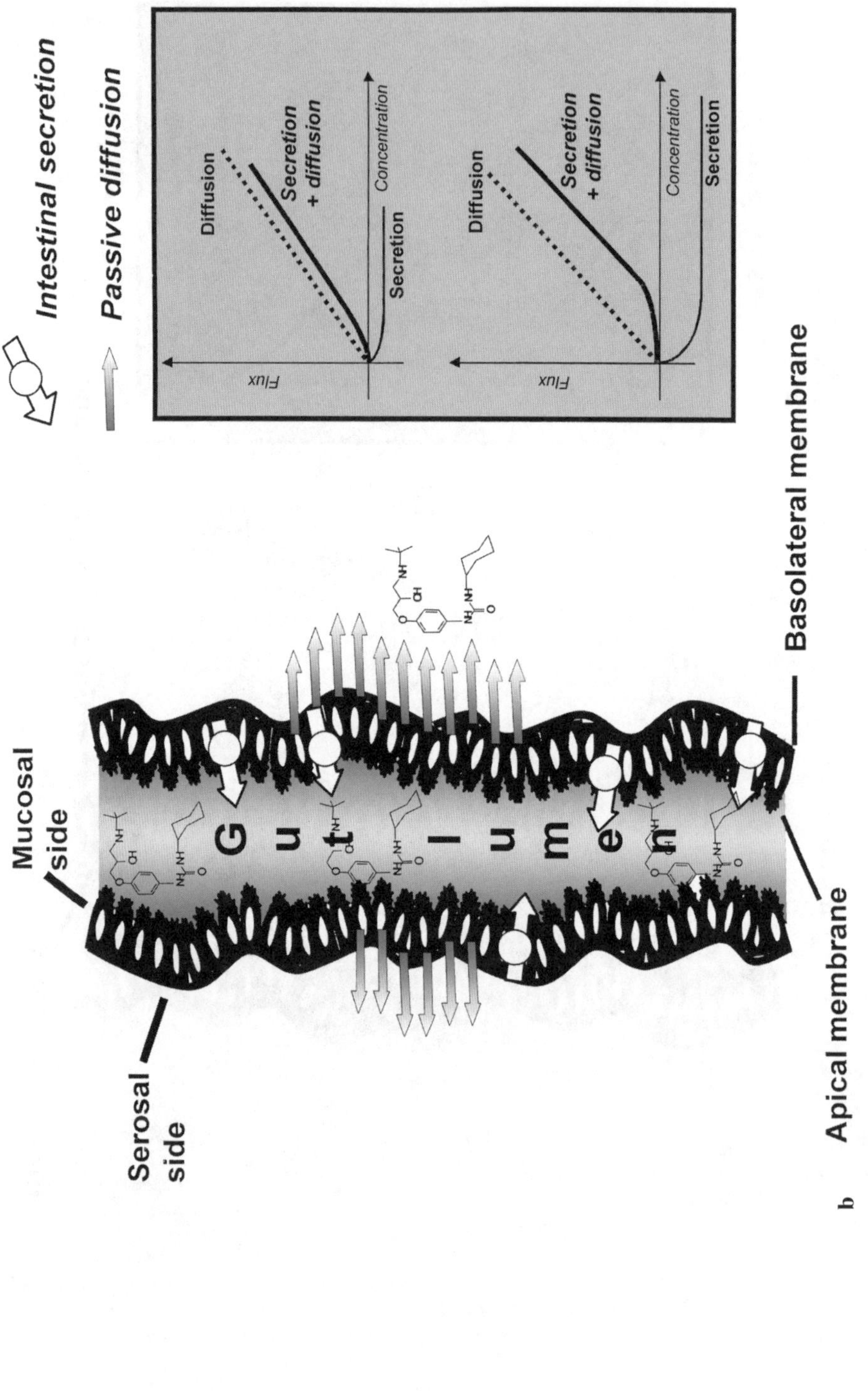

b

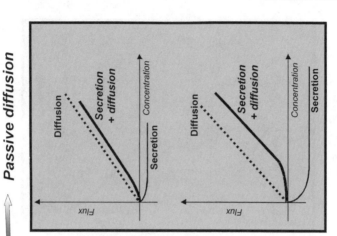

Fig. 1a,b. *Continued*

IV. Stereoisomers, Mutually Competing for Binding Sites at Receptors, Enzymes, and Transporters

The two enantiomers of a racemic drug/compound may or may not differ in their pharmacodynamic and/or pharmacokinetic properties. On the basis of their receptor affinity or pharmacological effect, it is possible to differentiate between the more active eutomer and the less active distomer. Both enantiomers are characterized by similar physicochemical properties, such as pK_a, molecular weight, and lipophilicity. Passive membrane passage is, hence, anticipated to be similar for both compounds, although the potential for differences has been discussed (FASSIHI 1993). Furthermore, solely on the basis of lipophilicity, which was found to be an important parameter for most kinetic processes upon investigation of structure-affinity relationships, the interaction with enzymes and transporters should not differ between enantiomers. There should only be a discrimination when the chiral center is contributing to the binding to enzyme or transporter.

Similar to stereo-/enantioselective drug action, transport enantioselectivity is only to be expected when a tight interaction between the chiral ligand and the transporter takes place. Similar to chromatography and to ligand-receptor interactions, a 3-point attachment at the binding area may be hypothesized for the enantiomer exhibiting higher affinity. Regarding drug action, the higher the affinity of the more affine enantiomer, the greater the enantiomeric difference in affinity (the so-called Pfeiffer's rule) (PFEIFFER 1956; LIEN et al. 1982). This is evident on the basis of the three-point attachment, since the loss of one binding point means considerable loss of binding forces. Therefore, it is hypothesized that lumen-to-blood directed transporters, which appear to exhibit a high substrate selectivity, have a higher potential for stereoselectivity than the enterocyte-to-lumen directed multidrug transporters responsible for intestinal secretion (blood-to-lumen transport).

Most absorption-related processes (e.g., gastrointestinal (GI) transit, pH variations, diffusion through membranes) do not appear to be relevant with respect to potential stereoselectivity. However, even when transport is not significantly stereoselective, other interesting aspects need to be considered: for carrier-mediated transport processes, an interaction between the two enantiomers is feasible, i.e., the eutomer might affect the transport of the distomer and vice versa, when both enantiomers have affinity to the relevant binding site of the respective transporter (see Sect. III).

V. Mucus Binding Prior to Interactions with the Brush-Border Membrane

Prior to the actual uptake into the enterocyte, the drug to be absorbed needs to pass the mucus layer present on the mucosal surface (LESUFFLEUR et al. 1990; MADARA and TRIER 1994; LANGGUTH et al. 1997). For various compounds, binding to the chiral sialic acid residues of gastrointestinal mucus has been

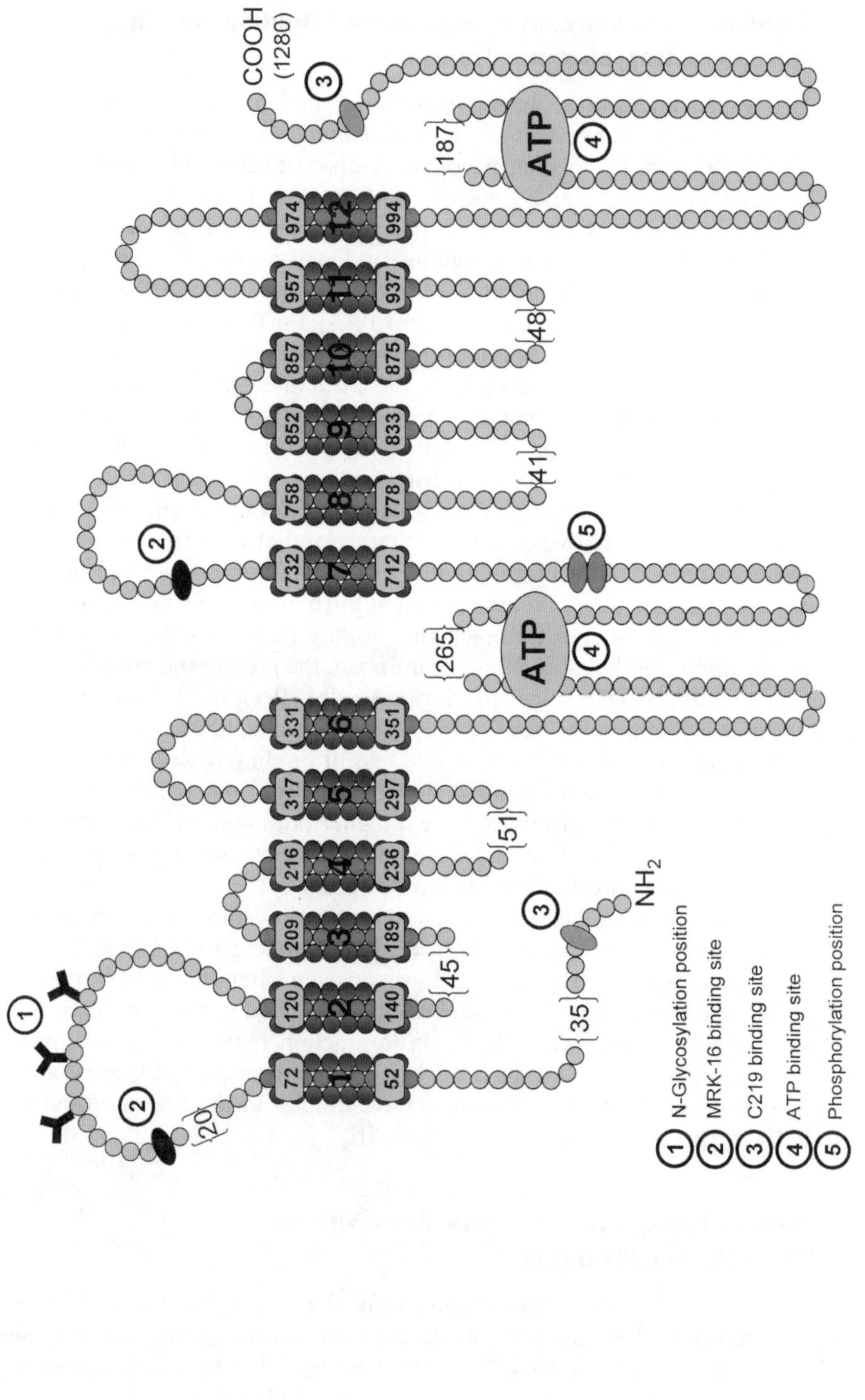
COOH
(1280)
NH2
ATP
ATP
72
52
120
140
209
189
216
236
317
297
331
351
732
712
758
778
852
833
857
875
957
937
974
994
20
35
45
51
265
41
48
187
1 N-Glycosylation position
2 MRK-16 binding site
3 C219 binding site
4 ATP binding site
5 Phosphorylation position

Fig. 2. A schematic representation of the ABC-transporter 170-P-glycoprotein (MDR1). [Besides the two ATP-binding sites, different drug binding sites were differentiated (vinca alkaloid-binding site, steroid-binding site) (CONSEIL et al. 1998)]. P-gp is composed of 12 transmembrane domains (TM) with the two nucleotide binding sites (NB) arranged in two symmetrical halves. This structural unit (6 TM, 1 NB) defines the ABC family of transporters. Glycosylation sites are located on the cell surface, i.e., on the outer loops, while ATP-binding sites are inside the cell. The two binding sites for the monoclonal P-gp antibody MRK-16 are located on the extracellular loops, those for C219 on the intracellular loops in proximity to the N- and the C-terminus. In mammals the ABC-transporter family includes transporters for peptides, chloride ions, acylated fatty acids, bile acids, and other substrates. Of particular interest is the multidrug resistant protein family, composed of at least six members in humans, with MRP1 and MRP2 capable of active drug transport when overexpressed in transfected cells. While P-gp can act on unmodified drug molecules, MRP1 and MRP2 can transport a variety of organic cations including glutathione and/or glucuronide adducts (JULIEN et al. 2000). JULIEN et al. purified P-gp following biotinylation under conservation of ATPase activity and tested a series of chiral 1,4-disubstituted piperazine derivatives, which were found to stimulate P-gp ATPase activity in a dose-dependent fashion, some of which in an *enantiospecific* fashion

detected. E.g., in vitro binding studies (equilibrium dialysis experiments) with isolated rat mucus revealed a 3:1 enrichment of the racemic model compound talinolol in the rat mucus-containing chambers when compared with controls. After the total dialysis time of 2h, the *S*/*R* ratio for talinolol was 1.07, although kinetic differences between the enantiomers had been observed indicating differences in the rate of association with rat mucus. Whether this is just a transient phenomenon or tight binding occurs, whether mucus binding leads to a decreased initial rate of absorption or a reduction of the amount of drug absorbed, i.e., whether it has any relevance in vivo, is unclear. However, Caco-2 transport studies with mucus-producing cocultures indicate that P_{eff} values remain virtually unaffected for most compounds when mucus is present (HILGENDORF et al. 2000).

VI. Dose-Dependent Stereoselectivity in Bioavailability Based on Active Transport and Countertransport

Active transport through membranes is known to play a role in bioavailability for a multitude of drugs. While high stereoselectivity is observed with respect to their carrier-mediated intestinal transport for nutritional constituents as well as for endogenous compounds that undergo enterohepatic cycling, the overall relevance of stereochemical aspects appears to be limited regarding drug absorption from the intestine. Although various drugs that are derived from endogenous compounds or nutrients utilize the respective stereoselective inside-directed transporters which is usually stereospecific, the other drug enantiomer appears to be quite frequently – less rapidly but

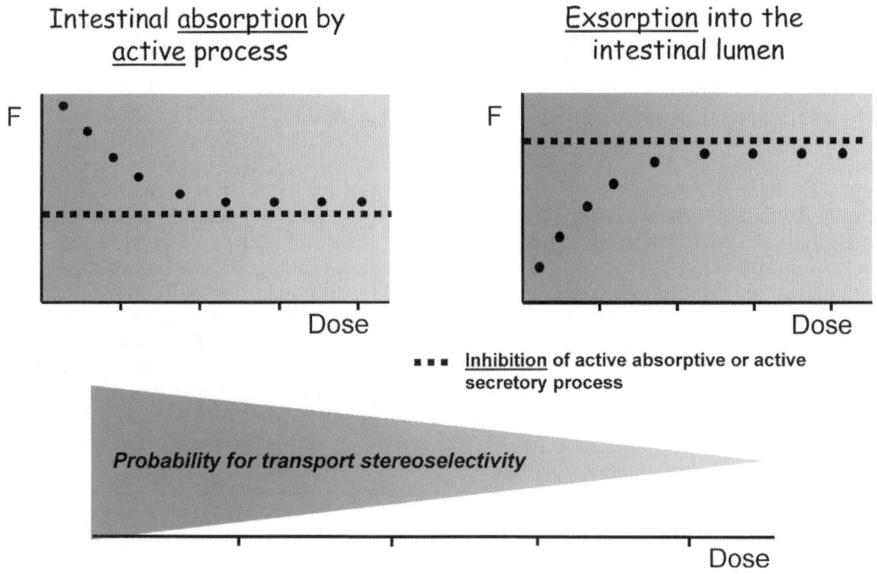

Fig. 3. Dose/concentration dependence of p.o. bioavailability (*F*) and probability for stereoselectivity when both passive and active processes contribute to absorption and when an active inside- or outside-directed transporter discriminates between two enantiomers and favors one of them. The probability for transport stereoselectivity depends on a number of factors. It decreases with decreasing impact of the active vs. a passive process, i.e., when the overall influence of this process is small and when the respective *active process* becomes *saturated* or when it is *inhibited*.

In different ways, net absorptive fluxes and bioavailabilities are dependent on luminal concentration or dose when active intestinal absorption or exsorption occurs [other bioavailability-limiting processes (e.g., metabolism) and alternative transport routes are not considered]:

1. For the *lumen-to-blood* directed active process, bioavailability decreases with dose, is determined by the maximum capacity of the transporter, and represents the sum of the saturable active process and – potentially – a simultaneous passive, dose-independent fraction. Maximum bioavailability may approximate 100% (F = 1) when the transporter operates with high efficiency and is not saturated. The smallest feasible value for F depends on the dose-dependent transport capacity and the fraction of the dose entering the blood via passive diffusion. The other extreme value equals 0 for *transport inhibition* with no passive fraction. (It approaches zero with increasing dose when active transport is saturated at no passive fraction, since the absolute amount actively transported reaches the value determined by its maximum capacity, and the relative contribution of this process decreases with increasing dose.)
2. For the *blood/enterocyte-to-lumen* directed active process (exsorption) bioavailability increases with dose (as observed for, e.g., talinolol; WETTERICH et al. 1996). Maximum bioavailability is reached by a complete blockade (inhibition) of exsorption. With saturation of exsorption the absolute amount transported per unit of time becomes independent of dose, i.e., the relative impact of exsorption decreases with increasing dose (asymptotically approaching zero). Upon inhibition or saturation of exsorption, bioavailability is mainly (or only) determined by the simultaneously occurring passive process

also often as completely – absorbed from the intestine by passive membrane passage [e.g., *S*-baclofen via the β-alanine-inhibitable amino acid transporter (Moll-Navarro et al. 1996)].

An important aspect for understanding the occurrence of transport stereoselectivities is the relationship between passive membrane permeation processes and active processes as indicated in Fig. 1 (inserts). The higher the (dose- or concentration-independent) passive permeability, the smaller the probability of inhibitable and saturable transport playing a role in overall absorption and the smaller the probability of net stereoselectivity in case of a stereoselective active absorptive transport or exsorption and of nonlinear bioavailability (Fig. 3).

VII. Transport and Affinity Assays and Related Problems

Today, numerous data about the relevance of particular transporters for compounds are available in the literature with most emphasis on the multidrug transporter P-glycoprotein. However, since neither a standard procedure nor a reference compound have been established, such data are not always comparable, as they are obtained with a variety of methods. For the establishment of structure-affinity- or structure-transport-relationships the generation of data with one or two defined and validated method(s) is desired (Scala et al. 1997).

1. Permeation Through Cell Monolayers

The numerous data and publications available for transport studies with colon carcinoma (Caco-2) cell monolayers (e.g., Artursson et al. 1996) represent an appropriate basis for the prediction of P-gp-mediated secretion and also of permeability (Sinko et al. 1991; Hilgendorf et al. 2000). In this cell line, different inside- and outside-directed transporters are expressed in addition to various enzymes. Other cell lines have also been proposed and used to some extent, such as HT29, HT29-MTX, and IEC-18, as well as genetically modified systems (Hilgendorf et al. 2000; Hilgendorf 1999).

Unless the fluxes originating from nonsaturable processes are predominant and active processes negligible, the presence of energy-consuming uptake or secretion mechanisms leads to one preferred transport direction through the membrane. In vitro, e.g., in transport studies through isolated tissues or cell monolayers, this leads to a *side-dependence* (*and transport against a concentration gradient*), resulting in a difference between apical-to-basolateral (a–b) and basolateral-to-apical (b–a) transport and b–a/a–b ratios (= efflux ratios) of >1 for compounds that are secreted through the apical microvilli membrane and <1 for compounds actively transported to the blood side. This side-dependence is a major characteristic of significant active transport processes through the enterocyte layer and is supplemented by two of the other characteristics for capacity-limited transport: saturability and (competitive or noncompetitive) inhibition. *Saturation of inside*-directed processes leads to decreasing permeabilities upon increasing donor concentrations (apical side), since the uptake mechanism present in addition to, e.g., passive diffusion works at maximum capacity and higher donor levels can no longer increase uptake. On the other hand, when active secretory processes (outside-directed) occur to a relevant extent, their saturation leads to higher a–b (and smaller b–a) permeabilities with reduced efflux ratios for higher donor concentrations. *Inhibition* of the respective active absorptive or active secretory process leads to decreased uptake when an inside-directed process occurs or an increased uptake when exsorption occurs (see also Fig. 3).

When isomers are to be assayed in one of the respective model systems or in vivo, particular demands need to be obeyed regarding assay specificity and validation. Chromatographic systems with separation of enantiomers on chiral stationary phases or as diastereomeric derivatives should be used when racemates are tested and the two isomers are assayed. Assay validation usually refers to both of the enantiomers separately. Systematic deviations of one of the calibration curves in particular concentrations ranges, however, may – artificially – create enantiomer differences.

2. Uptake Studies, Efflux Studies

As opposed to the transport through monolayers, a number of assays have been developed which are based on the uptake and/or extrusion of radioactively labeled or chromophoric marker compounds. Similar to the ^{3}H-vinblastine drug accumulation assay and the ^{14}C-adriamycine assay (e.g., FOJO et al. 1985; CRITCHFIELD et al. 1994), the rhodamine 123 efflux assay represents a well-known and frequently applied P-gp assay (e.g., LEE et al. 1994). It is based on the fluorescent rhodamine 123, which is a substrate for P-gp and other transporters, and is extruded from the P-gp expressing cell, i.e., the higher the P-gp content, the smaller the fluorescence of the cell. The calcein AM extrusion assay represents a more sophisticated P-gp assay, the principle of which is the affinity of calcein AM to P-gp and its extrusion from P-gp-containing cells, unless P-gp is inhibited. Calcein AM is an ester, which is hydrolyzed inside the cell by cytosolic hydrolases yielding the fluorescent calcein, which is less lipophilic and not a P-gp substrate and hence remains inside the cell (ENEROTH et al. 2001), where it is quantified.

The ^{14}C-tetraethylammonium (TEA) uptake assay using LLC-PK$_1$ cells represents a standard method to study the interaction with the organic cation transporter (OCT) (FAUTH et al. 1988; SAITO et al. 1992). A similar assay may be used to study the intestinal OCTs. The competitive inhibition of ^{14}C-TEA uptake by OCT substrates serves as measure of their affinity to the organic cation transporter. In addition, cytotoxicity assays have also been used (SKEHAN et al. 1990; SCALA et al. 1997).

3. Photoaffinity Labeling

Azidopine labeling was performed by BRUGGEMANN et al. (1989). These authors used plasma membranes of P-gp-expressing cells and photolabeled with ^{3}H-azidopine (under UV light) in the presence and absence of the test compound.

4. Radioligand Binding Assay

A competitive P-gp-radioligand binding assay (P-gp-RBA) was designed as a supplementary or alternative screening system to transport studies through monolayers of the human colon carcinoma cell line Caco-2. It is based on a Caco-2 cell preparation (that is overexpressing P-glycoprotein and is treated with lysolecithin to obtain porated cells) and ^{3}H-verapamil, -vinblastine or -talinolol as radioligands, and focuses on affinities to the respective binding site as opposed to the steroid-binding region (DÖPPENSCHMITT et al. 1999a; NEUHOFF et al. 2000). The data obtained for the β-adrenoceptor antagonist talinolol and other compounds revealed either no, negligible, or little stereoselectivity with respect to *binding* at the vinca alkaloid-binding site of intestinal P-gp (WAGNER et al. 2001). All radioligands, including bi-tritiated talinolol, are commercially available (e.g., www.isotopes.de).

5. ATPase Assay

RAO (1995) and others described quantification of ATPase activity (e.g., of isolated Sf9 cell membranes), where verapamil-induced activity may be included as reference (SARKADI et al. 1992). A test kit based on this principle is commercially available (ABEL et al. 2001).

6. Potential of Assays to Detect Enantiomer Differences

All assays mentioned above show some degree of variability, as generally occurs with biological systems. It may be difficult to detect small enantiomeric differences using such assays; hence the two enantiomers need to be *studied separately*, especially if no enantiomer-specific assay is used. The assays may well be suitable for detection of *marked* differences, for rankings, and for structure-affinity studies.

Only in transport studies combined with drug-specific assay are the two enantiomers of the respective compound tested simultaneously.

B. Chiral Drug Examples for Active Inside- and Outside-Directed Transport

I. Active Inside-Directed (Lumen-to-Blood) Transport

Differences in the rates and extents of absorption may be observed for compounds with affinities to amino acid or peptide carriers or other inside-directed transport systems located in the gut wall. With respect to amino acid- or peptide-type drugs, which are often used in a stereochemically pure form, stereoselective transport is assumed in many cases, yet not directly tested (TAYLOR and AMIDON 1995).

1. Methotrexate

The publication of HENDEL and BRODTHAGEN (1984) on the intestinal handling of methotrexate focuses on the differences between the D- and L-isomer. As opposed to L-methotrexate, which is actively and completely absorbed from the intestine (and reabsorbed after secretion into the bile) after oral doses of up to 30 mg/m^2 body surface, D-methotrexate is very poorly absorbed (<3%). In addition to saturability of D-methotrexate transport, competitive transport inhibition by folic acid was described (e.g., ZIMMERMAN 1992).

Because of methotrexate's widespread use in cancer chemotherapy, its transport through cell membranes was extensively studied and different influx and efflux carriers were detected in various cell systems, particularly since altered transport of methotrexate across the membrane of leukemic cells has resulted in resistance towards the drug. Because it inhibits the uptake of folic acid and 5-methyl tetrahydrofolate in preparations of rat intestinal tissues (SAID and REDHA 1987) via an electroneutral, Na^+-dependent, and DIDS-sensitive (DIDS = anion exchange inhibitor 4,4′-di-isothiocyanatostilbene-2,2′-disulfonic acid) transporter presumably located at the basolateral membrane, methotrexate was expected to share a common carrier in the intestinal luminal membrane with all folates. In the studies by ZIMMERMAN (1992), which used different pH levels and did not include stereochemical aspects, a 37% inhibition of methotrexate flux by folic acid was detected, indicating a proton-dependent active transport system in the human proximal small intestine.

Previous in vivo data on the stereoselectivity associated with methotrexate absorption via the folic acid transporter were recently confirmed by studies, in which rabbit intestinal brush-border membrane vesicles were used and studies performed in the presence of an H^+ gradient (ITOH et al. 2001). Methotrexate (L-isomer) and the antipode (D-amethopterin) were used as model substrates of the transporter. Initial uptake of folic acid and methotrexate, which acted as mutual competitive inhibitors apparently sharing the same transporter (Fig. 1A), was concentration-dependent with K_m values of 1.5 and 1.6 μM, respectively, and similar K_i values. Also, the D-isomer showed affinity to the folic acid transporter; however, the K_i value was 60-fold greater than that of the L-isomer.

2. Amino Acid-Related Structures

Early drug examples with active inside-directed transport include L-3,4-dihydroxyphenylalanine (L-DOPA). In 1973, two independent groups, WADE et al. and SHINDO et al., showed a competitive inhibition of the absorption of L-DOPA by other amino acids (L-alanine, L-phenylalanine, but not D-alanine, D-phenylalanine), accumulation of L-DOPA against a concentration gradient, and a higher uptake rate for L- than for D-DOPA. Furthermore, absorption followed saturation kinetics with decreasing absorption rates upon increasing doses, a phenomenon which is more significant for the L-isomer. These observations indicate that only L-DOPA is actively transported by a neutral amino acid transport system. In the in situ perfused rat intestine, disappearance from the lumen amounted to 55% (L) vs. 22% (D) after 10 min and 94% (L) vs. 55% (D) at 30 min (SHINDO et al. 1973). The respective fractions entering the blood compartment (corrected for metabolism) amounted to 41% (L) vs. 9% (D) at 10 min and 83% (L) vs. 29% (D) at 30 min post-dose.

Comparable behavior is assumed for the structurally related *α-methyldopa*. Its intestinal permeability in humans showed the expected pH- and concentration dependence (MERFELD et al. 1986).

Baclofen (*R*/*S*-4-amino-3-(4-chlorophenyl)butyric acid) is a centrally acting antispastic agent. The direct mimetic effect on $GABA_B$-receptors is enantioselective with the *R*-enantiomer being about 100 times more potent than the *S*-enantiomer (OLPE et al. 1978), yet pharmacodynamic antagonism between the two enantiomers was also found (SAWYNOK et al. 1985). However, unpublished transport studies with the everted-sac model (rat small intestine), the in situ perfused rat intestine, and Caco-2 cell monolayers indicated that both enantiomers are well absorbed with some preference for the *R*-enantiomer. In the studies using intestinal sacs and everted-sacs, a rapid equilibration was observed and the *R*/*S* ratio increased at the serosal (1.2) and decreased at the mucosal side (0.8) with time (90 min), which indicates active transport for *R*-baclofen in the absorptive direction (KRAUSS 1988; SPAHN et al. 1988; SPAHN 1989). Using the in situ perfused rat jejunum as experimental system, MOLL-NAVARRO et al. (1996) studied baclofen absorption with and without addition of inhibitors. For baclofen, these authors confirmed previously obtained in situ parameters: V_{max}, 27.73 ± 9.99 mmol/h; K_m, 8.06 ± 2.82 mmol; k_a (passive diffusion component), 0.40 ± 0.28 1/h. Transport was inhibited by the amino acid *β*-alanine (and the sulfonic *β*-amino acid taurine), and the data indicated that baclofen and the two amino acids share the same carrier. The most potent inhibitor was the *α*-amino acid leucine. The authors detected a residual absorption for baclofen, which was not inhibited. However, stereochemical aspects were not included in this publication, although presumably baclofen was studied as racemate and the chiral amino acids as pure L-form.

Transport studies through Caco-2-cell monolayers performed in the apical-to-basolateral direction using [^{3}H]-baclofen and [^{14}C]-rac-baclofen indicate a slightly higher absorptive flux for the *R*-enantiomer than for the racemate, which increased with time in culture [as an indication of higher (amino acid) transporter expression with increasing number of days in culture (DRESSLER et al. 2001)].

In studies in man with dosage of 20 mg p.o. and 10 mg i.v. as reference, the absolute bioavailability was 80% for racemate with no significant difference in absolute bioavailability between enantiomers, although that of the *R*-enantiomer – in spite of the higher C_{max} values – appeared slightly lower than that of its optical antipode. After p.o. administration C_{max} was always statistically significantly higher for *R*-baclofen (Fig. 4). Renal excretion rates showed an *R*/*S* ratio significantly different from unity for the first 2-h sampling interval only. The overall amount excreted into urine was slightly higher for *S*- than for *R*-baclofen (CL_R identical); however, during the first 2–4 h after administration of racemate, the amount excreted into urine was significantly higher for *R*- than for *S*-baclofen (KRAUSS 1988).

3. Peptide-Related Structures

In one of the early investigations that included stereochemical aspects of *cephalosporine* absorption (TAMAI et al. 1988), it was found that L-cephalexin does not reach the systemic circulation because of rapid degradation (hydrolysis), while D-cephalexin was well absorbed (yet its uptake was saturable) in rat intestine. Interestingly, L-cephalexin competitively inhibited the uptake of D-cephalexin, which suggested that both stereoisomers may be absorbed through the intestinal brush-border membrane via the same mechanism, presumably a dipeptide transport system, to which L-cephalexin appeared to have a higher affinity than its optical antipode. In a subsequent investigation

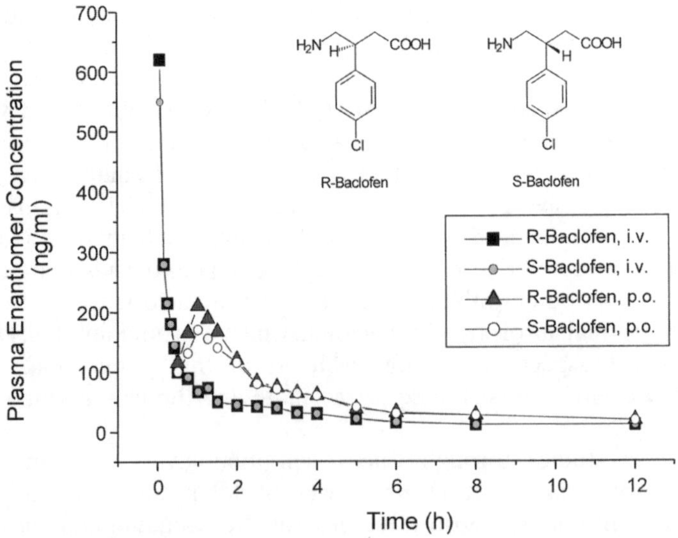

Fig. 4. Baclofen enantiomers in plasma. Average plasma concentration-time curves of baclofen enantiomers obtained in a clinical study with p.o. doses of 20 mg and i.v. doses of 10 mg administered to six healthy volunteers. (SPAHN 1989)

(KRAMER et al. 1992), these data were confirmed in radioaffinity studies, where the L-isomer reduced the extent of radiolabeling with [^{3}H]-benzylpenicillin to a higher extent than the D-isomer. However, simultaneously obtained results from a liposome-based reconstituted oligopeptide transport system from rabbit small intestine indicated that under the respective conditions only D-cephalexin and not L-cephalexin is actively transported.

WENZEL and coworkers (1995) included cephalexin enantiomers and loracarbef as well as dipeptides in their studies, which focused on affinities and transport rates of the enantiomers. They found higher affinities of the respective L-enantiomers to the oligopeptide/H^+-symporter in Caco-2 cells using the β-lactam [^{3}H]-cefadroxil as radiolabeled model substrate, as well as the cloned peptide transporter PepT1 from rabbit small intestine, that was expressed in *Xenopus laevis* oocytes. They demonstrated that the L-isomers are transported.

Studies on structure-activity relationships with respect to carbacephalosporins and cephalosporins included their antibacterial activity and their transport via the proton-dependent carrier in Caco-2 cells. Investigating a series of derivatives of cephalexin and loracarbef, particular structural features responsible for antimicrobial activity, transporter affinity, and deactivation via hydrolysis were illustrated (SNYDER et al. 1997).

Carrier-mediated processes have been implicated in determining the absorption characteristics of angiotensin-converting enzyme (*ACE*) *inhibitors* in experimental models (e.g., in situ luminal perfusion in rats with captopril: HU and AMIDON 1988) and after oral dosing in vivo (e.g., NICKLIN et al. 1996). Absorption of ACE inhibitors was inhibited by β-lactam antibiotics and dipeptides (cephradine, L-tyrosylglycine). Examples of drugs/prodrugs actively transported, besides the lead compound captopril, were enalapril (FRIEDMAN and AMIDON 1989a) and ceronapril (FRIEDMAN and AMIDON 1989b; NICKLIN et al. 1996), while one heavily substituted ACE inhibitor (fosinopril) was found to be absorbed from the intestinal lumen via a nonsaturable – presumably passive – route (FRIEDMAN and AMIDON 1989a).

Analogues of the natural renin substrate angiotensinogen, which represent peptide derivatives where the Leu-Val bond was replaced, were found to be potent renin inhibitors and also to be transported via the peptide transporter, since they reduced cephalexin uptake in a concentration-dependent manner (KRAMER et al. 1990).

4. Monocarboxylic Acid Transport

Although the absorption of weak organic acids is usually and mainly explained by the pH-partition theory, there is evidence that a carrier-mediated mechanism may contribute to the process to a varying extent.

Absorption of salicylic acid and benzoic acid, and a number of other carboxylic acids including chiral compounds was found to be absorbed through Caco-2 cell monolayers via a carrier-mediated intestinal transport mechanism

(TSUJI and TAMAI 1996). In addition, the inhibitory effect of lactic acid enantiomers, which was evident from investigations on intestinal membrane transport of nicotinic and acetic acid in different experimental models, e.g., the transport of D- and L-lactic acid, mandelic acid and ibuprofen enantiomers were examined (OGIHARA et al. 1996). A saturable as well as a nonsaturable process were revealed for both lactic acid isomers. No difference between the enantiomers was detected for the passive process; however, for the saturable process, a higher affinity and lower capacity was found for L- as opposed to D-lactic acid. In other words, at a low concentration, L-lactic acid is more efficiently transported, while at a high concentration, a preference for the D-enantiomer is evident. No transport stereoselectivity should be observed at an intermediate concentration of 1 mM. Under the experimental conditions used, differences between enantiomers were also observed for *S*- and *R*-mandelic acid with an approximately 30% higher permeability coefficient for the *S*-enantiomer, but not for ibuprofen enantiomers, for which the permeability coefficients were similar between stereoisomers (Tables 1, 2). Inhibition studies revealed a higher inhibitory effect of *S*-mandelic acid and *S*-ibuprofen on L-lactic acid transport than for *R*-mandelic acid and *R*-ibuprofen. The results of OGIHARA's studies clearly demonstrated that the detection of

Table 1. Permeability coefficients for [^{14}C]-lactic acid, mandelic and ibuprofen enantiomers (μl/min/mg protein) through Caco-2 cell monolayers at an apical pH of 6.0 and a basolateral pH of 7.3

Substrate	Concentration	Permeability
L-Lactic acid	1 μM	1.250 ± 0.004
D-Lactic acid	1 μM	0.626 ± 0.078
S-Mandelic acid	0.5 mM	0.144 ± 0.006
R-Mandelic acid	0.5 mM	0.106 ± 0.003
S-Ibuprofen	0.5 mM	0.106 ± 0.003
R-Ibuprofen	0.5 mM	9.57 ± 0.81

Table 2. Inhibitory effect on L/D-[^{14}C]-lactic acid transport (% of control) (data from OGIHARA et al. 1996)

Inhibitor	Concentration (mM)	Relative transport rate	
		L-[^{14}C]-lactic acid	D-[^{14}C]-lactic acid
L-Lactic acid	10	37.3 ± 3.11	42.6 ± 0.86
D-Lactic acid	10	76.7 ± 1.27	73.8 ± 2.93
S-Mandelic acid	10	35.7 ± 1.07	45.1 ± 0.55
R-Mandelic acid	10	78.7 ± 0.62	80.5 ± 1.23
S-Ibuprofen	1	70.6 ± 1.94	
R-Ibuprofen	1	73.3 ± 4.14	
S-Ibuprofen	5	24.0 ± 1.95	
R-Ibuprofen	5	35.2 ± 2.86	

stereoselectivity might depend on the selected substrate concentration and provided evidence that also for ibuprofen – and potentially related compounds – some transport stereoselectivity may be detected at appropriate experimental conditions.

Another compound of relevance due to its chiral character is L-ascorbic acid, for which the facilitative sugar transporters of the GLUT type can transport the oxidized form, dehydroascorbic acid, while 2 L-ascorbic acid transporters, SVCT1 and SVCT2, may mediate a concentrative, high-affinity L-ascorbic acid transport that is stereospecific and is driven by the Na^+-electrochemical gradient. Despite their close sequence homology and similar functions, the two isoforms are discretely distributed with SVCT1 located in the epithelial systems of the intestine as well as kidney and liver, while SVCT2 occurs in the metabolically active cells and membranes of specialized tissues, e.g., the brain and eye (Tsukaguchi et al. 1999).

Stereoselective uptake of carboxylic acids may also occur in other tissues and the relevance for the intestine has not yet been tested. MBCA (5-monomethyl-sulfamoyl-6,7-dichloro-2,3-dihydrobenzofuran-2-carboxylic acid) was studied with respect to its renal handling (Higaki et al. 1998). The uptake rate constant across the basolateral membrane and other processes were significantly larger for *R*-(+)-MBCA. The authors concluded that the uptake across the basolateral membrane and intracellular distribution were stereoselective processes.

II. Drug Examples for Active Outside-Directed (Blood-To-Lumen) Transport

1. Fluoroquinolones – Ofloxacin

Although previous studies (Prieto et al. 1988) had provided evidence for a saturable absorption when increasing luminal concentrations were used in perfusion studies in rats, carrier-mediated intestinal elimination of ofloxacin was detected in rats (Rabbaa et al. 1996, 1997), and appeared to favor the preferentially metabolized *R*-(+)-form (luminal perfusions with drug). Here, the initial portal blood concentrations and the portal AUCs of the *S*-(−)-enantiomer exceeded those of its optical antipode (C_{init} at 5 min: 0.17 [*S*] and 0.12 μg/ml [*R*]), whereas the intestinal clearances were determined as 0.23 (*S*) and 0.30 ml/min (*R*) following an i.v. dose of 20 mg/kg in rats. Intestinal elimination was maximal in the segments isolated from the upper gut (duodenum – proximal jejunum). Serum AUCs of both enantiomers increased linearly over the tested dose range (20–100 mg/kg as i.v. bolus). Transport inhibition studies in the in situ perfusion model with P-gp inhibitors, amino acids and peptides, and other fluoroquinolones indicated that more than one transporter is involved in the absorption of fluoroquinolones (Griffiths et al. 1994). At present, the literature data are not yet sufficient to attribute the observed stereoselectivity to an interaction with a particular transporter, such as P-gp or OCT (PIETIG et al. 2000).

2. Verapamil

Stereochemical aspects regarding verapamil and its interaction with P-glycoprotein have already been addressed by different authors. For example, SANDSTRÖM and coworkers (1998) performed jejunal absorption studies in man using the Loc-I-Gut model with different verapamil perfusate concentrations. They found evidence for saturability of exsorption, since the effective intestinal permeability increased with increasing perfusate concentration (4 mg/l (a), 40 mg/l (b)). The effective permeabilities were generally high, yielding values of 2.71 (a) and 4.74×10^{-4} cm/s (b) for *R*-(+)-verapamil and values of 2.25 (a) and 4.69×10^{-4} cm/s (b) for *S*-(–)-verapamil. These authors did not detect any difference in the jejunal P_{eff} between the two enantiomers and concluded that this indicates that P-glycoprotein in the human jejunal enterocyte does not discriminate between verapamil enantiomers. This conclusion is in agreement with data reported from in vitro studies using tumor cell lines (HAUSSERMANN et al. 1991) and data obtained in P-gp binding studies, such as those described by NEUHOFF et al. (2000).

3. β-Adrenoceptor Blockers

The different enantiomeric pairs of a group of β-adrenoceptor antagonists that were investigated in radioligand binding studies with ^{3}H-verapamil and/or ^{3}H-talinolol as radioligand (e.g., DÖPPENSCHMITT et al. 1999b; NEUHOFF et al. 2000; LEISEN et al. 2002) did not reveal any evidence for relevant binding stereoselectivity. Even for the compound exhibiting by far the highest efflux ratio, talinolol, the two IC_{50} values were similar.

Previous publications demonstrated for talinolol that its intestinal secretion affects concentration-time profiles and bioavailability, which was higher with higher doses (DE MEY et al. 1995; WETTERICH et al. 1996).

Similar results, including the occurrence of double peaks, were obtained with the structurally related celiprolol. For both compounds, secretion into the intestinal lumen was detected in the rat in situ intestinal perfusion model following intravenous administration of the respective compound and/or in intestinal perfusions in man (e.g., SPAHN-LANGGUTH et al. 1998; GRAMATTÉ et al. 1996).

In the studies of WETTERICH et al. (1996), some stereoselectivity was detected, mainly after p.o. dosage. Although the difference between the two enantiomers with higher levels of *R*- than *S*-talinolol was of small magnitude, it was statistically significant. Additional permeation studies with Caco-2 cell monolayers supported the hypothesis that transport through intestinal enterocytes may be one source of stereoselectivity, at least at low apical concentrations (WETTERICH et al. 1996). *S*/*R* ratios were slightly below unity, amounting to 0.85 at the lowest apical concentration of 0.2 mM rac-talinolol.

Table 3. Talinolol enantiomer exsorption in Caco-2 cells (grown without addition of antibiotics) is inhibited/reduced by the P-gp monoclonal antibody MRK-16 (HILGENDORF 1999)[a]

	Conditions		P_{eff} values		S/R P_{eff} ratio		Efflux ratio	
	$S_{a\text{-}b}$	$S_{b\text{-}a}$	$R_{a\text{-}b}$	$R_{b\text{-}a}$	$S/R_{a\text{-}b}$	$S/R_{b\text{-}a}$	S	R
Talinolol control	0.27	8.49	0.35	8.25	0.78	1.03	31.4	23.6
Talinolol + MRK-16	3.37	4.00	3.74	3.96	0.90	1.01	1.19	1.06

[a] Data represent average P_{eff} values of 2 × 3 representative studies; P_{eff} in 10^{-6} cm/s.

All data available supported the hypothesis that talinolol undergoes intestinal secretion (exsorption).

More extended studies with Caco-2 cell monolayers, where the P-gp specific monoclonal antibody MRK-16 was added, revealed that apical-to-basolateral vs. basolateral-to-apical transport differences were marked and stereoselective in a certain apical concentration range, but only when no inhibitor was added (Table 3).

Further studies were performed to support the hypothesis that outside-directed talinolol transport depends upon the extent of P-gp expression, e.g., by studying efflux ratios in different P-gp expressing- and nonexpressing-cell lines (IEC-18; HT29-MTX) and under different culture and transport conditions (DRESSLER et al. 2001) as well as in induced and genetically modified Caco-2 cells (MDR1-transfected vs. anti-MDR1-transfected cells; HILGENDORF et al. 1999, 2000; DÖPPENSCHMITT et al. 1999a).

Stereochemical aspects were also included in the clinical work of WESTPHAL et al. (2000b), who did not detect any difference between enantiomers in their population of healthy volunteers when talinolol was dosed alone as single or repetitive dose. They did detect a slight stereoselectivity following rifampicin pretreatment with higher levels of *R*-talinolol as observed in the previous studies in healthy volunteers, which had not been pretreated (WETTERICH et al. 1996).

III. The Distomer as Shoehorn or as Secretion Inhibitor for the Eutomer?

It is usually anticipated that the less active or inactive distomer is not relevant for the overall action of the racemate because of the lack of pharmacological activity when compared with the eutomer. The one restriction made is a potential competition at metabolizing enzymes or plasma proteins when both isomers have significant affinity. However, this assumption needs to be challenged from a different point of view when transporters are involved in drug absorption (and disposition).

Very recent studies with talinolol demonstrate that there may be a mutual influence between distomer and eutomer regarding distribution (HANAFY et al. 2002), transporter-mediated excretion, and also absorption, which was demonstrated in rodent studies (HANAFY 2001).

Table 4. The influence of the distomer *R*-talinolol on the intestinal absorption of *S*-talinolol[a]

	A[b] Net *secretory* flux	B[c] Effective permeability		C[d] Plasma AUC
		Ileum	Jejunum	
S (1+0)	100	100	100	100
S + 0.5 *R* (1+0.5)	81.6 ↓	–	–	–
S + *R* (1+1, rac)	55.1 ↓	283 ↑	255 ↑	111 ↑
S + 6 *R* (1+6)	26.3 ↓	506 ↑	446 ↑	205 ↑
S + V[e] (1+V)	10.2 ↓	121 ↑	114 ↑	156 ↑

[a] Intestinal transport studies with competitive inhibition of *S*-talinolol (*S*) exsorption by the distomer *R*-talinolol (*R*) as well as verapamil racemate (V) as positive control and p.o. bioavailability enhancement in vivo in rats (experimental conditions: *S*-talinolol donor concentration in Caco-2 cell monolayer transport studies: 250 μM; donor concentration in in situ intestinal perfusion studies: 5 μM; p.o. dose: 10 mg *S*-talinolol).
[b] Relative change of *net secretory flux* of *S*-talinolol (control value = 100%, 1.98 μmol h^{-1}) during presence of the competitive inhibitor *R*-talinolol or verapamil [0.5 mM (HILGENDORF 1999)].
[c] Relative change of *effective permeability* in in situ intestinal perfusion studies in rats [control value = 100%, 0.81 × 10^{-4} cm/s (ileum) and 0.94 × 10^{-4} cm/s (jejunum)] under presence of 30 μM *R*-talinolol or 500 μM verapamil (HANAFY 2001).
[d] Relative change of *plasma AUC* (0–8 h) of *S*-talinolol in rat following a 10 mg p.o. dose of *S*-talinolol without or with *R*-talinolol or verapamil [control AUC = 100%, 1.02 μg·ml^{-1} h; concomitant verapamil p.o. dose, 20 mg racemate (HANAFY 2001)].
[e] In all experimental studies, *R*-verapamil had a qualitatively similar influence on absorption as racemic verapamil.

With respect to absorption from the gastrointestinal tract, data obtained with enantiomers of talinolol in vitro, in situ, and in vivo indicate that the presence of the distomer *R*-(+)-talinolol (eudismic ratio approximately 40) at the absorption site reduces intestinal secretion of the active *S*-(−)-talinolol into the lumen and enhances the effective permeabilities for the pharmacologically active *S*-enantiomer. Hence, although basically inactive, the distomer does contribute to the overall effect of the eutomer, since it enhances its absorption from the intestine. These data were supported by in vitro and in situ data. Results were comparable for the Caco-2 model and the in situ intestinal perfusion model in rats (Table 4) (HANAFY 2001).

However, most inside-directed transporters show more stereoselectivity than the multidrug transporters responsible for exsorption. Under the condition of high stereoselectivity and preference for the eutomer, the distomer would be virtually irrelevant in this respect.

IV. Model Compounds for P-gp-Related Processes

1. Talinolol

Because of various advantageous kinetic properties, talinolol was proposed as a suitable model compound for P-gp-mediated transport in binding assays, in in vitro and in situ transport studies, as well as in vivo with respect to exsorptive transport-based drug–drug and drug–food interactions (Fig. 5) (SPAHN-LANGGUTH et al. 1998; SPAHN-LANGGUTH and LANGGUTH 2001; HANAFY 2001). This appears reasonable because of its broad therapeutic range (which represents a characteristic of drugs of this pharmacological group), a mainly unchanged renal and biliary clearance, the low protein binding (approximately 25%) and the sensitivity of its kinetics for changes in P-glycoprotein expression (e.g., HANAFY et al. 2001), but also to transporter function (inhibition by P-gp modulators). Evidence exists for an additional interaction with the organic cation transporter (OCT), and furthermore, from in vivo studies with MRP2 deficient rats, for a contribution of MRP2 to talinolol kinetics. Similar to the other compounds (see Sect. VIII.4.) talinolol has recently become available as tritium-labeled compound (www.isotopes.de).

Comprehensive work with talinolol as model compound has incorporated drug interaction studies (in vivo in rats and in the intestinal perfusion model)

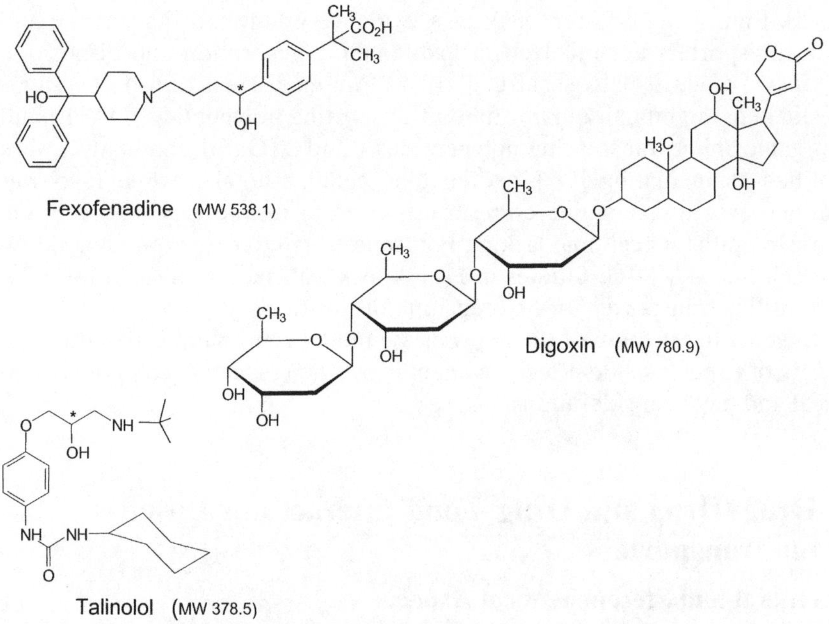

Fig. 5. Model compounds: Digoxin, fexofenadine, talinolol. Both in unlabeled and labeled form, these compounds may be used as reference or for investigational studies on P-glycoprotein-related processes, since they exhibit suitable physicochemical and kinetic properties to serve as P-gp model substrate. The graph depicts the structures and includes the molecular weights of the compounds (for example, talinolol: pK_a, 9.4; log P, 0.74; negligible binding to plasma protein and very low metabolic clearance (TRAUSCH et al. 1995; intermediate affinity to P-glycoprotein as shown in radioligand binding studies; high efflux ratio in Caco-2 cell studies, broad therapeutic range)

as performed by HANAFY (2001) and HANAFY et al. (2001), and cell culture work by HILGENDORF et al. (1999, 2000, 2001). Its recent acceptance in the clinical field is documented by the follow-up work of WESTPHAL et al. (2000a,b), ALTMANNSBERGER et al. (2000), LUDWIG et al. (2000), and SIEGMUND (2001); however, usually these authors did not include stereochemical aspects.

2. Digoxin

Owing to a favorable passive/active transport ratio, availability as radioactively labeled compound, and its widespread clinical use, but in spite of a fairly narrow therapeutic range, digoxin has been used as P-gp model substrate to investigate the influence of concomitantly administered drugs (Fig. 5) (CAVET et al. 1996; DRESCHER et al. 2000; MANNINEN et al. 1973; GREINER et al. 1999; FROMM et al. 1999).

3. Fexofenadine

Fexofenadine, a nonsedating antihistamine and metabolite of terfenadine, does – like talinolol – not undergo significant metabolic biotransformation (Fig. 5). Employing different cell lines, evidence was found that uptake and efflux transporters are involved in fexofenadine absorption and disposition (CVETKOVIC et al. 1999; RUSSELL et al. 1998). Among various transport systems investigated, the human organic anion transporting polypeptide (OATP) and rat organic anion transporting polypeptides 1 and 2 (Oatp1 and Oatp2) were identified as mediating [^{14}C]-fexofenadine cellular uptake, while P-gp was identified as a fexofenadine efflux transporter, using the LLC-PK_1 cell, the polarized epithelial cell line lacking P-gp, and the P-gp overexpressing derivative cell line L-MDR1. Studies in P-gp knock-out mice confirmed the relevance of this transporter for fexofenadine disposition.

In general, selection of the appropriate model compound is based on the severity of expected side-effects, availability of the respective compound, and ethical and legal considerations.

C. Drug–Drug and Drug–Food Interactions Based on Transporters

I. General and Stereochemical Aspects

Pharmacokinetic drug–drug interactions based on drug transport are feasible via an influence on the transporter function (e.g., competitive inhibition) or via an alteration of transporter expression.

Most relevant in this respect appears to be transport inhibition and induction of transporter expression. Transport inhibition in the intestine leads to a decreased bioavailability, when an inside-directed transport is reduced or inhibited, but to an increased bioavailability, when an outside-directed transport (exsorption) is reduced or inhibited. Stereoselectivity of a discriminative active transport should vanish upon its inhibition and increase upon induction (Figs. 6, 7).

Induction of secretory intestinal transporters has been shown to occur for P-glycoprotein in the intestine and to affect the absorption of talinolol (HANAFY et al. 2000, 2001; WESTPHAL et al. 2000a,b); transepithelial apical-to-basolateral fluxes decrease as soon as the expression of P-gp is increased. For induction of transporters responsible for an *absorptive* active process, increased apical-to-basolateral fluxes are to be expected.

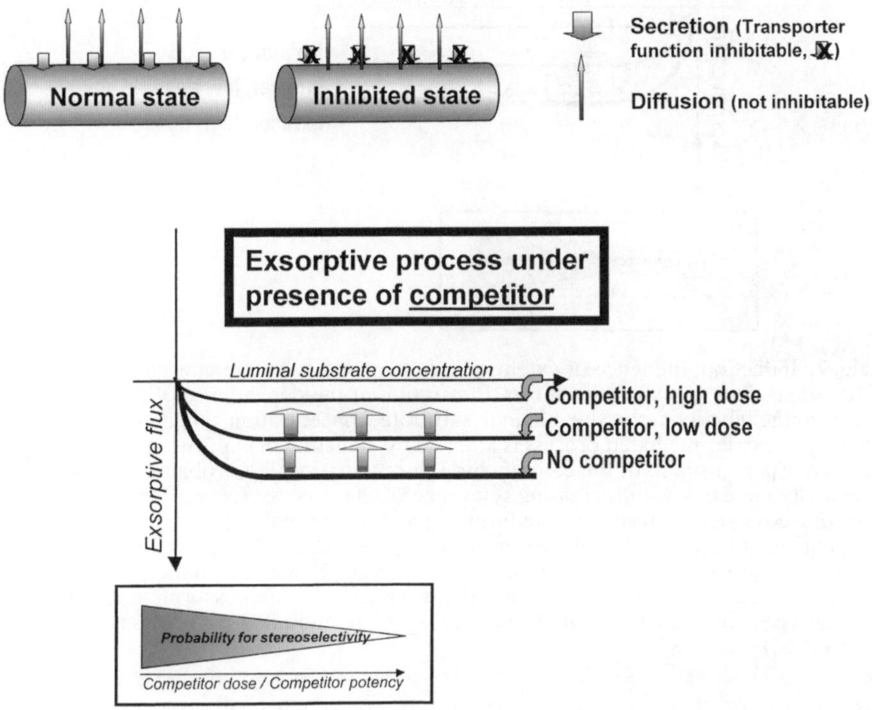

Fig. 6. Transport inhibition: Concentration- and competitor-dose-dependence of the stereoselectivity of transepithelial fluxes. Competitive inhibition or noncompetitive inhibition of potentially stereoselective (intestinal) transport leads to an increasing role of the passive process and a reduction of stereoselectivity (if present). With increasing competitor concentration, the probability for process stereoselectivity decreases in case of both, stereoselective active absorptive and stereoselective active exsorptive processes (if competitive inhibition is not stereoselective or -specific). Passive fluxes are linearly correlated with the donor concentration, passive diffusion does not exhibit significant stereoselectivity and is not affected by competitors. In the case of an active absorption process without passive diffusion, complete process inhibition at higher competitor concentration leads to a complete lack of absorption. For the combination of passive and active processes, the extent of nonlinearity in the flux-vs.-concentration/dose relationship (as well as the probability for stereoselectivity, if relevant) depends on the competitor dose. (Graphs modified from MARROUM et al., submitted)

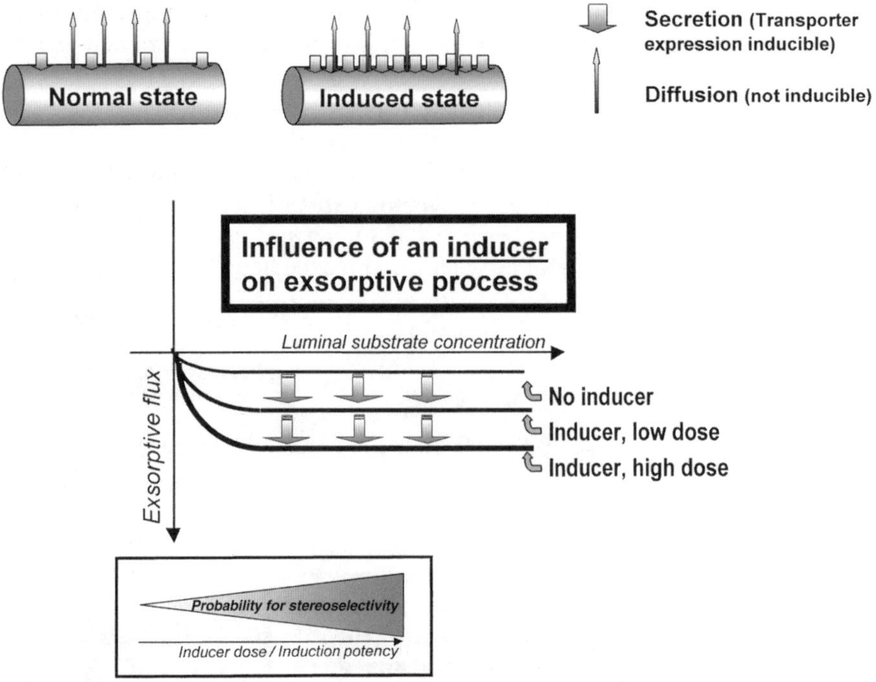

Fig. 7. Induction: Influence of extent of induction and substrate concentration on the stereoselectivity of exsorptive fluxes. Illustration of the dependence of the exsorptive transepithelial flux from the luminal substrate concentration when the respective pump- or carrier-mediated process is affected by induction of expression of the respective transport protein. In the case of chiral discrimination, the probability for stereoselectivity increases with increasing relevance of the process. Expression may depend on the dose of the respective inducer, hence, the probability for stereoselectivity appears greater, the higher the extent of induction is, i.e., with more potent inducers or higher inducer doses. In the upper part of the diagram, a schematic representation of the intestinal processes is given when passive diffusion and exsorption into the gut lumen occur in normal and induced state. (Graphs modified from MARROUM et al., submitted)

II. Competition (and Noncompetitive Effects at the Transporter)

Inhibition of absorptive active processes should lead to a decrease of transepithelial fluxes. Complete inhibition of the active component reduces the process to the fraction based on passive diffusion. In case of high stereoselectivity of the absorptive active process, stereoselectivity originating from this active transport process vanishes upon inhibition.

Radioligand binding data based on genetically modified cell lines permit screening for transporter affinities, but do not differentiate between compounds that are binding and transported and compounds that are binding and not transported (see Sect. A.VII.). Regarding permeation through membranes, the kinetic behavior of lipophilic drugs with P-gp affinity and with high passive fluxes may not be significantly affected by the presence of a second P-gp-ligand (DÖPPENSCHMITT et al. 1999b), since passive diffusion processes may readily compensate.

Studies performed to demonstrate the in vivo- and clinical relevance of drug–drug or drug–food interactions, which were hypothesized on the basis of in vitro binding or transport data, used different experimental setups.

HANAFY (2001) performed various drug–drug interaction studies with talinolol as model compound. In these studies, the in situ intestinal perfusion model was used and the relevance of the data tested in in vivo studies in rats. Cyclosporine A, rac- and *R*-verapamil, PSC833, vinblastine, and rifampicin were included. The IC_{50} values in perfusion studies were highest for cyclosporine A and PSC833, yet maximum inhibition was not obtained with PSC833 because of its poor solubility in perfusion buffer. With respect to talinolol AUC enhancement in vivo, following concomitant application of single p.o. doses, the most significant influence was detected for PSC833 followed by cyclosporine A, yet all modulators of P-gp-mediated secretion increased talinolol AUC to some extent. No significant enantiomeric difference was detected for talinolol in rats. Furthermore, the enantiomer ratio was not affected by the oral dosing of the modulators.

Drug bioavailability-enhancing ability has been attributed to a number of *grapefruit juice* flavonoids (GFJ; e.g., FUHR 1998). Yet, the underlying mechanism appeared to be largely related to inhibition of biotransformation (inhibition of CYP3A4) as opposed to inhibition of drug countertransport. However, evidence exists that transport modulation also plays a role. With respect to potential flavonoid effects, their cellular mechanism of action has best been elucidated in cancer cells. For example, CRITCHFIELD et al. (1994) identified flavonoids with increased ^{14}C-adriamycine (^{14}C-ADR) accumulation in P-glycoprotein-expressing HCT-15 colon cells, while a number of flavonoids reduced ADR accumulation and, hence, apparently upregulated and stimulated P-gp-mediated efflux. The most pronounced stimulation was found for galangin, kaempferol, and quercetin, which was blocked by MDR reversing agents (such as verapamil, vinblastine, and quinidine). The authors proposed allosteric effects, presumably at the "1,4-dihydropyridine-selective drug acceptor site" that was described by FERRY et al. (1992) to be allosterically coupled to the vinca alkaloid-selective site, as one feasible mechanistic explanation for the stimulation found.

In a GFJ-interaction study with the largely unmetabolized talinolol in the Caco-2 cell model, apical-to-basolateral talinolol transport in the Caco-2 model at 1 mM racemate concentration was increased almost threefold when GFJ was present (*S*-talinolol P_{eff}: 0.16×10^{-6} vs. 0.61×10^{-6} cm/s without vs. with GFJ; *R*-talinolol P_{eff}: 0.19×10^{-6} vs. 0.71×10^{-6} cm/s without vs. with GFJ). In vivo in rats, increased maximum plasma concentrations (C_{max} of *S*-talinolol: control, 77.5 ng/ml vs. GFJ, 163.6 ng/ml; C_{max} of *R*-talinolol: control, 79.5 ng/ml vs. GFJ, 163.0 ng/ml), enhanced AUC values, and decreased apparent oral clearances were found for both talinolol enantiomers when GFJ was administered together with a racemic 10 mg/kg p.o. dose. Furthermore, GFJ tended to accelerate the rate of talinolol input, but did not significantly affect terminal talinolol half-lives. It was therefore concluded that inhibition of intestinal secretion may contribute to bioavailability enhancement upon GFJ intake (SPAHN-LANGGUTH and LANGGUTH 2001). Stereochemical aspects were of minor relevance.

III. Induction

Most work performed with respect to induction (= the only feasible possibility for *increased* stereoselectivity of the respective active transport process) focuses on the expression of P-gp (MDR1). Its expression in tumors after chemotherapy is usually attributed to selection of preexisting multidrug resistant cells by P-glycoprotein-transported cytotoxic drugs. Expression of multidrug resistance gene-encoded P-glycoprotein and related carriers (e.g., BORST et al. 1997; SEELIG 1998; SEELIG et al. 2000) is known to be influenced by various factors under in vitro conditions. For example, P-gp expression in tissue culture can be increased by several types of stress-inducing treatments, including agents that activate protein kinase C. Some human tumor cell lines turn on the human multidrug resistance gene in response to heat and after exposure to toxic chemicals (e.g., CHAUDHARY and RONINSON 1993; VOLLRATH et al. 1994; PIRITY et al. 1996). In vivo, SCHRENK and coworkers (1993) detected the induction of multidrug resistance gene expression during cholestasis in rats and non-human primates and hypothesized that unidentified physiological inducers might mediate increased transcription of the *mdr* gene observed during cholestasis.

Various groups of chemicals have been described to influence the regulation of MDR1 expression including anthracyclines (HU et al. 1995), reserpine and yohimbine analogs (BHAT et al. 1995), and phenytoin, estramustine, paclitaxel (KAWAMURA et al. 1996; SPEICHER et al. 1994; SU et al. 1998), as well as vinblastine (ANDERLE et al. 1998). SCHÜTZ et al. (1996) and SCHUETZ and STROM (2001) described a coordinated upregulation of both P-glycoprotein and cytochrome P450–3A, e.g., in human colon carcinoma cells when cultured in the presence of modulators and substrates of P-glycoprotein.

As for competition studies, different experimental approaches were made by various authors to detect the in vivo relevance of P-gp induction.

1. In Situ Intestinal Perfusions in Rodents

Animals were pretreated with inducers and the effective permeability determined in studies in rats: a study on induction of intestinal secretion upon chronic pretreatment with different P-gp ligands (HANAFY et al. 2000, 2001) focused on talinolol as the model compound. In this study, intraindividual comparisons were performed by application of a two-step in situ intestinal perfusion, where the baseline effective permeability (consisting of passive diffusion plus exsorption) as well as the permeability upon inhibition of secretion (passive diffusion only) were determined. While repetitive verapamil and talinolol dosing had no statistically significant exsorption-inducing effect, vinblastine and rifampicin pretreatments resulted in decreased intestinal talinolol permeabilities in the three tested gut segments, duodenum, jejunum, and colon [e.g., *S*-talinolol in jejunum: control 2.50×10^{-4} cm/s; vinblastine induction, 1.48×10^{-4} cm/s ($p < 0.05$); rifampicin induction, 1.51×10^{-4} cm/s ($p < 0.05$)]. The inhibited permeability fractions were higher for vinblastine than for any other pretreatment and the difference from control pretreatment statistically significant for all intestinal segments (duodenum, 61.8%, colon, 63.1%; colon, 43.7%; *S*-talinolol). Statistically significant differences were also detected for rifampicin pretreatment in the perfused duodenum and jejunum (33.1% and 27.5% increase in inhibitable fraction for *S*-talinolol). These differences were explained by a significant induction of outside-directed transport in the intestinal enterocytes by vinblastine and rifampicin. The negligible stereoselectivity observed with respect to P_{eff} values in the rat was not enhanced upon induction of exsorption. Sandström and Lennernäs (1998) investigated the influence of oral rifampicin pretreatment on the jejunal permeability of *R*- and *S*-verapamil in the rat without observing an increase in stereoselectivity. In addition, the P_{eff} values decreased as a result of transporter induction, and formation and luminal occurrence of *R*- and *S*-norverapamil were enhanced upon rifampicin pretreatment, an observation which is in accordance with induction of CYP3A metabolism.

2. Clinical Studies

Under clinical conditions, in healthy volunteers, a study was performed where talinolol was administered both with and without rifampicin and steady-state talinolol kinetic parameters were measured (Westphal et al. 2000b). In addition, these authors quantified the expression of P-gp in the duodenum in biopsy samples obtained from the volunteers via reverse transcription-polymerase chain reaction (RT-PCR), immunohistochemistry, and Western-blot analysis before and after coadministration of rifampicin. Treatment with rifampicin resulted in lower talinolol AUC values following p.o. as well as i.v. talinolol and increased expression of P-gp in gut tissue. In these volunteers, differences between the kinetics of the talinolol enantiomers were detected upon rifampicin cotreatment only (Westphal et al. 1999, 2000b).

Rifampicin pretreatment was also studied under clinical conditions with the model compound fexofenadine (Hamman et al. 2001). The study showed that rifampicin effectively increased fexofenadine oral clearance and that this effect was independent of age and sex. Urinary recovery of fexofenadine within 24 h amounted to 3.9%–5.8% of the dose in the various groups without pretreatment and to 1.9%–2.7% with pretreatment. The amount of azacyclonol, a CYP3A-mediated metabolite of fexofenadine eliminated renally, increased on average twofold after rifampicin treatment; however, this pathway accounted for less than 0.5% of the dose (24-h excretion into urine: 0.1%–0.2% for the control treatment; 0.2%–0.3% for chronic rifampicin pretreatment).

D. Alternative Processes and Carriers, Alternative Species, and Variable Carrier Expression – Factors That May Complicate Data Interpretation

I. Studies of Intestinal Transport and p.o. Bioavailability in the Intact Organism

When studying drug absorption from the intestine, one may be focused on local processes and may not readily realize that the "attached organism" contributes to apparently local kinetics as well. For in situ perfusion studies, negative P_{eff} values with addition of an inhibitor of the respective transporter may indicate that the inhibitor is absorbed as well and affects drug disposition considerably. For most modulators, the increased tissue levels were explained by enhanced bioavailability, since the tissue-to-blood ratios were basically unaffected or just slightly altered. Racemic verapamil and *R*-verapamil coadministration, however, reduced the concentrations in most tissues in a similar way as *R*-talinolol (HANAFY 2001).

II. Are Data Obtained with Other P-gp-Expressing Cell Systems Representative of the Intestine of Animals and Man?

The relevance of carrier-mediated transport systems for tissue selectivity and the alteration of tissue selectivity plays an important role in drug targeting (TSUJI 1999). Recent investigations on stereoselective transport in rats have, for example, been performed with the noncompetitive NMDA-antagonist gacyclidine, which consists of (–)-1*S*, 2*R*- and (+)-1*S*, 2*R*-gacyclidine (HOIZEY et al. 2001). The concentrations of both enantiomers in the spinal cord extracellular fluid (ECF) were virtually identical in spite of higher total and unbound plasma concentrations of (+)-gacyclidine, indicating that an active transport system is responsible for drug uptake into the spinal cord.

Mefloquine was subject to studies which included a comparison of rat brain capillary endothelial cells (GPNT) and Caco-2 cells focusing on P-gp affinity. In GPNT cells, the (+)-stereoisomer was up to eightfold more effective than its antipode in increasing cellular accumulation of [^{3}H]-vinblastine, while in Caco-2 cells, both enantiomers were equally effective (presumably competitive) inhibitors. Stereoselective P-gp inhibition by mefloquine enantiomers was observed in rat brain cells, but not in human intestinal cells (PHAM et al. 2000).

III. P-gp and Other Transporters

Various P-gp substrates were found to exhibit affinity to other transporters as mentioned before (Sects. A.I., A.IV.). Hence, stereoselectivity may occur at various sites, or stereoselectivity at one site may be inverse to that at the other site resulting in no "apparent" stereoselectivity. Examples for important alternative transporters in this respect are MRP and OCT. MRP2 [canalicular multispecific organic anion transporter (cMOAT)] is, for example, responsible for intestinal secretion of glutathione conjugates (Gotoh et al. 2000).

Early work on the stereoselectivity of the renal OCT was performed by the group of Giacomini (e.g., Ott and Giacomini 1993; Hsyu and Giacomini 1985), who detected higher tetraethylammonium uptake inhibition for quinine, *S*-(–)-pindolol, *S*-(–)-verapamil, and *R*-(–)-disopyramide than for their optical antipodes. The relevance of these findings for the gastrointestinal tract is yet unknown and may be studied in experimental models on OCT-mediated transport in the intestine as summarized by Zhang et al. (1998).

IV. Variable Carrier Expression

In addition to observed species differences in stereoselectivity, variable carrier expression could lead to altered stereoselectivities in case of a stereoselective carrier-mediated transport. For P-gp, it was found that interindividual variability may be genetically defined (Kim et al. 2001). Hormonal influences were also detected (OCT2: Urakami et al. 2000; P-gp: Altmannsberger et al. 2000). It is further assumed that – as for metabolizing enzymes – developmental and environmental factors, diet, etc. should play a role and contribute to increased variability in kinetics (Lown et al. 1997), and affect stereoselectivity in overall drug absorption.

References

Abel S, Beaumont KC, Crespi CL, Eve MD, Fox L, Hyland R, Jones BC, Muirhead GJ, Smith DA, Venn RF, Walker DK (2001) Potential role for P-glycoprotein in the non-proportional pharmacokinetics of UK-343,664 in man. Xenobiotica 31:665–676

Altmannsberger S, Paneitz A, Franke G, Knoke M, Sperker B, Terhaag B (2000) Chronic thyroxine induces intestinal P-glycoprotein and elimination of talinolol in healthy volunteers (abstract). Eur J Clin Pharmacol 56:A16

Anderle P, Niederer E, Rubas W, Hilgendorf C, Spahn-Langguth H, Wunderli-Allenspach H, Merkle HP, Langguth P (1998) P-Glycoprotein (P-gp) mediated efflux in Caco-2 cell monolayers: The influence of culturing conditions and drug exposure on P-gp expression levels. J Pharm Sci 87:757–762

Artursson P, Karlsson J, Öcklind G, Schipper N (1996) Studying transport processes in absorptive epithelia. In: Shaw AJ (ed.), Epithelial cell culture: A practical approach, IRL Press at Oxford University Press, Oxford, New York, Tokyo, Chapter 6

Bhat UG, Winter MA, Pearce HL, Beck WT (1995) A structure-function relationship among reserpine and yohimbine analogues in their ability to increase expression of mdr1 and P-glycoprotein in a human colon carcinoma cell line. Mol Pharmacol 48:682–689

Borst P, Evers R, Kool M, Wjinholds J (1999) The multidrug resistance protein family. Biochim Biophys Act 1461:347–357

Borst P, Kool M, Evers R (1997) Do cMOAT (MRP2), other MRP homologues, and LRP play a role in MDR? Semin Cancer Biol 8:205–213

Bruggemann EP, Germann UA, Gottesman MM, Pastan I (1989) Two different regions of P-glycoprotein are photoaffinity labeled by azidopine. J Biol Chem 264:15483–15488

Cavet ME, West M, Simmons NL (1996) Transport and epithelial secretion of the cardiac glycoside, digoxin, by human intestinal epithelial (Caco-2) cells. Br J Pharmacol 118:1389–1396

Chaudhary PM, Roninson IB (1993) Induction of multidrug resistance in human cells by transient exposure to different chemotherapeutic drugs. J Nat Cancer Inst 85:632–639

Conseil G, Baubichon-Cortay H, Dayan G, Jault JM, Barron D, Di Pietro A (1998) Flavonoids: A class of modulators with bifunctional interactions at vicinal ATP- and steroid-binding sites on mouse P-glycoprotein. Proc Natl Acad Sci USA 95:9831–9836

Critchfield JW, Welsh CJ, Phang JM, Yeh GC (1994) Modulation of adriamycin accumulation and efflux by flavonoids in HCT-15 colon cells. Biochem Pharmacol 48:1437–1445

Cvetkovic M, Leake B, Fromm MF, Wilkinson G, Kim RB (1999) OATP and P-glycoprotein transporters mediate the cellular uptake and excretion of fexofenadine. Drug Metab Dispos 27:866–871

de Mey C, Schroeter V, Butzer R, Jahn P, Weisser K, Wetterich U, Terhaag B, Mutschler E, Spahn-Langguth H, Palm D, Belz GG (1995) Dose-effect and kinetic-dynamic relationships of the β-adrenoceptor blocking properties of various doses of talinolol in healthy humans. J Cardiovasc Pharmacol 26:879–888

Doherty MM, Pang KS (1997) First-pass effect: Significance of the intestine for absorption and metabolism. Drug Chem Toxicol 20:329–344

Döppenschmitt S, Langguth P, Regardh CG, Andersson TB, Hilgendorf C, Spahn-Langguth H (1999a) Characterization of binding properties to human P-glycoprotein: Development of a [^{3}H] verapamil radioligand-binding assay. J Exp Pharmacol Ther 288:348–357

Döppenschmitt S, Spahn-Langguth H, Regardh CG, Langguth P (1999b) Role of P-glycoprotein-mediated secretion in absorptive drug permeability: An approach using passive membrane permeability and affinity to P-glycoprotein. J Pharm Sci 88:1067–1072

Drescher S, Glaeser H, Hitzl M, Herrlinger H, van der Kuip H, Eichelbaum M (2000) Direct intestinal excretion is an important route of digoxin elimination in humans (abstract). Eur J Clin Pharmacol 56:A16

Dressler C, Neuhoff S, Andersson TB, Regardh CG, Leisen C, Langguth P, Spahn-Langguth H (2001) The duration of culturing may affect active transport in caco-2 cells: Amino acid- and peptide transporter vs. P-glycoprotein (abstract). Arch Pharm 334(suppl. 2):34

Eneroth A, Aström E, Hoogstraate J, Schrenk D, Conrad S, Kauffmann HM, Gjellan K (2001) Evaluation of a vincristine resistant Caco-2 cell line for use in calcein AM extrusion screening assay for P-glycoprotein interaction. Eur J Pharm Sci 12:205–214

Fassihi AR (1993) Racemates and enantiomers in drug development. Int J Pharm 92:1–14

Fauth C, Rossier B, Roch-Ramel F (1988) Transport of tetraethylammonium by a kidney epithelial cell line (LLC-PK1). Am J Physiol 254:F351–F357

Ferry DR, Russell MA, Cullen MH (1992) P-glycoprotein possesses a 1,4-dihydropyridine-selective drug acceptor site which is allosterically coupled to a vinca-alkaloid-selective binding site. Biochem Biophys Res Commun 188:440–445

Fojo A, Akiyama S, Gottesman MM, Pastan I (1985) Reduced drug accumulation in multiply drug-resistant human KB carcinoma cell lines. Cancer Res 45:3002–3007

Friedman DI, Amidon GL (1989a) Passive and carrier-mediated intestinal absorption components of two angiotensin-converting enzyme (ACE) inhibitor prodrugs in rats: Enalapril and fosinopril. Pharm Res 6:1043–1047
Friedman DI, Amidon GL (1989b) Intestinal absorption mechanism of dipeptide angiotensin converting enzyme inhibitors of the lysyl-proline type: Lisinopril and SQ29,852. J Pharm Sci 78:995–998
Fromm MF, Kim RB, Stein CM, Wilkinson GR, Roden DM (1999) Inhibition of P-glycoprotein-mediated drug transport: A unifying mechanism to explain the interaction between digoxin and quinidine. Circulation 99:552–557
Fuhr U (1998) Drug interactions with grapefruit juice. Extent, probable mechanism and clinical relevance. Drug Saf 18:251–272
Ganapathy V, Leibach FH (1986) Carrier-mediated reabsorption of small peptides in renal proximal tubule. Am J Physiol 251:F945–F953
Gotoh Y, Suzuki H, Kinoshita S, Hirohashi T, Kato Y, Sugiyama Y (2000) Involvement of an organic anion transporter (canalicular multispecific organic anion transporter/multidrug resistance-associated protein 2) in gastrointestinal secretion of glutathione conjugates in rats. J Pharmacol Exp Ther 292:433–439
Gramatté T, Oertel R, Terhaag B, Kirch W (1996) Direct demonstration of small intestinal secretion and site-dependent absorption of the beta-blocker talinolol in humans. Clin Pharmacol Ther 59:541–549
Greiner B, Eichelbaum M, Fritz P, Kreichgauer HP, von Richter O, Zundler J, Kroemer HK (1999) The role of intestinal P-glycoprotein in the interaction of digoxin and rifampin. J Clin Invest 104:147–153
Griffiths NM, Hirst BH, Simmons NL (1994) Active intestinal secretion of the fluoroquinolone antibacterials ciprofloxacin, norfloxacin and pefloxacin; a common secretory pathway? J Pharmacol Exp Ther 269:496–502
Hamman MA, Bruce MA, Haehner-Daniels BD, Hall SD (2001) The effect of rifampin administration on the disposition of fexofenadine. Clin Pharmacol Ther 69: 114–121
Hanafy A (2001) Transport inhibition and induction as sources for absorption- and disposition-related drug-drug-interaction: Talinolol as model substrate for the ABC-transporter P-glycoprotein. PhD Thesis, Department of Pharmacy, Martin-Luther-University Halle-Wittenberg
Hanafy A, Langguth P, Spahn-Langguth H (2000) Induction of P-glycoprotein (P-gp) expression in vivo: Intestinal permeabilities of talinolol in rats following pretreatment with P-gp substrates. European Graduate Student Meeting; DPhG and EUFEPS, Frankfurt/Main, Naunyn-Schmiedeberg´s Archives of Pharmacology, 361
Hanafy A, Langguth P, Spahn-Langguth H (2001) Pretreatment with potent P-glycoprotein ligands may increase intestinal secretion in rats. Eur J Pharm Sci 12:405–415
Hanafy A, Langguth P, Spahn-Langguth H (2002) Altered disposition into various tissues upon coadministration of P-glycoprotein substrates: Talinolol / verapamil. Die Pharmazie, in press
Hartmann C, Krauss D, Spahn H, Mutschler E (1989) Simultaneous determination of *R*- and *S*-celiprolol in human plasma and urine: High-performance liquid chromatographic assay on a chiral stationary phase with fluorimetric detection. J Chromatogr 496:387–396
Haussermann K, Benz B, Gekeler V, Schumacher K, Eichelbaum M (1991) Effects of verapamil enantiomers and major metabolites on the cytotoxicity of vincristine and daunomycin in human lymphoma cell lines. Eur J Clin Pharmacol 40:53–59
Hendel J, Brodthagen H (1984) Entero-hepatic cycling of methotrexate estimated by use of the D-isomer as a reference marker. Eur J Clin Pharmacol 26:103–107
Higaki K, Yukawa T, Takeuchi M, Nezasa K, Nakano M (1998) Stereoselective uptake of an organic anion across the renal basolateral membrane in isolated perfused rat kidney. Drug Metab Dispos 26:138–145

Hilgendorf C (1999) The intestinal epithelial barrier in vitro: Carrier-mediated and passive transport in normal and genetically modified Caco-2 cells and other intestinal cell lines. PhD Thesis, Department of Pharmacy, Martin-Luther-University Halle-Wittenberg

Hilgendorf C, Langguth P, Koggel A, Regardh CG, Spahn-Langguth H (2002) Identification of transporters involved in the intestinal secretion of selected β-adrenoceptor antagonists in Caco-2 cells: Relevance of P-glycoprotein and the organic cation transporter. Pharm Res (under revision)

Hilgendorf C, Spahn-Langguth H, Regardh CG, Lipka E, Amidon GL, Langguth P (2000) Caco-2 versus Caco-2/HT29-MTX co-cultured cell lines: Permeabilities via diffusion, inside- and outside-directed carrier-mediated transport. J Pharm Sci 89:63–75

Hoizey G, Kaltenbach ML, Dukic S, Lamiable D, Lallemand A, Millart H, D'Arbigny P, Vistelle R (2001) Distribution of gacyclidine enantiomers after experimental spinal cord injury in rats: possible involvement of an active transport system. J Pharm Sci 90:70–78

Hsyu PH, Giacomini KM (1985) Stereoselective renal clearance of pindolol in humans. J Clin Invest 76:1720–1726

Hu M, Amidon GL (1988) Passive and carrier-mediated intestinal absorption components of captopril. J Pharm Sci 77:1007–1011

Hu XF, Slater A, Wall DM, Kantharidis P, Parkin JD, Cowman A, Zalcberg JR (1995) Rapid up-regulation of mdr1 expression by anthracyclines in a classical multidrug-resistant cell line. Br J Cancer 71:931–936

Itoh T, Ono K, Koido KI, Li YH, Yamada H (2001) Stereoselectivity of the folate transporter in rabbit small intestine: Studies with amethopterin enantiomers. Chirality 13:164–169

Julien M, Kajiji S, Kaback RH, Gros P (2000) Simple purification of highly active biotinylated P-glycoprotein: Enantiomer-specific modulation of drug-stimulated ATPase activity. Biochemistry 39:75–85

Kawamura K, Grabowski D, Weizer K, Bukowski R, Ganapathi R (1996) Modulation of vinblastine cytotoxicity by dilantin (phenytoin) or the protein phosphatase inhibitor okadaic acid involves the potentiation of antimitotic effects and induction of apoptosis in human tumour cells. Br J Cancer 73:183–188

Kim R, Leake BF, Choo EF, Dresser GK, Kubba SV, Schwarz UI, Taylor A, Xie HG, McKinsey J, Zhou S, Lan LB, Schütz JD, Schütz EG, Wilkinson GR (2001) Identification of functionally variant MDR1 alleles among European Americans and African Americans. Clin Pharmacol Ther 70:189–199

Kramer W, Girbig F, Gutjahr U, Kleemann HW, Leipe I, Urbach H, Wagner A (1990) Interaction of renin inhibitors with the intestinal uptake system for oligopeptides and β-lactam antibiotics. Biochim Biophys Acta 1027:25–30

Kramer W, Girbig F, Gutjahr U, Kowalewski S, Adam F, Schiebler W (1992) Intestinal absorption of beta-lactam antibiotics. Functional and stereospecific reconstitution of the oligopeptide transport system from rabbit small intestine. Eur J Biochem 204:923–930

Krauss D (1988) Investigations on the kinetics of the racemic baclofen, its enantiomers and its fluoro analog. PhD Thesis, School of Pharmacy, Johann Wolfgang Goethe-University Frankfurt/M

Langguth P, Amidon G, Lipka E, Spahn-Langguth H (1997) Gastrointestinal transport processes: Potentials for stereoselectivities at substrate-specific and -nonspecific epithelial transport systems. In: Aboul-Enein H, Wainer IW (eds) "Stereochemistry in Drug Development and Use", John Wiley & Sons New York, 611–632

Lee JS, Paull K, Alvarez M, Hose C, Monks A, Grever M, Fojo AT, Bates SE (1994) Rhodamine efflux patterns predict P-glycoprotein substrates in the National Cancer Institute drug screen. Mol Pharmacol 46:627–638

Leisen C, Dressler C, Herber B, Langguth P, Spahn-Langguth H (2002) Lipophilicities of baclofen ester prodrugs correlate with affinities to the ATP-dependent efflux pump P-glycoprotein. Pharm Res (under revision)
Lesuffleur T, Barbat A, Dussaulx E, Zweibaum A (1990) Growth adaptation to methotrexate of HT-29 human colon carcinoma cells is associated with their ability to differentiate into columnar absorptive and mucus-secreting cells. Cancer Res 50:6334–6343
Lien EJ, Rodriguez de Miranda JF, Ariens EJ (1982) Quantitative structure-activity correlation of optical isomers: A molecular basis for Pfeiffer´s rule. Mol Pharmacol 12:598–604
Lown KS, Mayo RR, Leichtman AB, Hsiao HL, Turgeon DK, Schmiedlin-Ren P, Brown MB, Guo W, Rossi SJ, Benet LZ, Watkins PB (1997) Role of intestinal P-glycoprotein (mdr1) in interpatient variation in the oral bioavailability of cyclosporine. Clin Pharmacol Ther 62:248–260
Ludwig K, Franke G, Popowski K, Sperber B, Warzok R, Schroeder E (2000) MDR1 genotype affects disposition of the P-glycoprotein substrate talinolol in man (abstract) Eur J Clin Pharmacol 56:A15
Madara JL, Trier JS (1994) The functional morphology of the mucosa of the small intestine. In: Johnson R (ed.) "Physiology of the gastrointestinal tract", 3rd Edition, Raven Press, New York, Chapter 45
Manninen V, Apajalahti A, Melin J, Karesoja M (1973) Altered absorption of digoxin in patients given propantheline and metoclopramide. Lancet 1:398–400
Marroum P, Spahn-Langguth H, Langguth P (submitted) Metabolism, Transport and Drug Interactions. Chapter 9 in: Sahajwalla C. "New Drug Application: Regulatory and Scientific principles for Clinical Pharmacology and Biopharmaceutics"
Merfeld AE, Mlodozeniec AR, Cortese MA, Rhodes JB, Dressman JB, Amidon GL (1986) The effect of pH and concentration on α-methyldopa absorption in man. J Pharm Pharmacol 38:815–822
Moll-Navarro MJ, Merino M, Casabó VG, Nácher A, Polache A (1996) Interaction of taurine on baclofen intestinal absorption: A nonlinear mathematical treatment using differential equations to describe kinetic inhibition models. J Pharm Sci 85:1248–1254
Murer H, Hopfer U, Kinne R (1976) Sodium/proton antiport in brush-border-membrane vesicles isolated from rat small intestine and kidney. Biochem J 154: 597–604
Neuhoff S, Langguth P, Dressler C, Andersson TB, Regardh CG, Spahn-Langguth H (2000) Affinities at the verapamil binding site of MDR1-encoded P-glycoprotein: Drugs and analogs, stereoisomers and metabolites. Int J Clin Pharmacol Ther 38:168–179
Nicklin PL, Irwin WJ, Timmins P, Morrison RA (1996) Uptake and transport of the ACE-inhibitor ceronapril (SQ 29852) by monolayers of human intestinal absorptive (Caco-2) cells in vitro. Int J Pharmaceut 140:175–183
Ogihara T, Tamai I, Takanaga H, Sai Y, Tsuji A (1996) Stereoselective and carrier-mediated transport of monocarboxylic acids across Caco-2 cells. Pharm Res 13:1828–1832
Olpe HR, Demieville H, Baltzer V, Bentze WL, Koella WP, Wolf P, Haas HL (1978) The biological activity of D- and L-baclofen (Lioresal™). Eur J Pharmacol 52:133–136
Ott RJ, Giacomini KM (1993) Stereoselective interactions of organic cations with the organic cation transporter in OK cells. Pharm Res 10:1169–1173
Pfeiffer CC (1956) Optical isomerism and pharmacological action, a generalization. Science 124:29–31
Pham YT, Regina A, Farinotti R, Couraud P, Wainer IW, Roux F, Gimenez F (2000) Interactions of racemic mefloquine and its enantiomers with P-glycoprotein in an immortalised rat brain capillary endothelial cell line, GPNT. Biochim Biophys Acta 1524:212–219

Pietig G, Gerhardt U, Schlatter E, Hohage H (2000) Does cyclosporine A inhibit the renal organic cation transport system? Eur J Clin Pharmacol 56:A15 (51)
Pirity M, Hever-Szabo A, Venetianer A (1996) Overexpression of P-glycoprotein in heat- and/or drug-resistant hepatoma variants. Cytotechnology 19:207–214
Prieto JG, Barrio JP, Alvarez AI, Gómez G (1988) Kinetic mechanism for the intestinal absorption of ofloxacin. J Pharm Pharmacol 40:211–212
Rabbaa L, Dautrey S, Colas-Linhart N, Carbon C, Farinotti R (1996) Intestinal elimination of ofloxacin enantiomers in the rat: Evidence of a carrier-mediated process. Antimicrob Agents Chemother 40:2126–2130
Rabbaa L, Dautrey S, Colas-Linhart N, Carbon C, Farinotti R (1997) Absorption of ofloxacin isomers in the rat small intestine. Antimicrob Agents Chemother 41:2274–2277
Rao US (1995) Mutation of glycine 185 to valine alters the ATPase function of the human P-glyoprotein expressed in Sf9 cells. J Biol Chem 270:6686–6690
Russell T, Stoltz M, Weir S (1998) Pharmacokinetics, pharmacodynamics, and tolerance of single- and multiple-dose fexofenadine hydrochloride in healthy male volunteers. Clin Pharmacol Ther 64:612–621
Said HM, Redha R (1987) A carrier-mediated transport for folate in basolateral membrane vesicles of rat small intestine. Biochem J 247:141–146
Saito H, Yamamoto M, Inui K, Hori R (1992) Transcellular transport of organic cation across monolayers of kidney epithelial cell line LLC-PK-1. Am J Physiol 262: C59–C66
Sandström R, Karlsson A, Knutson L, Lennernäs H (1998) Jejunal absorption and metabolism of *R*/*S*-verapamil in humans. Pharm Res 15:856–862
Sandström R, Lennernäs H (1999) Repeated oral rifampicin decreases the jejunal permeability of *R*/*S*-verapamil in rats. Drug Metab Dispos 27:951–955
Sarkadi B, Price EM, Boucher RC, Germann UA, Scarborough GA (1992) Expression of the human multidrug resistance cDNA in insect cells generates a high activity drug-stimulated membrane ATPase. J Biol Chem 267:4854–4858
Sawynok J, Dickson C (1985) D-Baclofen is an antagonist at baclofen receptors mediating antinociception in the spinal cord. Pharmacology 31:248–259
Scala S, Akhmed N, Rao US, Paull K, Lan LB, Dickstein B, Lee JS, Elgemeie GH, Stein WD, Bates SE (1997) P-Glycoprotein substrates and antagonists cluster into two distinct groups. Mol Pharmacol 51:1024–1033
Schrenk D, Gant TW, Preisegger KH, Silverman JA, Marino PA, Thorgeirsson SS (1993) Induction of multidrug resistance gene expression during cholestasis in rats and nonhuman primates. Hepatology 17:854–860
Schütz EG, Beck WT, Schütz JD (1996) Modulators and substrates of P-glycoprotein and cytochrome P4503 A coordinately up-regulate these proteins in human colon carcinoma cells. Mol Pharmacol 49:311–318
Schuetz E, Strom S (2001) Promiscuous regulator of xenobiotic removal. Nat Med 7:536–537
Seelig A (1998) A general pattern for substrate recognition by P-glycoprotein. Eur J Biochem 251:252–261
Seelig A, Blatter XL, Wohnsland F (2000) Substrate recognition by P-glycoprotein and the multidrug resistance-associated protein MRP1: a comparison. Int J Clin Pharmacol Ther 38:111–121
Shindo H, Komai T, Kawai K (1973) Studies on the metabolism of D- and L-isomers of 3,4-dihydroxyphenylalanine (DOPA). V. Mechanism of intestinal absorption of D- and L-Dopa-^{14}C in rats. Chem Pharm Bull 21:2031–2038
Siegmund W (2001) Reply. Clin Pharmacol Ther 69:80
Sinko PJ, Leesman GD, Amidon GL (1991) Predicting fraction dose absorbed in humans using a macroscopic mass balance approach. Pharm Res 8:979–988
Skehan P, Storeng R, Scudiero D, Monks A, McMahon J, Vistica D, Warren JT, Bokesch H, Kenney S, Boyd MR (1990) New colorimetric cytotoxicity assay for anticancer-drug screening. J Natl Cancer Inst 82:1107–1112

Snyder NJ, Tabas LB, Berry DM, Duckworth DC, Spry DO, Dantzig AH (1997) Structure-activity relationship of carbacephalosporins and cephalosporins: Antibacterial activity and interaction with the intestinal proton-dependent dipeptide transport carrier of Caco-2 cells. Antimicrob Agents Chemother 41: 1649–1657

Spahn H (1989) Characterization of stereoselective processes in drug metabolism and pharmaco-kinetics. Habilitation Thesis, Dept Pharmacol Toxicol, Johann Wolfgang Goethe-University, Frankfurt/M

Spahn H, Krauss D, Mutschler E (1988) Enantiospecific high-performance liquid chromatographic (HPLC) determination of baclofen and its fluoro analogue in biological material. Pharm Res 5:107–12

Spahn-Langguth H, Benet LZ, Möhrke W, Langguth P (1997) First-pass phenomena: Sources of stereoselectivities and variabilities of concentration-time profiles after oral dosage. In: Aboul-Enein H, Wainer IW (eds) "Stereochemistry in Drug Development and Use", John Wiley & Sons New York, 573–610

Spahn-Langguth H, Baktir G, Radschuweit A, Okyar A, Terhaag B, Ader P, Hanafy A, Langguth P (1998) P-Glycoprotein transporters and the gastrointestinal tract: Evaluation of the potential in vivo relevance of in vitro data employing talinolol as model compound. Int J Clin Pharmacol Ther 36:16–24

Spahn-Langguth H, Langguth P (2001) Grapefruit juice enhances intestinal absorption of the P-glycoprotein substrate talinolol. Eur J Pharm Sci 12:361–367

Speicher LA, Barone LR, Chapman AE, Hudes GR, Laing N, Smith CD, Tew KD (1994) P-glycoprotein binding and modulation of the multidrug-resistant phenotype by estramustine. J Natl Cancer Inst 86:688–694

Su GM, Davey MW, Davey RA (1998) Induction of broad drug resistance in small cell lung cancer cells and its reversal by paclitaxel. Int J Cancer 76:702–708

Tamai I, Ling HY, Timbul SM, Nishikido J, Tsuji A (1988) Stereospecific absorption and degradation of cephalexin. J Pharm Pharmacol 40:320–324

Taylor MD, Amidon GL (eds.) (1995) Peptide-based drug design: Controlling transport and metabolism. Washington DC: ACS

Trausch B, Oertel R, Richter K, Gramatté T (1995) Disposition and bioavailability of the β_1-adrenoceptor antagonist talinolol in man. Biopharm Drug Dispos 16:403–414

Tsuji A (1999) Tissue selective drug delivery utilizing carrier-mediated transport systems. J Control Release 62:239–244

Tsuji A, Tamai I (1996) Carrier-mediated intestinal transport of drugs. Pharm Res 13:963–977

Tsukaguchi H, Tokui T, Mackenzie B, Berger UV, Chen XZ, Wang Y, Brubaker RF, Hediger MA (1999) A family of mammalian Na^+-dependent L-ascorbic acid transporters. Nature 399:70–75

Ullrich KJ, Rumrich G, David C, Fritzsch G (1993) Bisubstrates: Substrates that interact with both, renal contraluminal organic anion and organic cation transport systems II. Pflüger's Arch 425:300–312

Urakami Y, Okuda M, Saito H, Inui K (2000) Hormonal regulation of organic cation transporter OCT2 expression in rat kidney. FEBS Lett 473:173–176

Vollrath V, Wielandt AM, Acuna C, Duarte I, Andrade L, Chianale J (1994) Effect of colchicine and heat shock on multidrug resistance gene and P-glycoprotein expression in rat liver. J Hepatol 21:754–763

Wacher VJ, Wu CY, Benet LZ (1995) Overlapping substrate specificities and tissue distribution of cytochrome P450 3 A and P-glycoprotein: Implications for drug delivery and activity in cancer chemotherapy. Mol Carcinog 13:129–134

Wade DN, Mearrick PT, Morris JL (1973) Active transport of L-Dopa in the intestine. Nature 242:463–464

Wagner D, Spahn-Langguth H, Hanafy A, Koggel A, Langguth P (2001) Intestinal drug efflux: Formulation and food effects. Adv Drug Delivery 50 [Suppl 21]:S13–31

Wenzel U, Thwaites DT, Daniel H (1995) Stereoselective uptake of β-lactam antibiotics by the intestinal peptide transporter. Br J Pharmacol 116:3021–3027

Westphal K, Weinbrenner A, Zschiesche M, Franke G, Knoke M, Hachenberg T, Greiner B, Fritz P, Oertel R, Terhaag B, Kroemer HK, Siegmund W (1999) Talinolol – rifampicin interaction: potential role of drug transporters. Naunyn-Schmiedeberg's Arch Pharmacol 359 (Suppl.), R123

Westphal K, Weinbrenner A, Giessmann T, Stuhr M, Franke G, Zschiesche M, Oertel R, Terhaag B, Kroemer HK, Siegmund W (2000a) Oral bioavailability of digoxin is enhanced by talinolol: evidence for involvement of intestinal P-glycoprotein. Clin Pharmacol Ther 68:6–12

Westphal K, Weinbrenner A, Zschiesche M, Franke G, Knoke M, Oertel R, Fritz P, von Richter O, Warzok R, Hachenberg T, Kauffmann HM, Schrenk D, Terhaag B, Kroemer HK, Siegmund W (2000b) Induction of P-glycoprotein by rifampin increases intestinal secretion of talinolol in human beings: A new type of drug/drug interactions. Clin Pharmacol Ther 68:345–355

Wetterich U, Spahn-Langguth H, Mutschler E, Terhaag B, Rösch W, Langguth P (1996) Evidence for intestinal secretion as an additional clearance pathway of talinolol enantiomers: Concentration- and dose-dependent absorption in vitro and in vivo. Pharm Res 13:514–522

Wu CY, Benet LZ, Hebert MF, Gupta SK, Rowland M, Gomez DY, Wacher VJ (1995) Differentiation of absorption and first-pass gut and hepatic metabolism in humans: Studies with cyclosporine. Clin Pharmacol Ther 58:492–497

Wu X, Whitfield LR, Stewart BH (2000) Atorvastatin transport in the Caco-2 cell model: Contributions of P-glycoprotein and the proton-monocarboxylic acid co-transporter. Pharm Res 17:209–215

Zhang L, Brett CM, Giacomini KM (1998) Role of organic cation transporters in drug absorption and elimination. Annu Rev Pharmacol Toxicol 38:431–460

Zimmerman J (1992) Methotrexate transport in the human intestine. Evidence for heterogeneity. Biochem Pharmacol 43:2377–2383

CHAPTER 12

Enantioselective Plasma and Tissue Binding

P.J. HAYBALL and D. MAULEÓN

A. Introduction

Many drugs, chiral or otherwise, interact with plasma (Fig. 1) or tissue (Fig. 2) proteins or with other macromolecules to form a drug-macromolecular complex. The formation of such a complex with protein is generally referred to as "drug–protein binding." Drug–protein binding may be either reversible or irreversible. A frequently cited example of irreversible binding is the covalent chemical attachment of a drug to plasma or other protein via its acyl-glucuronide metabolite (see reviews: SPAHN-LANGGUTH and BENET 1992; HAYBALL 1995). In this chapter discussion will be largely restricted to reversible chiral drug–protein associations.

Reversible drug–protein binding implies that the drug (ligand) binds with the aid of intermolecular bond forces such as those involved in salt-bridge, hydrogen, and van der Waals bonds. Hydroxyl, carboxyl, amino, and phenyl groups of specific amino acid constituents of binding proteins are available to interact with analogous functional groups of drugs. The N–H and C=O groups of the protein backbone are able to establish relatively strong hydrogen bonds on regions not involved with α-helices or β-sheets. At physiological pH, carboxyl and amino functional groups will predominately exist in their respective ionized states and the highly polar guanidine group in arginine residues are collectively capable of forming salt-bridge links with oppositely charged ligand groups. In addition to lipophilic phenyl groups, indole groups in tryptophan are important π-donors. Moreover, lipophilic pockets or surfaces in proteins are frequently constituted by a combination of aliphatic side-chains (for example, leucine and isoleucine), aryl groups and disulphide bridges where the principle binding force is hydrophobic through water exclusion. Clearly, since these groups will be aligned in specific three-dimensional shapes for both proteins and chiral drugs, the potential exists for drug–protein binding to differ depending on the enantiomeric form of the chiral ligand, giving rise to so-called enantioselective protein-binding. It is the inherent chirality of the α-amino acid subunits of proteins which ultimately leads to enantioselective binding of chiral drugs. However, it is the three-dimensional shape of the protein which is really responsible for the experimentally observed enantioselectivity.

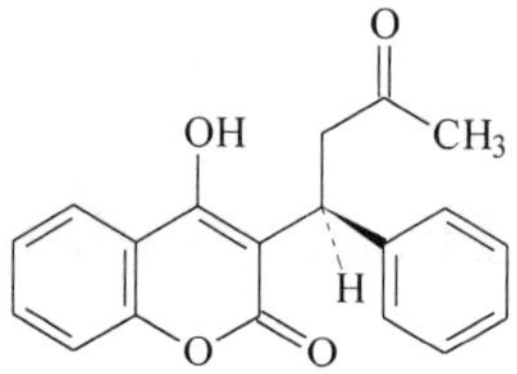

S-Warfarin

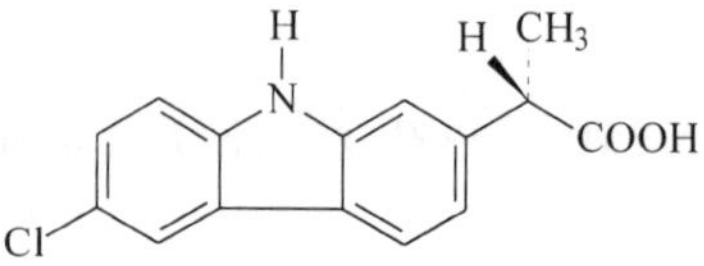

S-Carprofen

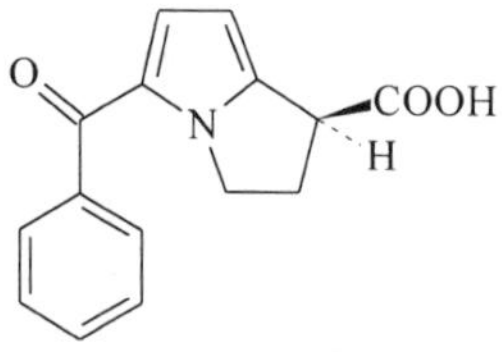

S-Ketorolac

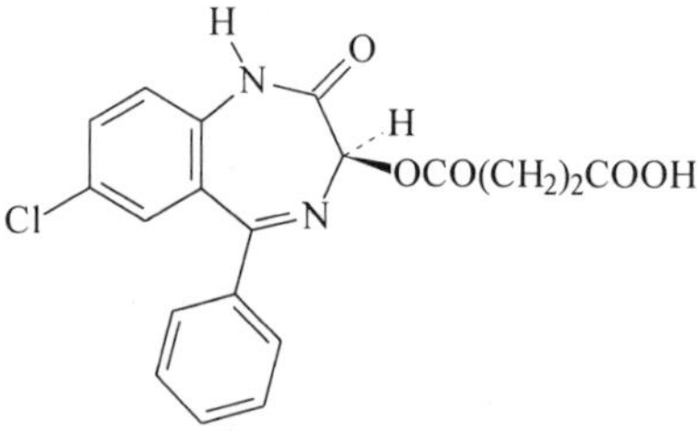

R-Oxazepam hemisuccinate

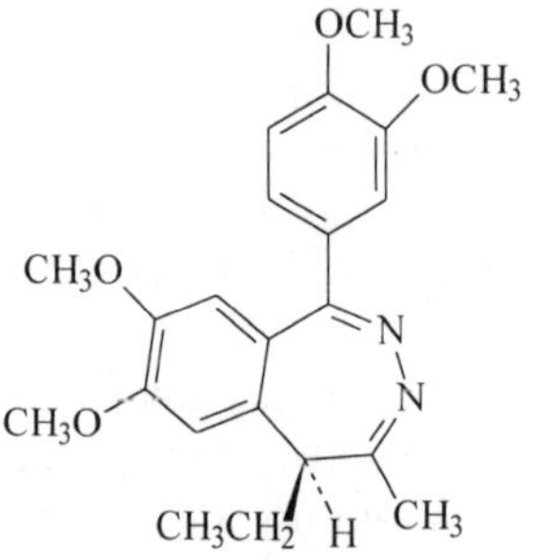

S-Tofisopam

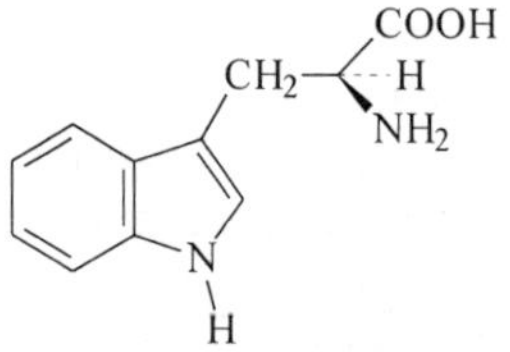

S-Tryptophan

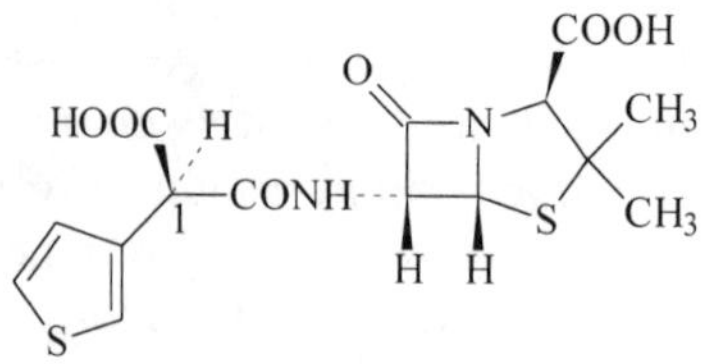

1R-Ticarcillin

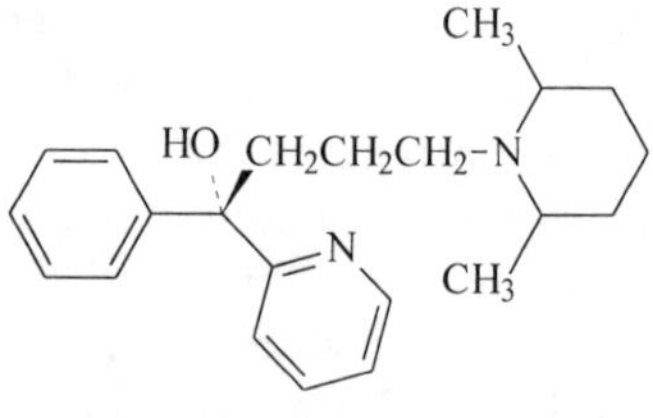

S-Pirmenol

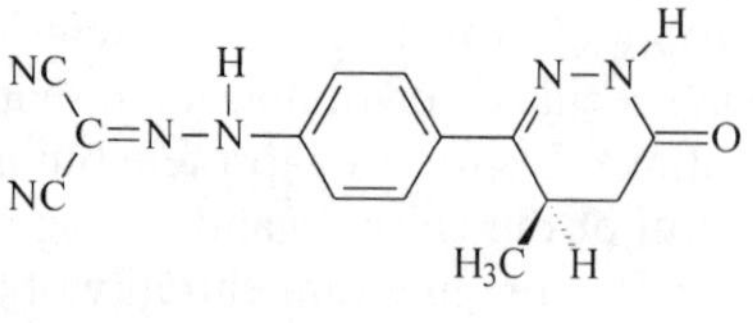

R-Levosimendan

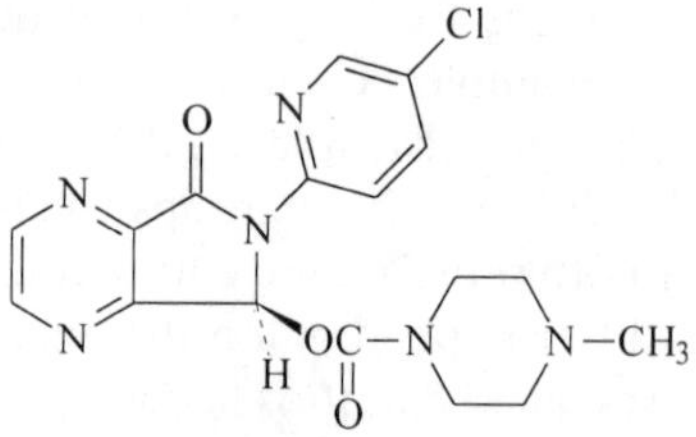

S-Zopiclone

Fig. 1. Chemical structures of one of the enantiomers from a group of representative chiral drugs which exhibit enantioselective binding to plasma protein

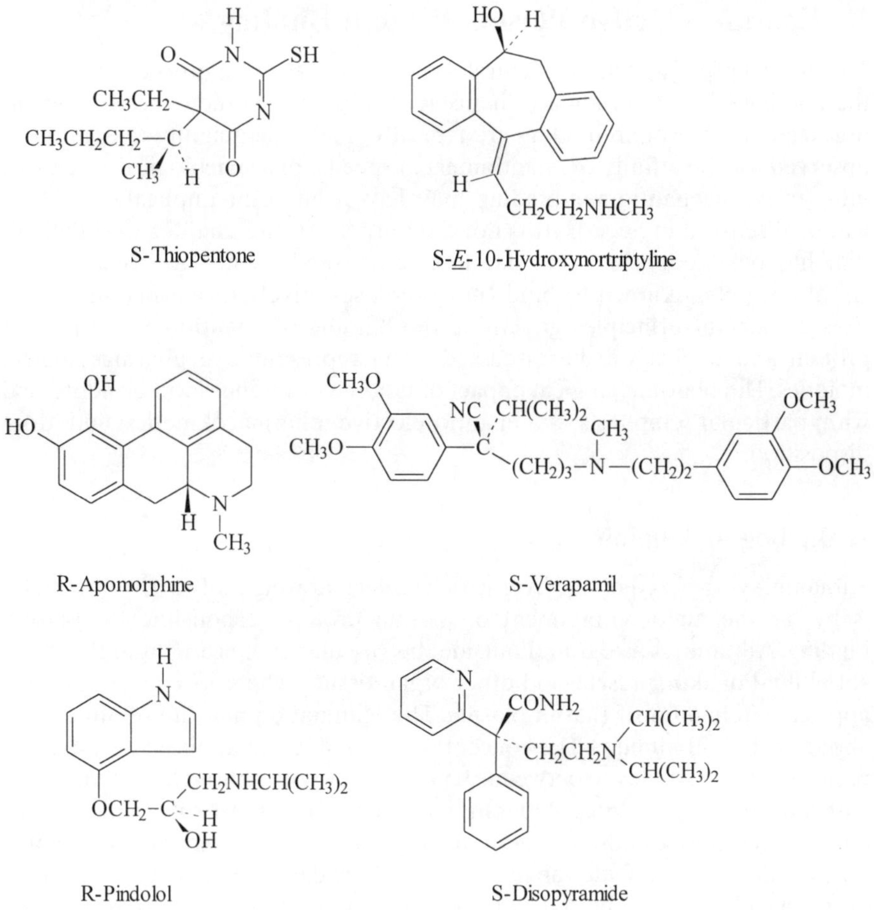

Fig. 2. Chemical structures of enantiomers of representative chiral drugs which are reported to bind to tissue components enantioselectively

Protein binding of drugs in the vascular compartment, rather than in the tissue compartment, is the primary objective for review here since there have been relatively few controlled studies that have examined the latter. Virtually all such reversible binding of drugs occurs with plasma proteins; however, there have been a very limited number of studies which have examined, for example, the uptake of isomers of chiral drugs into red blood cells. Also limited are the number of studies which have examined the blood–brain barrier passage of chiral drugs and the selective partitioning of enantiomers into other tissues.

B. Enantioselective Plasma Protein Binding

The plasma protein binding of chiral drugs is potentially enantioselective since the binding sites in plasma are themselves chiral. While the degrees of enantioselectivity in protein binding are typically smaller in magnitude than those observed for the affinity of enantiomers at specific pharmacological receptors, differences in enantiomer binding may have significant implications which will be discussed in Sect. E. It is not the purpose of this chapter to collate all that has been published on enantioselective binding, nor to discuss every chiral drug class known to bind (non)stereoselectively to plasma and tissue sites. Important principles governing the binding of enantiomers to plasma protein and tissues will be discussed using representative pharmacological moieties. The pharmacological impact of binding differences will be addressed with particular emphasis on enantioselective pharmacokinetics and drug disposition.

I. Binding to Albumin

Albumin, synthesized by the liver with a molecular weight of 65000–69000 Da, is by far the major component of plasma protein responsible for plasma binding. Albumin is also found outside the vascular compartment in the interstitial fluid of skin, muscle, and other organ tissues where its concentration is approximately 60% of that in plasma. The elimination half-life of albumin is about 17 days in humans. The concentration of albumin in plasma is held at a relatively constant level between 3.5%–5.5% (w/v) although this can vary with various disease states. Albumin has a major transport role in plasma for endogenous compounds (for example, free fatty acids, hormones, and bilirubin) in addition to a wide range of drugs, particularly those with acidic properties, which may be bound very avidly. It is now apparent that albumin associates reversibly with drugs at specific sites and accordingly, may be regarded as a silent receptor. The number of specific binding sites on the albumin molecule is yet to be fully characterized with upwards of six sites having been proposed (ASCOLI et al. 1998; YOSHIKAWA et al. 1999). The two main binding regions are the warfarin-azapropazone region (region I) and the indole-benzodiazepine region (region II). It is now generally accepted that both of these regions are able to discriminate between enantiomers and that enantioselectivity may arise from differing affinities of enantiomers towards a common binding region. In addition, preferences of individual enantiomers to different binding regions may result in enantioselective binding to this protein.

II. Specific Albumin Binding Regions

Binding region I (FEHSKE et al. 1981) has been the most extensively studied, since it is occupied by a diverse range of therapeutically useful drugs. It would appear that two specific binding sites exist in region I, the azapropazone and the warfarin binding sites (FEHSKE et al. 1981); however, recently three distinct regions were identified capable of specifically binding dansyl-L-asparagine, *n*-butyl-*p*-aminobenzoate, and acenocoumarol (YAMASAKI et al. 1996). Recent studies of the prototypical binding of warfarin enantiomers to region I of HSA have indicated that the *S*-enantiomer is a direct competitor for *R*-warfarin, while *R*-warfarin was an indirect competitor for its antipode (BERTUCCI et al. 1999) (see Sect. B.VI.); however, differences in the unbound fraction are not great with values of 1.2% and 0.9% reported for the *R*- and *S*-enantiomers, respectively (YACOBI and LEVY 1977).

Binding region II is capable of substantial chiral discriminatory capacity. The most widely studied ligands have been the chiral 1,4-benzodiazepines (BDZs). In fact, this specific, high-affinity BDZ site on HSA shows stereospecificity of binding putatively similar in terms of which enantiomer binds preferentially (at least for some congeners) to BDZ receptors in the central nervous system (CNS), and has been proposed as a model to elucidate structure-activity relationships in the CNS (SALVADORI et al. 1997). The binding to this HSA region by the hemisuccinate esters of the chiral benzodiazepine oxazepam has been reported to occur with a degree of enantioselectivity of approximately two orders of magnitude (MULLER and WOLLERT 1975a), as has the binding of lorazepam acetate enantiomers (FITOS et al. 1999). Similarly, the binding of the essential amino acid (L)-tryptophan to region II occurs in a highly stereoselective fashion being circa two orders of magnitude more avidly bound than the unnatural (D)-enantiomer (MULLER and WOLLERT 1975b).

In summary, the precise origin of the chiral discrimination process is often difficult to gauge since conformational alterations caused by the chiral ligands are likely. Far more useful is the study of the individual binding behavior of the enantiomers. To date enantioselective binding to albumin binding sites, other than regions I and II, have yet to be demonstrated, possibly since specific chiral markers for these alternative sites remain to be identified.

III. Binding to α_1-Acid Glycoprotein

α_1-Acid glycoprotein (AGP), or orosomucoid, is a plasma globulin with a molecular weight of approximately 44000 Da. AGP, present at a much lower concentration (0.4%–1.0% w/v) than albumin, is an acidic protein shown to be important in the binding of basic drugs. Unlike albumin, it is thought to possess a single binding site (MULLER and STILLBAUER 1983), and enantioselective binding by AGP will presumptively occur solely due to differences in affinities between enantiomers. Interestingly, there are at least two genetic variants of AAG and this binding protein exists as a mixture of these variants: the A variant and the F1 and/or S variant or variants (HERVE et al. 1998) which individually (and hence collectively) are capable of discriminating between enantiomers (EAP et al. 1990). Enantioselective binding to AGP has been reported for a range of chiefly basic drugs such as rac-verapamil (EICHELBAUM et al.

1984), rac-methadone (ROMACH et al. 1981), rac-disopyramide (LIMA et al. 1984; VALDIVIESCO et al. 1988), rac-gallopamil (GROSS et al. 1993), and rac-zopiclone (FERNANDEZ et al. 1999). In general, while AGP is a qualitatively important plasma protein in the binding of basic drugs, it typically exhibits low enantioselectivity, substantially less than has been observed for HSA-drug binding.

IV. Binding to Other Plasma Proteins

Other plasma proteins, including lipoproteins, potentially exhibit enantioselective binding because of their inherent chirality. And while the concentrations of these proteins may be minor compared to albumin and AGP, in situations where their production is increased (for example, hyperlipoproteinemia) or when plasma ligand concentrations have lead to saturation of the aforementioned binding sites, chiral recognition by other plasma proteins may become important. To date, there have been no such reported studies.

V. Species Differences in Plasma Protein Binding

Differences in the stereoselectivity of plasma protein binding can depend on the animal species selected. It has been observed (LIMA 1988) that *S*-disopyramide was more highly bound to plasma proteins than its antipode in humans, gorilla, and pig serum, whereas the reverse was observed in serum harvested from the blood of cow and sheep. Similarly, species (human and rat) differences have been noted for the protein binding of propranolol enantiomers (ALBANI et al. 1984; TABKAHSHI et al. 1990) and verapamil enantiomers (GROSS et al. 1988). However, these differences are typically small and a consequence of very similar binding affinities of various animal binding proteins for each enantiomer. The pharmacological significance of this phenomenon would be expected to be minor.

VI. Enantiomer Interactions at Plasma Protein Binding Sites

Consequences of enantiomer-enantiomer interactions at plasma protein binding sites are discussed below (Sects. E.I. and E.II.). A number of specific mechanisms for such interactions have been proposed (ORAVCOVA et al. 1996).

Competition between enantiomers (competitive inhibition) has been observed for enantiomers of 2-phenylpropionic acid (JONES et al. 1986) and flurbiprofen (KNADLER et al. 1989), where in each case inhibition and mutual displacement occurred at one site on the albumin molecule. It has been suggested that competitive binding of ligands to albumin might occur at different parts on the macromolecule, a so-called induced-fit model, where conformational change accompanies the binding of one enantiomer which subsequently modifies the binding of its antipode (HONORE 1990).

An alternative mechanism involves allosteric displacement and has been proposed to operate when *S*-warfarin displaces benzodiazepines from human serum albumin (Fitos and Simonyi 1992). Allosteric displacement is thought to be sensitive to the concentration of displacing ligand following studies of a model fatty acid displacer, octanoic acid (Noctor et al. 1992). Suprofen and ketoprofen enantiomers, which bind predominately to region II of HSA (see Sect. B.II.), were affected by the presence of octanoic acid using an affinity chromatographic technique (see Sect. C.III.).

Enantioselective site-to-site displacement of protein bound drug has also been examined (Rahman et al. 1993). Upon calculation of binding affinity parameters for enantiomers of carprofen and ibuprofen to specific HSA binding regions, the investigators concluded that in the presence of *R*- and *S*-ibuprofen, carprofen enantiomers were displaced from their high affinity site (region II) to their low affinity site (region I). When the latter region was blocked by racemic warfarin, substantial displacement of carprofen enantiomers ensued and a fourfold increase in their unbound fraction was observed.

C. Methods for Determining Plasma Protein Binding of Enantiomers

I. Classical Methods: Isolation of Unbound Drug

Typically, the amount of drug bound reversibly to protein is not measured directly, rather it is calculated as the difference between unbound drug and total (bound plus unbound) drug. Classical techniques such as equilibrium dialysis, ultrafiltration and gel filtration involve the separation of a protein-free solution containing unbound drug from the protein-drug solution. With equilibrium dialysis, unbound drug diffuses from the plasma (or other protein solution) into protein-free buffer outside a semi-impermeable membrane. Using this method, the composition of the buffer, particularly its ionic strength, may greatly influence the movement of unbound drug into the protein-free solution, although this influence would be expected to be nonenantioselective. Another limitation of the method is with highly lipophilic drugs which can form micellar dispersions in the protein-free buffer thereby leading to an overestimate of the unbound fraction; again, this is unlikely to be enantioselective.

Filtration methods involve the centrifugal generation of a protein-free filtrate containing unbound drug. Subsequently, the protein-free solution is assayed to determine drug concentration. For a chiral drug, when both enantiomers are present or when interconversion is likely, enantioselective analytical methods need to be used. This may create significant methodological problems since not only are unbound concentrations often much lower than total (bound plus unbound) concentrations, but the additional problem of analyzing one enantiomer when its antipode is also present in the sample needs to be overcome.

Circular dichroism (CD) is a physicochemical technique which has been used to identify binding sites of drug enantiomers on proteins. When a chiral ligand binds to a protein, the asymmetric complex formed can be detected using CD (CHIGNELL 1973). Asymmetrical complex formation can be analyzed quantitatively as a measure of the ellipticity induced, and accordingly, provides information on binding site(s) affinities. CD investigations have been performed on numerous chiral drugs, including the binding of *R*- and *S*-warfarin to HSA (BERTUCCI et al. 1999) and carprofen enantiomers (and their modulation by ibuprofen) to HSA (RAHMAN et al. 1993).

II. Quantification Using Radiolabeled Ligands

For chiral drugs, which are extensively bound to protein, the use of radiolabeled enantiomers may facilitate determination of binding. These may be sourced from pharmaceutical companies or radiochemical suppliers or alternatively, synthesized in-house. However, the use of radiolabeled ligand in assessing protein binding requires an extremely pure radiochemical. Relatively minor amounts of poorly bound radiochemical impurities can lead to spuriously elevated estimates of the unbound fraction (BJORNSSON et al. 1981), and typically, a clean-up (often chromatographic) of radioligand is required immediately prior to its use. Much less of a problem is the presence of optical impurity in the radioligand source since, at the low ligand to protein ratios at which radiolabeled drug is typically added, competitive displacement reactions between enantiomers are likely to be minimal. As long as one is capable of resolving the unbound radiolabeled isomers prior to their quantification, an accurate description of enantioselective binding behavior is possible. For example, when the plasma protein binding of ketoprofen enantiomers was assessed (HAYBALL et al. 1991) [1-^{14}C]-labeled racemic drug was synthesized and successfully used as unresolved radioligand to elucidate the binding of the separate enantiomers. Similarly, labeled racemate was used to determine the plasma unbound fractions of ibuprofen enantiomers, including their concentration-dependence and mutually competitive nature (EVANS et al. 1989).

The potential problem of resolving the typically small amounts of radiolabeled drugs available to investigators may be circumvented by semipreparative HPLC using a suitable chiral stationary phase. A novel approach when investigating the enantiomer-protein binding of the (putatively) extensively bound nonopiate analgesic ketorolac involved facile tritium exchange between racemate and ^3H_2O followed by direct chiral HPLC resolution to give separate labeled isomers (HAYBALL et al. 1994a).

III. Binding Studies Using Immobilized Plasma Protein

The binding of drug enantiomers to immobilized protein HPLC columns has been used to assess enantioselective binding. Information on the affinity of individual enantiomers and interactions between ligands (including enantiomer-enantiomer interactions) may be obtained. The binding properties of enantiomers are directly related to the chromatographic parameters, column capacity (retention) being a measure of the affinity to the protein, and the enantioselectivity factor (ratio of enantiomer capacity factors) indicating the degree of enantioselective protein binding. This occurs because in some cases the immobilized protein retains the binding specificity (DOMENICI et al. 1990; ASCOLI et al. 1998) and the conformational mobility (CHOSSON et al. 1993) of the native protein. By adding competing ligands to the mobile phase, the chromatographic displacement of an enantiomer can be readily studied (BERTUCCI et al. 1999).

HSA- and α1-acid glycoprotein (AGP)-based protein columns, given the predominance of these proteins in plasma, are the most widely used. There have been numerous attempts to correlate HPLC column enantioselectivity with native protein binding enantioselectivity to both albumin (LOUN and HAGE 1996; BERTUCCI et al. 1999) and AGP (FITOS et al. 1995). Immobilized albumin, in particular, has been a useful tool for rapidly probing enantioselective drug binding, as discussed below for chiral NSAIDs, in addition to providing detailed information about the nature of binding to specific binding sites on the albumin molecule. In particular, this method has been used to distinguish between competitive, cooperative, noncooperative, and independent binding, following established definitions (HONORE 1990).

Correlation of chromatographic resolution by AGP-based protein columns and binding enantioselectivity has been achieved with some but not all representatives of chiral 1,4-benzodiazepines thus studied (FITOS et al. 1995). ORAVCOVA and coworkers (ORAVCOVA et al. 1996), in an extensive review of binding methodologies, concluded that unlike albumin-based columns, AGP columns are not suitable as a screening tool to assess the binding of ligands to native AGP even though the system has been widely used to separate enantiomers. It needs to be recognized that the nonphysiological conditions often used chromatographically (pH differences and/or presence of organic modifiers) may dramatically alter protein conformation and binding behavior from that of the native protein. And clearly, if a number of the binding regions of native protein have become associated with the linkages to the silica chromatographic support, then differences in binding behavior would be expected.

D. Enantioselective Tissue Binding and Partitioning

Compared to the number of studies that have examined the binding of enantiomers to plasma protein, tissue binding studies are very limited. Tissue sequestration of drugs is determined by the physicochemical properties of the drug and of the site at which binding takes place in the tissue. Theoretically, a specific tissue constituent to which a drug binds preferentially, may represent an important pharmacokinetic determinant. However, relatively few specific tissue binding sites have been identified which make such an impact on the dispositional properties of a drug, particularly those capable of discriminating between enantiomers. Tissue phosphatidylserine has been proposed (YATA et al. 1990) to play a role in the tissue distribution of weakly basic drugs. This acidic phosphatide possesses a chiral amino acid substituent and accordingly, may bind drugs enantioselectively particularly in tissues, such as the lung (presumptively including pulmonary surfactant), where significant amounts of this binding component reside. It has been hypothesized to account, at least partially, for the enantioselective tissue distribution of disopyramide and verapamil (HANADA et al. 1998). Binding of chiral drugs to specific tissues (brain, red blood cells, adipose tissue, and synovial fluid and tissue) are explored below (Sect. D.I–IV.).

I. Blood–Brain Barrier Passage of Drugs: Impact of Chirality

Drug transfer across cellular membranes such as the blood–brain barrier is dominated by the lipophilicity of the drug and by its degree of ionization at physiological pH, which in turn is related to its acid-base properties. These factors are not affected by chirality and accordingly, transport across the blood–brain barrier should be nonenantioselective even though cell membranes do form a chiral environment by virtue of their chiral constituents. Chiral drugs may, rather than being passively transported, traverse with the aid of active membrane-bound transporters and, since these processes involve more specific molecular interactions, potentially experience chiral discrimination.

Indeed, in the few cases which have examined transport mechanisms for chiral drugs, mostly their passage is not modulated by their stereoconfiguration. ROCHAT and coworkers (ROCHAT et al. 1999) evaluated the in vitro transport of enantiomers of the selective serotonin reuptake inhibitor (SSRI) citalopram, using rat brain microvessel cells; while transport was saturable and temperature-dependent (suggesting the involvement of a carrier mechanism) passage was nonstereoselective and bidirectional without influences of active efflux mechanisms or monoamine oxidases. Most notably, lack of stereoselectivity was noted before carrier-protein saturation was reached; clearly, once saturation has occurred for a membrane-bound transporter, passive diffusion (predictably nonenantioselective) would predominate as the transport mechanism. Similarly, the uptake and distribution of the new racemic SSRI (rac-NS2381) into living porcine brain, using positron emission tomographic study of the ^{11}C-labeled analogue, showed drug accumulation occurred in regions rich in serotonin uptake sites and this was largely nonenantioselective (SMITH et al. 1999). Ethopropazine enantiomers have been shown not to exhibit differences in either their plasma protein binding nor in their uptake by various rat brain tissues (MABOUDIAN-ESFAHANI and BROCKS 1999).

An in vivo microdialysis study of the blood–brain equilibration of thiopental enantiomers in the rat was recently conducted (Mather et al. 2000). Concentrations of *R*- and *S*-thiopental were measured in plasma, tissue, and brain microdialysate of rats after intravenous infusion of racemate and tissue:plasma distribution coefficients of *R*-thiopental were observed to be greater than those of *S*-thiopental, when calculated from either total (bound plus unbound) or unbound plasma concentrations. However, the investigators could find no pharmacokinetic evidence to support enantioselectivity in the rates of equilibration across the blood–brain barrier.

The concentration and distribution of apomorphine enantiomers in plasma, brain tissue, and striatal extracellular fluid was assessed with microdialysis following subcutaneous infusion of the individual enantiomers to steady state (Sam et al. 1997). Unbound apomorphine in the extracellular fluid collected in the striatum was much higher than corresponding unbound plasma concentrations. However, since this extracellular tissue compartment is in equilibrium with the intracellular compartment which was rich in apomorphine, the authors were unable to conclude whether this was solely due to an active transport across the blood–brain barrier or a redistribution from the cellular compartment. Since the concentration of *R*-apomorphine was double that of its optical antipode in brain microdialysate, a possible stereoselective uptake could not be ruled out. This apparent trapping of apomorphine in the striatum cells may be artifactual due to overloading by the drug administration method. Since unbound apomorphine is not in fast equilibration between brain fluid and plasma, this means that exit of this compound from the CNS should be slower than plasma clearance by other routes; in that case, the ratio of compound inside the CNS versus compound in plasma should change with time. Alternatively, the distribution of enantiomers in the brain could be confounded by the reported higher affinity of *R*-apomorphine for dopamine receptors located on the cell membranes in the striatum (Seeman and Van Tol 1993). Taken together, the situation for apomorphine above is complicated by the simultaneous operation of numerous dynamic equilibria. The use of specific transport cell cultures would be needed to narrow down the different interacting mechanisms operating in vivo.

The classical nonselective β-adrenoreceptor antagonist rac-pindolol is used clinically as the racemate for the treatment of hypertension and more recently, given its reported 5-$HT_{1A/1B}$ antagonist activity (Newman-Tancredi et al. 1998), as a potential augmentor of numerous antidepressants including selective serotonin reuptake inhibitors (SSRIs) (Perez et al. 1997; Zanardi et al. 1998). Yan and Lewander (Yan and Lewander 1999) investigated the enantiomer concentration ratio in various tissues of the rat at 90 min following subcutaneous bolus dosing with the racemate and found total brain levels of pindolol were fourfold higher than in plasma and moreover, the S:R ratio (for bound plus unbound compound) was 1.5 in the brain, a value significantly higher than that observed in plasma and other tissues. Since the enantiomers have identical lipophilicity, this enantioselectivity might conceivably be due to a higher free concentration of *S*-pindolol since differences in the binding of enantiomers ($f_{uS} > f_{uR}$) has been shown for rat α_1-acid glycoprotein (MURAI-KUSHIYA et al. 1993). The investigators have suggested that these data support a role for the stereoselective passage of pindolol across brain endothelial cells as has been shown for this drug when its transport by renal tubular cells was assessed (HSYU and GIACOMINI 1985). However, the time course of the concentration ratio for pindolol enantiomers may be relevant. Brain and plasma levels are related through the equilibrium established by enantiomers crossing the blood–brain barrier. If the plasmatic clearance is relatively fast, it may be possible to find opposite isomer ratios at different sampling times. At short postadministration times (absorption phase), plasma concentrations may be higher than corresponding brain tissue concentrations, since the compound is still entering the central compartment and the entry into the CNS is only starting. During the distribution phase, CNS concentrations may be rising, while the drug in plasma is decaying due to metabolic and other elimination processes and possibly trapping in other peripheral compartments. Consequently, CNS levels might be higher than those in plasma during this time. Later, one could observe a progressive lowering of CNS drug levels, but when the plasmatic clearance is very efficient this can act as a concentration sink and the brain/plasma ratio is maintained above unity. Clearly, this underscores the limitations of experiments performed when drug concentrations are determined at limited sampling times and accordingly, the pitfalls in the interpretation of comparative concentration ratios of enantiomers between blood and tissue determined over a single-dose time profile. This can lead to an apparent enantioselectivity which is not due to intrinsic stereoselectivity at the blood–brain barrier; rather, it is more likely due to enantioselective metabolism as has been noted for many of the chiral β-adrenoreceptor antagonists (WALLE et al. 1988).

The involvement of enantioselective efflux processes (such as *p*-glycoprotein) have been proposed to also operate with the enantiomers of E-10-hydroxynortriptyline; evidence has implicated a stereoselective, active transport of drug out of cerebrospinal fluid (BERTILSSON et al. 1991) after drug concentrations have putatively reached steady-state. Similarly, the time course may have a strong influence on the interpretation of these data.

Conflicting data are reported for the quinoline-methanol antimalarial rac-mefloquine. Administration of the racemate and individual enantiomers to rats yielded plasma concentrations (bound plus unbound) for the (+)-isomer 2–3 times higher than those for its antipode, whereas the opposite was reported in all tissues of the brain examined (BAUDRY et al. 1997). In contrast, human brain levels of (+)-mefloquine were substantially higher than (–)-mefloquine in two postmortem cerebral biopsies performed on patients with fatal cerebral malaria treated with racemate (PHAM et al. 1999). In one of those cases the plasma ratio of (–)/(+) for bound plus unbound drug was 3.5. Given the absence of plasma protein binding data for mefloquine enantiomers and evidence for differential enantiomer binding of the structural analogues [rac-chloroquine (OFORI-ADJEI et al. 1986) and rac-hydroxychloroquine (MCLACHLAN et al. 1993)], it remains to be determined whether the apparent enantioselectivity in blood–brain transport is simply related to the relative availability of unbound drug isomers in plasma. Also underscored by these studies of mefloquine enantiomers is the question of whether the pathophysiological status of the blood–brain barrier plays a role in enantioselective drug transport. Furthermore, the comparison of these two sets of data is complicated by the differences in doses used in the rat study compared with those in humans. It is possible that the postmortem drug analyses in humans were conducted after a high-dose steady-state had been achieved with the antimalarial in an attempt to load the brain as much as possible to cure the disease. This means that linearity cannot be presumed here and protein binding might be saturated.

II. Partitioning of Chiral Drugs into Red Blood Cells

The distribution of drugs through the red cell membrane has been well documented (SCHANKER 1964) and is usually governed by an equilibrium with unbound drug concentrations in serum. Accordingly, one would predict that partitioning of chiral drugs into these cells should be nonenantioselective assuming equal availability of unbound enantiomers. Mostly, this has been examined where unbound enantiomer concentrations in plasma are very similar; red cell partitioning, after equilibrium has been attained, has shown no differences for enantiomers of mexiletine (KWOK et al. 1995), ketoprofen (HAYBALL et al. 1992), and propranol (BAI et al. 1983; OLANOFF et al. 1984). However, the antirheumatic and antimalarial drug rac-hydroxychloroquine has been shown (DUCHARME et al. 1994) to accumulate stereoselectively in red blood cells (R > S), which could be explained by the higher plasma protein binding of *S*-hydroxychloroquine (MCLACHLAN et al. 1993).

III. Uptake into Adipose Tissue

Adipose tissue uptake of drugs is largely a result of lipophilicity, a chemical property independent of enantiomer stereoconfiguration. However, members of the chiral 2-arylpropionic acid nonsteroidal antiinflammatory drug (NSAID) class, commonly referred to as profens, are subject to enantiospecific conjugation with coezyme A (CoA) leading to the formation of an intermediate CoA thioester, which in turn may lead to its incorporation into hybrid triglycerides. This is a different process from the reversible equilibrium resulting simply from drug lipophilicity, since triglyceride formation is due to covalent (irreversible) attachment of an NSAID. The 2-arylpropionates undergo metabolic activation to reactive coenzyme A thioesters, which in turn have been involved in the formation of hybrid triglycerides of these xenobiotics (CARABAZA et al. 1996). This process is specific for enantiomers of the (*R*)-

configuration and is thought to occur via a similar mechanism to the metabolic chiral inversion process (HUTT and CALDWELL 1983). Some have suggested that adipose tissue deposition of such compounds may have toxicological consequences (CALDWELL and MARSH 1983) and while this is still speculative, it highlights the potential for drug toxicity resulting from enantioselective drug distribution. The molecular aspects of 2-arylpropionate thioester generation have been recently reviewed (KNIGHTS 1998).

IV. Sequestration into Synovial Fluid and Articular Tissue

Since rheumatic diseases involve local articular pathology, it is generally assumed that NSAIDs, the commonly used prototypical chiral drug class, need to partition into the joint cavity to be efficacious. Trans-synovial passage of NSAIDs appears to be a diffusion-mediated process, regulated principally by the physico-chemical properties of the drug and the pathophysiological status of the synovial membrane (NETTER et al. 1989). Accordingly, the individual enantiomers of a chiral NSAID should exhibit comparable transport kinetics across the synovial membrane with respect to the unbound species, although enantioselective protein binding within the plasma and/or synovial fluid compartments might confound the interpretation of data generated for total species.

Relatively few studies have monitored chiral NSAID enantiomer concentrations in synovial fluid. DAY and coworkers (DAY et al. 1988) monitored the time-course of total *R*- and *S*-ibuprofen in patients with rheumatoid arthritis who were prescribed racemic drug. Synovial fluid levels of *S*-ibuprofen exceeded those of its antipode across the dose interval, yet in plasma its relative enrichment was significantly less, although this isomer still predominated at most time points. Compartmental analysis showed a faster rate of entry of the *S*-enantiomer into synovial fluid, most likely as a result of enantioselective plasma protein binding. It should be noted that since synovial fluid contains albumin, albeit at a substantially lower concentration than in plasma, at steady state the enantiomeric ratio of total species in synovial fluid will also be modulated by enantioselective binding to synovial albumin and other proteins.

More recently, the distribution of tiaprofenic acid isomers in synovium and cartilage was assessed in osteoarthritic patients dosed with racemate for 2 days prior to surgical arthroplasty (LAPICQUE et al. 1996). Rather than synovial fluid, these tissues are the true local targets for NSAID therapy. Plasma and synovium concentrations of the pharmacologically active *R*-enantiomer were higher than those of its antipode, while negligible amounts of drug were found in cartilage. Various studies (MULLER et al. 1992; MIGNOT et al. 1996) have established that enantioselective binding of this NSAID to HSA occurred in vitro; at a physiological protein concentration the *S*-enantiomer was bound to a greater extent.

E. Pharmacological Ramifications of Enantioselective Plasma and Tissue Binding

Reversible drug–protein binding is a primary determinant of the pharmacokinetic and pharmacodynamic properties of the drug. Taking the case of the latter first, it is only unbound drug which is capable of eliciting a receptor-mediated pharmacological effect; protein-bound drug is generally assumed to be pharmacologically inactive. Clearly, this has ramifications when exploring enantioselective in vitro activities of chiral drugs which may be probed in the absence of protein-binding compared to the situation in vivo.

I. Enantioselective Plasma Protein Binding: Implications for Interpretation of Pharmacokinetics of Chiral Drugs

In terms of the pharmacokinetic properties of drug enantiomers, the situation is more complex. For some drugs that are highly extracted by the liver, the degree of plasma protein binding does not significantly alter the rate of metabolism, since the drug is efficiently removed from its plasma binding sites during circulation through the liver. These drugs, such as rac-propranolol and rac-verapamil, are classified kinetically as nonrestrictively cleared drugs (Wilkinson and Shand 1975). An analogous situation is seen with the renal tubular secretion of drugs; secretion into the tubular lumen occurs in spite of extensive protein binding. However, the majority of drugs are restrictively cleared, for example, the enantiomers of warfarin, and in this case the hepatic clearance of total drug is directly proportional to the fraction unbound in plasma (f_u). Consequently, for restrictively cleared drug enantiomers which are highly protein bound, a small difference between enantiomers in the fraction bound to plasma protein will not only result in a relatively large difference in f_u, but also in a proportional difference in plasma clearance. This will have potentially important consequences in the interpretation of the relationship between pharmacokinetic parameters of enantiomers of chiral drugs and their observed pharmacological response. Cognizance of enantioselective plasma protein binding will also be important when interpreting mechanisms for changes to total clearance or total drug concentrations, for example, as a result of disease states. In theoretical treatments, Evans and coworkers (Evans et al. 1988) and Tucker and Lennard (Tucker and Lennard 1990) have illustrated, among other variables, the impact of enantioselective plasma protein binding on the apparent, but incorrect, pharmacokinetics of the racemic drug.

Changes in the plasma protein binding of drugs arising from multiple-drug therapy is well-documented; it is still used as an explanation for drug-drug interactions. However, plasma is an open compartment and any drug which is displaced from plasma proteins will rapidly distribute into tissue compartments spreading out to increase its volume of distribution. Accordingly, the transient increase in the unbound level of a drug, due to the presence of a displacing drug, is usually of little therapeutic importance (Sansom and Evans 1995).

The pharmacokinetics of the chiral antiarrhythmic drug rac-pirmenol was recently examined in dogs following intravenous administration of the racemate (Janiczek et al. 1997). The (–)-enantiomer had 47% lower clearance and 33% lower steady-state volume of distribution than (+)-pirmenol and these differences were mostly attributed to stereoselective plasma protein binding reflected in a 58% higher unbound fraction for the dextrorotatory isomer. Similar trends were observed when separate enantiomers were administered, which suggested a lack of interaction between enantiomers.

II. Pharmacokinetic Implications of Enantioselective Plasma Protein Binding: Case Studies of Chiral NSAIDs

Chiral NSAIDs, most notably of the 2-arylpropionate chemical class, have been extensively studied in terms of their plasma protein binding properties. Of all papers cited in ISI Current Contents (1993 to 2000) which have examined protein binding of drug enantiomers, either as a subject in itself or as part of a pharmacokinetic/dispositional study, more than a third of these have examined chiral NSAID binding behavior. Without exception, they are all highly bound to plasma protein (principally to albumin) with f_u values typically less than 1%. Given the extent to which these drug enantiomers are bound to albumin, only a small difference in the degree of binding is necessary to lead to significant enantioselectivity in f_u. Further, since these drugs are kinetically characterized as restrictively cleared, low hepatic extraction ratio drugs, their total plasma clearance will be dependent on f_u and substantial differences may arise between the total plasma concentration profiles of the enantiomers. In studies examining the disposition and protein binding in humans of the chiral arylalkanoic acid NSAID rac-ketorolac (Hayball et al. 1994a,b), the investigators were able to conclude that the twofold higher total plasma clearance of *S*-ketorolac was solely attributed to the concordant higher unbound fraction of this isomer in plasma (and in buffer solutions containing physiological concentrations of human serum albumin). By relying solely on total plasma concentrations of enantiomers, incorrect assignment of mechanisms (for example: metabolic, enzymic factors) could be postulated to explain enantioselective drug clearance.

For compounds which are extensively bound to plasma proteins, such as enantiomers of chiral NSAIDs, the apparent volume of distribution (V) of each enantiomer can be represented by the following equation (Lin et al. 1987):

$$V = V_P(1 + R_{E/I}) + V_T(f_u/f_{uT})$$

where f_{uT} represents the unbound fraction of the enantiomer extravascularly, $R_{E/I}$ is the ratio of the amount of binding protein (in the case of NSAIDs this will be albumin) in the extracellular fluid outside the plasma to that in plasma,

V_P is plasma volume, and V_T is the aqueous volume outside the extracellular fluid into which the drug distributes. If one assigns physiological values to $R_{E/I}$, V_P, and V_T of 1.4, 3 l, and 30 l, respectively (OIE and TOZER 1979) then this equation simplifies to:

$$V = 7.2 + 30(f_u / f_{uT})$$

It follows that the enantiomers of a chiral NSAID will have different volumes of distribution if they differ in the relative extent to which they bind to plasma and tissue components. Given that distribution volumes of NSAID enantiomers are typically very similar and of the order 7–14 l in humans, this means that their tissue binding is appreciably less than their plasma protein binding (the fraction f_u/f_{uT} is small in magnitude). Accordingly, plasma protein binding, rather than tissue protein binding, will usurp the greater role when attempting to understand mechanisms of enantioselective drug distribution for drugs of this and related classes.

Since chiral NSAIDs are extensively bound to plasma albumin and they have relatively low molecular weights, saturation of binding sites may occur when enantiomer concentrations in plasma exceed available binding sites; this situation may be exacerbated when both enantiomers are present in plasma. Insight into the nonlinear human pharmacokinetics of ibuprofen isomers was achieved when it was recognized that concentration-dependent binding was occurring when escalating doses of racemate were administered (EVANS et al. 1989, 1990). Evidence of displacement of each enantiomer by its antipode has been noted by other workers including more recent studies in rats (ITOH et al. 1997) and humans (PALIWAL et al. 1993; SMITH et al. 1994).

III. Enantioselective Tissue Binding: Implications for Interpretation of Pharmacokinetics of Chiral Drugs

With regard to the extent of tissue distribution of enantiomers, a tissue-to-plasma drug concentration ratio (K_p), more precisely probes enantioselective tissue distribution than the plasma volume of distribution term as discussed in Sect. E.II. At steady state, K_p can be expressed by the following equation, provided a specific membrane transport system is not in operation:

$$K_p = C_T^{ss} / C_P^{ss}(Q_T + f \cdot Cl_{int}) / Q_T)$$

where C_T^{ss} and C_P^{ss} are the steady-state tissue and plasma drug concentrations, respectively; and Q_T and Cl_{int} are the tissue plasma flow and intrinsic clearance of drug by the tissue. In a tissue of a noneliminating organ Cl_{int} is zero and the above equation collapses (CHEN and GROSS 1979) to:

$$K_p = C_T^{ss} / C_P^{ss}$$

It is generally assumed that in the distribution equilibrium state, the concentration of unbound drug in plasma and tissue are very similar (assuming no active transport) and thus:

$$f_u \times C_P^{ss} = f_{uT} \times C_T^{ss}$$

and therefore, K_{pf}, the tissue-to-plasma unbound drug concentration ratio, is an inverse function solely of the unbound fraction in tissue:

$$K_{pf} = K_p / f_u = 1/f_{uT}$$

In the case of a chiral drug, it is independent of enantioselective plasma protein binding unlike volumes of distribution terms.

Clearly, plasma protein binding assumes a far greater importance than tissue binding when attempting to assess pharmacokinetic properties of drugs, chiral or achiral.

F. Conclusions

In conclusion, the major plasma proteins, by virtue of their inherent chirality, are capable of discriminating between enantiomers of chiral drugs. For enantiomers which are extensively bound, relatively minor differences in affinities for plasma protein can result in significant enantioselectivity when calculating unbound fractions. This can have significant pharmacokinetic implications when mechanistically assessing plasma concentration-time profiles expressed for total (bound plus unbound) enantiomers, most notably, when the drug in question is restrictively cleared by the liver. In many instances, thc clinical consequences of enantioselectivity in the plasma protein binding of drugs will be relatively minor; however, in some cases there will be a significant impact such as outlined for congeners of some chiral NSAIDs. Enantioselective tissue binding is poorly recognized by comparison, and significant pitfalls may arise when documenting blood and tissue concentrations of individual enantiomers when steady-state has not been achieved (see Sect. D.I.). Essentially, enantioselective sequestration of drugs into tissue, when correctly assessed under steady-state conditions, can be explained by differences in the availability of unbound drug in the vascular compartment, that is, by differences in plasma protein binding.

References

Albani F, Riva R, Contin C, Baruzzi A (1984) Stereoselective Binding of Propranolol Enantiomers to human α-acid glycoprotein and human plasma. British Journal of Clinical Pharmacology 18:244–246

Ascoli GA, Bertucci C, Salvadori P (1998) Ligand binding to a human serum albumin stationary phase: use of same-drug competition to discriminate pharmacologically relevant interactions. Biomedical Chromatography 12:248–254

Bai SA, Walle UK, Wilson MJ, Walle T (1983) Stereoselective binding of the (–)-enantiomer of propranolol to plasma and extravascular binding sites in the dog. Drug Metabolism & Disposition 11:394–395

Baudry S, Pham YT, Baune B, Vidrequin S, Crevoisier C, Gimenez F, Farinotti R (1997) Stereoselective passage of mefloquine through the blood-brain barrier in the rat. Journal of Pharmacy & Pharmacology 49:1086–1090

Bertilsson L, Otani K, Dahl ML, Nordin C, Aberg-Wistedt A (1991) Stereoselective efflux of (*E*)-10-hydroxynortriptyline enantiomers from the cerebrospinal fluid of depressed patients. Pharmacology & Toxicology 68:100–103

Bertucci C, Canepa A, Ascoli GA, Guimaraes LFL, Felix G (1999) Site I on human albumin: Differences in the binding of (*R*)- and (*S*)-warfarin. Chirality 11:675–679

Bjornsson TD, Brown JE, Tschanz C (1981) Importance of radiochemical purity of radiolabeled drugs used for determining plasma protein binding of drugs. Journal of Pharmaceutical Sciences 70:1372–1373

Caldwell J, Marsh MV (1983) Interrelationships between xenobiotic metabolism and lipid biosynthesis. Biochemical Pharmacology 32:1667–1672

Carabaza A, Suesa N, Tost D, Pascual J, Gomez M, Gutierrez M, Ortega E, Montserrat X, Garcia AM, Mis R, Cabre F, Mauleon D, Carganico G (1996) Stereoselective metabolic pathways of ketoprofen in the rat: incorporation into triacylglycerols and enantiomeric inversion. Chirality 8:163–172

Chen HSG, Gross JF (1979) Estimation of tissue-to-plasma partition coefficients used in physiological pharmacokinetic models. Journal of Pharmacokinetics and Biopharmaceutics 7:117–125

Chignell CF (1973) Drug-protein binding: recent advances in methodology: spectroscopic techniques. Annals of the New York Academy of Sciences 226:44–59

Chosson E, Uzan S, Gimenez F, Wainer IW, Farinotti R (1993) Influence of specific albumin ligand markers used as modifiers on the separation of benzodiazepine enantiomers by chiral liquid chromatography on a human serum albumin column. Chirality 5:71–77

Day RO, Williams KM, Graham GG, Lee EJ, Knihinicki RD, Champion GD (1988) Stereoselective disposition of ibuprofen enantiomers in synovial fluid. Clinical Pharmacology & Therapeutics 43:480–487

Domenici E, Bertucci C, Salvadori P, Motellier S, Wainer IW (1990) Immobilized serum albumin: rapid HPLC probe of stereoselective protein-binding interactions. Chirality 2:263–268

Ducharme J, Wainer IW, Parenteau HI, Rodman JH (1994) Stereoselective Distribution of Hydroxychloroquine in the Rabbit Following Single and Multiple Oral Doses of the Racemate and the Separate Enantiomers. Chirality 6:337–346

Eap CB, Cuendet C, Baumann P (1990) Binding of d-methadone, l-methadone, and dl-methadone to proteins in plasma of healthy volunteers: role of the variants of α1-acid glycoprotein. Clinical Pharmacology and Therapeutics 47:338–346

Eichelbaum M, Mikus G, Vogelgesang B (1984) Pharmacokinetics of (+)-, (–)- and (+/–)-verapamil after intravenous administration. British Journal of Clinical Pharmacology 17:453–458

Evans AM, Nation RL, Sansom LN, Bochner F, Somogyi AA (1988) Stereoselective drug disposition: potential for misinterpretation of drug disposition data. British Journal of Clinical Pharmacology 26:771–780

Evans AM, Nation RL, Sansom LN, Bochner F, Somogyi AA (1989) Stereoselective plasma protein binding of ibuprofen enantiomers. European Journal of Clinical Pharmacology 36:283–290

Evans AM, Nation RL, Sansom LN, Bochner F, Somogyi AA (1990) The relationship between the pharmacokinetics of ibuprofen enantiomers and the dose of racemic ibuprofen in humans. Biopharmaceutics & Drug Disposition 11:507–518

Fehske KJ, Muller WE, Wollert U (1981) The location of drug binding sites in human serum albumin. Biochemical Pharmacology 30:687–692

Fernandez C, Gimenez F, Thuillier A, Farinotti R (1999) Stereoselective Binding of Zopiclone to Human Plasma Proteins. Chirality 11:129–132

Fitos I, Simonyi M (1992) Stereoselective effect of phenprocoumon enantiomers on the binding of benzodiazepines to human serum albumin. Chirality 4:21–23

Fitos I, Visy J, Simonyi M, Hermansson J (1995) Separation of enantiomers of benzodiazepines on the Chiral-AGP column. Journal of Chromatography A 709:265–273
Fitos I, Visy J, Simonyi M, Hermansson J (1999) Stereoselective allosteric binding interaction on human serum albumin between ibuprofen and lorazepam acetate. Chirality 11:115–120
Gross AS, Eser C, Mikus G, Eichelbaum M (1993) Enantioselective Gallopamil Protein Binding. Chirality 5:414–418
Gross AS, Heuer B, Eichelbaum M (1988) Stereoselective Protein Binding of Verapamil Enantiomers. Biochemical Pharmacology 37:4623–4627
Hanada K, Akimoto S, Mitsui K, Mihara K, Ogata H (1998) Enantioselective Tissue Distribution of the Basic Drugs Disopyramide, Flecainide and Verapamil in Rats – Role of Plasma Protein and Tissue Phosphatidylserine Binding. Pharmaceutical Research 15:1250–1256
Hayball PJ (1995) Formation and reactivity of acyl glucuronides: the influence of chirality. Chirality 7:1–9
Hayball PJ, Holman JW, Nation RL, Massy-Westropp RA, Hamon DP (1994a) Marked enantioselective protein binding in humans of ketorolac in vitro: elucidation of enantiomer unbound fractions following facile synthesis and direct chiral HPLC resolution of tritium-labelled ketorolac. Chirality 6:642–648
Hayball PJ, Nation RL, Bochner F (1992) Enantioselective pharmacodynamics of the nonsteroidal antiinflammatory drug ketoprofen: in vitro inhibition of human platelet cyclooxygenase activity. Chirality 4:484–487
Hayball PJ, Nation RL, Bochner F, Newton JL, Massy-Westropp RA, Hamon DP (1991) Plasma protein binding of ketoprofen enantiomers in man: method development and its application. Chirality 3:460–466
Hayball PJ, Wrobel J, Tamblyn JG, Nation RL (1994b) The pharmacokinetics of ketorolac enantiomers following intramuscular administration of the racemate. British Journal of Clinical Pharmacology 37:75–78
Herve F, Caron G, Duche J-C, Gallard P, Rahman NA, Tsantili-Kakoulidou A, Carrupt P-A, D'Anthis P, Tillement J-P, Testa B (1998) Ligand specificity of the genetic variants of human α1-acid glycoprotein: generation of a three-dimensional quantitative structure activity relationship model for drug binding to the a variant. Molecular Pharmacology 54:129–138
Honore B (1990) Conformational changes in human serum albumin induced by ligand binding. Pharmacology & Toxicology 66:7–26
Hsyu PH, Giacomini KM (1985) Stereoselective renal clearance of pindolol in humans. Journal of Clinical Investigation 76:1720–1726
Hutt AJ, Caldwell J (1983) The metabolic chiral inversion of 2-arylpropionic acids – a novel route with pharmacological consequences. Journal of Pharmacy & Pharmacology 35:693–704
Itoh T, Maruyama J, Tsuda Y, Yamada H (1997) Stereoselective Pharmacokinetics of Ibuprofen in Rats – Effect of Enantiomer-Enantiomer Interaction in Plasma Protein Binding. Chirality 9:354–361
Janiczek N, Smith DE, Chang T, Sedman AJ, Stringer KA (1997) Pharmacokinetics of Pirmenol Enantiomers and Pharmacodynamics of Pirmenol Racemate in Patients With Premature Ventricular Contractions. Journal of Clinical Pharmacology 37: 502–513
Jones ME, Sallustio BC, Purdie YJ, Meffin PJ (1986) Enantioselective disposition of 2-arylpropionic acid nonsteroidal antiinflammatory drugs. II. 2-Phenylpropionic acid protein binding. Journal of Pharmacology & Experimental Therapeutics 238:288–294
Knadler MP, Brater DC, Hall SD (1989) Plasma protein binding of flurbiprofen: Enantioselectivity and influence of pathophysiological status. Journal of Pharmacology & Experimental Therapeutics 249:378–385
Knights KM (1998) Role of hepatic fatty acid:coenzyme A ligases in the metabolism of xenobiotic carboxylic acids. Clinical & Experimental Pharmacology & Physiology 25:776–782

Kwok DW, Kerr CR, McErlane KM (1995) Pharmacokinetics of Mexiletine Enantiomers in Healthy Human Subjects – a Study of the in Vivo Serum Protein Binding, Salivary Excretion and Red Blood Cell Distribution of the Enantiomers. Xenobiotica 25:1127–1142

Lapicque F, Mainard D, Gillet P, Payan E, Bannwarth B, Netter P (1996) Stereoselective Distribution of Tiaprofenic Acid in Synovium and Cartilage in Osteoarthritic Patients. European Journal of Clinical Pharmacology 50:283–287

Lima JL (1988) Species-Dependent Binding of Disopyramide Enantiomers. Drug Metabolism and Disposition 16:563–567

Lima JL, Jungbluth GL, Devine T, Robertson LW (1984) Stereoselective Binding of Disopyramide to Human Plasma Protein. Life Sciences 35:835–839

Lin JH, Cocchetto DM, Duggan DE (1987) Protein binding as a primary determinant of the clinical pharmacokinetic properties of non-steroidal antiinflammatory drugs. Clinical Pharmacokinetics 12:402–432

Loun B, Hage DS (1996) Chiral Separation Mechanisms in Protein-Based Hplc Columns .2. Kinetic Studies of (*R*)- and (*S*)-Warfarin Binding to Immobilised Human Serum Albumin. Analytical Chemistry 68:1218–1225

Maboudian-Esfahani M, Brocks DR (1999) Disposition of ethopropazine enantiomers in the rat: Tissue distribution and plasma protein binding. Journal of Pharmacy & Pharmaceutical Sciences 2:23–29

Mather LE, Edwards SR, Duke CC, Cousins MJ (2000) Microdialysis study of the blood-brain equilibration of thiopental enantiomers. British Journal of Anaesthesia 84:67–73

McLachlan AJ, Cutler DJ, Tett SE (1993) Plasma Protein Binding of the Enantiomers of Hydroxychloroquine and Metabolites. European Journal of Clinical Pharmacology 44:481–484

Mignot I, Presle N, Lapicque F, Monot C, Dropsy R, Netter P (1996) Albumin Binding Sites For Etodolac Enantiomers. Chirality 8:271–280

Muller N, Lapicque F, Monot C, Payan E, Dropsy R, Netter P (1992) Stereoselective binding of etodolac to human serum albumin. Chirality 4:240–246

Muller WE, Stillbauer AE (1983) Characterization of a common binding site for basic drugs on human α1-acid glycoprotein (orosomucoid). Naunyn-Schmiedebergs Archives of Pharmacology 322:170–173

Muller WE, Wollert U (1975a) High stereospecificity of the benzodiazepine binding site on human serum albumin. Studies with d- and l-oxazepam hemisuccinate. Molecular Pharmacology 11:52–60

Muller WE, Wollert U (1975b) Influence of various drugs on the binding of L-tryptophan to human serum albumin. Research Communications in Chemical Pathology & Pharmacology 10:565–568

Murai-Kushiya M, Okada S, Kimura T, Hasegawa R (1993) Stereoselective binding of β-blockers to purified rat α1-acid glycoprotein. Journal of Pharmacy & Pharmacology 45:225–228

Netter R, Bannworth B, Royer-Marrot MJ (1989) Recent findings on the pharmacokinetics of non-steroidal antiinflammatory drugs in synovial fluid. Clinical Pharmacokinetics 17:145–162

Newman-Tancredi A, Chaput C, Gavaudan S, Verriele L, Millan MJ (1998) Agonist and antagonist actions of (–)pindolol at recombinant, human serotonin1 A (5-HT1 A) receptors. Neuropsychopharmacology 18:395–398

Noctor TA, Wainer IW, Hage DS (1992) Allosteric and competitive displacement of drugs from human serum albumin by octanoic acid, as revealed by high-performance liquid affinity chromatography, on a human serum albumin-based stationary phase. Journal of Chromatography 577:305–315

Ofori-Adjei D, Ericsson O, Lindstrom B, Sjoqvist F (1986) Protein binding of chloroquine enantiomers and desethylchloroquine. British Journal of Clinical Pharmacology 22:356–358

Oie S, Tozer TN (1979) Effect of altered plasma protein binding on apparent volume of distribution. Journal of Pharmaceutical Sciences 68:1203–1205

Olanoff LS, Walle T, Walle UK, Cowart TD, Gaffney TE (1984) Stereoselective clearance and distribution of intravenous propranolol. Clinical Pharmacology & Therapeutics 35:755–761

Oravcova J, Bohs B, Lindner W (1996) Drug-Protein Binding Studies. New Trends in Analytical and Experimental Methodology. Journal of Chromatography B: Biomedical Applications 677:1–28

Paliwal JK, Smith DE, Cox SR, Berardi RR, Dunnkucharski VA, Elta GH (1993) Stereoselective, Competitive, and Nonlinear Plasma Protein Binding of Ibuprofen Enantiomers As Determined Invivo in Healthy Subjects. Journal of Pharmacokinetics & Biopharmaceutics 21:145–161

Perez V, Gilaberte I, Faries D, Alvarez E, Artigas F (1997) Randomised, double-blind, placebo-controlled trial of pindolol in combination with fluoxetine antidepressant treatment. Lancet 349:1594–1597

Pham YT, Nosten F, Farinotti R, White NJ, Gimenez F (1999) Cerebral uptake of mefloquine enantiomers in fatal cerebral malaria. International Journal of Clinical Pharmacology & Therapeutics 37:58–61

Rahman MH, Maruyama T, Okada T, Yamasaki K, Otagiri M (1993) Study of Interaction of Carprofen and Its Enantiomers With Human Serum Albumin .1. Mechanism of Binding Studied By Dialysis and Spectroscopic Methods. Biochemical Pharmacology 46:1721–1731

Rochat B, Baumann P, Audus KL (1999) Transport mechanisms for the antidepressant citalopram in brain microvessel endothelium. Brain Research 831:229–236

Romach MK, Piafsky KM, Abel JG, Khouw V, Sellers EM (1981) Methadone binding to orosomucoid (α1-acid glycoprotein): determinant of free fraction in plasma. Clinical Pharmacology & Therapeutics 29:211–217

Salvadori P, Bertucci C, Ascoli G, Uccellobarretta G, Rossi E (1997) Direct Resolution, Characterization, and Stereospecific Binding Properties of an Atropisomeric 1,4-Benzodiazepine. Chirality 9:495–505

Sam E, Sarre S, Michotte Y, Verbeke N (1997) Distribution of apomorphine enantiomers in plasma, brain tissue and striatal extracellular fluid. European Journal of Pharmacology 329:9–15

Sansom LN, Evans AM (1995) What is the true clinical significance of plasma protein binding displacement interactions? Drug Safety 12:227–233

Schanker LS (1964) Physiological transport of drugs. Advances in Drug Research 1:71–106

Seeman P, Van Tol HH (1993) Dopamine D4 receptors bind inactive (+)-aporphines, suggesting neuroleptic role. Sulpiride not stereoselective. European Journal of Pharmacology 233:173–174

Smith DE, Paliwal JK, Cox SR, Berardi RR, Dunnkucharski VA, Elta GH (1994) The Effect of Competitive and Nonlinear Plasma Protein Binding On the Stereoselective Disposition and Metabolic Inversion of Ibuprofen in Healthy Subjects. Biopharmaceutics & Drug Disposition 15:545–561

Smith DF, Gee AD, Hansen SB, Moldt P, Nielsen EO, Scheel-Kruger J, Gjedde A (1999) Uptake and distribution of a new SSRI, NS2381, studied by PET in living porcine brain. European Neuropsychopharmacology 9:351–359

Spahn-Langguth H, Benet LZ (1992) Acyl glucuronides revisited: Is the glucuronidation process a toxification as well as a detoxification mechanism. Drug Metabolism Reviews 24:5–48

Tabkahshi H, H. O, Kanno S, Takeuchi H (1990) Plasma Protein Binding of Propranolol Enantiomers as a Major Determinant of Their Stereoselective Distribution in Rats. Journal of Pharmacology & Experimental Therapeutics 252:272–278

Tucker GT, Lennard MS (1990) Enantiomer specific pharmacokinetics. Pharmacology & Therapeutics 45:309–329

Valdiviesco L, Giacomini KM, Nelson WL, Pershe R, Blaschke TF (1988) Stereoselective Binding of Disopyramide to Plasma Proteins. Pharmaceutical Research 5: 316–318

Walle T, Webb JG, Bagwell EE, Walle UK, Daniell HB, Gaffney TE (1988) Stereoselective delivery and actions of β receptor antagonists. Biochemical Pharmacology 37:115–124
Wilkinson GR, Shand DG (1975) Commentary: a physiological approach to hepatic drug clearance. Clinical Pharmacology & Therapeutics 18:377–390
Yacobi A, Levy G (1977) Protein binding of warfarin enantiomers in serum of humans and rats. Journal of Pharmacokinetics & Biopharmaceutics 5:123–131
Yamasaki K, Maruyama T, Kraghhansen U, Otagiri M (1996) Characterization of Site I On Human Serum Albumin – Concept About the Structure of a Drug Binding Site. Biochimica et Biophysica Acta – Protein Structure & Molecular Enzymology 1295:147–157
Yan H, Lewander T (1999) Differential tissue distribution of the enantiomers of racemic pindolol in the rat. European Neuropsychopharmacology 10:59–62
Yata N, Toyoda T, Murakami T, Nishiura A, Higashi Y (1990) Phosphatidylserine as a determinant for the tissue distribution of weakly basic drugs in rats. Pharmaceutical Research 7:1019–1025
Yoshikawa T, Oguma T, Ichihashi T, Kinoshita H, Hirano K, Yamada H (1999) Epimerization of moxalactam by albumin and simulation of in vivo epimerization by a physiologically based pharmacokinetic model. Chirality 11:309–315
Zanardi R, Franchini L, Gasperini M, Lucca A, Smeraldi E, Perez J (1998) Faster onset of action of fluvoxamine in combination with pindolol in the treatment of delusional depression: a controlled study. Journal of Clinical Psychopharmacology 18:441–446

CHAPTER 13

Stereoselective Drug Metabolism and Drug Interactions

A.S. Gross, A. Somogyi, and M. Eichelbaum

A. Introduction

The pharmacokinetics of many chiral drugs are stereoselective. In the majority of cases it is stereoselective metabolism that makes a major contribution to the difference in disposition of the individual stereoisomers. Stereochemistry influences not only the disposition and activity of the parent drug but also the nature and potential pharmacological activity of the metabolites formed. Many recent developments are furthering our understanding of drug metabolism and the implications of stereochemistry. The enzymes catalyzing drug metabolism are being identified and characterized in increasing detail with the genes encoding these proteins being cloned and sequenced, and variants identified and studied in different populations. The factors that regulate the expression, levels, and activity of the enzymes are being determined and the structural requirements for substrate binding are being understood as the profile of substrates and their binding affinities become known. Furthermore, the potential for clinically significant interactions between coadministered drugs in vivo which associate with or are metabolized by the same enzyme can be assessed and predicted from in vitro data.

The in vitro techniques utilizing human liver microsomes, heterologously expressed enzymes, primary hepatocytes, and tissue slices are now being used to characterize the metabolic fate of new drugs during development. Knowledge of the enzymes catalyzing biotransformation in vivo and their kinetics enables the prediction of the factors that will contribute to inter- and intraindividual variation in metabolism and hence pharmacokinetics as well as drug–drug interactions. Developments in stereospecific analytical techniques which now enable individual stereoisomers to be measured in biological fluids at very low concentrations have not only facilitated our understanding of the metabolic fate of chiral drugs, but continue to provide valuable mechanistic insights into the three-dimensional interaction between substrates and enzymes. Induction and inhibition of drug-metabolizing enzymes have been identified as two of the important mechanisms responsible for clinically significant drug–drug interactions. A consideration of stereoselectivity can provide additional understanding of the clinical significance of the interaction of drugs used as mixtures of stereoisomers.

The majority of drugs derived from natural sources are chiral, as they are the products of the stereospecificity of biological synthesis. Chiral drugs synthesized in the laboratory will be produced as racemic mixtures unless stereospecific synthetic techniques are deliberately used. Recent developments in synthetic chemistry have facilitated the large-scale production of single stereoisomers and therefore many newly introduced chiral drugs are being used clinically as single stereoisomers. Nevertheless, established drugs in widespread use today are used as mixtures of stereoisomers and it remains important to have a clear understanding of the implications of stereoselectivity for drug metabolism and interactions. This chapter will describe relevant and contemporary concepts with limited examples but will not provide a comprehensive catalogue of the extensive literature relating to stereoselective drug metabolism. General concepts relating to stereoselective drug metabolism are discussed in a number of reviews (TESTA 1988; TRAGER and JONES 1987; MASON and HUTT 1997).

B. Stereoselective Metabolism

When administered to man, drugs interact with the chiral environment which is characteristic of all life. Production of pharmacological and toxicological effects as well as distribution in the body and elimination via excretion or metabolism are usually a consequence of an interaction with chiral macromolecules such as proteins. Metabolism, the conversion of the parent drug to another chemical species, which is often more polar and thus readily excreted from the body via the kidneys, bile, and gut, makes a major contribution to the elimination of most drugs. As metabolism is a consequence of the interaction of a drug substrate with a chiral drug-metabolizing enzyme, it is not surprising that there can be stereochemical consequences.

I. Substrate and Product Stereoselectivity

The concepts of substrate and product stereoselectivity must be distinguished when discussing stereoselective metabolism (JENNER and TESTA 1973). Differential metabolism of the stereoisomers of a chiral molecule is known as substrate stereoselectivity. Although individual pathways of metabolism may display a high degree of substrate stereoselectivity, the net metabolic clearance of two enantiomers often differs only two- to fivefold as a number of enzymes, each with varying substrate stereoselectivities, contribute to the overall metabolism of each stereoisomer.

Product stereoselectivity refers to the formation of metabolites with a center of asymmetry from a prochiral substrate. These reactions, catalyzed by a single enzyme, often display a high degree of stereoselectivity or even stereospecificity, and only one of two possible enantiomeric metabolites is usually formed. For example, the formation of 4-hydroxydebrisoquine from debrisoquine in vivo, a reaction catalyzed by CYP2D6, overwhelmingly favors the production of the (*S*)-(+)- enantiomer (Eichelbaum et al. 1988). In extensive metabolizers (EMs) who express CYP2D6, the enantiomeric excess (a value of 0 indicates no stereoselectivity and 100 stereospecificity) of this reaction is greater than 96. In poor metabolizers (PMs) who lack active CYP2D6, the enantiomeric excess ranges from 28% to 90% (mean 70%). An additional example is the stereospecific reduction of naltrexone to the corresponding 6β- and not 6α-alcohol (Malpeis et al. 1975).

Many enzymes have now been identified as contributing to the biotransformation of drugs used clinically. Stereoselective metabolism has been reported both for enzymes that introduce or unmask functional groups (phase 1 metabolism such as oxidation, reduction, or hydrolysis) as well as those which conjugate the molecule with an endogenous moiety (phase 2 metabolism such as glucuronidation or sulfation). Some examples of stereoselective metabolism catalyzed by well-characterized drug-metabolizing enzymes are given in Table 1.

The liver is the major organ responsible for drug metabolism; however, each enzyme has a characteristic distribution throughout the body between the liver and other tissues. Hence substantial extrahepatic metabolism may occur for some drugs. In some cases extrahepatic metabolism can have important clinical consequences. For example, CYP3A enzymes are localized on enterocytes on the intestinal lumen (Watkins et al. 1987; Lown et al. 1994),

Table 1. Examples of stereoselective pathways of metabolism catalyzed by identified enzymes

Enzyme	Metabolic pathway	Stereoselectivity	Reference
FMO3	Chlorpheniramine *N*-oxidation	(*R*) > (*S*)	Cashman et al. 1992
CYP2A6	Losigamone phenolic oxidation	(+) > (–)	Torchin et al. 1996
CYP2B6	Ifosfamide 4-hydroxylation	(*S*) > (*R*)	Roy et al. 1999
CYP2D6	Fluoxetine *N*-demethylation	(*R*) > (*S*)	Margolis et al. 2000
CYP2C9	Phenytoin 4′hydroxylation	(*S*) > (*R*)	Yasumori et al. 1999
CYP2C19	Omeprazole hydroxylation	(*R*) > (*S*)	Äbelö et al. 2000
	Mephobarbital 4-hydroxylation	(*R*) > (*S*)	Kobayashi et al. 2001
CYP3A4	Omeprazole sulfoxidation	(*S*) > (*R*)	Äbelö et al. 2000
Phenolsulfotransferase	Salbutamol sulfation	(*R*) > (*S*)	Eaton et al. 1996
Reduction	Warfarin keto-reduction	(*R*) > (*S*)	Caldwell et al. 1988
UDP-glucuronosyl-transferases	Zileuton glucuronidation	(*S*) > (*R*)	Sweeny and Nellans 1995

and when drugs which are substrates of CYP3A are administered orally substantial metabolism can occur during absorption, e.g., nifedipine (HOLTBECKER et al. 1996) and cyclosporine (LOWN et al. 1997).

1. Different Rates and Routes of Metabolism

Substrate stereoselectivity can be a consequence of metabolism by the same enzyme at different rates for each stereoisomer due to differences in the three dimensional interaction between the drug and the active site of the enzyme. Metabolism at different rates via different enzymes may also occur. For example, in human liver microsomes the intrinsic clearance of omeprazole to the sulfone metabolite, a reaction catalyzed by CYP3A4 for both enantiomers, is fourfold greater for the (*S*)-(3.87 ± 1.66 μl/min/mg) (mean ± SD) than the (*R*)-(0.84 ± 0.17 μl/min/mg) enantiomer. By contrast, the intrinsic clearance of omeprazole to the hydroxy-metabolite is tenfold greater for the (*R*)-enantiomer (39.9 ± 22.5 μl/min/mg), a reaction catalyzed predominantly by CYP2C19, than the (*S*)-enantiomer (4.00 ± 1.59 μl/min/mg), a reaction catalyzed by both CYP3A4 and CYP2C19 (ÄBELÖ et al. 2000).

The enantioselectivity of the metabolism of the calcium antagonist verapamil, used clinically as a racemic mixture, has also been studied in man in vivo and in vitro using human liver microsomes (KROEMER et al. 1992). The enantioselectivity of the metabolism to norverapamil, D-617, D-703, and D-702 differs (Table 2). This is a consequence of different enzymes catalyzing the various pathways of verapamil metabolism. For example, CYP3A4 catalyzes the formation of norverapamil, and both CYP3A4 and CYP1A2 contribute to

Table 2. Parameters (mean ± SD) describing the kinetics of the metabolism of (*S*)- and (*R*)-verapamil [Michaelis-Menten kinetics (K_m, V_{max} and intrinsic clearance, Cl_{int}) in vitro in human liver microsomes ($n = 10$) and clearance to the metabolite (Cl_{met}) in vivo in healthy volunteers] (KROEMER et al. 1992)

Parameter	Norverapamil		D-702		D-703		D-617	
	S	R	S	R	S	R	S	R
K_m (μM)	52.8 ± 41.3[a]	63.8 ± 50.4	159 ± 142	329 ± 382	59.4 ± 29.3[a]	40.9 ± 25.6	44.7 ± 18.5	92.4 ± 95.3
V_{max} (pmol/mg/min)	809 ± 611	817 ± 701	112 ± 64	146 ± 54	309 ± 182[a]	174 ± 107	1068 ± 978	1158 ± 871
Cl_{int} (ml/min/g protein)	18.4 ± 16.1[a]	14.7 ± 12.4	1.02 ± 0.75	0.84 ± 0.66	7.0 ± 4.3[a]	5.4 ± 3.6	27 ± 24	20 ± 14
Cl_{met} (ml/min)	851[b] ± 215[a]	213[b] ± 45	n.d.	n.d.	531 ± 141[a]	9.6 ± 3.5	1857[c] ± 534[a]	568[c] ± 185

[a] Significant difference between (*R*)- and (*S*)-enantiomers ($P < 0.05$).
[b] Sum of metabolites norverapamil + D-715 + D-620.
[c] Sum of metabolites D-617 + D-717.
n.d., not determined.

the formation of D617 (Kroemer et al. 1993). A further interesting example is the stereoselective glucuronidation of zileuton as shown in Fig. 1. The rate of glucuronidation of (*S*)-zileuton is 3.6- to 4.3-fold greater than (*R*)-zileuton at all concentrations measured, with both enantiomers having a similar apparent K_m for the metabolizing enzyme but 3.4-fold greater V_{max} for the (*S*)- than for the (*R*)- enantiomer (Sweeny and Nellans 1995).

Alternatively, stereoselective metabolism can occur when the different spatial orientation of the stereoisomers results in only one stereoisomer interacting with a particular active site of a drug-metabolizing enzyme. The biotransformation of the stereoisomers is consequently catalyzed by different drug-metabolizing enzymes, or the same enzyme may catalyze biotransformation at different sites on the molecule. Different pathways of metabolism for the two enantiomers can therefore occur. A classic example is that of the racemic anticonvulsant mephenytoin. The (*S*)-enantiomer is principally oxidized to 4′-hydroxymephenytoin by CYP2C19, whereas the (*R*)-enantiomer is principally *N*-dealkylated by CYP2C9 and CYP2B6 to form phenylethylhydantoin (De Morais et al. 1994; Goldstein et al. 1994; Ko et al. 1998). Stereoselective mephenytoin pharmacokinetics result in vivo as the 4-hydroxylation of (*S*)-mephenytoin is more rapid than the *N*-demethylation of (*R*)-mephenytoin (Wedlund et al. 1985). An interesting recent in vitro study reports that the same enzyme, CYP2C19, preferentially metabolizes each enantiomer of omeprazole via a different pathway: 5-hydroxylation of the pyridine group of (*R*)-omeprazole and 5-*O*-demethylation of (*S*)-omeprazole (Äbelö et al. 2000).

In general, the closer the center of asymmetry to the site of metabolism, the greater the potential for stereoselective metabolism. For example, modifi-

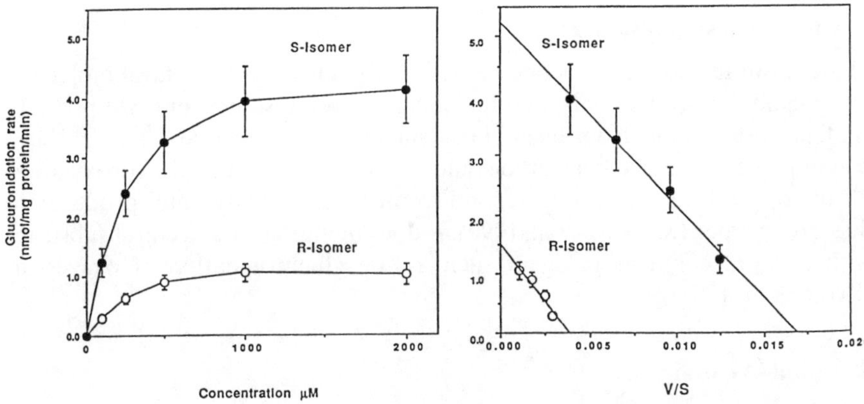

Fig. 1. Stereoselective glucuronidation of zileuton. Influence of isomer concentration on the rate of microsomal glucuronidation and Eadie-Hofstee plots of (*R*)-isomer and (*S*)-isomer glucuronidation. Results represent mean ± SE of hepatic microsomes from four human livers. (Sweeny and Nellans 1995)

cation of the amino acid at the active site of metabolism can influence the regioselectivity of metoprolol oxidation but does not influence the enantioselectivity of oxidation at sites distant from the active enzyme site (Ellis et al. 1996). However, this is not observed for all the metabolic pathways of chiral drugs, e.g., methadone *N*-demethylation (Foster et al. 1999).

Stereoselective metabolism must therefore consider the chirality of both the parent drug and metabolite and five scenarios can be considered.

2. Achiral–Chiral

A center of chirality is introduced into an achiral molecule during metabolism and a chiral metabolite is produced. For example, tacrine metabolism to the 1-hydroxy metabolite is highly stereoselective, preferentially producing the (+)-isomer in both rat and man (Hooper et al. 1994).

3. Chiral–Chiral

Metabolism of a chiral molecule occurs at a site separate from the center of asymmetry. Consequently, the chiral center is retained in the metabolite. For example, the antidepressant citalopram is metabolized by *N*-demethylation to demethyl- and didemethylcitalopram, retaining the center of chirality and the absolute configuration of the parent enantiomer in each metabolite (Sidhu et al. 1997). Similarly the chiral alkylating agent cyclophosphamide is metabolized to the noncytotoxic *N*-dechloroethyl metabolite with retention of the center of chirality and absolute configuration (Williams et al. 1999a). If the racemate is administered the rates of formation of the metabolite from each enantiomer can differ and therefore the ratio of the metabolites in vivo may not be unity, as observed for citalopram (Sidhu et al. 1997).

4. Chiral–Diastereoisomer

Metabolism introduces a second center of asymmetry into a chiral molecule. The metabolic product will therefore be a diastereoisomer and two possible configurations result from each chiral substrate. Both phase 1 and phase 2 reactions can produce diastereoisomers. For example, the α-hydroxylation of metoprolol introduces a second center of chirality and produces a diastereoisomer (Murthy et al. 1990) as does metabolism of a chiral substrate with a chiral conjugating agent such as the glucuronidation of oxazepam (Patel et al. 1995).

5. Chiral–Achiral

Metabolism occurs at the site of chirality and results in the loss of that chiral center. For example, aromatization of the dihydropyridine ring of chiral calcium channel blockers such as nitrendipine produces the achiral pyridine analogue (Mast et al. 1992).

II. Chiral Inversion

Metabolism of one enantiomer to its optical antipode has been termed chiral inversion (CALDWELL et al. 1988) and has been described for a number of antiinflammatory chiral 2-arylpropionic acid derivatives. The inversion of (*R*)- to (*S*)-enantiomers of compounds such as ibuprofen is considered to be attributable to the initial formation of an acyl coenzyme-A-thioester of the (*R*)-enantiomers, which is partially racemized by a CoA racemase to an intermediate, which is then hydrolyzed to free the (*S*)-enantiomer (HALL and XIAOTAO 1994). Chiral inversion is discussed in Chap. 14, this volume.

Nonenzymatic chiral inversion can occur for some enantiomer pairs in aqueous solution, or with albumin as a catalyst as reported for thalidomide (ERIKSSON et al. 1995, 1998), resulting in exposure to both enantiomers even when an enantiomerically pure formulation of a single enantiomer is administered. In this case there will be no advantage to the administration of a single enantiomer rather than the racemic mixture.

III. Species Differences

Animals are widely used in pharmacological research and play a pivotal role in toxicology testing during drug development. It is well established that there are species differences in the suite of drug-metabolizing enzymes and consequently the rates and pathways of metabolism observed in the rat, rabbit, mouse, monkey, or dog can differ from those observed in man. For example, the opioid antagonist naltrexone is metabolized in man to an active metabolite 6-β-naltrexol and this metabolite is not formed in the rat (MALPEIS et al. 1975). It is therefore not surprising that stereoselectivity of chiral drug metabolism can also demonstrate species differences. For example, in man in vivo (*R*)-warfarin is preferentially reduced to (*R,S*)-warfarin alcohol. In the rat, in vivo, this reaction is also stereoselective but favors (*S*)-warfarin, producing (*S,S*)-warfarin alcohol (LEWIS et al. 1974; POHL et al. 1976; HERMANS and THIJSSEN 1989). This can have important implications for the interpretation of the exposure of animal species to metabolites of pharmacological and toxicological significance during chronic administration of racemic drugs.

C. Stereoisomerism and Metabolic Drug Interactions

Pharmacokinetic interactions between drugs can have a number of mechanisms targeted at plasma binding sites, carrier-mediated transporters and metabolic enzymes, each of which have the potential to display stereoselectivity. Most clinically significant drug interactions are due to drug metabolism. When a racemic drug interacts with another, coadministered compound three different molecules are involved and often only measurement of the concentrations of the individual stereoisomers can be used to fully explain the clinical scenarios observed. This is especially important when the enantiomers of a racemic drug differ in potency for clinically important pharmacological effects and the interaction is stereoselective. Metabolic drug–drug interactions may involve inhibition or induction of the enzymes catalyzing biotransformation.

I. Inhibition

When two drugs compete for interaction with the active site of a drug-metabolizing enzyme which has a finite capacity, the drug with a higher affinity will displace that with a lower affinity. If the displaced compound is a substrate for the enzyme, its metabolism will be diminished, or inhibited. The inhibition of each enantiomer of a racemic drug by an interacting drug will depend on the affinities of the enzyme for the putative inhibitor relative to those of the enantiomers of the racemic drug. Hence, even if the enantiomers are metabolized by the same enzyme, each enantiomer may be differentially affected. In the extreme case the two enantiomers may be metabolized by different enzymes and hence the interaction may affect only one optical antipode. Stereoselective interactions of clinical significance for the racemic anticoagulant warfarin have been extensively studied (for review see Eichelbaum and Gross 1996) and can be illustrated using the well-characterized example of the interaction of warfarin with sulfinpyrazone (Toon et al. 1986).

(*S*)-Warfarin is a more potent anticoagulant than (*R*)-warfarin and the rates and routes of metabolism of the two enantiomers differ (Park 1988). (*S*)-Warfarin is metabolized to (*S*)-7- and some (*S*)-6-hydroxywarfarin while (*R*)-warfarin is metabolized by both oxidation to (*R*)-6- and (*R*)-7-hydroxywarfarin and reduction to (*R*,*S*)-warfarin alcohol (Fig. 2). Coadministration of warfarin with sulfinpyrazone results in prolongation of the prothrombin time, while total plasma warfarin concentrations (R+S) fall or remain unchanged. Stereospecific analysis reveals that sulfinpyrazone inhibits the clearance by metabolism of (*S*)-warfarin, with the resultant higher (*S*)-warfarin plasma concentrations responsible for the increased pharmacological response (Toon et al. 1986). Sulfinpyrazone increases the clearance of (*R*)-warfarin via displacement from plasma protein binding sites, and total (*R*)-warfarin plasma concentrations fall. The sum of (*R*)- and (*S*)-warfarin plasma concentrations therefore remains almost unchanged. By contrast, the inhibitor of leukotriene biosynthesis, zileuton, decreases the mean clearance of (*R*)-warfarin, but does not affect (*S*)-warfarin plasma clearance and no major changes in prothrombin time occur (Awni et al. 1995). These examples highlight the need to measure the unbound concentrations of individual enantiomers when the interactions of highly plasma bound racemic drugs are being investigated.

An interesting recent example of the in vivo and in vitro investigation of stereoselective drug interactions has been reported by Hemeryck et al. (2000; 2001). In vivo, paroxetine inhibits the clearance of both (*R*)- and (*S*)-metoprolol and abolishes the stereoselectivity of metoprolol pharmacokinet-

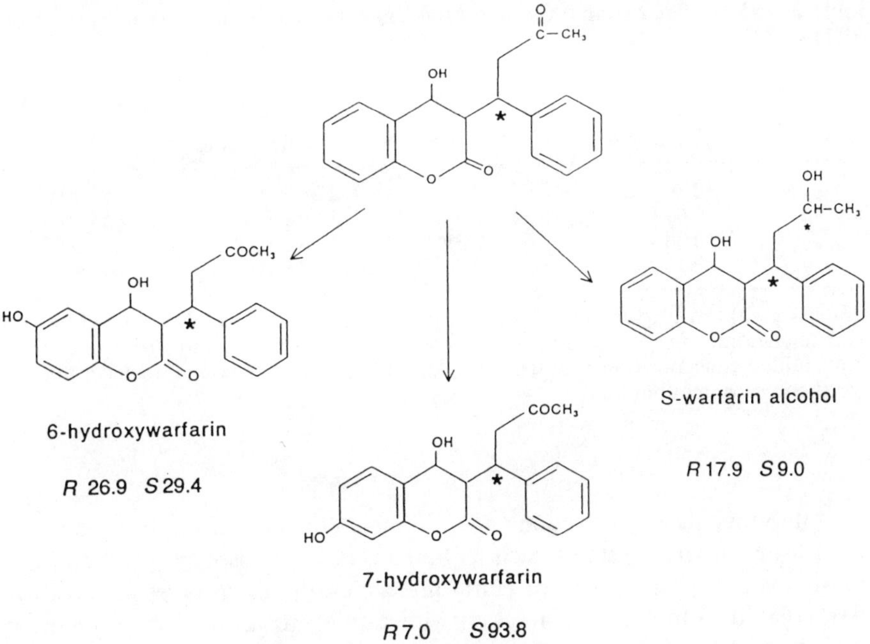

Fig. 2. Metabolic pathways of warfarin. The formation clearances (ml/min) from (*R*)- and (*S*)-warfarin for each metabolite in healthy volunteers are given (NIOPAS et al. 1991). The formation of 7-hydroxywarfarin is the major pathway of metabolism of (*S*)-warfarin, whereas reduction to the diastereoisomeric alcohol is the major pathway of metabolism of (*R*)-warfarin. Clearance to 6-hydroxywarfarin is the same for both enantiomers. (EICHELBAUM and GROSS 1996)

ics. In vitro studies confirmed that this was a consequence of preferential inhibition of metoprolol *O*-demethylation, which favors the (*R*)-enantiomer, by paroxetine. This route of metabolism is catalyzed by polymorphic CYP2D6. At high concentrations of paroxetine, the stereoselectivity of metoprolol metabolism in vitro is lost, indicating that metabolism via other pathways catalyzed by other enzymes does not display net stereoselectivity (see Table 3). Further examples of stereoselective inhibitory drug interactions include the inhibition of propranolol by quinidine (ZHOU et al. 1990), verapamil by cimetidine (MIKUS et al. 1990), nitrendipine by cimetidine but not grapefruit juice (SOONS et al. 1991), acenocoumarol by piroxicam (BONNABRY et al. 1996), flecainide by quinidine (BIRGERSDOTTER et al. 1992), oral nicardipine by grapefruit juice (UNO et al. 2000), and metoprolol by cimetidine (TOON et al. 1988).

Inhibitory drug interactions can also result when two drugs interact with different binding sites on the same enzyme, with changes in the conformation of the enzyme leading to noncompetitive inhibition of metabolism. Stereoselectivity in noncompetitive inhibition must also be considered if racemic drugs are involved.

Table 3. Effect of paroxetine on metoprolol metabolism in vitro[a] (HEMERYCK et al. 2001)

Paroxetine μM		Cl_{int} μl/min/mg		(*R*)/(*S*)-ratio
Total	Unbound	(*R*)-metoprolol	(*S*)-metoprolol	
0	0	3.94 ± 0.56	2.54 ± 0.23	1.55 ± 0.16
0.5	0.04	2.24 ± 0.32	1.89 ± 0.52	1.25 ± 0.43
1	0.08	1.57 ± 0.31	1.48 ± 0.12	1.07 ± 0.27
2	0.16	1.15 ± 0.20	1.22 ± 0.20	0.98 ± 0.31

[a] Mean (±SD) intrinsic clearance (Cl_{int}) of (*R*)- and (*S*)-metoprolol in pooled human liver microsomes in the absence and in the presence of 0.5, 1, and 2 μM paroxetine (total added concentrations; substrate depletion experiments taking into account non-specific microsomal binding).

Inhibitory drug interactions are routinely studied during drug development using in vitro systems such as human liver microsomes and heterologously expressed enzymes. For chiral molecules, the in vitro results must be extrapolated to the in vivo situation with caution if stereoselective pharmacokinetics results in a different ratio of concentrations of stereoisomers in vivo than that evaluated in vitro.

II. Induction

Drug-metabolizing enzymes are differentially affected by inducing agents. If the enantiomers of a racemic drug are metabolized by different enzymes, exposure to inducing agents can preferentially influence the activity of the enzymes metabolizing only one optical antipode and a change in the net stereoselectivity of metabolism can result. For example, there is stereoselective induction of the metabolism of the prodrug cyclophosphamide, used in cancer chemotherapy, by the anticonvulsant phenytoin (WILLIAMS et al. 1999b). Phenytoin significantly increases the formation of the *N*-dechloroethylated metabolite of (*S*)- and not (*R*)-cyclophosphamide, while also inducing the formation of the active 4-hydroxymetabolite of both (*R*)- and (*S*)-cyclophosphamide. Rifampin induces the metabolism of (*R*)-verapamil to a greater degree than (*S*)-verapamil after both oral and intravenous administration (FROMM et al. 1996). A far greater induction of verapamil metabolism is observed after oral rather than intravenous administration, consistent with a greater effect of rifampin on the drug-metabolizing enzymes of the intestine than of the liver, and intestinal enzymes make a major contribution to verapamil's first-pass metabolism after oral administration (VON RICHTER et al. 2001). A further interesting example of stereoselective induction is that of ibuprofen by clofibrate (SCHEUERER et al. 1998). Not only is the clearance of both *R*- and *S*-ibuprofen increased by clofibrate, there is also an increase in the fractional inversion of the *R*- to the *S*-enantiomer mediated via increased formation of *R*-ibuprofenoyl-coenzyme A rather than the oxidative metabolism of ibuprofen.

III. Enantiomer–Enantiomer Interaction

As two compounds are coadministered when a racemic mixture is given, it is not surprising that interactions between the enantiomers of chiral drugs have been observed. If the enantiomers of a racemic drug interact with the same active site of a drug-metabolizing enzyme, three scenarios for enantiomer–enantiomer interactions are possible.

1. Mutually competitive inhibition of the metabolism of both enantiomers is observed as they are metabolized by the same enzyme and compete for catalysis.
2. Although both enantiomers are metabolized by the same enzyme, one enantiomer has a higher affinity for the enzyme and acts as a competitive inhibitor of its optical antipode. In this case the metabolism of one enantiomer remains unchanged and that of the optical antipode is inhibited.
3. Only one enantiomer is a substrate for the enzyme; however, the optical antipode can inhibit the enzyme and hence the metabolism of the other enantiomer.

Enantiomer–enantiomer interactions have been reported for an increasing number of chiral drugs used as racemates including flurbiprofen, nitrendipine, disopyramide, methadone, and propafenone. The interaction between the enantiomers of the antiarrhythmic propafenone, used clinically as the racemate, has been investigated in detail. In vitro (*R*)-propafenone is a potent competitive inhibitor of the 5-hydroxylation of (*S*)-propafenone, catalyzed by CYP2D6 (Kroemer et al. 1991), with (*S*)-propafenone being a weaker inhibitor of (*R*)-propafenone metabolism. In vivo the clearance of (*R*)-propafenone is similar when administered as the single enantiomer or the racemate, whereas the clearance of (*S*)-propafenone is significantly lower after administration of the racemate (920 ± 300 ml/min) compared with the (*S*)-enantiomer alone (2521 ± 1450 ml/min). As (*S*)- but not (*R*)-propafenone has β-blocking properties, the higher plasma concentrations of the (*S*)-enantiomer following administration of the racemate can lead to β-blockade in patients (Kroemer et al. 1989).

A recent example of a competitive enantiomer–enantiomer interaction characterized in vitro is shown in Fig. 3. The apparent K_m for (*R*)-zileuton glucuronidation is increased 1.8- to 3.4-fold by the (*S*)-isomer and the apparent K_m for (*S*)-isomer glucuronidation is increased 1.3- to 1.8-fold by the (*R*)-enantiomer. For both enantiomers there is minimal effect on the maximal

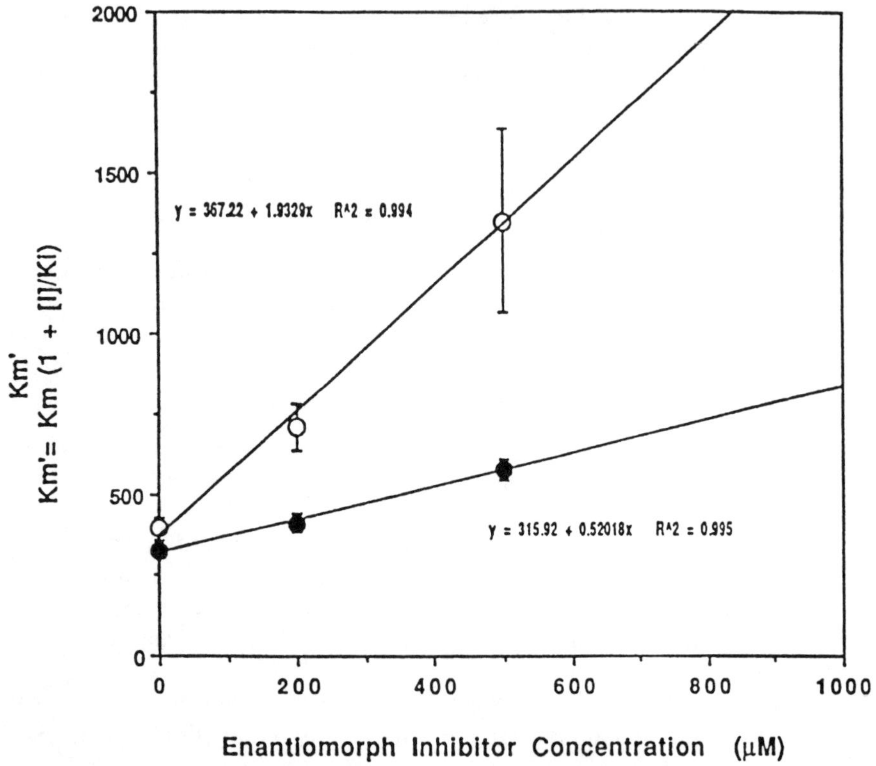

Fig. 3. Enantiomer–enantiomer interaction for zileuton. Relationship between enantiomer inhibitor concentration and K_m of *R*-isomer (○) and (*S*)-isomer (●) glucuronidation. The apparent K_m for *R*-zileuton glucuronidation is increased 1.8- to 3.4 fold by the *S*-isomer and the apparent K_m for *S*-isomer glucuronidation is increased 1.3- to 1.8-fold by the *R*-enantiomer. For both enantiomers there is minimal effect on the maximal velocity of glucuronidation. (SWEENY and NELLANS 1995)

velocity of glucuronidation (SWEENY and NELLANS 1995). Interestingly the ratio of the apparent K_m values ((*R*)/(*S*) = 1.2) is smaller than the ratio of the K_i values (2.4). These results may be consistent with the involvement of multiple UDP-glucuronosyltransferase isoforms in the enantioselective glucuronidation of zileuton.

If enantiomer–enantiomer interactions occur, a racemic mixture does not behave as the simple sum of the individual enantiomers. The disposition of each enantiomer observed when administered individually may not reflect the pharmacokinetics of the enantiomers when given as the racemic mixture. If chiral drugs are used as racemic mixtures, appropriate investigations using stereospecific analytical techniques will be necessary to establish whether there are clinical implications of enantiomer–enantiomer interactions observed in vitro.

D. Additional Consequences of Stereoselective Metabolism

I. In Vivo and In Vitro Pharmacological Potency

The relative potency of the stereoisomers of a drug for pharmacological effects related to efficacy or safety is usually assessed in vitro at defined concentrations. If the pharmacological effects display stereoselectivity, the more potent enantiomer is termed the eutomer, and the less potent enantiomer the distomer (ARIENS 1986). If stereoselective metabolism resulting in stereoselective pharmacokinetics occurs in vivo, the plasma concentrations of the two enantiomers will differ when equal doses are administered. If the pharmacological effects are stereoselective, the relative potency of the enantiomers in vivo may differ from that observed in vitro as the relative concentrations of the eutomer and distomer are not the same in vitro and in vivo. Depending on whether the eutomer or distomer is preferentially metabolized, the in vitro/in vivo potency may be attenuated or amplified. For example, in vitro (*S*)-acenocoumarol is three to five times more potent an oral anticoagulant than (*R*)-acenocoumarol (MEINERTZ et al. 1978). At the same unbound plasma concentration in vivo the (*S*)-enantiomer will be three to five times more potent than the (*R*)-enantiomer. However, in vivo the metabolic clearance of (*R*)-acenocoumarol is 14 times lower than that of the (*S*)-enantiomer. Therefore, when given at equal molar doses the plasma concentrations of (*R*)-acenocoumarol are far higher than those of the (*S*)-enantiomer. The higher plasma concentrations of (*R*)-acenocoumarol produce a greater pharmacological response than the lower concentrations of (*S*)-acenocoumarol. If only the same doses and not the different plasma concentrations are considered, the (*R*)-enantiomer could appear to be more potent than the (*S*)-enantiomer in vivo (THIJSSEN et al. 1986; HERMANS and THIJSSEN 1993). This potential discrepancy highlights the importance of relating pharmacological effects in vivo to plasma concentrations of the pharmacologically active agent rather than the dose administered.

II. Active Chiral Metabolites

Metabolism does not always modify a drug molecule in a way that leads to abolition of pharmacological activity (bioinactivation). There are many examples where drug metabolites produce pharmacological effects – either those desired effects associated with efficacy and/or the undesired effects associated with adverse reactions and toxicity. Substrate stereoselectivity therefore may result in one enantiomer of a racemic drug preferentially forming an active metabolite associated with efficacy or toxicity. For example, the alkylating agent ifosfamide is used clinically as a racemic mixture for the treatment of cancer (WILLIAMS and WAINER 1999). Both enantiomers are metabolized via a number of different pathways with 4-hydroxylation favoring the (*R*)-enantiomer. In studies using the single enantiomers in animals, (*R*)-ifosfamide produces greater cytotoxicity, the desired pharmacological effect, which has been attributed to higher concentrations of the 4-hydroxy metabolite. Patients with ifosfamide-related neurotoxicity have been observed to have high concentrations of the metabolite (*R*)-3-*N*-dechloroethylifosfamide, which is formed from (*S*)-rather than (*R*)-ifosfamide. Administration of the single enantiomer, (*R*)-ifosfamide, has therefore been advocated to minimize the dose-limiting neurotoxicity that can compromise the cytotoxic efficacy of racemic ifosfamide (WILLIAMS and WAINER 1999).

Some drugs have been developed as prodrugs, where metabolism in vivo transforms the inactive molecule administered (the prodrug) to a metabolite responsible for the pharmacological effects desired, a process called bioactivation. For example, enalapril, given orally as an antihypertensive, requires deesterification to enalaprilat before exhibiting angiotensin-converting enzyme inhibition. Stereoselective metabolism of a prodrug in vivo can theoretically have important consequences if the active metabolite is chiral and displays enantioselectivity for the pharmacological effects produced.

III. First-Pass Metabolism

1. Intestinal and Hepatic Metabolism

Following oral administration, a drug traverses the intestinal mucosa and is transported to the liver via the portal circulation before it enters the systemic circulation. High concentrations of drug-metabolizing enzymes in the liver and the intestinal enterocytes are consequently able to metabolize the drug prior to systemic distribution throughout the body. The sum of the presystemic intestinal and hepatic metabolism is termed first pass metabolism. If the intestinal and hepatic enzymes display substrate stereoselectivity for a racemic drug, the first pass metabolism and hence bioavailability of the chiral drug will be stereoselective (SPAHN-LANGGUTH et al. 1997). The ratio of the enantiomers in plasma in vivo following oral administration will also differ from that observed following intravenous administration and consequently the total concentration of the two enantiomers required to elicit a pharmacological effect may be route dependent. The stereoselective first-pass metabolism of verapamil is a clear example of this scenario (EICHELBAUM and GROSS 1996). The systemic metabolism of verapamil is stereoselective, resulting in a higher clearance rate of the (*S*)- than the (*R*)-enantiomer and an (*R*)/(*S*) plasma concentration ratio following intravenous administration of 2. The first-pass metabolism of verapamil is also stereoselective, favoring the *S*-enantiomer, with the bioavailability of (*S*)- and (*R*)-verapamil being 20% and 50%, respectively (VOGELGESANG et al. 1984). Consequently, after oral administration plasma concentrations of the (*R*)-enantiomer are far greater than those of the (*S*)-enantiomer (ratio (*R*)/(*S*) = 5); that is, total concentrations after intravenous therapy are enriched with the *S*-enantiomer relative to oral administration. As the effect of verapamil on AV-node conduction is stereoselective, favoring the (*S*)-enantiomer (ECHIZEN et al. 1985, 1988), equieffective (*S*)-verapamil concentrations are attained at higher total concentrations after oral rather than intravenous therapy. Therapeutic efficacy is attained at oral doses of 80–160 mg and intravenous doses of 5–10 mg racemic verapamil. Further examples of stereoselective first pass metabolism include propranolol (LINDNER et al. 1989), nitrendipine (MAST et al. 1992), and salbutamol (WARD et al. 2000). Figure 4 shows the plasma concentration-time profiles of the enantiomers of salbutamol following oral and intravenous administration, clearly demonstrating net stereoselective first-pass metabolism.

2. Bioequivalence

The implications of stereoselectivity for the assessment of the bioequivalence of chiral drugs deserves mention (MEHVAR and JAMALI 1997; NERURKAR et al. 1992). If stereospecific analytical techniques are not used, plasma concentrations of the sum of the individual enantiomers will be measured and used as a basis for assessment. If the racemic drug is subject to significant enan-

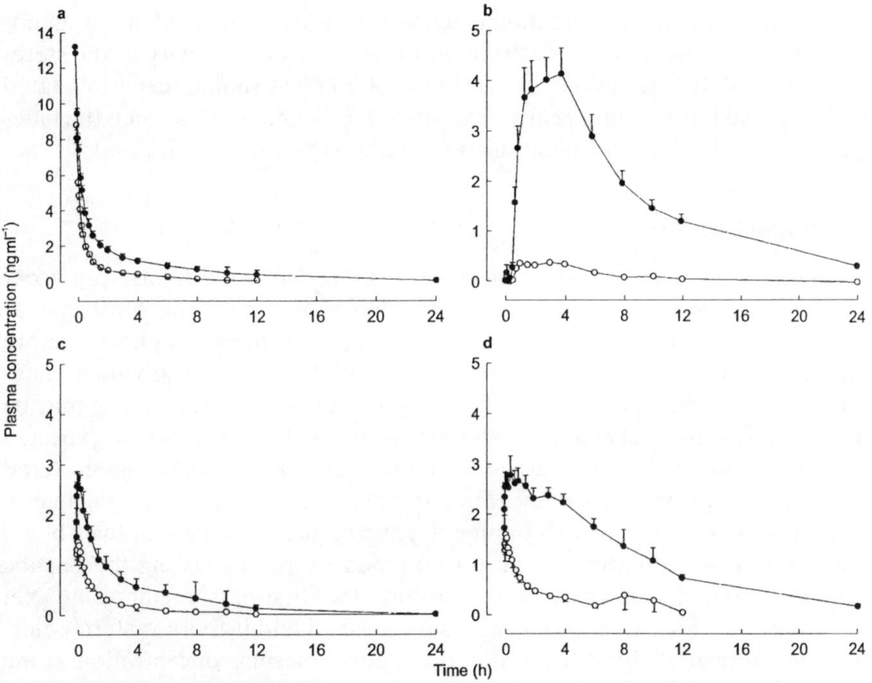

Fig. 4. Plasma concentration-time profiles of (*R*)-(○) and (*S*)-(●) salbutamol after administration of (**a**) 500 μg intravenously (**b**) 2 mg orally (**c**) 1200 μg by inhalation with oral charcoal and (**d**) 1200 μg by inhalation. Mean ± SE in 15 healthy subjects. Net stereoselective first-pass metabolism after oral administration clearly favors the *R*-enantiomer. (WARD et al. 2000)

tioselective first-pass metabolism and clearance, the pharmacokinetics of the sum of the enantiomers will not represent the profile of the individual enantiomers. This may be especially significant if one enantiomer makes a greater contribution to drug pharmacological effects.

For two formulations of a racemic drug, the first pass metabolism of both enantiomers should be identical if the rates of dissolution and rates of absorption are similar. However, if the formulation has a chiral matrix there is the potential for differential release of the enantiomers from two different formulations. The enantiomer which is more slowly released can be more extensively metabolized during absorption and the bioavailabilities of the enantiomers in the two formulations can differ. It is possible, therefore, for the results of the assessment of bioequivalence to differ for the sum of the enantiomers and each of the individual enantiomers, especially when immediate and sustained release formulations are compared. Bioequivalence of the total drug may therefore not reflect bioequivalence of the individual enantiomers. For example, the bioequivalence of two formulations of the racemic antihistamine chlorpheniramine has recently been studied. Bioequivalence was demonstrated for total ((*R*)- plus (*S*)-)chlorpheniramine as well as for the (*S*)-enantiomer. However, although the difference between the formulations was not large, based on (*R*)-enantiomer concentrations, bioequivalence was not observed, and this result was attributed to the limit of sensitivity of the stereospecific analytical technique used (HIEP et al. 2000). A similar scenario related to the optical purity of the analyte measured has been reported when the bioequivalence of nadolol formulations was studied (SRINIVAS et al. 1996).

IV. Impact of Disease

Changes in the activity of drug-metabolizing enzymes with disease can occur which can alter the stereoselective metabolism of chiral drug substrates. A greater response than expected from total drug concentrations may therefore occur. If only the total drug concentration rather than the individual enantiomers are considered, a shift in the concentration-effect relationship may be observed. The greatest change in stereoselective metabolism may be expected to occur for high clearance drugs with low oral bioavailability administered to patients with hepatic disease. For example, cirrhosis leads to a decline in hepatic mass with loss of functioning hepatic tissue, a decrease in liver blood flow as well as a shunting of the portal blood supply, bypassing the hepatic drug-metabolizing enzymes and therefore the first-pass metabolism. Net metabolism of high clearance drugs is diminished and if there is stereoselective metabolism of chiral substrates, the relative plasma concentration of the enantiomers may be altered relative to healthy subjects. Systemic concentrations will not only be higher in patients with cirrhosis, they will be enriched in the enantiomer which is usually preferentially metabolized during first pass. For example, the bioavailability of (*R*)-nimodipine is increased 3- to 4-fold to 60%–65% in patients with cirrhosis, while that of (*S*)-nimodipine is increased

17-fold to 40%–45% (SPORKMANN 1992). As the cardiovascular effects of the *S*-enantiomer are at least 10-fold greater than those of the *R*-enantiomer, enrichment of the eutomer as well as the higher plasma concentrations achieved, which are substantially greater than the increase in total (*R*)- and (*S*)- concentrations, must both be considered during dosage adjustment in cirrhotic patients.

V. Genetics

1. Genetic Polymorphisms in Drug Metabolism

Genetic polymorphisms in drug-metabolizing enzymes have been widely studied in recent years. Within populations there are subgroups of individuals carrying mutations in the genes coding for the synthesis of drug-metabolizing enzymes. These different alleles can result in loss of enzymatic activity, reduced or increased catalytic activity or increased activity, the latter due to much higher enzyme expression. At the level of phenotype these mutations give rise to distinct subgroups in the population such as poor metabolizers (PM), intermediate metabolizers (IM), extensive metabolizers (EM), and ultrarapid metabolizers (UM), as reported for CYP2D6.

The major allelic variants of many clinically significant drug-metabolizing enzymes have been identified. The frequencies of these alleles have been determined in different ethnic groups and the consequences of the changes in the genetic code for substrate specificity as well as enzyme activity in vitro and in vivo have been investigated. Genotype has been shown to contribute to variability in the activity of many enzymes catalyzing drug biotransformation, including the well-characterized examples of genetic polymorphisms in CYP2D6, CYP2C19, and *N*-acetyltransferase 1 and 2, but also CYP1A1, CYP2E1, CYP2A6, CYP2C9, glutathione-*S*-transferases (WORMHOUDT et al. 1999), flavin containing monooxygenase 3 (FMO3) (SACHSE et al. 1999), and UDP-glucuronosyltransferase (BHASKER et al. 2000) among others. If a polymorphic enzyme predominantly metabolizes chiral substrates stereoselectively, the stereoselective pathways will be absent in individuals who lack the active enzyme. For example, CYP2D6 contributes to the oxidative metabolism of many chiral basic drugs including fluoxetine (MARGOLIS et al. 2000), dexfenfluramine (HARITOS et al. 1998), metoprolol (LENNARD et al. 1983), propafenone (KROEMER et al. 1989), flecainide (GROSS et al. 1989), and mexiletine (ABOLFATHI et al. 1993; VANDAMME et al. 1993), and CYP2C19 contributes to the stereoselective metabolism of omeprazole (ÄBELÖ et al. 2000), mephobarbital (KOBAYASHI et al. 2001), pantoprazole (TANAKA et al. 1997), and mephenytoin (KÜPFER et al. 1981). Stereoselective metabolism of these drugs is observed in extensive metabolizers (EM) who express the active form of CYP2D6 or CYP2C19. For these two enzymes, 7% and 3% of the Caucasian population respectively inherit two nonfunctional alleles, do not form the active enzyme, and these individuals are termed poor metabolizers (PM). Not only is the net clearance via metabolism decreased in PMs relative to EMs, but the stereoselectivity of metabolism is altered. For example, omeprazole, used for the treatment of acid-related gastrointestinal disorders, is stereoselectively metabolized by CYP2C19. In PMs the plasma concentrations of (*R*)-omeprazole are 7.5-fold higher than in EMs (Fig. 5), and plasma concentrations of (*R*)-hydroxyomeprazole formed from (*R*)-omeprazole are 3.8-fold lower. (*S*)-omeprazole concentrations are raised in PMs relative to EMs, but only by 3.1-fold, and no difference in (*S*)-hydroxyomeprazole plasma concentrations are observed. CYP2C19-catalyzed metabolism of omeprazole therefore favors the (*R*)-enantiomer, and when absent in PMs a greater change in the metabolism and hence pharmacokinetics of (*R*)-omeprazole than of (*S*)-omeprazole is observed (Tybring et al. 1997). An additional interesting example of stereoselective metabolism in relation to genotype is with the chiral antidepressant fluoxetine. In CYP2D6 EMs, the apparent oral clearances of (*R*)- and (*S*)-fluoxetine are similar (40 and 36 l/h, respectively). In CYP2D6 PMs the apparent oral clearance of (*R*)-fluoxetine (17 l/h) is sixfold higher than that of *S*-fluoxetine (3 l/h). The plasma concentrations of both enantiomers are greater in CYP2D6 PMs than in EMs; however, (*S*)-

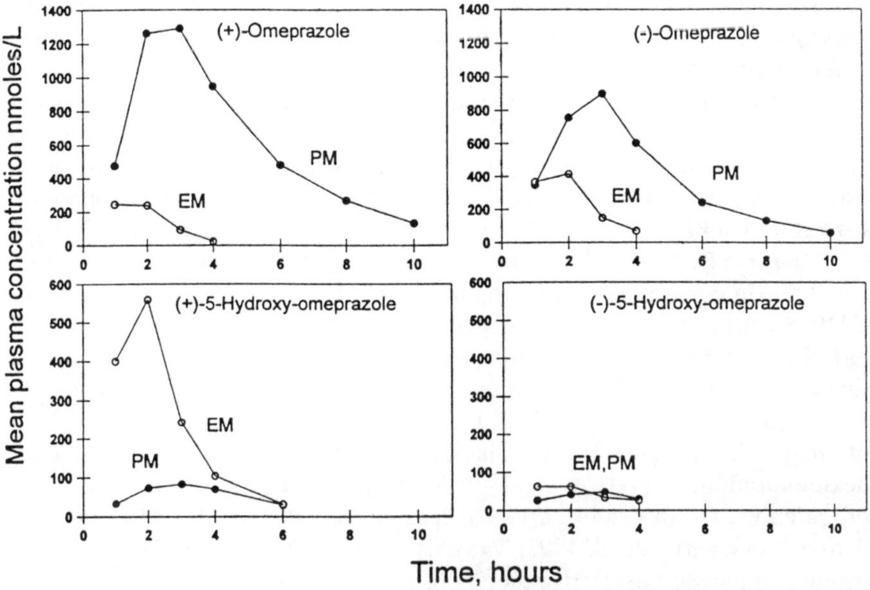

Fig. 5. Mean plasma concentration versus time curve for the enantiomers of omeprazole and 5-hydroxyomeprazole in five EM and five PM of CYP2C19 after a single oral dose of 20 mg racemic omeprazole. Plasma concentrations of (*R*)- omeprazole are 7.5-fold higher in CYP2C19 EMs than PMs and plasma concentrations of (*R*)-hydroxyomeprazole formed from (*R*)-omeprazole are 3.8-fold lower. (*S*)-omeprazole concentrations are raised in PMs compared to EMs, but only by 3.1-fold, and no difference in (*S*)-hydroxyomeprazole levels is observed. (Tybring et al. 1997)

fluoxetine plasma concentrations are increased 11.5-fold whereas (*R*)-fluoxetine concentrations are only increased 2.5-fold. When CYP2D6 is absent in PMs, the stereoselectivity of non-CYP2D6-catalyzed biotransformation is clearly revealed. Stereoselectivity is not observed in EMs because metabolic pathways of opposite stereoselectivity balance the stereoselective CYP2D6-catalyzed metabolism (Fjordside et al. 1999).

Investigations of polymorphic enzyme substrate specificity have used the stereoselectivity of chiral substrate metabolism as a tool to provide mechanistic insights into the substrate-enzyme interaction. For example, changing the amino acid residue at position 374 from the N-terminus of CYP2D6 from valine to methionine in in vitro expression systems alters the regioselectivity of metoprolol metabolism, placing this residue at the active site of the enzyme. This residue, however, is sufficiently distant from the chiral center of metoprolol for amino acid changes at position 374 not to influence the enantioselectivity of metoprolol oxidation (Ellis et al. 1996). In yeast cells, this same amino acid change from valine to methionine at position 374 alters the enantioselectivity of the 4-hydroxylation of bunitrolol from (+) > (−), which is consistent with human liver microsomes, to (−) > (+) (Narimatsu et al. 1999). Modeling based on similar experiments supports the observation that the valine at 374 in CYP2D6 is oriented towards the active site for substrate oxidation. Interestingly, additional experiments suggest that the amino acid at position 304, serine, is a determinant of CYP2D6 catalyzed metoprolol enantioselectivity.

2. Inter-Ethnic Differences

The frequencies of alleles encoding for polymorphic drug-metabolizing enzymes can vary between ethnic groups (Xie et al. 2001). Consequently, ethnic differences in the proportion of individuals in a population with altered enzyme activities occur. For example, the inactive CYP2C19 alleles *CYP2C19*2* and *CYP2C19*3* are more frequent in Asian than Caucasian populations and therefore CYP2C19 PMs comprise 15%–20% of Asian populations but only 3% of Caucasian populations (Bertilsson 1995). Stereoselective metabolism per se will be similar in individuals from different ethnic groups of the same CYP2C19 genotype. However, different average metabolic capacities – and average stereoselectivities of the metabolism of chiral drug substrates, can result. For example, the stereoselective pharmacokinetics of hexobarbital has been studied in Chinese and Caucasian EMs and PMs of CYP2C19 (Adedoyin et al. 1994). The stereoselectivity of hexobarbital apparent oral clearance, principally attributable to metabolism, in EMs ((*R*)/(*S*) = 6) differs from that in PMs ((*R*)/(*S*) = 0.5), irrespective of ethnic group. Because a greater proportion of the Chinese than the Caucasian population are CYP2C19 PMs, a greater proportion of the Chinese than the Caucasian population will have altered stereoselectivity of hexobarbital metabolism. The population *average* stereoselective hexobarbital apparent oral clearance will therefore differ between Chinese and Caucasians populations.

VI. Influence of Age

In utero and during growth to adulthood, individual drug-metabolizing enzymes develop at different rates. If the enantiomers of a racemic drug are metabolized by different enzymes, it is possible that the stereoselectivity of metabolism will change as the activities of the individual enzymes develops. For example, developmental changes in the pharmacokinetics of the enantiomers of warfarin attributable to metabolism have been studied in prepubertal, pubertal, and adult patients on long-term warfarin therapy (TAKAHASHI et al. 2000). The unbound oral clearances of (*S*)- and (*R*)-warfarin with age are shown in Fig. 6, and are mediated principally via CYP2C9 for (*S*)-warfarin and via a number of enzymes for (*R*)-warfarin. The differential effect of age on the oral clearance of the two enantiomers is related to estimated liver weight changes for (*S*)-warfarin, but not for (*R*)-warfarin. This enantiomeric difference is postulated to be attributable to the multiple enzymes catalyzing (*R*)-warfarin biotransformation maturing and developing at different rates during childhood.

Although the activity of some drug-metabolizing enzymes declines with old age, it is unlikely that stereoselectivity of metabolism would change. For example, a decrease in the metabolism of both (*R*)- and (*S*)-verapamil is observed in elderly subjects (ABERNETHY et al. 1993; SASAKI et al. 1993; SCHWARTZ et al. 1994). However, if the enantiomers of a racemic drug are metabolized by various enzymes which are differentially affected by aging, dif-

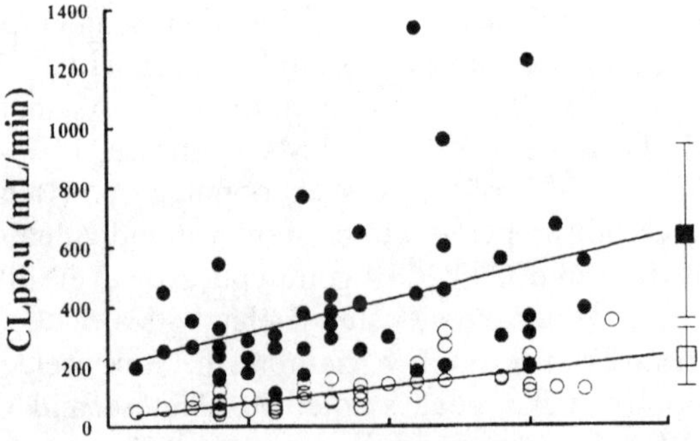

Fig. 6. Relationship between age and unbound oral clearance values (*Clpo,u*) for (*S*)-(●) and (*R*)-(○) warfarin from 52 pediatric patients. The *squares* represent average adult data. Developmental changes in warfarin metabolism differ for the two enantiomers and hence the stereoselectivity of metabolism and unbound oral clearance changes with age. (TAKAHASHI et al. 2000)

ferences in the net stereoselectivity of metabolism between young adults and the elderly may occur. It has been reported that the intrinsic clearance of oral (*S*)-propranolol, an index of hepatic propranolol metabolic activity, is 30% lower in elderly versus young adult volunteers and that there is no significant age-related difference in the intrinsic clearance of (*R*)-propranolol (Gilmore et al. 1992). However, other authors have not observed a change in (*R*)- and (*S*)-propranolol metabolism with age (Lalonde et al. 1990; Colangelo et al. 1992; Zhou et al. 1992). Changes in stereoselective metabolism with age have also been reported for hexobarbital (Chandler et al. 1988) and mephobarbital (Hooper and Qing 1990).

VII. Influence of Sex

Differences in the activities of some drug-metabolizing enzymes have been observed in males and females (Harris et al. 1995). It is therefore not surprising that sex differences in the stereoselective metabolism of some chiral drugs such as mephobarbital (Hooper and Qing 1990), metoprolol (Luzier et al. 1999), and verapamil (Gupta et al. 1995) have been observed.

E. Conclusion

Stereoselective pharmacokinetics have been observed for many chiral drugs, and stereoselective metabolism is frequently responsible. However, the underlying metabolic mechanism is often not fully understood. When a racemic drug is used clinically, the two different enantiomers administered can be eliminated from the body by metabolism via different pathways and at different rates. As enantiomer–enantiomer metabolic interactions can also occur, the fate of the racemate may not be a simple mean of the fate of the single enantiomers. Stereoselective aspects of chiral drug metabolism therefore must be considered in order to fully characterize and understand the mechanisms responsible for, and the clinical consequences of, drug biotransformation and drug–drug interactions. Further research is required to increase knowledge on the structural requirements for substrate interaction and catalysis for individual drug-metabolizing enzymes. Such data will improve the accuracy of predictions of chiral drug metabolism and metabolic interactions with coadministered drugs in vivo.

References

Äbelö A, Andersson TB, Antonsson M, Naudot AK, Skånberg I, Weidolf L (2000) Stereoselective metabolism of omeprazole by human cytochrome P450 enzymes. Drug Metabolism and Disposition 28:966–972

Abernethy DR, Wainer IW, Longstreth JA, Andrawis NS (1993) Stereoselective verapamil disposition and dynamics in aging during racemic verapamil administration. Journal of Pharmacology and Experimental Therapeutics 266:904–911

Abolfathi Z, Fiset C, Gilbert M, Moerike K, Belanger PM, Turgeon J (1993) Role of polymorphic debrisoquin 4-hydroxylase activity in the stereoselective disposition of mexiletine in humans. Journal of Pharmacology and Experimental Therapeutics 266:1196–1201

Adedoyin A, Prakash C, O'Shea D, Blair IA, Wilkinson GR (1994) Stereoselective disposition of hexobarbital and its metabolites: relationship to the *S*-mephenytoin polymorphism in Caucasian and Chinese subjects. Pharmacogenetics 4:27–38

Ariëns EJ (1986) Chirality in bioactive agents and its pitfalls. Trends in Pharmacological Sciences 7:200–205

Awni WM, Hussein Z, Granneman GR, Patterson KJ, Dube LM, Cavanaugh JH (1995) Pharmacodynamic and stereoselective pharmacokinetic interactions between zileuton and warfarin in humans. Clinical Pharmacokinetics 29 [Suppl 2]:67–76

Bertilsson L (1995) Geographic/interracial differences in polymorphic drug oxidation: current state of knowledge of cytochromes P450 (CYP) 2D6 and 2C19. Clinical Pharmacokinetics 29:192–209

Bhasker CR, McKinnon W, Stome A, Lo ACT, Kubota T, Ishizaki T, Miners JO (2000) Genetic polymorphism of UDP-glucuronosyltransferase 2B7 (UGT2B7) at amino acid 268: ethnic diversity of alleles and potential clinical significance. Pharmacogenetics 10:679–685

Birgersdotter UM, Wong W, Turgeon J, Roden DM (1992) Stereoselective genetically-determined interaction between chronic flecainide and quinidine in patients with arrhythmias. British Journal of Clinical Pharmacology 33:275–280

Bonnabry P, Desmeules J, Rudaz S, Leeman T, Veuthey JL, Dayer P (1996) Stereoselective interaction between piroxicam and acenocoumarol. British Journal of Clinical Pharmacology 41:525–530

Caldwell J, Hutt AJ, Fournel-Gigleux S (1988) The metabolic chiral inversion and dispositional enantioselectivity of the 2-arylpropionic acids and their biological consequences. Biochemical Pharmacology 37:105–114

Cashman JR, Celestial JR, Leach AR (1992) Enantioselective *N*-oxygenation of chlorpheniramine by the flavin-containing monoxygenase from hog liver. Xenobiotica 22:459–469

Chandler MH, Scott SR, Blouin RA (1988) Age-associated stereoselective alterations in hexobarbital metabolism. Clinical Pharmacology and Therapeutics 43:436–441

Colangelo PM, Blouin RA, Steinmetz JE, McNamara PJ, DeMaria AN, Wedlund PJ (1992) Age and propranolol stereoselective disposition in humans. Clinical Pharmacology and Therapeutics 51:489–494

de Morais SM, Wilkinson GR, Blaisdell J, Nakamura K, Meyer UA, Goldstein JA (1994) The major genetic defect responsible for the polymorphism of *S*-mephenytoin metabolism in humans. Journal of Biological Chemistry 269: 15419–15422

Eaton EA, Walle UK, Wilson HM, Aberg G, Walle T (1996) Stereoselective sulphate conjugation of salbutamol by human lung and bronchial epithelial cells. British Journal of Clinical Pharmacology 41:201–206

Echizen H, Brecht T, Niedergesäss S, Vogelgesang B, Eichelbaum M (1985) The effect of dextro-, levo-, and racemic verapamil on atrioventricular conduction in humans. American Heart Journal 109:210–217

Echizen H, Manz M, Eichelbaum M (1988) Electrophysiologic effects of dextro- and levo-verapamil on sinus node and AV node function in humans. Journal of Cardiovascular Pharmacology 12:543–546

Eichelbaum M, Bertilsson L, Kupfer A, Steiner E, Meese CO (1988) Enantioselectivity of 4-hydroxylation in extensive and poor metabolizers of debrisoquine. British Journal of Clinical Pharmacology 25:505–508

Eichelbaum M, Gross AS (1996) Stereochemical aspects of drug action and disposition. Advances in Drug Research 28:1–64

Ellis SW, Rowland K, Ackland MJ, Rekka E, Simula AP, Lennard MS, Wolf CR, Tucker GT (1996) Influence of amino acid residue 374 of cytochrome P-450 2D6

(CYP2D6) on the regio- and enantio-selective metabolism of metoprolol. Biochemical Journal 316:647–654

Eriksson T, Björkman S, Roth B, Fyge Å, Höglund P (1995) Stereospecific determination, chiral inversion in vitro and pharmacokinetics in humans of the enantiomers of thalidomide. Chirality 7:44–52

Eriksson T, Björkman S, Roth B, Fyge Å, Höglund P (1998) Enantiomers of thalidomide: blood distribution and the influence of serum albumin on chiral inversion and hydrolysis. Chirality 10:223–228

Fjordside L, Jeppesen U, Eap CB, Powell K, Baumann P, Brøsen K (1999) The stereoselective metabolism of fluoxetine in poor and extensive metabolizers of sparteine. Pharmacogenetics 9:55–60

Foster DJR, Somogyi AA, Bochner F (1999) Methadone *N*-demethylation in human liver microsomes: lack of stereoselectivity and involvement of CYP3A4. British Journal of Clinical Pharmacology 47:403–412

Fromm MF, Busse D, Kroemer HK, Eichelbaum M (1996) Differential induction of prehepatic and hepatic metabolism of verapamil by rifampin. Hepatology 24:796–801

Gilmore DA, Gal J, Gerber JG, Nies AS (1992) Age and gender influence the stereoselective pharmacokinetics of propranolol. Journal of Pharmacology and Experimental Therapeutics 261:1181–1186

Goldstein JA, Faletto MB, Romkes-Sparks M, Sullivan T, Kitareewan S, Raucy JL, Lasker JM, Ghanayem BI (1994) Evidence that CYP2C19 is the major (*S*)-mephenytoin 4′-hydroxylase in humans. Biochemistry 33:1743–1752

Gross AS, Mikus G, Fischer C, Hertrampf R, Gundert-Remy U, Eichelbaum M (1989) Stereoselective disposition of flecainide in relation to the sparteine/debrisoquine metaboliser phenotype. British Journal of Clinical Pharmacology 28:555–566

Gupta SK, Atkinson L, Tu T, Longstreth JA (1995) Age and gender related changes in stereoselective pharmacokinetics and pharmacodynamics of verapamil and norverapamil. British Journal of Clinical Pharmacology 40:325–331

Hall SD, Xiaotao Q (1994) The role of coenzyme A in the biotransformation of 2-arylpropionic acids. Chemical and Biological Interactions 90:235–251

Haritos VS, Ching MS, Ghabrial H, Gross AS, Taavitsainen P, Pelkonen O, Battaglia SE, Smallwood RA, Ahokas JT (1998) Metabolism of dexfenfluramine in human liver microsomes and by recombinant enzymes: role of CYP2D6 and 1A2. Pharmacogenetics 8:423–432

Harris RZ, Benet LZ, Schwartz JB (1995) Gender effects in pharmacokinetics and pharmacodynamics. Drugs 50:222–239

Hemeryck A, Lefebvre RA, De Vriendt C, Belpaire FM (2000) Paroxetine affects metoprolol pharmacokinetics and pharmacodynamics in healthy volunteers. Clinical Pharmacology and Therapeutics 67:283–291

Hemeryck A, De Vriendt CA, Belpaire FM (2001) Metoprolol-paroxetine interaction in human liver microsomes: stereoselective aspects and prediction of the in vivo interaction. Drug Metabolism and Disposition 29:656–663

Hermans JJR, Thijssen HHW (1989) The in vitro ketone reduction of warfarin and analogues. Substrate stereoselectivity, product stereoselectivity and species differences. Biochemical Pharmacology 38:3365–3370

Hermans JJ, Thijssen HH (1993) Human liver microsomal metabolism of the enantiomers of warfarin and acenocoumarol: P450 isozyme diversity determines the differences in their pharmacokinetics. British Journal of Pharmacology 110:482–490

Hiep BT, Fernandez C, Khanh V, Hung NK, Thuillier A, Farinotti R, Arnaud P, Gimenez F (2000) Stereospecific versus nonstereospecific assessments for the bioequivalence of two formulations of racemic chlorpheniramine. Chirality 12:599–605

Holtbecker N, Fromm MF, Kroemer HK, Ohnhaus EE, Heidemann H (1996) The nifedipine-rifampin interaction. Evidence for induction of gut wall metabolism. Drug Metabolism and Disposition 24:1121–1123

Hooper WD, Pool WF, Woolf TF, Gal J (1994) Stereoselective hydroxylation of tacrine in rats and humans. Drug Metabolism and Disposition 22:719–724
Hooper WD, Qing MS (1990) The influence of age and gender on the stereoselective metabolism and pharmacokinetics of mephobarbital in humans. Clinical Pharmacology and Therapeutics 48:633–640
Jenner P, Testa B (1973) The influence of stereochemical factors on drug disposition. Drug Metabolism Reviews 2:117–184
Ko JW, Desta Z, Flockhart DA (1998) Human *N*-demethylation of (*S*)-mephenytoin by cytochrome P450 s 2C9 and 2B6. Drug Metabolism and Disposition 26:775–778
Kobayashi K, Kogo M, Tani M, Shimada N, Ishizaki T, Numazawa S, Yoshida T, Yamamoto T, Kuroiwa Y, Chiba K (2001) Role of CYP2C19 in stereoselective hydroxylation of mephobarbital by human liver microsomes. Drug Metabolism and Disposition 29:36–40
Kroemer HK, Funck-Brentano C, Silberstein DJ, Wood AJJ, Eichelbaum M, Woosley RL, Roden DM (1989) Stereoselective disposition and pharmacologic activity of propafenone enantiomers. Circulation 79:1068–1076
Kroemer HK, Fischer C, Meese CO, Eichelbaum M (1991) Enantiomer/enantiomer interaction of (*S*)- and (*R*)-propafenone for cytochrome P450IID6-catalyzed 5-hydroxylation: in vitro evaluation of the mechanism. Molecular Pharmacology 40:135–142
Kroemer HK, Echizen H, Heidemann H, Eichelbaum M (1992) Predictability of the in vivo metabolism of verapamil from in vitro data: contribution of individual metabolic pathways and stereoselective aspects. Journal of Pharmacology and Experimental Therapeutics 260:1052–1057
Kroemer HK, Gautier JC, Beaune P, Henderson C, Wolf CR, Eichelbaum M (1993) Identification of P450 enzymes involved in metabolism of verapamil in humans. Naunyn Schmiedebergs Archives of Pharmacology 348:332–337
Küpfer A, Roberts RK, Schenker S, Branch RA (1981) Stereoselective metabolism of mephenytoin in man. Journal of Pharmacology and Experimental Therapeutics 218:193–199
Lalonde RL, Tenero DM, Burlew BS, Herring VL, Bottorff MB (1990) Effects of age on the protein binding and disposition of propranolol stereoisomers. Clinical Pharmacology and Therapeutics 47:447–455
Lennard MS, Tucker GT, Silas JH, Freestone S, Ramsay LE, Woods HF (1983) Differential stereoselective metabolism of metoprolol in extensive and poor debrisoquine metabolizers. Clinical Pharmacology and Therapeutics 34:732–737
Lewis RJ, Trager WF, Chan KK, Breckenridge AM, Orme ME, Rowland M, Shary W (1974) Warfarin. Stereochemical aspects of its metabolism and the interaction with phenylbutazone. Journal of Clinical Investigation 53:1607–1617
Lindner W, Rath M, Stoschitzky K, Semmelrock HJ (1989) Pharmacokinetic data of propranolol enantiomers in a comparative human study with (*S*)- and (R,S)-propranolol. Chirality 1:10–13
Lown KS, Kolars JC, Thummel KE, Barnett JL, Kunze KL, Wrighton SA, Watkins PB (1994) Interpatient hetereogeneity in expression of CYP3A4 and CYP3A5 in small bowel: lack of prediction by the erythromycin breath test. Drug Metabolism and Disposition 22:947–955
Lown KS, Mayo RR, Leichtman AB, Hsiao HL, Turgeon DK, Schmiedlin-Ren P, Brown MB, Guo W, Rossi SJ, Benet LZ, Watkins PB (1997) Role of intestinal P-glycoprotein (mdr1) in interpatient variability in the oral bioavailability of cyclosporine. Clinical Pharmacology and Therapeutics 62:248–260
Luzier AB, Killian A, Wilton JH, Wilson MF, Forrest A, Kazierad DJ (1999) Gender-related effects on metoprolol pharmacokinetics and pharmacodynamics in healthy volunteers. Clinical Pharmacology and Therapeutics 66:594–601
Malpeis L, Bathala MS, Ludden TM, Bhat HB, Frank SG, Sokoloski TD, Morrison BE and Reuning RH (1975) Metabolic reduction of naltrexone I. Synthesis, separa-

tion and characterization of naloxone and naltrexone reduction products and qualitative assay of urine and bile following administration of naltrexone, α-naltrexol, or β-naltrexol. Research Communications in Chemical Pathology and Pharmacology 12:43–65

Margolis JM, O'Donnell JP, Mankowski DC, Ekins S, Obach RS (2000) (*R*)-, (*S*)-, and racemic fluoxetine *N*-demethylation by human cytochrome P450 enzymes. Drug Metabolism and Disposition 28:1187–1191

Mason JP, Hutt AJ (1997) Stereochemical aspects of drug metabolism. Chapter 3. In: Aboul-Enien HY, Wainer IW (eds) The Impact of Stereochemistry on Drug Development and Use. John Wiley and Sons, Inc Chemical Analysis Series Vol 142, pp 45–105

Mast V, Fischer C, Mikus G, Eichelbaum M (1992) Use of pseudoracemic nitrendipine to elucidate the metabolic steps responsible for stereoselective disposition of nitrendipine enantiomers. British Journal of Clinical Pharmacology 33:51–59

Mehvar R, Jamali F (1997) Bioequivalence of Chiral drugs. Clinical Pharmacokinetics 33:122–141

Meinertz T, Kasper W, Karl C, Jähnchen E (1978) Anticoagulant activity of the enantiomers of acenocoumarol. British Journal of Clinical Pharmacology 5:187–188

Mikus G, Eichelbaum M, Fischer C, Gumulka S, Klotz U, Kroemer HK (1990) Interaction of verapamil and cimetidine: stereochemical aspects of drug metabolism, drug disposition and drug action. Journal of Pharmacology and Experimental Therapeutics 253:1042–1048

Murthy SS, Shetty HU, Nelson WL, Jackson PR, Lennard MS (1990) Enantioselective and diastereoselective aspects of the oxidative metabolism of metoprolol. Biochemical Pharmacology 40:1637–1644

Narimatsu S, Kato R, Horie T, Ono S, Tsutsui M, Yabusaki Y, Ohmori S, Kitada M, Ichioka T, Shimada N, Kato R, Ishikawa T (1999) Enantioselectivity of bunitrolol 4-hydroxylation is reversed by the change of an amino acid residue from valine to methionine at position 374 of cytochrome P450–2D6. Chirality 11:1–9

Nerurkar SG, Dighe SV, Williams RL (1992) Bioequivalence of racemic drugs. Journal of Clinical Pharmacology 32:935–943

Niopas I, Toon S, Rowland M (1991) Further insight into the stereoselective interaction between warfarin and cimetidine in man. British Journal of Clinical Pharmacology 32:508–511

Park BK (1988) Warfarin: Metabolism and mode of action. Biochemical Pharmacology 37:19–27

Patel M, Tang BK, Grant DM, Kalow W (1995) Interindividual variability in the glucuronidation of (*S*) oxazepam contrasted with that of (*R*) oxazepam. Pharmacogenetics 5:287–97

Pohl LR, Bales R, Trager WF (1976) Warfarin. Stereochemical aspects of its metabolism in vivo in the rat. Research Communications in Chemical Pathology and Pharmacology 15:233–256

Roy P, Tretyakov O, Wright J, Waxman DJ (1999) Stereoselective metabolism of ifosfamide by human P-450 s 3A4 and 2B6. Drug Metabolism and Disposition 27:1309–1318

Sachse C, Ruschen S, Dettling M, Schley J, Bauer S, Müller-Oerlinghausen B, Roots I, Brockmöller J (1999) Flavin monooxygenase 3 (FMO3) polymorphism in a white population: allele frequencies, mutation linkage, and functional effects on clozapine and caffeine metabolism. Clinical Pharmacology and Therapeutics 66:431–438

Sasaki M, Tateishi T, Ebihara A (1993) The effects of age and gender on the stereoselective pharmacokinetics of verapamil. Clinical Pharmacology and Therapeutics 54:278–285

Scheuerer S, Hall SD, Williams KM, Geisslinger G (1998) Effect of clofibrate on the chiral inversion of ibuprofen in healthy volunteers. Clinical Pharmacology and Therapeutics 64:168–176

Schwartz JB, Capili H, Wainer IW (1994) Verapamil stereoisomers during racemic verapamil administration: effects of aging and comparisons to administration of individual stereoisomers. Clinical Pharmacology and Therapeutics 56:368–376

Sidhu J, Priskorn M, Poulsen M, Segonzac A, Grollier G, Larsen F (1997) Steady-state pharmacokinetics of the enantiomers of citalopram and its metabolites in humans. Chirality 9:686–692

Soons PA, Vogels BAPM, Roosemälen MCM, Schoemaker HC, Uchida E, Edgar E, Lundahl J, Cohen AF, Breimer DD (1991) Grapefruit juice and cimetidine inhibit stereoselective metabolism of nitrendipine in humans. Clinical Pharmacology and Therapeutics 50:394–403

Spahn-Langguth H, Benet LZ, Möhrke W, Langguth P (1997) First-pass phenomena: sources of stereoselectivities and variabilities of concentration-time profiles after oral dosage. In: Aboul-Enein, Wainer IW (eds) The Impact of Stereochemistry on Drug Development and Use. Chemical Analysis Series Vol 142; John Wiley and Sons Inc New York, pp 573–610

Sporkmann K (1992) MD Thesis University of Tübingen

Srinivas NR, Barr WH, Shyu WC, Mohandoss E, Chow S, Staggers J, Balan G, Belas FJ, Blair IA, Barbhaiya RH (1996) Bioequivalence of two tablet formulations of nadolol using single and multiple dose data: assessment using stereospecific and nonstereospecific assays. Journal of Pharmaceutical Sciences 85:299–303

Sweeny DJ, Nellans HN (1995) Stereoselective glucuronidation of zileuton isomers by human hepatic microsomes. Drug Metabolism and Disposition 23:149–153

Takahashi H, Ishikawa S, Nomoto S, Nishigaki Y, Ando F, Kashima T, Kimura S, Kanamori M, Echizen H (2000) Developmental changes in pharmacokinetics and pharmacodynamics of warfarin enantiomers in Japanese children. Clinical Pharmacology and Therapeutics 68:541–555

Tanaka M, Yamazaki H, Hakusui H, Nakamichi N, Sekino H (1997) Differential stereoselective pharmacokinetics of pantoprazole, a proton pump inhibitor in extensive and poor metabolisers of pantoprazole – a preliminary study. Chirality 9:17–21

Testa B (1988) Substrate and product stereoselectivity in monooxygenase-mediated drug activation and inactivation. Biochemical Pharmacology 37:85–92

Thijssen HHW, Janssen GMJ, Baars LGM (1986) Lack of effect of cimetidine on pharmacodynamics and kinetics of single oral doses of R- and *S*-acenocoumarol. European Journal of Clinical Pharmacology 30:619–623

Toon S, Low LK, Gibaldi M, Trager WF, O'Reilly RA, Motley CH, Goulart DA (1986) The warfarin-sulfinpyrazone interaction: stereochemical considerations. Clinical Pharmacology and Therapeutics 39:15–24

Toon S, Davidson EM, Garstang FM, Batra H, Bowes RJ, Rowland M (1988) The racemic metoprolol H_2-antagonist interaction. Clinical Pharmacology and Therapeutics 43:283–289

Torchin CD, McNeilly PJ, Kapetanovic IM, Strong JM, Kupferberg HJ (1996) Stereoselective metabolism of a new anticonvulsant drug candidate, losigamone, by human liver microsomes. Drug Metabolism and Disposition 24:1002–1008

Trager WF, Jones JP (1987) Stereochemical considerations in drug metabolism. In: Bridges JW, Chasseaud LF, Gibson GG (eds) Progress in Drug Metabolism. Vol 10. Taylor and Francis Ltd, pp 55–83

Tybring G, Böttiger Y, Widén J, Bertilsson L (1997) Enantioselective hydroxylation of omeprazole catalyzed by CYP2C19 in Swedish white subjects. Clinical Pharmacology and Therapeutics 62:129–137

Uno T, Ohkubo T, Sugawara K, Higashiyama A, Motomura S, Ishizaki T (2000) Effects of grapefuit juice on the stereoselective disposition of nicardipine in humans: evidence for dominant presystemic elimination at the gut site. European Journal of Clinical Pharmacology 56:643–649

Vandamme N, Broly F, Libersa C, Courseau C, Lhermitte M (1993) Stereoselective hydroxylation of mexiletine in human liver microsomes: implication of P450IID6 – a preliminary report. Journal of Cardiovascular Pharmacology 21:77–83

Vogelgesang B, Echizen H, Schmidt E, Eichelbaum M (1994) Stereoselective first-pass metabolism of highly cleared drugs: studies of the bioavailability of L- and D-verapamil examined with a stable isotope technique. British Journal of Clinical Pharmacology 18:733–740

Von Richter O, Greiner B, Fromm MF, Fraser R, Omari T, Barclay ML, Dent J, Somogyi AA, Eichelbaum M (2001) Determination of in vivo absorption, metabolism, and transport of drugs by the human intestinal wall and liver with a novel perfusion catheter. Clinical Pharmacology and Therapeutics 70:217–227

Ward JK, Dow J, Dallow N, Eynott P, Milleri S, Ventresca GP (2000) Enantiomeric disposition of inhaled, intravenous and oral racemic-salbutamol in man – no evidence of enantioselective lung metabolism. British Journal of Clinical Pharmacology 49:15–22

Watkins PB, Wrighton SA, Schuetz EG, Guzelian PS (1987) Identification of glucocorticoid-inducible cytochromes P-450 in the intestinal mucosa of rats and man. Journal of Clinical Investigation 80:1029–1030

Wedlund PJ, Aslanian WS, Jacqz E, McAllister CB, Branch RA, Wilkinson GR (1985) Phenotypic differences in mephenytoin pharmacokinetics in normal subjects. Journal of Pharmacology and Experimental Therapeutics 234:662–669

Williams ML, Wainer IW (1999) Cyclophosphamide versus ifosfamide: "To use ifosfamide or not to use, that is the three-dimensional question." Current Pharmaceutical Design 5:665–672

Williams ML, Wainer IW, Granvil CP, Gehrcke B, Bernstein ML, Ducharme MP (1999a) Pharmacokinetics of (*R*)- and (*S*)-cyclophosphamide and their dechloroethylated metabolites in cancer patients. Chirality 11:301–308

Williams ML, Wainer IW, Embree L, Barnett M, Granvil CL, Ducharme MP (1999b) Enantioselective induction of cyclophosphamide metabolism by phenytoin. Chirality 11:569–574

Wormhoudt LW, Commandeur JNM, Vermeulen NPE (1999) Genetic polymorphisms of human *N*-acetyltransferase, cytochrome P450, glutathione-*S*-transferase, and epoxide hydrolase enzymes: relevance to xenobiotic metabolism and toxicity. Critical Reviews in Toxicology 29:59–124

Xie HG, Kim RB, Wood AJJ, Stein CM (2001) Molecular basis of ethnic differences in drug disposition and response. Annual Review of Pharmacology and Toxicology 41:815–850

Yasumori T, Chen LS, Li QH, Ueda M, Tsuzuki T, Goldstein JA, Kato R, Yamazoe Y (1999) Human CYP2C-mediated stereoselective phenytoin hydroxylation in Japanese: difference in chiral preference of CYP2C9 and CYP2C19. Biochemical Pharmacology 57:1297–1303

Zhou HH, Anthony LB, Roden DM, Wood AJJ (1990) Quinidine reduces clearance of (+)-propranolol more than (−)-propranolol through marked reduction in 4-hydroxylation. Clinical Pharmacology and Therapeutics 47:686–693

Zhou HH, Whelan E, Wood AJJ (1992) Lack of effect of ageing on the stereochemical disposition of propranolol. British Journal of Clinical Pharmacology 33:121–123

CHAPTER 14

Metabolic Chiral Inversion of 2-Arylpropionic Acids

I. TEGEDER, K. WILLIAMS, and G. GEISSLINGER

A. Introduction

I. 2-Arylpropionic Acids

The 2-arylpropionic acid derivatives (2-APAs) represent a major group of nonsteroidal antiinflammatory drugs (NSAIDs) which are used in the treatment of inflammatory diseases, particularly rheumatoid arthritis and osteoarthritis (marketed 2-APAs are listed in Table 1). They have also found much wider application as over-the-counter formulations for the self-medication of headache and minor pain as well as pain associated with dysmenorrhea. Moreover, they are used in perisurgical and cancer pain management. 2-APAs exist in two enantiomeric forms due to the presence of an asymmetric carbon atom. With the exception of naproxen and – in a few countries – ibuprofen and ketoprofen, they are marketed as racemates. Their anti-inflammatory efficacy is primarily based on the inhibition of cylooxygenase isoenzymes resulting in the inhibition of prostaglandin biosynthesis. This activity resides almost exclusively in the (*S*)-(+)-enantiomers. A change in the chirality of these drugs from the *R*-configuration to the *S*-configuration results in a dramatic increase of this inhibitory potency. The incidence of adverse effects of racemates (e.g., *R*,*S*-ibuprofen) is not greater than that of pure *S*-enantiomers (e.g., *S*-naproxen). Thus, the features of the 2-arylpropionic acids that are the focus of this review are:

- They are asymmetric.
- Only *S*-enantiomers inhibit cyclooxygenase activity (COX-1 and COX-2).
- With the exception of naproxen, and recently ibuprofen and ketoprofen, they have been administered as racemates.
- The *R*-enantiomer of some APAs is unidirectionally inverted to the prostaglandin synthesis inhibiting *S*-enantiomer.
- Inversion is accomplished by enzymes involved in lipid metabolism with implications for toxicity and efficacy.

Table 1. Marketed chiral 2-arylpropionic acid antiinflammatory drugs

Carprofen	Ketoprofen[b]
Fenoprofen	Naproxen[a]
Flurbiprofen	Pirprofen
Ibuprofen[b]	Tiaprofenic acid
Indoprofen	

[a] *S*-enantiomer only.
[b] *S*-enantiomer available in some countries.

II. Inversion in Man

The first evidence of metabolic inversion came from observations on the chirality of urinary metabolites of ibuprofen in humans. These metabolites were dextrorotatory regardless of whether the racemate, the *R*-enantiomer or the *S*-enantiomer had been administered (ADAMS et al. 1967). More detailed pharmacokinetic studies clearly illustrated the unidirectional nature of the inversion process (Fig. 1) (GEISSLINGER et al. 1990; LEE et al. 1985; MILLS et al. 1973). Following administration of the pure *R*-enantiomer of ibuprofen, about 60% of an administered dose was stereospecifically inverted to the *S*-enantiomer. There was no measurable inversion of *S*- to *R*-ibuprofen (GEISSLINGER et al. 1990). The pharmacokinetics of the individual enantiomers were altered by concurrent administration of the respective optical antipode (LEE et al. 1985). It was suggested that this change reflects an interaction between the enantiomers at plasma protein binding sites or glucuronidation (LAPICQUE et al. 1993). The use of isotopically labeled *S*-ibuprofen in combination with the *R*-enantiomer as pseudoracemate allowed an estimation of the fractional inversion of *R*-ibuprofen in the form in which it is usually administered, i.e., the racemate (RUDY et al. 1991, 1995; SCHEUERER et al. 1998a). Using this methodology the fractional inversion was found to be of the order of 75% (SCHEUERER et al. 1998a).

In addition to ibuprofen, chiral inversion has been demonstrated unequivocally in humans for two other 2-arylpropionates, fenoprofen (RUBIN et al. 1985) and benoxaprofen (BOPP et al. 1979). No significant inversion was detected in man for indoprofen (TAMASSIA et al. 1984) or flurbiprofen (GEISSLINGER et al. 1994; JAMALI et al. 1988). Other 2-arylpropionates which are probably not inverted in man are tiaprofenic acid (SINGH et al. 1986), ketoprofen (FOSTER et al. 1988a,b; GEISSLINGER et al. 1995; SALLUSTIO et al. 1988) and carprofen (STOLTENBORG et al. 1981). These observations clearly indicate the selectivity of the isomerization, and the relatively narrow group of drugs, even within the class of 2-arylpropionic acids, for which inversion is an important feature.

III. Inversion in Animals: Models of Inflammation

Comparative in vivo and in vitro data provided further evidence that there was chiral inversion of some 2-arylpropionic acids. It was shown that *R*-

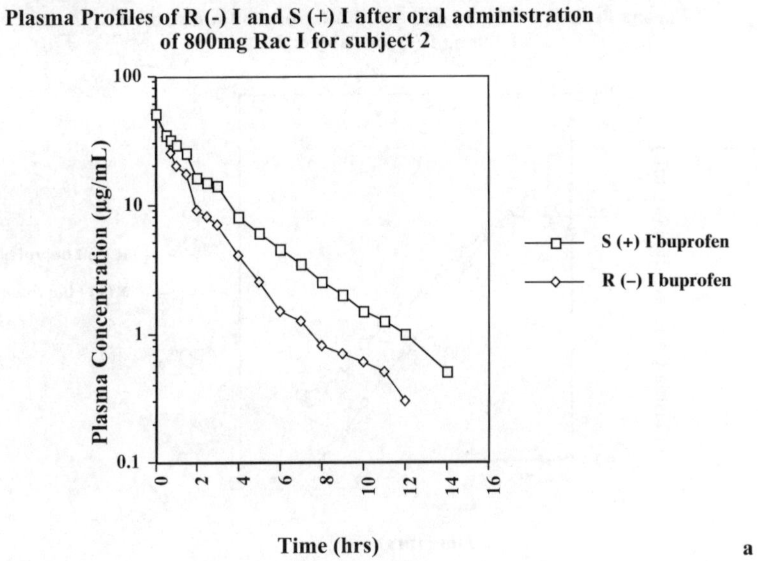

a

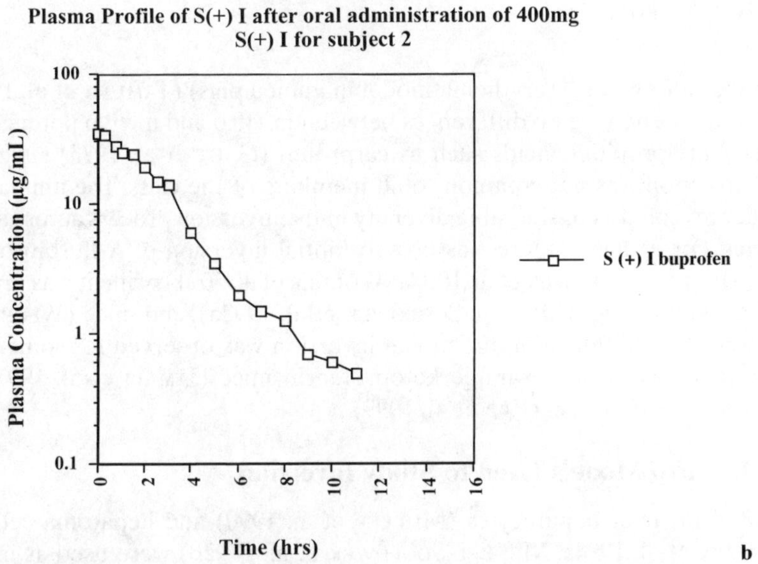

b

Fig. 1. Plasma concentration-time profiles of the *R*- and *S*-enantiomers of ibuprofen in healthy volunteers after oral administration of solutions of: **a** 800 mg *RS*-ibuprofen; **b** 400 mg *S*-ibuprofen; **c** 400 mg *R*-ibuprofen. (Lee et al. 1985)

ibuprofen did not inhibit prostaglandin synthesis in vitro, but was active in vivo in animal models of inflammation. This was reflected by a change in the potency ratio (S/R) of the enantiomers from 160 in vitro to 1.3 in vivo (Adams et al. 1976). Similarly, it was found that clidanac had an *S/R* potency ratio of 1000 in vitro (model: microsomal preparations) (Tamura et al. 1981a) but in

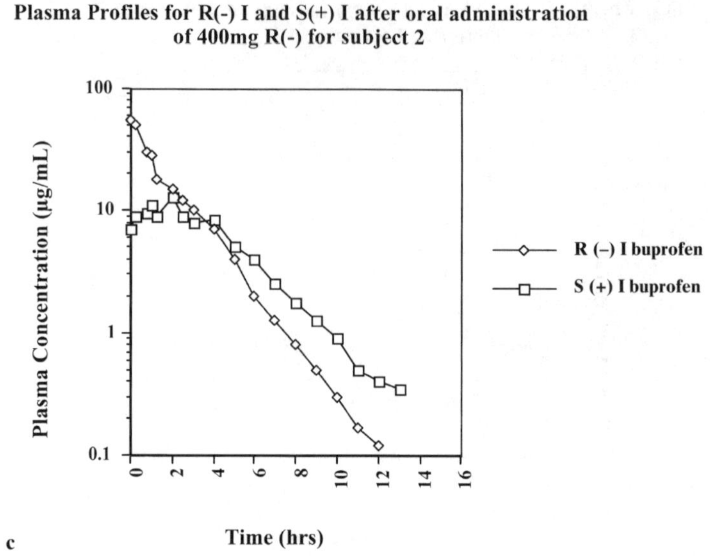

Fig. 1. *Continued*

vivo the ratio was 1.0 (erythema model in guinea pigs) (TAMURA et al. 1981b). However, there were no differences between in vitro and in vivo potencies for other 2-arylpropionic acids such as carprofen (GAUT et al. 1975) suggesting that inversion was not common to all members of the class. The animal data further revealed a considerable diversity in the inversion process across animal species. For example, there was no substantial inversion of *R*-flurbiprofen in rats (MENZEL-SOGLOWEK et al. 1992a; TEGEDER et al. 2001) while it was inverted in dogs, guinea pigs (MENZEL-SOGLOWEK et al. 1992a), and mice (WECHTER et al. 1997). In addition, a bi-directional inversion was observed for some APAs in certain species, for example ketoprofen in mice (JAMALI et al. 1997) and tiaprofenic acid in rats (ERB et al. 1999).

IV. In Vitro Models Used to Study Inversion

Isolated rat liver hepatocytes (MULLER et al. 1990) and hepatoma cell lines (MENZEL et al. 1994a; MENZEL-SOGLOWEK et al. 1992b) were used as in vitro models to study chiral inversion and stereoselective metabolism. The inversion of *R*-ibuprofen to *S*-ibuprofen in isolated rat hepatocytes was shown to obey apparent first order kinetics. In hepatoma cells, inversion occurred slowly. Nevertheless, the tumor cells were able to mimic qualitatively the species and substance specificity of inversion of 2-arylpropionic acids in vivo (MENZEL-SOGLOWEK et al. 1992b). Therefore, both models were suitable for studying the underlying mechanisms. In addition to cell culture experiments, the process of inversion and particularly the contribution and regulation of involved enzymes was studied using different subcellular preparations from liver homogenates. Finally, inversion was studied in the perfused rat intestine (JEFFREY et al. 1991), liver (ROY-DE VOS et al. 1996) and lung (HALL et al. 1992). These experiments provided some insight into the site of inversion.

V. The In Vivo Site of Inversion

Following a single oral dose of pure *R*-ibuprofen, approximately 60% has been observed to be inverted to *S*-ibuprofen (LEE et al. 1985). Following intravenous administration of racemic ibuprofen; however, the differences between plasma concentrations of *R*- and *S*-ibuprofen were less prominent, and it was suggested that there was a presystemic site of inversion (COX et al. 1987). The rate of inversion was found to depend not only on the route of administration but also on the duration of gastrointestinal absorption (SATTARI et al. 1994). Thus, first-pass inversion was thought to occur in the gut. Supporting this view, rat and human whole small intestinal homogenates were shown to invert *R*-ibuprofen and *R*-benoxaprofen in vitro (SIMMONDS et al. 1980). Studies using rat in situ-perfused intestine and/or liver revealed that perfusion of the liver alone or combined perfusion of intestine and liver with the *R*-enantiomer resulted in the formation of the *S*-isomer (JEFFREY et al. 1991; ROY-DE VOS et al. 1996). Furthermore, inversion of *R*-ibuprofen was observed in liver homogenates (KNIHINICKI et al. 1989), isolated rat hepatocytes (MULLER et al. 1990), and subcellular preparations (mitochondria and microsomes) from rat and human liver homogenates (BRUGGER et al. 1996; KNADLER et al. 1990) suggesting that the liver is the predominant site of the presystemic and systemic inversion. *R*-ibuprofen and *R*-fenoprofen were also found to be inverted in other organs such as kidney, lung, and heart, but to a much lesser extent than in the liver (HALL et al. 1992; KNADLER et al. 1990). Thus, metabolic chiral inversion may take place in several organs correlating with the pattern of expression of the responsible enzymes (discussed in Sect. B.). However, the quantitatively most important organ with respect to inversion is the liver.

B. Mechanism of Inversion

I. Formation of Coenzyme A Thioesters

Since 2-arylpropionic acid enantiomers per se are isomerically stable at physiological pH, it was assumed that an enzymatically mediated activation was required for the inversion. Since xenobiotic carboxylic acids were known to form coenzyme A (CoA) thioesters in the course of conjugation reactions, it was suggested that the formation of *R*-ibuprofenoyl-CoA by an acyl CoA-synthetase was the first step of inversion (WECHTER et al. 1974). This was supported by the finding that the inversion of *R*-ibuprofen by rat liver homogenate required CoA and ATP as cofactors. Later, numerous studies have demonstrated the formation of the CoA-thioester by ibuprofen and other enantiomers of 2-arylpropionic acids in rat and human liver preparations supplemented with CoA, ATP and Mg^{2+} (Brugger et al. 1996; Knadler et al. 1990; Knights et al. 1988, 1989; Sevoz et al. 2000). In these experiments, CoA thioester formation and inversion was only observed with *R*- but not *S*-ibuprofen. However, while *S*-ibuprofen was not inverted, *S*-ibuprofenoyl-CoA was inverted. These experiments suggested that the stereospecificity of inversion is controlled by the acyl-CoA synthetase. The specificity of acyl-CoA formation was further evaluated by passing solutions of the *R*- and *S*-enantiomers of fenoprofen and ibuprofen through an agarose affinity column into which long chain acyl-CoA synthetase from rat liver had been absorbed. Formation of the respective CoA thioesters was only observed with the *R*-enantiomers and not the *S*-enantiomers, confirming the stereospecificity of long chain CoA synthetase (Knights et al. 1988; Knihinicki et al. 1989). Since the coupling of other carboxylic acids to CoA involves the formation of an intermediary adenylate, a similar mechanism was suggested for ibuprofen. This hypothesis was addressed by incubation of rat liver mitochondria with either chemically synthesized *R*- and *S*-ibuprofenoyl-adenylates or with the free acids. Interestingly, formation of the respective optical antipodes and the CoA thioester occurred with both adenylates and with *R*-ibuprofen, but not with *S*-ibuprofen (Menzel et al. 1994b). Thus, the formation of *R*-ibuprofenoyl-adenylate appears to be the first stereoselective step of chiral inversion. It is not known whether the formation of the adenylate ester is mediated by a specific adenylase or also by acyl-CoA synthetase.

Numerous acyl-CoA synthetases have been described in mammalian cells but, at least in the case of *R*-ibuprofenoyl-CoA formation, it appears that the long chain acyl CoA synthetase (LACS) is the most important one with respect to chiral inversion. Strong evidence in favor of this hypothesis is the competitive inhibition of *R*-ibuprofenoyl-CoA formation by palmitic acid, a prototypical long chain fatty acid (Brugger et al. 1996; Tracy et al. 1993). The involvement of other acyl CoA synthetases in inversion is probably not significant considering the weak or absent inhibition of *R*-ibuprofenoyl-CoA formation by octanoic acid or propionic acid (Tracy et al. 1993). The copurification of palmitoyl-CoA synthetase and *R*-ibuprofenoyl CoA synthetase from rat liver and the subsequent recombinant expression of this enzyme in *E. coli* leaves little doubt that long chain acyl CoA synthetase and *R*-ibuprofenoyl CoA synthetase are identical enzymes (Brugger et al. 2001; Sevoz et al. 2000). Five different isoenzymes of long chain acyl CoA synthetase (LACS) have been characterized in the rat. Each LACS appears to have a "limited tissue-specific" tissue distribution and distinct regulation (Suzuki et al. 1995). LACS 1 is abundant in liver, heart, and adipose tissue (Suzuki et al. 1990), while LACS 2 and 3 are predominant in brain (Fujino et al. 1996; Fujino et al. 1992). LACS 4 and 5 are primarily found in steroid-generating tissues and small intestine (Cho et al. 2000; Oikawa et al. 1998). The differences between the LACS enzymes may have contributed to the suggestion that organs other than the liver were quantitatively significant sites of inversion.

II. Racemization (Epimerization) of the Coenzyme A Thioesters

Following the formation of the CoA thioester, the next step of inversion is racemization. The thioesters are activated molecular species in that the acidity is greatly increased by conversion of the carboxylic acid to its thioester (see above). The question was thus raised whether CoA thioesters racemize spontaneously in aqueous solution as had been observed for other drugs which have labile hydrogens at their asymmetric centers. In dilute alkaline solution, i.e., under aggressive biochemical conditions, some racemization may occur but at physiological pH ibuprofenoyl CoA thioesters were chirally stable (KNIGHTS et al. 1988). In the presence of human or rat liver homogenate, however, both *R*-ibuprofenoyl-CoA and *S*-ibuprofenoyl CoA thioesters were readily epimerized (KNIHINICKI et al. 1991) indicating an enzymatic but not stereospecific contribution to this event. A 42-kDa 2-arylpropionyl CoA epimerase was isolated from the cytosolic and mitochondrial fractions of rat liver. The purified enzyme catalyzed the epimerization of various 2-arylpropionyl CoAs with no obvious stereochemical preference (REICHEL et al. 1995; SHIEH et al. 1993). Expression of 2-arylpropionyl CoA epimerase was found predominantly in the liver and to a lesser extent in kidney, heart, and brain (REICHEL et al. 1995, 1997). Analysis of the genetic structure of rat epimerase revealed a significant sequence homology to the carnitine dehydratase from several species suggesting that the epimerase is involved in lipid metabolism (REICHEL et al. 1997). However, an essentially identical sequence was found for α-methylacyl-CoA racemase, suggesting that 2-aryl-propionoyl-CoA epimerase and α-methylacyl-CoA racemase are identical enzymes (SCHMITZ et al. 1997). α-methylacyl-CoA racemase rapidly epimerizes CoA *R*- and *S*-esters of a variety of α-methyl-branched fatty acids (SCHMITZ et al. 1997), including bile acid intermediates such as di- and trihydroxycoprostanoyl-CoA, suggesting that racemization is an essential step in bile acid formation.

III. Hydrolysis of Coenzyme A Thioesters

The final step of inversion is the enzymatic hydrolysis of the CoA thioesters to release the free 2 arylpropionic acids. Incubation of synthetically prepared *R*- or *S*-ibuprofenoyl CoA thioesters with rat liver homogenates or subcellular preparations resulted in rapid epimerization followed by much slower hydrolysis (TRACY et al. 1991). No stereoselectivity of hydrolysis was noted for any of the enzyme preparations (TRACY et al. 1991). The specific identity of the 2-aryl-propionoyl-CoA hydrolase is unknown. Enzymes such as palmitoyl-CoA hydrolase are relatively nonselective in that they are active on a range of fatty acids of varying degrees of chain length. However, other hydrolases such as the one which cleaves *S*-methylmalonyl-CoA is enantiospecific. Whether the hydrolase which cleaves *R*- and *S*-ibuprofenoyl-CoA is a distinct enzyme or identical with one of the fatty acid acyl-CoA hydrolases has yet to be determined (Fig. 2).

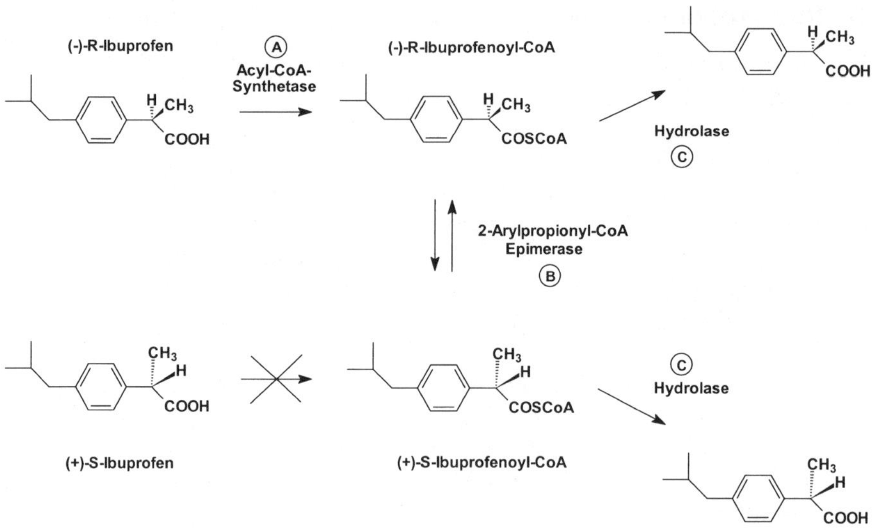

Fig. 2. Mechanism of inversion of ibuprofen and related antiinflammatory drugs as first proposed by NAKAMURA et al. (REICHEL et al. 1997). Stereospecificity of inversion is controlled by the formation of an adenylate which is enantiospecific for the *R*-enantiomer followed by formation of the respective CoA-thioester. Racemization of the CoA-thioester by a racemase produces *S*-ibuprofenoyl-CoA. Finally, hydrolysis of the CoA-thioester releases *S*-ibuprofen. The same or a similar hydrolase also cleaves *R*-ibuprofenoyl-CoA. (From REICHEL et al. 1997)

C. Consequences of Chiral Inversion

The most obvious consequence of the stereoselective formation of 2-arylpropionoyl CoA is that additional quantities of the potent antiinflammatory *S*-enantiomer are produced which contribute to the desired effects but also to the toxicity of the drug. In respect to inversion, the *R*-enantiomer can be considered to be a prodrug. As a consequence, the inter-subject variability in the degree of inversion and thus production of the active drug may contribute to the variability with which different patients respond to the same dose of the racemic drug. Moreover, concomitant treatment with other drugs and reduced liver function may affect the rate of inversion and thus the availability of the active enantiomer. This uncertainty could be avoided by administering the pure *S*-enantiomer. Clinically, however, there is as yet no clear distinction between the response variability with *S*-naproxen or *S*-ibuprofen and racemic nonsteroidal antiinflammatory agents (GEISSLINGER et al. 1993).

In addition to these pharmacokinetic implications, the CoA thioester intermediates per se might affect cellular functions, particularly those involving fatty acid metabolism. In this respect, it has been shown that the CoA thioesters of ibuprofen and fenoprofen inhibit acetyl-CoA carboxylase, thereby inhibiting the rate limiting step in fatty acid biosynthesis (KEMAL et al. 1992) with *S*-ibuprofenoyl-CoA being 15-fold more potent as an inhibitor of acetyl-CoA carboxylase than *R*-ibuprofenoyl-CoA (KEMAL et al. 1992). Furthermore, there is some evidence that ibuprofen stereoselectively inhibits *β*-oxidation of palmitic acid (FRENEAUX et al. 1990). This long chain fatty acid needs to be activated to palmitoyl-CoA thioester on the outer mitochondrial membrane before it is converted to an acylcarnitine that can penetrate the inner mitochondrial membrane. This is a prerequisite for *β*-oxidation. Formation of *R*-ibuprofenoyl-CoA may partly sequester extramitochondrial coenzyme A, thereby inhibiting the formation of the long chain CoA thioester. When CoA concentrations are increased the stereoselective inhibition of *β*-oxidation is partially reversed. This suggests that *R*-ibuprofen competes with palmitic acid for available CoA. In addition to the stereoselective inhibition of *β*-oxidation by *R*-ibuprofen, flurbiprofen and ibuprofen enantiomers were found to inhibit *β*-oxidation nonstereoselectively (BROWNE et al. 1999; ZHAO et al. 1992). Since mitochondrial respiration was moderately inhibited by the drugs, uncoupling of oxidative phosphorylation was suggested to contribute to this effect (BROWNE et al. 1999). Thus, inhibition of *β*-oxidation may involve stereoselective CoA-dependent and nonstereoselective non-CoA-dependent mechanisms. Ibuprofenoyl-CoA thioesters were also found to inhibit the induction of cyclooxygenase-2 expression in LPS treated blood monocytes, which was associated with a reduction of stimulated prostaglandin release (NEUPERT et al. 1997). Thus, ibuprofenoyl-CoA thioesters might directly influence gene regulation.

In addition to the possible direct effects of 2-arylpropionyl-CoAs on cellular function, there may be indirect effects resulting from the incorporation of these intermediates into glycerolipids such as triglycerides and phospholipids. A rather surprising finding and an aspect overlooked in original studies of the distribution of radiolabeled ibuprofen into tissues, was the very slow elimination of the drug from fat tissue (ADAMS et al. 1969). The estimated elimination half-life from this latter site was of the order of 7 days, contrasting markedly with the very rapid elimination from blood (half-life of about 2 h). This could not be explained by simple aqueous-lipid partition coefficients. The reason for the accumulation of ibuprofen in fat tissue was later discovered to be due to incorporation of the drug into triglycerides, thereby retaining the drug in fat tissue (FEARS 1985). The resulting unusual product, where one or more of the fatty acids was replaced by the drug was termed a "hybrid" triglyceride (WILLIAMS et al. 1986). The formation of such "hybrid" triglycerides only occurred after thioesterification of the drug, suggesting that the uptake into lipids should be stereospecific. This hypothesis was confirmed by chronic administration of pure ibuprofen enantiomers to rats. While *R*-ibuprofen accumulated in fat tissue, this phenomenon was not observed with *S*-ibuprofen (WILLIAMS et al. 1986). Qualitatively similar accumulation occurs in man after chronic treatment with racemic ibuprofen (K. Williams et al., unpublished data). Although the consequences of the formation of "hybrid" lipids are uncertain, one could speculate that formation of hybrid phospholipids, in particular, may cause changes in membrane characteristics and function.

I. Factors That May Modulate Inversion

Pretreatment of rat hepatocytes with clofibric acid was found to increase the rate of inversion of *R*-ibuprofen in vitro (MAYER 1996). In contrast, clofibric acid did not elicit such a reaction with flurbiprofen (ROY-DE VOS et al. 1996). The enhanced inversion of ibuprofen by clofibric acid was linked to an increase in hepatic microsomal long-chain CoA synthetase expression, a finding consistent with the proposed mechanism of inversion and the importance of the synthetase.

As expected with regard to the in vitro effects of clofibric acid on CoA synthetase, pretreatment of rats with clofibric acid for 3 days followed by a single intravenous dose of pseudoracemic ibuprofen (13C-*S*-ibuprofen + *R*-ibuprofen) resulted in a considerable increase in the clearance of *R*-ibuprofen by inversion while the clearance of *S*-ibuprofen remained almost unaltered (SCHEUERER et al. 1998b). In addition, there was a fourfold increase in the volume of distribution for *R*-ibuprofen while that of *S*-ibuprofen was unchanged (SCHEUERER et al. 1998b). In an attempt to explain these alterations of drug distribution it was hypothesized that clofibric acid, by inducing CoA synthetase, was increasing the proportion of *R*-ibuprofen being incorporated into triglycerides and thus being retained in adipose tissue. Western blot analyses confirmed that clofibric acid indeed induced the synthetase, while not altering the racemase. Furthermore, studies of radiolabeled ibuprofen demonstrated a significant increase in the concentrations of the label in tissue samples (SCHEUERER et al. 1998b). It was subsequently demonstrated that clofibrate similarly modified the pharmacokinetics of ibuprofen in humans (SCHEUERER et al. 1998a).

D. Conclusions

Chiral inversion of enantiomers is not a common phenomenon in man, even for the 2-arylpropionic acids. However, for drugs such as ibuprofen where it does occur, it is an intriguing phenomenon with potentially important clinical and toxicological considerations. The current understanding is that the 2-arylpropionic acids are inverted in a multistage process. This includes firstly the enantiospecific formation of a CoA thioester via an adenylate intermediate that is catalyzed by long chain acyl-CoA synthetase. Secondly, the acyl-CoA thioester is racemized by a specific, but not enantioselective, racemase. Lastly, the acyl-CoA thioester is hydrolized by hydrolases of unknown specificity.

References

Adams SS, Bough RG, Cliffe EE, Lessel B, Mills RF (1969) Absorption, distribution and toxicity of ibuprofen. Toxicol Appl Pharmacol 15:310–330

Adams SS, Bresloff P, Mason CG (1976) Pharmacological differences between the optical isomers of ibuprofen: evidence for metabolic inversion of the (–)-isomer. J Pharm Pharmacol 28:256–257

Adams SS, Cliffe EE, Lessel B, Nicholson JS (1967) Some biological properties of 2-(4-isoburylphenyl)-propionic acid. J Pharm Sci 56:1686

Bopp RJ, Nash JF, Ridolfo AS, Shepard ER (1979) Stereoselective inversion of (*R*)-(−)-benoxaprofen to the (*S*)-(+)-enantiomer in humans. Drug Metab Dispos 7:356–359

Browne GS, Nelson C, Nguyen T, Ellis BA, Day RO, Williams KM (1999) Stereoselective and substrate-dependent inhibition of hepatic mitochondria beta-oxidation and oxidative phosphorylation by the nonsteroidal antiinflammatory drugs ibuprofen, flurbiprofen, and ketorolac. Biochem Pharmacol 57:837–844

Brugger R, Garcia Alia B, Reichel C, Waibel R, Menzel S, Brune K, Geisslinger G (1996) Isolation and characterization of rat liver microsomal R-ibuprofenoyl-CoA synthetase. Biochem Pharmacol 52:1007–1013

Brugger R, Reichel C, Garcia Alia B, Brune K, Yamamoto T, Tegeder I, Geisslinger G (2001) Expression of rat liver long-chain acyl-CoA synthetase and characterization of its role in the metabolism of R-ibuprofen and other fatty acid-like xenobiotics. Biochem Pharmacol 61:651–656

Cho YY, Kang MJ, Ogawa S, Yamashita Y, Fujino T, Yamamoto TT (2000) Regulation by adrenocorticotropic hormone and arachidonate of the expression of acyl-CoA synthetase 4, an arachidonate-preferring enzyme expressed in steroidogenic tissues. Biochem Biophys Res Commun 274:741–745

Cox SR, VanderLugt JT, Gumbleton TJ, Smith RB (1987) Relationships between thromboxane production, platelet aggregability, and serum concentrations of ibuprofen or flurbiprofen. Clin Pharmacol Ther 41:510–521

Erb K, Brugger R, Williams K, Geisslinger G (1999) Stereoselective disposition of tiaprofenic acid enantiomers in rats. Chirality 11:103–108

Fears R (1985) Lipophilic xenobiotic conjugates: the pharmacological and toxicological consequences of the participation of drugs and other foreign compounds as substrates in lipid biosynthesis. Prog Lipid Res 24:177–195

Foster RT, Jamali F, Russell AS, Alballa SR (1988a) Pharmacokinetics of ketoprofen enantiomers in healthy subjects following single and multiple doses. J Pharm Sci 77:70–73

Foster RT, Jamali F, Russell AS, Alballa SR (1988b) Pharmacokinetics of ketoprofen enantiomers in young and elderly arthritic patients following single and multiple doses. J Pharm Sci 77:191–195

Freneaux E, Fromenty B, Berson A, Labbe G, Degott C, Letteron P, Larrey D, Pessayre D (1990) Stereoselective and nonstereoselective effects of ibuprofen enantiomers on mitochondrial beta-oxidation of fatty acids. J Pharmacol Exp Ther 255:529–535

Fujino T, Kang MJ, Suzuki H, Iijima H, Yamamoto T (1996) Molecular characterization and expression of rat acyl-CoA synthetase 3. J Biol Chem, 271:16748–16752

Fujino T, Yamamoto T (1992) Cloning and functional expression of a novel long-chain acyl-CoA synthetase expressed in brain. J Biochem (Tokyo) 111:197–203

Gaut ZN, Baruth H, Randall LO, Ashley C, Paulsrud JR (1975) Stereoisomeric relationships among antiinflammatory activity, inhibition of platelet aggregation, and inhibition of prostaglandin synthetase. Prostaglandins 10:59–66

Geisslinger G, Lotsch J, Menzel S, Kobal G, Brune K (1994) Stereoselective disposition of flurbiprofen in healthy subjects following administration of the single enantiomers. Br J Clin Pharmacol 37:392–394

Geisslinger G, Menzel S, Wissel K, Brune K. (1995) Pharmacokinetics of ketoprofen enantiomers after different doses of the racemate. Br J Clin Pharmacol 40:73–75

Geisslinger G, Schuster O, Stock KP, Loew D, Bach GL, Brune K (1990) Pharmacokinetics of S(+)- and R(−)-ibuprofen in volunteers and first clinical experience of S(+)-ibuprofen in rheumatoid arthritis. Eur J Clin Pharmacol 38:493–497

Geisslinger G, Stock KP, Loew D, Bach GL, Brune K (1993) Variability in the stereoselective disposition of ibuprofen in patients with rheumatoid arthritis. Br J Clin Pharmacol 35:603–607

Hall SD, Hassanzadeh-Khayyat M, Knadler MP, Mayer PR (1992) Pulmonary inversion of 2-arylpropionic acids: influence of protein binding. Chirality 4:349–352
Jamali F, Berry BW, Tehrani MR, Russell AS (1988) Stereoselective pharmacokinetics of flurbiprofen in humans and rats. J Pharm Sci 77:666–669
Jamali F, Lovlin R, Aberg G (1997) Bi-directional chiral inversion of ketoprofen in CD-1 mice. Chirality 9:29–31
Jeffrey P, Tucker GT, Bye A, Crewe HK, Wright PA (1991) The site of inversion of R(−)-ibuprofen: studies using rat in-situ isolated perfused intestine/liver preparations. J Pharm Pharmacol 43:715–720
Kemal C, Casida JE (1992) Coenzyme A esters of 2-aryloxyphenoxypropionate herbicides and 2-arylpropionate antiinflammatory drugs are potent and stereoselective inhibitors of rat liver acetyl-CoA carboxylase. Life Sci 50:533–540
Knadler MP, Hall SD (1990) Stereoselective arylpropionyl-CoA thioester formation in vitro. Chirality 2:67–73
Knights KM, Drew R, Meffin PJ (1988) Enantiospecific formation of fenoprofen coenzyme A thioester in vitro. Biochem Pharmacol 37:3539–3542
Knihinicki RD, Day RO, Williams KM (1991) Chiral inversion of 2-arylpropionic acid nonsteroidal anti inflammatory drugs–II. Racemization and hydrolysis of (*R*)- and (*S*)- ibuprofen-CoA thioesters. Biochem Pharmacol 42:1905–1911
Knihinicki RD, Williams KM, Day RO (1989) Chiral inversion of 2-arylpropionic acid nonsteroidal antiinflammatory drugs–1. In vitro studies of ibuprofen and flurbiprofen. Biochem Pharmacol 38:4389–4395
Lapicque F, Muller N, Payan E, Dubois N, Netter P (1993) Protein binding and stereoselectivity of nonsteroidal antiinflammatory drugs. Clin Pharmacokinet 25:115–123
Lee EJ, Williams K, Day R, Graham G, Champion D (1985) Stereoselective disposition of ibuprofen enantiomers in man. Br J Clin Pharmacol 19:669–674
Mayer JM (1996) Ibuprofen enantiomers and lipid metabolism. J Clin Pharmacol 36:27S–32S
Menzel S, Sauernheimer C, Brune K, Geisslinger G (1994a) Is the inversion from R- to *S*-ketoprofen concentration dependent? Investigations in rats in vivo and in vitro. Biochem Pharmacol 47:1267–1270
Menzel S, Waibel R, Brune K, Geisslinger G. (1994b) Is the formation of R-ibuprofenyl-adenylate the first stereoselective step of chiral inversion? Biochem Pharmacol 48:1056–1058
Menzel-Soglowek S, Geisslinger G, Beck WS, Brune K (1992a) Variability of inversion of (*R*)-flurbiprofen in different species. J Pharm Sci 81:888–891
Menzel-Soglowek S, Geisslinger G, Mollenhauer J, Brune K (1992b) Metabolic chiral inversion of 2-arylpropionates in rat H4IIE and human Hep G2 hepatoma cells. Relationship to in vivo metabolism. Biochem Pharmacol 43:1487–1492
Mills RF, Adams SS, Cliffe EE, Dickinson W, Nicholson JS (1973) The metabolism of ibuprofen. Xenobiotica 3:589–598
Muller S, Mayer JM, Etter JC, Testa B (1990) Metabolic chiral inversion of ibuprofen in isolated rat hepatocytes. Chirality 2:74–78
Neupert W, Brugger R, Euchenhofer C, Brune K, Geisslinger G (1997) Effects of ibuprofen enantiomers and its coenzyme A thioesters on human prostaglandin endoperoxide synthases. Br J Pharmacol 122:487–492
Oikawa E, Iijima H, Suzuki T, Sasano H, Sato H, Kamataki A, Nagura H, Kang MJ, Fujino T, Suzuki H, Yamamoto TT (1998) A novel acyl-CoA synthetase, ACS5, expressed in intestinal epithelial cells and proliferating preadipocytes. J Biochem (Tokyo) 124:679–685
Reichel C, Bang H, Brune K, Geisslinger G, Menzel S (1995) 2-Arylpropionyl-CoA epimerase: partial peptide sequences and tissue localization. Biochem Pharmacol 50:1803–1806
Reichel C, Brugger R, Bang H, Geisslinger G, Brune K (1997) Molecular cloning and expression of a 2-arylpropionyl-coenzyme A epimerase: a key enzyme in the inversion metabolism of ibuprofen. Mol Pharmacol 51:576–582

Roy-de Vos M, Mayer JM, Etter JC, Testa B (1996) Clofibric acid increases the undirectional chiral inversion of ibuprofen in rat liver preparations. Xenobiotica 26:571–582

Rubin A, Knadler MP, Ho PP, Bechtol LD, Wolen RL (1985) Stereoselective inversion of (*R*)-fenoprofen to (*S*)-fenoprofen in humans. J Pharm Sci 74:82–84

Rudy AC, Knight PM, Brater DC, Hall SD (1995) Enantioselective disposition of ibuprofen in elderly persons with and without renal impairment. J Pharmacol Exp Ther 273:88–93

Rudy AC, Knight PM, Brater DC, Hall SD (1991) Stereoselective metabolism of ibuprofen in humans: administration of R-, *S*- and racemic ibuprofen. J Pharmacol Exp Ther 259:1133–1139

Sallustio BC, Purdie YJ, Whitehead AG, Ahern MJ, Meffin PJ (1988) The disposition of ketoprofen enantiomers in man. Br J Clin Pharmacol 26:765–770

Sattari S, Jamali F (1994) Evidence of absorption rate dependency of ibuprofen inversion in the rat. Chirality 6:435–439

Scheuerer S, Hall SD, Williams KM, Geisslinger G (1998a) Effect of clofibrate on the chiral inversion of ibuprofen in healthy volunteers. Clin Pharmacol Ther 64:168–176

Scheuerer S, Williams KM, Brugger R, McLachlan AJ, Brune K, Day RO, Geisslinger G (1998b) Effect of clofibrate on the chiral disposition of ibuprofen in rats. J Pharmacol Exp Ther 284:1132–1138

Schmitz W, Helander HM, Hiltunen JK, Conzelmann E (1997) Molecular cloning of cDNA species for rat and mouse liver alpha-methylacyl-CoA racemases. Biochem J 326:883–889

Sevoz C, Benoit E, Buronfosse T (2000) Thioesterification of 2-arylpropionic acids by recombinant acyl-coenzyme A synthetases (ACS1 and ACS2). Drug Metab Dispos 28:398–402

Shieh WR, Chen CS (1993) Purification and characterization of novel "2-arylpropionyl-CoA epimerases" from rat liver cytosol and mitochondria. J Biol Chem 268:3487–3493

Simmonds RG, Woodage TJ, Duff SM, Green JN (1980) Stereospecific inversion of (*R*)-(−)-benoxaprofen in rat and man. Eur J Drug Metab Pharmacokinet 5:169–172

Singh NN, Jamali F, Pasutto FM, Russell AS, Coutts RT, Drader KS (1986) Pharmacokinetics of the enantiomers of tiaprofenic acid in humans. J Pharm Sci 75:439–442

Stoltenborg JK, Puglisi CV, Rubio F, Vane FM (1981) High-performance liquid chromatographic determination of stereoselective disposition of carprofen in humans. J Pharm Sci 70:1207–1212

Suzuki H, Kawarabayasi Y, Kondo J, Abe T, Nishikawa K, Kimura S, Hashimoto T, Yamamoto T (1990) Structure and regulation of rat long-chain acyl-CoA synthetase. J Biol Chem 265:8681–8685

Suzuki H, Watanabe M, Fujino T, Yamamoto T (1995) Multiple promoters in rat acyl-CoA synthetase gene mediate differential expression of multiple transcripts with 5′-end heterogeneity. J Biol Chem 270:9676–9682

Tamassia V, Jannuzzo MG, Moro E, Stegnjaich S, Groppi W, Nicolis FB (1984) Pharmacokinetics of the enantiomers of indoprofen in man. Int J Clin Pharmacol Res 4:223–230

Tamura S, Kuzuna S, Kawai K (1981a) Inhibition of prostaglandin biosynthesis by clidanac and related compounds: structural and conformational requirements for PG synthetase inhibition. J Pharm Pharmacol 33:29–32

Tamura S, Kuzuna S, Kawai K, Kishimoto S (1981b) Optical isomerization of R(−)-clidanac to the biologically active S(+)-isomer in guinea-pigs. J Pharm Pharmacol 33:701–706

Tegeder I, Niederberger E, Israr E, Guhring H, Brune K, Euchenhofer C, Grosch S, Geisslinger G (2001) Inhibition of NF-{kappa}B and AP-1 activation by R- and *S*-flurbiprofen. Faseb J 15:2–4

Tracy TS, Hall SD (1991) Determination of the epimeric composition of ibuprofenyl-CoA. Anal Biochem 195:24–29

Tracy TS, Wirthwein DP, Hall SD (1993) Metabolic inversion of (*R*)-ibuprofen. Formation of ibuprofenyl-coenzyme A. Drug Metab Dispos 21:114–120

Wechter WJ, Kantoci D, Murray ED, Jr Quiggle DD, Leipold DD, Gibson KM, McCracken JD (1997) R-flurbiprofen chemoprevention and treatment of intestinal adenomas in the APC(Min)/+ mouse model: implications for prophylaxis and treatment of colon cancer. Cancer Res 57:4316–4324

Wechter WJ, Loughhead DG, Reischer RJ, VanGiessen GJ, Kaiser DG (1974) Enzymatic inversion at saturated carbon: nature and mechanism of the inversion of R(−) *p*-iso-butyl hydratropic acid. Biochem Biophys Res Commun 61:833–837

Williams K, Day R, Knihinicki R, Duffield A (1986) The stereoselective uptake of ibuprofen enantiomers into adipose tissue. Biochem Pharmacol 35:3403–3405

Zhao B, Geisslinger G, Hall I, Day RO, Williams KM (1992) The effect of the enantiomers of ibuprofen and flurbiprofen on the beta-oxidation of palmitate in the rat. Chirality 4:137–141

CHAPTER 15

Stereoselective Renal Elimination

C.M. Brett, R.J. Ott, and K.M. Giacomini

A. Introduction

Recent progress has been made in understanding the role of membrane transporters in drug absorption and elimination. In particular, we are gaining an understanding of membrane transporters in drug elimination by both the kidney and the liver. In the past 5 years, many membrane transporters relevant to drug elimination have been cloned, sequenced and their functional characteristics in heterologous expression systems have been evaluated. The emerging paradigm is that like enzymes involved in drug metabolism, there are multiple and redundant transporters that may be responsible for the elimination of particular drugs.

The field of stereoselective drug disposition has grown largely around the area of stereoselective drug metabolism. In 1985, the first example of stereoselective renal elimination of a drug was published (Hsyu and Giacomini 1985); however, the mechanisms responsible for the stereoselective elimination were not identified. With the cloning, identification, and characterization of multiple membrane transporters involved in drug elimination, it is now possible to examine interactions of enantiomeric compounds with transporters involved in renal elimination of drugs and to ascertain mechanisms that may be responsible for stereoselective renal elimination. In this review, we first provide an overview of renal handling of compounds (Sect. B.), followed by a brief discussion of transporters in the kidney that may play a role in drug elimination (Sect. C.). In Sect. D., we focus on examples of stereoselective interactions of drugs with membrane transporters in isolated renal preparations or in whole animals. Finally, in Sect. E., we present clinical examples of stereoselective drug elimination by the kidney.

B. Overview of Renal Handling: Filtration, Secretion, Reabsorption

Ensuring minimal fluctuation in electrolyte and water content, conserving critical nutrients such as amino acids and sugars, and eliminating endogenous metabolic wastes and exogenous toxins or xenobiotics summarize the critical role of the kidney in maintaining physiologic homeostasis. In part, the integration of these complex and diverse functions is dependent on tightly regulated transport processes at the basolateral and apical surfaces of the renal tubules. Secretion and reabsorption of compounds are the physiologic effect of net transport at the tubular membrane. For any compound, the net balance of three processes determines the rate of renal excretion: filtration at the glomeruli, tubular reabsorption, and tubular secretion. To discuss these three functions, the anatomy of the kidney must be understood.

Microscopic examination of renal tissue reveals that the functional unit, the nephron, consists of two main structures: the glomerulus and tubules. The glomerulus is composed of interconnected capillary loops inside a space called Bowman's capsule. The structure of the glomerulus surrounded by Bowman's capsule is ideal for the primary process of filtration. The blood in the capillary network is separated from the Bowman's space by a series of thin membranes (vascular endothelium, basement membrane, capsular epithelium), but the glomerular membrane is 100 to 500 times as permeable as the usual capillary. Macromolecules (up to 5 kDa) are easily filtered through the glomerulus.

Protein-bound substances of any size (e.g., protein-bound calcium) are not filtered. Filtration at the glomerulus results in a cell-free, protein-free filtrate of plasma, which enters the Bowman's capsule and eventually the renal proximal tubule. This "bulk flow" filtrate of the blood exiting the glomerulus is drastically different from the urine, which results after further "processing" along the tubules, the other main structure of the nephron. The volume of filtration per time is called the glomerular filtration rate (GFR). In addition to the net filtration pressure, GFR is dependent on the surface area and the permeability of the glomeruli. In a normal 70-kg human, the GFR is 180 l/day or 125 ml/min; urine flow is 1 ml/min.

Approximately 20% of the plasma that enters the glomerulus via the afferent arteriole filters from capillaries into Bowman's space. From there, the glomerular filtrate passes into a series of tubules (proximal, descending and ascending Henle's loop, distal, and collecting-ducts), each with unique anatomy reflecting function at that site. The 80% of the plasma which enters the glomerulus but is not filtered into Bowman's space proceeds into the efferent arterioles, which eventually become the peritubular capillaries that are associated with the tubules. Since the tubules are closely associated with the peritubular capillaries, substances can be transferred from the tubule to the capillary (reabsorption) and from the capillary to the tubule (secretion). The processes of reabsorption and secretion occur along the tubular epithelia. The peritubular capillaries are highly permeable compared to capillaries in other areas of the body, so that flow from the interstitium is rapid.

The tubules concentrate the glomerular filtrate to avoid water, electrolyte, and nutrient loss. The large volume of the initial filtrate is reduced by 99% via activity of various channels and transporters along the renal tubules. The processes of reabsorption and secretion are selective systems, which is in marked contrast to filtration, a bulk flow phenomenon where all material that is filtered moves together. Tubular epithelial cells are much less penetrable than glomerular epithelial cells. Although in some cases, molecules move across the epithelial barrier via a paracellular route, even the "leaky" tight junctions of the proximal tubule generally prohibit bulk flow movement between cells. Thus, the primary route across these epithelial barriers is transcellular. The transport mechanisms of the renal epithelium are similar to those of any epithelial cell: diffusion, facilitated diffusion, primary active transport, and secondary active transport.

Transepithelial renal reabsorption (movement from tubular fluid into blood) is a selective "salvage" pathway during which a substance must cross the luminal membrane (apical membrane) and then the basolateral membrane. After these two steps, the substance is in the interstitial space and the final phase of reabsorption involves movement into the capillary, which is generally by bulk flow or diffusion. In contrast, secretion moves a substance in the opposite direction – from the blood into the tubular space. In this case, a substance must cross the basolateral membrane to enter the epithelial cell and then exit via the apical membrane. In both reabsorption and secretion, net movement is possible because of differences in the transport characteristics of the two membranes of the epithelial cell.

The excretion rate and clearance of a compound can be summarized by the following equations:

$$\text{renal excretion rate} = \text{rate of filtration} + \text{rate of secretion} - \text{rate of reabsorption}$$

$$\text{renal clearance} = \frac{\text{renal excretion rate}}{\text{plasma concentration}}$$

$$\text{renal clearance} = \frac{\text{rate of filtration} + \text{rate of secretion-rate of re-absorption}}{\text{plasma concentration}}$$

The rate of filtration is equal to GFR*fu*C, where fu is the unbound fraction of a substance, and C is the plasma concentration. GRF can be determined in a number of ways. For example, the renal clearance of inulin is equal to GFR, since this exogenous polysaccharide is not protein bound, is not reabsorbed, secreted, or metabolized. By comparing the clearance of a substance to its clearance by filtration, one can determine if net reabsorption or secretion is occurring. If the clearance is greater than fu*GFR, then net secretion occurs; if clearance is less than fu*GFR, net reabsorption occurs. Since 99% of the water of the glomerular filtrate is reabsorbed, a substance that is not reabsorbed at any site along the tubule is concentrated 99-fold. Similarly, a substance that is highly reabsorbed will decrease in concentration as it flows from proximal to convoluted tubule.

The absorptive and secretary functions vary from site to site along the renal tubule and correlate well with differences in anatomic characteristics. The proximal tubule has an extensive apical membrane (brush border) and

Table 1. Physiologic parameters in the kidney

Site	ml/min flow		
Glomerulus	125		
Loops of Henle	45		
Distal tubules	25		
Collecting tubules	12		
Urine	1		
	Amount Filtered/day	% Excreted	Amount Reabsorbed
Water (l)	180	1.8	99.0
Sodium (g)	630	3.2	99.5
Glucose (g)	180	0	100

many mitochondria, consistent with rapid and active membrane transport. About 65% of the glomerular filtrate is reabsorbed at this site. In particular, organic nutrients such as glucose and amino acids are reabsorbed along the proximal tubule. Typically, these nutrients are reabsorbed against a concentration gradient at the luminal membrane by a secondary active process. Movement across the basolateral membrane is via facilitated diffusion. Similarly, the proximal tubule secretes both endogenous and exogenous organic cations (e.g., choline, creatinine, dopamine, histamine, serotonin, atropine, cimetidine, and quinine) and anions (e.g., bile salts, fatty acids, prostaglandins, penicillin, salicylates, and chlorothiazide). Many of these substances are also filtered at the glomerular membrane. If a substance is protein bound, glomerular filtration is limited and secretion is the critical mechanism for excretion. Because water and endogenous organic nutrients are reabsorbed and/or secreted and exogenous organic cations and anions are secreted at the proximal tubule, stereoselective aspects of transport are most relevant at this site (Table 1).

C. Transporters Involved in Active Secretion and Reabsorption

A number of membrane transporters are involved in active secretion and/or reabsorption of clinically used drugs. These transporters are generally located in the proximal tubule of the kidney and may be sorted to either the basolateral or brush border membrane. Cellular models of secretory or reabsorptive flux require two distinct types of transporters, one localized to the basolateral membrane and one to the brush border membrane. These transporters work in concert to mediate transepithelial flux of compounds in either the reabsorptive or secretory direction (Fig. 1).

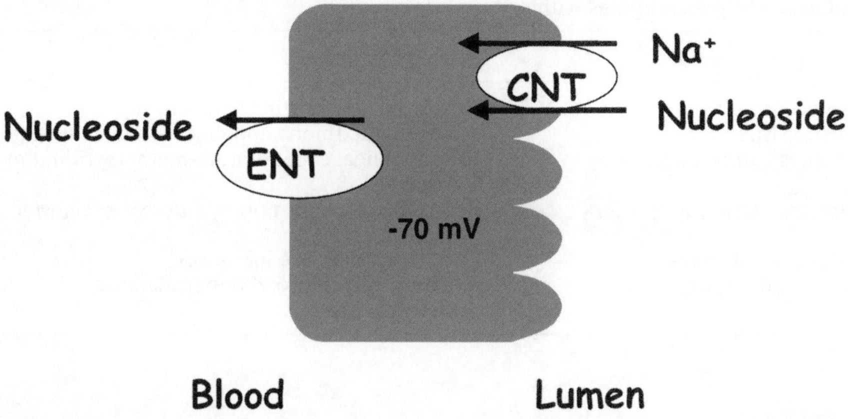

Fig. 1. Model of reabsorptive flux of nucleoside transport in a proximal tubule epithelial cell. A concentrative (Na^+-coupled) nucleoside transporter is on the apical (lumen facing) membrane and an equilibrative nucleoside transporter is on the basolateral membrane.

Transporters involved in active secretion, termed xenobiotic transporters in this chapter, generally mediate the transport of structurally diverse molecules. In contrast, transporters involved in reabsorption of drugs are generally nutrient transporters that translocate essential nutrients or structural analogs of nutrients (e.g., oligopeptides or nucleosides). Below we provide a description of six membrane transporter families that are involved in secretion or reabsorption. Along with the general characteristics of the transporters, we provide information on the tissue distribution and the molecular properties. Representative drugs that are substrates of each of the transporters are shown in Table 2.

I. P-Glycoprotein or MDR1

Multidrug resistance protein or P-glycoprotein (also termed MDR1 protein or P-gp) is a membrane transporter which is overexpressed in cancer cells and confers resistance to a variety of structurally diverse anticancer drugs (Kartner et al. 1985). The MDR genes are members of the larger family of transport proteins known as ABC (ATP-Binding Cassette) transporters (Germann 1996; Schinkel 1997).

1. Tissue Distribution

P-gp is widely expressed in normal and cancerous tissues with the highest levels in tissues with excretory function, consistent with its proposed role in the protection of cells against the toxic effects of xenobiotics (Schinkel 1997; Hall et al. 1999; Wacher et al. 1998). In the kidney, P-gp is localized to the

Table 2. Drug interactions with selected transporter

Transporter Family	Selected Substrates
P-glycoprotein	Vinblastine, vincristine, digoxin
Mrp1, Mrp2	Anionic glutathione and glucuronide conjugates
Organic cation transporters	Procainamide, cimetidine, *N*-methylpyridinium (Mpp+)
Organic anion transporters	Penicillins, cephalosporins, dideoxy-nucleoside analogs
Oligopeptide transporters	Cephalosporins, ace inhibitors
Nucleoside transporters	Adenosine, 2-chloro-2′-deoxyadenosine, 5-fluorouridine

Table 3. Molecular characteristics of multidrug resistance proteins

Transporter Family	MDR1 ABC transporter	MRP1 ABC transporter	MRP2 ABC transporter
Protein length	1279 a.a.	1531 a.a.	1545 a.a.
Predicted TMD[a]	12	11 + 6	11 + 6

[a] Transmembrane domains.

brush border membrane of the proximal tubule and acts to move drugs from cell to tubule lumen.

2. Molecular Characteristics

The molecular characteristics of MDR1 are described in Table 3.

II. Multidrug Resistance Associated Proteins (MRP1 and MRP2)

The MRP proteins are also members of the ABC protein family and are associated with resistance of cancer cells to multiple drugs. Like P-gp, MRP1 and MRP2 are also ATP-dependent drug transporters with broad substrate selectivity. MRP1 functions as the glutathione *S*-conjugate pump and may transport sulfate and glucuronide conjugates (COLE and DEELEY 1998). MRP2 or cMOAT was first described in mutant rats defective in the excretion of glucuronide, glutathione, and sulfate conjugates (JANSEN et al. 1985; TAKIKAWA et al. 1991).

1. Tissue Distribution

MRP1 is ubiquitously expressed in major organ systems including the kidney; MRP2 is expressed primarily in the liver (COLE and DEELEY 1998).

2. Molecular Characteristics of MRP1 and MRP2

The molecular characteristics of MRP2 are described in Table 3.

III. Organic Cation Transporters

Organic cation transporters in kidney, liver, and intestine secrete a variety of xenobiotics including many clinically used drugs (ZHANG et al. 1998; DRESSER et al. 1999). The transporters, OCT1, OCT2 and OCT3, mediate the uptake of a wide array of structurally diverse, hydrophilic organic cations including a variety of drugs, endogenous amines and foreign substances (ZHANG et al. 1997, 1998; GRUNDEMANN et al. 1998; DRESSER et al. 2000). Although there is considerable overlap in the substrate selectivities of OCT1, OCT2, and OCT3, recent studies suggest that there are distinct differences in the substrate and inhibitor selectivities of the transporters (DRESSER et al. 2000). More recently, OCTN1 and OCTN2 have been identified as members of the OCT superfamily (TAMAI et al. 1997, 1998; WU et al. 1998). In general, the OCT1–3 transporters appear to be localized to the basolateral membrane of epithelial cells and function in series with brush border membrane transporters to mediate the transepithelial secretion of organic cations (MEYER-WENTRUP et al. 1998; URAKAMI et al. 1998).

1. Tissue Distribution

In human tissues, OCT1 is localized primarily in liver; OCT2 is found primarily in kidney but is also present in brain and placenta; and OCT3 is more broadly distributed in intestine, kidney, placenta, brain, and liver (ZHANG et al. 1998; GRUNDEMANN et al. 1998; GORBOULEV et al. 1997). Functional evidence and studies with antibodies suggest that the transporters are localized to the basolateral membrane.

2. Molecular Characteristics

The molecular characteristics of OCT1–3 are described in Table 4.

IV. Organic Anion Transporters

Multiple organic anion transport mechanisms have been identified in the basolateral and brush border membranes of the kidney (PRITCHARD and MILLER 1993, 1996). Organic anion transporters are involved in renal and hepatic secretion of a diverse array of organic anions including many clinically used drugs. The prototypic substrate, is para-aminohippurate and the prototypic inhibitor is probenecid. The organic anion transporters which have been are either members of the OAT or OATP/OAT-K gene family. Organic anion

Table 4. Molecular characteristics of organic cation transporters (OCT)

Transporter Family	hOCT1 OCT, SLC22A1[a]	hOCT2 OCT, SLC22A2	hOCT3 OCT, SLC22A3
Protein Length	554	555	
Predicted TMD[b]	12	12	12

[a] SLC is solute carrier family.
[b] Transmembrane domains.

Table 5. Molecular characteristics of organic anion transporters

Transporter Family	hOAT1 OAT	rOAT2 OAT	rOAT3 OAT	rOATK-1 OATP	rOATK-2 OATP/OATK	hOATP1 OATP/OATK	rOATP2 OATP/OATK	rOATP3 OATP/OATK
Protein length		?	536 a.a.	669 a.a.	498 a.a.	670 a.a.	661 a.a.	670 a.a.
Predicted TMD[a]	7–12	?	12	12	8	10–12	12	12

[a] Transmembrane domains.

transporters have been cloned from rat primarily; however, some human homologs (namely, OAT1 and OATP) have been cloned.

1. Tissue Distribution

Organic anion transporters are expressed primarily in one or more epithelial tissues such as kidney, liver, and intestine, where they play roles in drug absorption, distribution, and elimination.

2. Molecular Characteristics

The molecular characteristics of the organic anion transporters are shown in Table 5.

V. Nucleoside Transporters

The two major classes of nucleoside transporters are equilibrative and concentrative transporters (BELT et al. 1993; CASS 1995; GRIFFITH and JARVIS 1996). The equilibrative nucleoside transporters (ENT1 and ENT2) mediate the facilitated diffusion of nucleosides across the plasma membrane. This class of transporters is broadly selective, accepting both purine and pyrimidine nucleosides and a variety of nucleoside analogs (GRIFFITHS et al. 1997; CRAWFORD et al. 1998). ENT1 is inhibited by nanomolar concentrations of the thiopurine, NBMPR, whereas ENT2 is not. The concentrative nucleoside transporters (CNT) mediate the active transport of nucleosides into the cell by coupling to the Na^+-gradient (BELT et al. 1993; CASS 1995; WANG et al. 1997a). Two major

Table 6. Molecular characteristics of nucleoside transporters

Transporter Family	cNT1 CNT SLC28A1[a]	SPNT1 CNT SLC28A2	ENT1 ENT	ENT2 ENT
Protein length	650 a.a.	659 a.a.	456 a.a.	456 a.a.
Predicted TMD[b]	12–14	12–14	11	11

[a] Solute carrier family 28.
[b] Transmembrane domains.

subtypes of concentrative nucleoside transporters belonging to the CNT family have been characterized and cloned, SPNT (or cNT2) and cNT1 (Wang et al. 1997; Huang et al. 1994; Che et al. 1995; Ritzel et al. 1997). SPNT is mainly purine selective but also accepts uridine, whereas cNT1 is mainly pyrimidine selective but also accepts adenosine.

1. Tissue Distribution

The equilibrative nucleoside transporters are widely distributed in most tissues (Belt et al. 1993; Griffith and Jarvis 1996; Crawford et al. 1998). Based on Northern analysis, cNT1 appears to be highly localized to the renal and intestinal epithelium (Huang et al. 1994; Ritzel et al. 1997) whereas SPNT appears to be distributed in a variety of human tissues including kidney, heart, liver, intestine, skeletal muscle, and pancreas (Wang et al. 1997). Functional studies suggest that concentrative nucleoside transporters are localized to the brush border membrane in the renal tubules although to date, studies with specific antibodies are lacking.

2. Molecular Characteristics

The molecular characteristics of nucleoside transporters are shown in Table 6.

VI. Oligopeptide Transporters

The proton-dependent oligopeptide transporters (POT or PTR or H^+/dipeptide transporters) move peptides and diverse peptide-like drugs across membranes and usually function in the absorptive direction. This class of secondary active transporters uses the pH gradient to drive small peptides into cells against their concentration gradient. hPepT1 is thought to play a critical role in oral bioavailability of peptide-like drugs (Liang et al. 1995; Bretschneider et al. 1999). hPepT2 is thought to play roles in renal reabsorption (Liu et al. 1995). Substrates for both transporters include drugs such as β-lactam antibiotics, antineoplastic agents, and ACE inhibitors (Bretschneider et al. 1999; Amidon and Lee 1994).

Table 7. Molecular characteristics of oligopeptide transporters

Transporter	hPepT1	hPepT2
Family	POT SLC15A1*[a]	POT SLC15A2
Protein length	708 a.a.	729 a.a.
Predicted TMD[b]	12	12

[a] Solute carrier family 15.
[b] Transmembrane domains.

1. Tissue Distribution

In normal human tissues, hPepT1 is localized primarily in intestine and hPepT2 in kidney.

2. Molecular Characteristics

The molecular characteristics of oligopeptide transporters are shown in Table 7.

D. Stereoselective Interactions of Drugs with Transporters in the Kidney

Stereoselective elimination of a compound can result from its specific interaction with a transporter at either the apical or basolateral membrane, from the process of metabolism intracellularly, or at the step of protein binding. This discussion focuses on the stereoselectivity that is evident at the level of the transporters, although examples of stereoselective renal metabolism are presented. In some cases the interactions have been specifically studied using a cloned transporter in a heterologous expression system. In other cases, however, the interactions have been studied in cells or tissues containing multiple and redundant transporters. In such cases, it is difficult to attribute the stereoselective interaction to a particular transporter.

I. Stereoselective Interactions with Organic Cation Transporters

The secretion of organic cations is common, since many therapeutic agents (e.g., cimetidine, procainamide, quinine, quinidine, and pindolol) are members of this chemical class. Stereoselective renal secretion of organic cations has widespread pharmacologic implications related to maintaining effective concentrations of a drug as well as avoiding toxicity.

To date, multiple organic cation transporters have been identified, characterized, and cloned. The driving forces, substrate specificity, and tissue distribution vary among these transporters (see Sect. C.). By reviewing the data related to the renal secretion of several specific drugs, the role of stereoselectivity in secretion via the organic cation transporters will become apparent.

1. Pindolol

The renal clearance of pindolol is stereoselective in humans (see below); however, in vivo renal clearance studies cannot pinpoint the site of stereoselectivity that occurs at the level of a transporter (e.g., brush border or basolateral membrane) or the transporter per se. To address this, studies of the inhibition of uptake of a nonmetabolized organic cation, *N*-methylnicotinamide (NMN) by pindolol, were performed in renal brush border membrane vesicles (BBMV) of rabbits (OTT et al. 1991). Both enantiomers inhibited NMN transport, but the IC_{50}s were similar ($140 \pm 20\,\mu M$, (*R*)-(+)-pindolol; $120 \pm 20\,\mu M$, (*S*)-(−)-pindolol). In BBMV from rats, similar results were found when effects of both enantiomers of pindolol ($90\,\mu M$) on the transport of tetraethylammonium (TEA) were studied ($52.3 \pm 7.3\%$ of control, (*R*)-(+)-pindolol; $60.5 \pm 14.9\%$ of control, (*S*)-(−)-pindolol) (GROSS and SOMOGYI 1994). Modest differences were observed in a study of TEA transport in opossum kidney (OK) cells, which showed that the IC_{50} for (*R*)-(+)-pindolol ($30 \pm 4\,\mu M$) was somewhat greater than for (*S*)-(−)-pindolol ($23 \pm 4\,\mu M$) (OTT and GIACOMINI 1993).

2. Quinine and Quinidine

The diastereomers quinine and quinidine are both secreted in the kidney (clearance greater than inulin clearance) and exhibit stereoselective renal clearance in humans (see below). In studies using BBMV from rabbit (OTT et al. 1991) and rat (GROSS and SOMOGYI 1994), stereoselective inhibition of model organic cation transport was not observed for quinine and quinidine. In contrast, in BBMV from dogs, quinidine ($K_I = 0.7\,\mu M$) was found to be a more potent inhibitor of NMN uptake compared to quinine ($K_I = 7.0\,\mu M$) (BENDAYAN et al. 1990). In contrast to these data in BBMV, several reports provide support that quinine is a more potent inhibitor of organic cation transport than is quinidine. For example, in renal cortical slices from either rats or humans, quinine [$K_I = 261 \pm 44\,\mu M$ (human); $288 \pm 21\,\mu M$, (rat)] inhibited amantadine transport more than quinidine did [$K_I = 586 \pm 68\,\mu M$ (human); $861 \pm 79\,\mu M$, (rat)]. Data in cortical slices reflect basolateral membrane transport. In the proximal tubule, quinine (female: $IC_{50} = 45 \pm 9\,\mu M$; male: $42 \pm 7\,\mu M$) more potently inhibited amantadine, compared to quinidine (female: $IC_{50} = 95 \pm 6\,\mu M$; male: $112 \pm 8\,\mu M$) (WONG et al. 1993). No differences were identified in male versus female rats. In the distal tubule, no difference in IC_{50} was identified (quinine, female/male: $97 \pm 8/105 \pm 10\,\mu M$; quinidine, $99 \pm 6/92 \pm 10\,\mu M$) and no gender differences were apparent. Collectively, the data with quinidine and quinine are difficult to simplify. Namely, the studies were carried out in multiple tissue preparations from various species and most likely reflect interactions of the diastereomers with various organic cation transporters. Studies with cloned transporters in heterologous expression systems will be useful to determine which isoform of OCT exhibits stereoselectivity for quinine and quinidine.

3. Other Organic Cation Drugs

a) NS-49

[*R*-(−)3′-(2-amino-1-hydroxyethyl)-4′fluoromethanesulphonanilide hydrochloride)] is an α_{1A}-adrenoceptor agonist that was developed to treat stress incontinence, since its cardiovascular effects are minimal at the same time that it increases intraurethral pressure. The renal clearance of N*S*-49, the R-enantiomer, and PNO-49C, the *S*-enantiomer, were both approximately threefold greater than GFR during an in vivo study in rats. Inhibition of uptake of each by quinine and quinidine was measured in rat renal BBMV. Quinine interacted more potently with N*S*-49 and quinidine more potently with PNO-49C. This difference in inhibitor potency with different enantiomers is consistent with the existence of many subsystems for organic cation transport with different, but overlapping stereospecific characteristics (Mukai and Morino 1999).

b) Fluorinated Quinolone Derivatives (e.g., Ciprofloxacin, Levofloxacin)

These agents are clinically important antibiotics, known to be secreted into the urine unmetabolized. Several of these antibiotics (e.g., enoxacin, ciprofloxacin, and levofloxacin) have been shown to permeate LLC-PK_1 monolayers in the basal-to-apical (secretory) direction (Sasaya et al. 1997). This transport is saturable and inhibited by quinidine, but not TEA.

Levofloxacin is the *S*-(−) isomer of ofloxacin; the *R*-(+) isomer is pharmacologically inactive. The movement of these isomers across LLC-PK_1 cells is characteristic of quinolone derivatives (Sasaya et al. 1999). The levofloxacin flux ratio is 1.6:1 (basal-to-apical:apical-to-basal) compared to its isomer's ratio of (1.0–1.2):1. Guanidine, enoxacin, and L-arginine inhibited the basal-to apical movement of levofloxacin but not the *R*-(+) isomer. TEA and D-arginine had no effect on either compound. The permeation of the *R*-(+) isomer was unaffected by any of the compounds tested. In addition, apically-added guanidine enhanced the efflux of levofloxacin from the cells; cimetidine did not.

The transport of these fluorinated quinolone compounds is notable in that they seem to utilize a guanidine-selective transport system, which is distinct from the TEA and NMN transport mechanism. Of note, cyclosporine A, a potent modulator of P-glycoprotein, did not affect the permeation of levofloxacin, eliminating this as a mechanism of transport in the LLC-PK_1 monolayers. Of particular significance to a discussion of stereoselectivity is that the isomer of levofloxacin did not exhibit strong secretory movement, indicating a stereoselective transport difference between these two isomers. The location of the chiral ring in the fluorinated quinolones may be significant in determining their stereoselectivity. For example, the chiral ring in levofloxacin is located away from the piperazine ring whereas the chiral center of grepafloxacin, which is not stereoselectively secreted, involves the piperazine ring. Similarly, only L-arginine (not D-arginine or D- or L-lysine) had an inhibitory effect.

c) Carnitine

Na^+-dependent organic cation transport has been reported and at least two human transporters have been cloned: hOCTN1 and hOCTN2. Carnitine, a substrate of OCTN2, is important in the β-oxidation of fatty acids, and is associated with a specific drug-induced (e.g., valproate) deficiency syndrome. At least in part, the drug-induced carnitine deficiency syndrome has been attributed to an interaction at the level of transporter. In characterizing this transporter, OHASHI et al. (1999) established that OCTN2 is stereoselective. HEK293 cells (human embryonic kidney cells) were transfected with hOCTN2. The expressed hOCTN2 transported D-carnitine with lower affinity than L-carnitine (K_m = 10.9 μM vs 4.3 μM). Consistent with this, in placental choriocarcinoma cell line (JAR cells), PRASAD et al. (1996) also reported stereospecific carnitine transport, with a lower affinity noted for the D-isomer.

II. Stereoselective Interactions with Oligopeptide Transporters

The effect of structure, including stereospecificity, of β-lactam antibiotics on their affinity for the oligopeptide transporter has been addressed in at least two studies. In BBMV from rats (DANIEL and ADIBI 1993), various substitutions at three sites were systematically analyzed by measuring and comparing affinities for β-lactam antibiotics. Of relevance to this discussion on stereoselectivity, the K_i for D- and L-isomers of cephalexin and loracarbef were evaluated. For L-cephalexin, the apparent affinity (43.7 ± 1.2 μM) was higher than that of its D-isomer (63.1 ± 1.1 μM). The stereospecificity noted with loracarbef was more striking, in that the affinity of the L-isomer (198.7 ± 2.0 μM) was more than four times that determined for the D-isomer (84.1 ± 1.1 μM).

In another study, the stereospecificity of the rPEPT 2 transporter was examined in *Xenopus laevis* oocytes, which had been injected with the cRNA isolated from rabbit kidney. The transport of dipeptides by rPEPT2 required an L-alanine in the C-terminal position (BOLL et al. 1994). Peptides with D-alanine in the N-terminal position maintained some inhibitory effect on ^{3}H-cefadroxil transport. However, with D-alanine in the C-terminal position or with D-alanine in both the N- and C-terminal positions, ^{3}H-cefadroxil transport was not inhibited.

III. Stereoselective Interactions with Nucleoside Transporters

Although a vast amount of information is available about the distinct mechanisms involved in nucleoside transport, little attention has been given to the topic of stereoselectivity of this group of transporters. Gati et al. (1989) studied the transport of D- and L-adenosine in mouse red blood cells (RBCs) and L1210 cells. In the RBCs that express ENT transporters, the inward flux of D-adenosine (5 μM) was 20-fold that of L-adenosine (5 μM). L-Adenosine (100–200 μM) flux was inhibited by D-adenosine (200 μM) and by NBMPR (5 μM). In the L1210 cells which express a purine-selective concentrative nucleoside transporter, SPNT, in the presence of a sodium gradient, D-adenosine flux (2.9 $\pm$ 0.1 pmol/s/μl of cell water) occurred at a rate approximately 20-fold greater than L-adenosine flux (0.13 $\pm$ 0.01 pmol/s/μl of cell water). In the presence of 20 μM dipyridamole (inhibits equilibrative nucleoside transport), findings were similar, suggesting that the Na^+-dependent nucleoside transport is stereoselective for adenosine.

Plasmodium falciparum, the parasite that causes malaria, infects red blood cells where it maintains an asexual life cycle. When mRNA from developing *P. falciparum* parasites was injected into *Xenopus* oocytes, transport systems distinct from those of the host are expressed. Relevant to this discussion is that D-adenosine uptake was greater in the mRNA-injected oocytes than in the water-injected controls. L-adenosine did not inhibit the transport of D-adenosine, suggesting that in this system, the uptake of this purine is stereoselective (Penny et al. 1998).

IV. Stereoselective Interactions with Organic Anion Transporters

1. DBCA

Several reports suggest stereoselective transport of the organic anion, 5-dimethylsulfamoyl-6,7-dichloro-2,3-dihydrobenzofuran-2-carboxylic acid (DBCA), by organic anion transporters in the proximal tubule. DBCA is a uricosuric antihypertensive diuretic in which the (*S*)-(−)-enantiomer is responsible for the diuretic and antihypertensive activity (Nakumura et al. 1990). As with several other diuretics, DBCA is actively secreted by a proximal tubule organic anion transport system in the rat (Higaki and Nikano 1992), rabbit, dog (Nakumura et al. 1990) and monkey (Nakano and Kawahara 1992). In the monkey, the unbound drug renal clearance was 14–29 times the creatinine clearance (Nakano and Kawahara 1992; Nakano et al. 1993). In addition, the active secretion of DBCA is inhibited by probenecid (Nakumura et al. 1990).

DBCA appears to be handled stereoselectively in protein binding, hepatic extraction, metabolism (Higaki and Nikano 1992; Higaki et al. 1992), and active renal secretion at the organic anion transport system (Nakano and Kawahara 1992; Higaki et al. 1994). In the monkey (where 78% of the drug is excreted unchanged in the urine), probenecid preferentially inhibited excretion of the (*S*) enantiomer by 53% and the (*R*) enantiomer by only 14% (Nakano and Kawahara 1992; Nakano et al. 1993). The unbound renal clearance of the (*S*) enantiomer was 1.3–1.6 times that of the (*R*) enantiomer (Nakano and Kawahara 1992). Excluding the possible effects of probenecid on renal metabolism of DCBA, these data suggest that there is stereoselective renal transport of DCBA via an organic anion transporter in the kidney.

When racemic DBCA was perfused in a rat kidney, the unbound renal clearance of the (*R*) enantiomer was 3.3 times that of the (*S*) enantiomer (Higaki et al. 1994). In contrast, when the enantiomers of DBCA were perfused individually, the unbound renal clearances were not significantly different. This species difference between stereoselective renal clearance in rat and monkey may be due to species differences in renal metabolism of DBCA. Notably, the rat selectively *N*-demethylates the *R*-enantiomer, whereas there is little renal metabolism of DBCA by the monkey.

2. Ofloxacin

Ofloxacin is a clinically marketed racemic antibiotic, where the *S*-(−)-enantiomer is 8–128 times more potent than the (*R*)-(+)-enantiomer. In monkeys and humans this compound is excreted unchanged in the urine, while in rats ofloxacin is extensively metabolized. Studies in rats determined that differences in the disposition of the ofloxacin enantiomers were due to stereoselective glucuronidation. However, when the stereoselective disposition differences were examined in the monkey (which does not extensively metabolize ofloxacin), it was determined that the stereoselective differences were due to competition between the two enantiomers for renal excretion, especially for renal secretion (Okazaki et al. 1992), probably by organic anion transporters.

3. Ibuprofen

Ahn et al. (1991) demonstrated that *R*(−) ibuprofen is cleared more rapidly in the isolated perfused rat kidney than the *S*(+) enantiomer, suggesting differences in tubular transport of unchanged or conjugated drug species, differences in renal drug metabolism (formation of the glucuronide) or differences in cellular uptake.

E. Clinical Examples

I. Carbenicillin

Recently, the stereoselective disposition of carbenicillin, an organic anion marketed clinically as a racemic mixture, was studied in healthy volunteers (Itoh et al. 1993). The data demonstrated that the fraction unbound in the plasma (fu) of the (*R*)-carbenicillin was significantly higher than that of the (*S*)-carbenicillin (the ratio averaged 1.4). The ratio of the renal clearance was 1.2, which was reasonably close to the fu ratio. The investigators interpreted the differences between the two ratios as a true stereoselective renal secretion, with the (*S*)-enantiomer being secreted preferentially in the renal tubule ($Cl_{int\ sec}$ differed between the two enantiomers). Interestingly, following coadministration of probenecid, which inhibits the tubular secretion of organic anions, the ratio of renal clearances was 1.4, reflecting the stereoselective plasma protein binding of the two enantiomers. Similar data are obtained when sulbenicillin is co-administered with probenecid in humans (Itoh et al. 1998).

II. Pindolol

Pindolol, a β-adrenoreceptor blocking agent, is actively secreted (unbound renal clearance is five times greater than GFR) by the organic cation transport system. In addition, greater than 50% of a dose is excreted unchanged by the kidneys. Clinical studies have suggested that it is stereoselectively eliminated. After oral administration of 20 mg of racemic pindolol to humans, the renal clearance of (*S*)-(−)-pindolol was 240 ± 55 ml/min, significantly higher that that of (*R*)-(+)-pindolol (200 ± 51 ml/min). The net clearance by secretion of (*S*)-pindolol was 196 ± 47 ml/min (30% higher than that of (*R*)-pindolol at 157 ± 48 ml/min) (Hsyu and Giacomini 1985). Possible mechanisms responsible for this stereoselective renal clearance of pindolol appear to be stereoselective renal metabolism or renal secretion (stereoselective binding to plasma proteins was not observed). These data are consistent with a stereoselective renal elimination mechanism for pindolol, with the S enantiomer being preferentially cleared. Somogyi et al. (1992) investigated the effect of coadministration of cimetidine on the renal clearance of the two enantiomers of pindolol. In their study, the renal clearance of (*S*)-pindolol was 222 ± 66 ml/min, significantly higher that that of (*R*)-pindolol (170 + 55 ml/min). Cimetidine significantly reduced the renal clearance of both enantiomers, but reduced the renal clearance of the R enantiomer to a greater extent than the S enantiomer. Renal clearance was 155 ± 38 ml/min for (*S*)-pindolol (a reduction of 26%) and 104 ± 18 ml/min for (*R*)-pindolol (a reduction of 34%) upon coadministration with cimetidine. These studies suggest that pindolol is stereoselectively secreted in the kidney and that cimetidine has a stereoselective inhibitory effect on the renal tubular secretion of pindolol.

Other β-blockers [e.g., sotalol (Carr et al. 1992) and pafenolol (Regardh et al. 1990)] have been shown to be actively secreted by the kidney, but no evidence for renal stereoselectivity has been clearly identified.

III. Quinine and Quinidine

Although diastereoisomers, quinine, and quinidine have similar physical properties, the unbound renal clearances of these two compounds are both greater than inulin clearance, an indication of net renal tubular secretion. In clinical studies, the renal clearance of quinidine was fourfold greater than that of quinine (Notterman et al. 1986; Gaudry et al. 1993). No stereoselective differences in plasma protein binding were observed, although both are highly protein bound (>90%). The renal filtration and passive reabsorption of these two compounds should be similar since they have similar octanol-water partition coefficients and pK_a values. Therefore, stereoselective active renal secretion may be the mechanism responsible for the observed differences in the renal clearance of quinine and quinidine. However, the potential concomitant stereoselective renal metabolism of these compounds was not addressed in these studies.

The effect of coadministration of quinine and quinidine on the biliary and renal elimination of digoxin was determined (HEDMAN et al. 1990). A significant reduction was found in the steady state biliary clearance of digoxin from 134 ± 57 ml/min to 87 ± 39 ml/min during treatment with quinine and from 95 ± 24 ml/min to 55 ± 27 ml/min during treatment with quinidine. Conversely, only quinidine significantly reduced the renal clearance of digoxin (from 155 ± 26 ml/min to 100 ± 21 ml/min). Upon coadministration of quinine, the renal clearance of digoxin was unchanged (from 177 ± 40 ml/min to 185 ± 53 ml/min). This study not only confirmed the clinically documented interaction seen between quinidine and digoxin, but also attempted to quantitate the contributions of quinidine inhibition to the different digoxin clearance pathways. This study suggests that the stereoselectivity for quinidine and quinine is more pronounced in the kidney than in the liver. Further studies will need to be conducted to determine which renal transporter is the underlying mechanism of this drug–drug interaction.

IV. Metabolite of Verapamil

The kidney (MIKUS et al. 1990) actively secretes a major metabolite of verapamil, D-617. Upon coadministration of cimetidine and verapamil, the renal clearance of the *S*-D-617 isomer was significantly decreased, whereas the clearance of the *R*-D-617 metabolite was unaffected by cimetidine administration. Stereoselective renal secretion was suggested as the mechanism of this effect. However, it is not known if this metabolite is actually secreted by the organic cation transport system or by MDR1.

V. Carprofen Glucuronides

Carprofen is a nonsteroidal antiinflammatory agent in which most of the pharmacological activity resides in the (*S*)-configuration. Development of a direct stereoselective HPLC method to measure carprofen and glucuronide conjugates permitted the investigation of its disposition in humans. No significant differences between the renal clearance and metabolic clearance values of the drug enantiomers themselves were detected. However, the investigators found that the renal clearance of the (*S*)-glucuronide in each human volunteer was higher than that of the (*R*)-glucuronide. Although these metabolites are diastereoisomers, not enantiomers, these data support a stereoselective renal process in the handling of carprofen glucuronides (IWAKAWA et al. 1988). It is not known at this time which transport system within the kidney transports carprofen and the glucuronide metabolites.

References

Ahn H-Y, Jamali F, Cox SR, Kittayanond D, Smith DE (1991) Stereoselective disposition of ibuprofen enantiomers in the isolated perfused rat kidney. Pharm Res 8:1520–1524

Amidon GL, Lee HJ (1994) Absorption of peptide and peptidomimetic drug, Annu Rev Pharmacol Toxicol 34:321–341

Belt JA, et al (1993) Nucleoside transport in normal and neoplastic cells. Adv Enzyme Regul 33:235–252

Bendayan R, Sellers EM, Silverman M (1990) Inhibition kinetics of cationic drugs on N'-methylnicotinamide uptake by brush border membrane vesicles from the dog kidney cortex. Can J Physiol Pharmacol 68:467–475

Boll M, Markovich D, Weber W-M, Korte H, Daniel H, Murer H (1994) Expression cloning of a cDNA from rabbit small intestine related to proton-coupled transport of peptides, lactam antibiotics and ACE-inhibitors. Pflugers Arch 429:146–149

Bretschneider B, Brandsch M, Neubert R (1999) Intestinal transport of beta-lactam antibiotics: analysis of the affinity at the H+/peptide symporter (PEPT1), the uptake into Caco-2 cell monolayers and the transepithelial flux. Pharm Res 16:55–61

Carr RA, Pasutto FM, Forster RT (1996) Influence of cimetidine co-administration on the pharmacokinetics of sotalol enantiomers in an anaesthetized rat model: evidence supporting active renal excretion of sotolol. Biopharm Drug Disposit 17:55–69

Carr RA, Pasutto FM, Foster RT (1996) Renal, biliary, and intestinal clearance of sotalol enantiomers in rat model: evidence of intestinal exsorption. Biopharm Drug Disposit 17:725–735

Carr RA, Foster RT, Lewanczuk RZ, Hamilton PG (1992) Pharmacokinetics of sotalol enantiomers in humans. J Clin Pharmacol 32:1105–1109

Cass CE (1995) Nucleoside transport. In: Georgopapadakou NH (ed.) Drug transport in antimicrobial and anticancer chemotherapy. Marcel Dekker, New York, pp 403–451

Che M, Ortiz DF, Arias IM (1995) Primary structure and functional expression of a cDNA encoding the bile canalicular, purine-specific Na(+)-nucleoside cotransporter. J Biol Chem 270:13596–13599

Cole SP, Deeley RG (1998) Multidrug resistance mediated by the ATP-binding cassette transporter protein MRP. Bioessays 20:931–940

Crawford CR, et al (1998) Cloning of the human equilibrative, nitrobenzylmercaptopurine riboside (NBMPR)-insensitive nucleoside transporter ei by functional expression in a transport-deficient cell line. J Biol Chem 273:5288–5293

Daniel H, Adibi SA (1993) Transport of β-lactam antibiotics in kidney brush border membrane: Determinants of their affinity for the oligopeptide/H+ symporter. J Clin Invest 92:2215–2223

Dresser MJ, Zhang L, Giacomini KM (1999) Molecular and functional characteristics of cloned human organic cation transporters. In: Amidon G, Sadée W (eds) Membrane transporters as drug targets. Plenum Publishing Corporation, New York, pp 441–469

Dresser MJ, Gray AT, Giacomini KM (2000) Kinetic and selectivity differences between rodent, rabbit, and human organic cation transporters (OCT1). J Pharmacol Exp Ther 292:1146–1152

Fiset C, Philippon F, Gilbert M, Turgeon J (1993) Stereoselective disposition of (±)-sotalol at steady-state conditions. Br J Clin Pharmacol 36:75–77

Ganapathy ME, Brandsch M, Prasad PD, Ganapathy V, Leibach FH (1995) Differential recognition of β-lactam antibiotics by intestinal and renal peptide transporters, PEPT 1 and PEPT 2. J Biol Chem 270:25672–25677

Gati WP, Dagnini L, Paterson ARP (1989) Enantiomeric selectivity of adenosine transport systems in mouse erythrocytes and L1210 cells. Biochem J 263:957–960

Gaudry SE, Sitar DS, Smyth DD, McKenzie JK, Aoki FY (1993) Gender and age as factors in the inhibition of renal clearance of amantadine by quinine and quinidine. Clin Pharmacol Ther 54:23–27

Germann UA (1996) P-glycoprotein – a mediator of multidrug resistance in tumour cells. Eur J Cancer 32 A:927–44

Gorboulev V, et al (1997) Cloning and characterization of two human polyspecific organic cation transporters. DNA Cell Biol 16:871–881

Griffith DA, Jarvis SM (1996) Nucleoside and nucleobase transport systems of mammalian cells. Biochim Biophys Acta 1286:153–181

Griffiths M, et al (1997) Cloning of a human nucleoside transporter implicated in the cellular uptake of adenosine and chemotherapeutic drugs. Nat Med 3:89–93

Gross AS, Somogyi AA (1994) Interaction of the stereoisomers of basic drugs with the uptake of tetraethylammonium by rat renal brush-border membrane vesicles. J Pharmacol Exp Ther 268:1073–1080

Grundemann D, et al (1998) Molecular identification of the corticosterone-sensitive extraneuronal catecholamine transporter. Nat Neurosci 1:349–351

Hall SD, et al (1999) Molecular and physical mechanisms of first-pass extraction. Drug Metab Dispos 27:161–166

Hedman A, Angelin B, Arvidsson A, Dahlqvist R, Nilsson B (1990) Interactions in the renal and biliary elimination of digoxin: Stereoselective difference between quinine and quinidine. Clin Pharmacol Ther 47:20–26

Higaki K, Nakano M (1992) Stereoselective disposition of *S*-8666, a novel uricosuric antihypertensive diuretic and its *N*-monodemethylated metabolite in a perfused rat liver preparation: Effect of protein binding on the kinetics of *S*-8666. Drug Metab Dispos 20:350–355

Higaki K, Kadono K, Goto S, Nakano M (1994) Stereoselective renal tubular secretion of an organic anion in the isolated perfused rat kidney. J Pharmacol Exp Ther 270:329–335

Higaki K, Kadono K, Nakano M (1992) Stereoselective pharmacokinetics a novel uricosuric antihypertensive diuretic in rats: Pharmacokinetic interactions between enantiomers. J Pharm Sci 81:935–939

Hsyu P-H, Giacomini KM (1985) Stereoselective renal clearance of pindolol in humans. J Clin Invest 76:1720–1726

Hsyu P-H, Giacomini KM (1987) The pH gradient-dependent transport of organic cations in the renal brush border membrane. J Biol Chem 262:3964–3968

Huang QQ, et al (1994) Cloning and functional expression of a complementary DNA encoding a mammalian nucleoside transport protein. J Biol Chem 269:17757–17760

Iwakawa S, Suganuma T, Lee *S*-F, Spahn H, Benet LZ, Lin ET (1988) Direct determination of diastereomeric carprofen glucuronides in human plasma and urine and preliminary measurements of stereoselective metabolic and renal elimination after oral administration of carprofen in man. Drug Metab and Dispos 17:414–419

Itoh T, Ishida M, Onuki Y, Tsuda Y, Shimada H, Yamada H (1993) Stereoselective renal tubular secretion of carbenicillin. Antimicrob Agents Chemother 37:2327–2332

Itoh T, Watanabe N, Ishida M, Tsuda Y, Koyano S, Tsunoi T, Shimada H, Yamada H (1998) Stereoselective disposition of sulbenicillin in humans. Antimicrob Agents Chemother 42:325–331

Jansen P, Peters WH, Lamers WH (1985) Hereditary chronic conjugated hyperbilirubinemia in mutant rats caused by defective hepatic anion transport. Hepatology 5:573–579

Kartner N, et al (1985) Detection of P-glycoprotein in multidrug-resistant cell lines by monoclonal antibodies. Nature 316:820–823

Liang R, et al (1995) Human intestinal H^+/peptide cotransporter: Cloning, functional expression, and chromosomal localization. J Biol Chem 270:6456–6463

Liu W, et al (1995) Molecular cloning of PEPT 2, a new member of the H^+/peptide cotransporter family, from human kidney. Biochim Biophys Acta 1235:461–466

Meyer-Wentrup F, et al (1998) Membrane localization of the electrogenic cation transporter rOCT1 in rat liver. Biochem Biophys Res Commun 248:673–678
Mikus G, Eichelbaum M, Fisher C, Gumulka S, Klotx U, Kroemer HK (1990) Interactions of verapamil and cimetidine: Stereoselective aspects of drug metabolism, drug disposition and drug action. J Pharmacol Exp Ther 253:1042–1048
Mukai H, Morino A (1999) Renal excretion mechanism of N*S*-49, a phenylethylamine class alpha 1A-adrenoceptor agonist. Biopharm Drug Dispos 20:207–215
Nakano M, Kawahara S (1992) Stereoselective renal tubular secretion of a new uricosuric diuretic, 6,7-dichloro-5-(N, *N*-dimethylsulfamoyl)-2,3-dihydro-2-benzofurancarboxylic acid (*S*-8666) in cynomolgus monkeys. Drug Metab Dispos 20:179–185
Nakano M, Higaki K, Kawahara S (1993) Enantiomer-enantiomer interaction of a uricosuric antihypertensive diuretic (DBCA) in renal tubular secretion and stereoselective inhibition by probenecid in the cynomolgus monkey. Xenobiotica 23:525–536
Nakumura M, Kawabata T, Itoh T, Muyata K, Harada H (1990) Stereoselective saluretic effect and localization of renal tubular secretion of enantiomers of *S*-8666, a novel uricosuric antihypertensive diuretic. Drug Dev Res 19:23–36
Notterman DA, Drayer DE, Metakis L, Reidenberg MM (1986) Stereoselective renal tubular secretion of quinidine and quinine. Clin Pharmacol Ther 40:511–517
Ohashi R, Tamai I, Yabuuchi H, Nezu J-I, Oku A, Sai Y, Shimane M, Tsuji A (1999) Na^+-dependent carnitine transport by organic cation transporter (OCTN2): Its pharmacological and toxicological relevance. J Pharmacol Exp Ther 291:778–784
Okazaki O, Kurata T, Hakusui H, Tachizawa H (1992) Species-related stereoselective disposition of ofloxacin in the rat, dog and monkey. Xenobiotica 22:439–450
Ott RJ, Giacomini KM (1993) Stereoselective interactions of organic cations with the organic cation transporter in OK cells. Pharm Res 10:1169–1173
Ott RJ, Hui AC, Wong FM, Hsyu P-H, Giacomini KM (1991) Interactions of quinidine and quinine and (+)- and (−)- pindolol with the organic cation/proton antiporter in renal border membrane vesicles. Biochem Pharmacol 41:142–145
Penny JI, Hall ST, Woodrow CJ, Cowan GM, Gero AM, Kreshna S (1998) Expression of substrate-specific transporters encoded *by Plasmodium falciparum* in *Xenopus laevis* oocytes. Mol Biochem Parisotol 93:81–89
Prasad PD et al (1996) Sodium-dependent carnitine transport in human placental choriocarcinoma cells. Biochim Biophys Acta 1284:109–117
Pritchard JB, Miller DS (1993) Mechanisms mediating renal secretion of organic anions and cations. Physiol Rev 73:765–796
Pritchard JB, Miller DS (1996) Renal secretion of organic anions and cations. Kidney Int 49:1649–1654
Regardh CG, Heggelund A, Kylberg-Hanssen K, Lundeborg P (1990) Pharmacokinetics of pafenolol after IV and oral administration of three separate doses of different strength to man. Biopharm Drug Dispos 11:607–617
Ritzel MW, et al (1997) Molecular cloning and functional expression of cDNAs encoding a human Na^+-nucleoside cotransporter (hCNT1). Am J Physiol 272(2 Pt 1):C707–C714
Sasaya M, Oda M, Endo T, Saitoh H, Takada M (1997) The transport of ciprofloxacin in cultured kidney epithelial cells LLC-PK1. Biol Pharm Bull 20:887–891
Sasaya M, Hatakeyama Y, Saitoh H, Takada M (1999) Stereoselective permeation of new fluorinated quinolone derivatives across LLC-PK1 cell monolayers. Biol Pharm Bull 22:707–712
Schinkel AH (1997) The physiological function of drug-transporting P-glycoproteins. Semin Cancer Biol 8:161–170
Somogyi AA, Bochner F, Sallustio BC (1992) Stereoselective inhibition of pindolol renal clearance by cimetidine in humans. Clin Pharmacol Ther 51:379–387
Takikawa H, et al (1991) Biliary excretion of bile acid conjugates in a hyperbilirubinemic mutant Sprague-Dawley rat. Hepatology 14:352–360
Tamai I, et al (1997) Cloning and characterization of a novel human p*H*-dependent organic cation transporter, OCTN1. FEBS Lett 419:107–111

Tamai I, et al (1998) Molecular and functional identification of sodium ion-dependent, high affinity human carnitine transporter OCTN2. J Biol Chem 273:20378–20382
Urakami Y, et al (1998) Functional characteristics and membrane localization of rat multispecific organic cation transporters, OCT1 and OCT2, mediating tubular secration of cationic drugs. J. Pharmacol Exp Ther 287:800–805
Vander AJ (19) In: Renal Physiology, 4th edn. McGraw-Hill, Inc, New York, pp
Wacher VJ, et al (1998) Role of P-glycoprotein and cytochrome P450 3 A in limiting oral absorption of peptides and peptidomimetics. J Pharm Sci 87:1322–1330
Wang J, et al (1997) Functional and molecular characteristics of Na(+)-dependent nucleoside transporters. Pharm Res 14:1524–1532
Wang J, et al (1997) Na(+)-dependent purine nucleoside transporter from human kidney: cloning and functional characterization. Am J Physiol 273(6 Pt 2): F1058–F1065
Wong LTY, Escobar MR, Smyth DD, Sitar DS (1993) Gender-associated differences in rat renal tubular amantadine transport and absence of stereoselective transport inhibition by quinine and quinidine in distal tubules. J Pharmacol Exp Ther 267:1440–1444
Wu, X, et al. cDNA sequence, transport function, and genomic organization of human OCTN2, a new member of the organic cation transporter family. Biochem Biophys Res Commun 246:589–595
Zhang L, et al (1997) Cloning and functional expression of a human liver organic cation transporter. Mol Pharmacol 51:913–921
Zhang L, Brett CM, Giacomini KM (1998) Role of organic cation transporters in drug absorption and elimination. Annu Rev Pharmacol Toxicol 38:431–460
Zhang L, Schaner ME, Giacomini KM (1998) Functional characterization of an organic cation transporter (hOCT1) in a transiently transfected human cell line (HeLa). J Pharmacol Exp Ther 286:354–361

Section IV
Implications for Drug Development and Therapy

CHAPTER 16

Regulatory Requirements for the Development of Chirally Active Drugs

R.R. SHAH and S.K. BRANCH

The views expressed in this chapter are those of the authors and do not necessarily represent the views or the opinions of Medicines Control Agency, other regulatory authorities or any of their advisory committees.

A. Introduction

The evolution of regulatory guidelines following advances in scientific and technical methods is best illustrated by regulation of chiral drugs. The significance of stereoselectivity in drug action is well established; for example, it was known as long ago as 1926 that the biological activity of atropine resided in only one stereoisomer (CUSHNY 1926). However, the absence of suitable methods for either the large-scale preparation of pure enantiomers or for stereoselective analysis meant that the majority of stereochemically pure drugs on the market were of only natural origin. Whether enantiomers of a chirally active drug should be developed was largely an academic question and historically, synthetic chiral drugs were mainly presented as the racemate.

I. Historical Background

The late 1980s saw the introduction of new methods for the large-scale separation or asymmetric syntheses of single enantiomers accompanied by advances in chiral analytical procedures. As the new synthetic, preparative and analytical methods evolved, there emerged a better understanding of the pharmacological significance of chirality. Stereoselectivity in pharmacokinetics and pharmacodynamics was already acknowledged by the 1970s (JENNER and TESTA 1973), but it was not until almost a decade later that the wider implications of chirality for clinical efficacy and safety of drugs began to be recognized. By late 1980s, prompted largely by many acrimonious academic debates, a stage had been reached when the issue needed to be addressed by both the pharmaceutical industry and the regulatory authorities responsible for licensing medicinal products.

Regulators and representatives from the pharmaceutical industry in the European Union (EU), Japan and the US debated issues arising from chirality in relation to quality, safety and efficacy – the three principles which form the basis for approval of a medicinal product. A concern emerged within the industry on the possible adverse regulatory attitude towards, and more stringent requirements for, the future development of racemates. An additional concern was the extent of harmonization in requirements by different regulatory authorities.

II. Regional Evolution of Regulatory Requirements

Regulatory control of chiral drugs was first considered by Sweden in the early 1980s, although they now fully subscribe to the EU guidance note. Formal guidance on the development of chiral drugs was published first in the US in 1992 and then shortly thereafter in the EU in 1994. A similar attitude prevails in Japan, although a formal guidance note has not been issued. The Canadian guidance is also based on the same principles (RAUWS and GROEN 1994).

As early as 1987, the US Food and Drug Administration (FDA) had required the sponsors to prepare individual stereoisomers and include in the New Drug Application the documentation concerning their physical, chemical, pharmacological, preclinical and clinical properties. Later in 1989, a special Stereoisomeric Committee was established to determine the requirements to be imposed formally. Entitled the *Policy Statement for the Development of New Stereoisomeric Drugs*, it was formally published in the Federal Register in January 1992 with amendments made in January 1997.

Common legislation governs the criteria for the approval of human medicines throughout the Member States of the European Union whether applications for marketing authorizations are made through the centralized or national routes. This is set out in a number of volumes of the *Rules Governing Medicinal Products in the European Community* (EUDRALEX 1998). The note for guidance of primary interest for drugs which may exist as optical isomers is set out in volume III and is entitled *Investigation of Chiral Active Substances* (CPMP/III/3501/91). Discussion on this guideline started in 1991; it was adopted in October 1993 and came into force in April 1994.

The Canadian Health Protection Branch set up a Working Group on Drug Stereochemistry in 1990 and adopted a guideline in 1994. This guideline on Stereochemical Issues in Chiral Drug Investigation was updated in 1996.

All these guidelines set out requirements for studies to justify the chosen strategy in the areas corresponding to the three technical parts (pharmaceutical, preclinical and clinical) of the dossier accompanying the application.

The Japanese Ministry of Health and Welfare (MHW) and the Pharmaceutical and Medical Safety Bureau (PSMB), responsible for the promulgation of national and international guidelines in the form of Notifications, have not issued specific guidance on the development of chiral drugs but have nonetheless responded to the "enantiomer-versus-racemate" scientific debate. The attitude of the MHW and its advisory body, the Central Pharmaceutical Affairs Council (CPAC) has been previously discussed (SHINDO and

CALDWELL 1991, 1995). Although there is a lack of formal guidance in Japan, it is apparent that there is a considerable degree of concordance with the regulatory principles established elsewhere. While the lack of official guidance on investigation of chiral drugs offers limited or no information on requirements, there is considerable discussion on individual cases either with individual applicants or pharmaceutical industry associations and the resulting ad hoc decisions assume the force of "case laws" based on precedence.

These authorities have recognized the various scenarios possible with respect to enantioselectivity in pharmacokinetics and in primary and secondary pharmacodynamics as well as the potential beneficial or adverse interactions between the two enantiomers. It is further acknowledged that (a) the common practice of developing racemates has resulted in (*only*) a few recognized adverse consequences, (b) despite the problems identified with a few racemates, the development of racemates may continue to be appropriate, especially when one enantiomer is inactive, (c) there are examples of stereoselective toxicity in which the toxicity has resided in one of the pair and (d) although some enantiomeric pairs have useful complementary therapeutic activities, there are no reasons to believe that a 50:50 mixture represents clinically the most ideal mix.

All the regulatory authorities have adopted essentially the same attitude, with slight differences in emphasis, with respect to the development of chiral drugs. It has been considered desirable to adopt a pragmatic approach to the regulation of chiral drugs based on scientific data relating to quality, safety, efficacy and risk-benefit. The decision as to whether to develop a racemate or a single enantiomer of an optically active drug is therefore left to the applicant for the marketing authorization or sponsor of the product, as long as the decision is justified.

III. Scope of Regulatory Control on Chiral Drugs

It may also be emphasized that all the pharmaceutical, preclinical and clinical regulatory requirements and guidelines (either national or supranational) which apply to nonchiral drugs continue to apply equally to chiral drugs – either racemic forms or enantiomers. The guidance notes on chirality promulgated by these authorities aim merely to highlight the features unique to, and additional documentation required for, chirally active drugs.

In addition to guidelines promulgated by national authorities, there are other guidelines developed through The International Conferences on Harmonization of Technical Requirements for the Registration of Pharmaceuticals for Human Use (ICH). This important tripartite initiative has been sponsored by regulatory authorities and research-based industry representatives from the European Union, Japan and the United States. In addition, the ICH Steering Committee includes observers from the World Health Organization (WHO), the Canadian Health Protection Branch and the European Free Trade Association (EFTA).

It is the aim of the ICH initiative to promote international harmonization of regulatory requirements. Such harmonization avoids the duplication of the development work required for registering new medicinal products, and is of importance to the pharmaceutical industry, which is becoming increasingly globalized. The ICH guidelines supersede national guidelines developed by the signatories to the ICH process. At present there is no ICH guideline specifically addressing issues directly and exclusively related to chiral drugs, but a number of safety, quality and efficacy guidelines have emerged through this process. These include references to chirality when appropriate and the following discussion on the regulatory requirements also includes the requirements referred to in the ICH guidelines.

The three important regulatory considerations in the development of chirally active drugs are early development of enantioselective assays, the possibility of interenantiomeric interactions and consequently, careful planning of bridging studies required. The latter concept applies when the development changes from racemate to a single enantiomer and vice versa in order to minimize duplication of studies. Most authorities are cautiously content to allow pharmacological/toxicological bridging studies. However, only the EU considers this possibility explicitly for clinical documentation when it is proposed to develop a single enantiomer from an approved racemate. The extent of bridging studies should be defined on a case-by-case basis.

IV. Justifying the Risk/Benefit

The recommendations from various regulatory authorities have in turn stimulated further the development of single enantiomers by the pharmaceutical industry. The progress made in this area has been such that, in retrospect, it is difficult to understand why the biological significance of chirality was not initially more widely appreciated and why there was so much controversy when the need for regulating chiral drugs was first debated. Nowadays, it is widely accepted that enantiomers should be treated as separate compounds from the point of view of their pharmacological action as lessons continue to be learned with respect to enantiospecific drug safety (SHAH et al. 1998).

The presentation of a drug as either a racemate or a single enantiomer now requires full justification before a marketing authorization can be granted, which means that studies with individual enantiomers using chiral analytical procedures are a necessity during product development even if it is the racemate that is finally marketed. The key to a successful outcome for an application for a marketing authorization is proper justification of the decisions made concerning a product during development. When there are doubts or the sponsor decides to deviate from the guidelines, it is best to discuss the issue as early as possible with the appropriate regulatory authority. Discussions between the sponsor and the regulatory authority are the norm as a way of guidance in Japan. The FDA is also happy to discuss any cases where questions arise regarding the definition of "significant toxicity" in the context of an unexpected preclinical toxicity finding and the possible role of an individual enantiomer and invites discussion with sponsors on whether to pursue development of the racemate or single enantiomer. In the EU, discussions are also possible with individual Member States in addition to requesting a formal Scientific Advice from the Committee for Proprietary Medicinal Products of the EU.

The remainder of this chapter outlines from a stereochemical perspective the regulatory requirements for the investigation and the development of chiral drugs in the three regions of Europe, the US and Japan. They aim to provide a strategic framework of considerations which should guide the development of a chiral drug. Since there is a great degree of concordance, no attempt is made to make any regional distinctions.

B. Pharmaceutical Requirements

Active substances with one or more stereogenic centers can be produced as a single enantiomer, an equimolar mixture of enantiomers (a racemate), a non-equimolar mixture of enantiomers (nonracemic mixture) or, in the case of multiple centers, a mixture of diastereoisomers. An application to market a medicinal product containing a new chemical or molecular entity must include information both on the drug substance itself and on the finished product or dosage form. For the drug substance, details of the route by which it is synthesized, evidence of its molecular structure, physicochemical properties, control by an appropriate specification, description of any potential impurities and investigation of stability must all be provided. In the case of chiral drug substances, special attention must be given to the identity and the stereoisomeric purity of the active ingredient.

The need for determination of optical purity is well illustrated by the D-thyroxine episode. Thyroxine increases metabolic rate and heart rate and suppresses the release of thyroid-stimulating hormone in addition to lowering serum cholesterol levels.

L-thyroxine is the naturally occurring stereoisomer of thyroxine. On a per microgram basis, this natural isomer (with its half-life of 7–11 days) is 15–40 times more potent as a thyromimetic (effects on metabolism and the release of thyroid-stimulating hormone) than is D-thyroxine (which has a half-life of 3–4 days). On the basis of earlier studies, D-thyroxine was assumed to be more effective in decreasing serum cholesterol levels than L-thyroxine per unit of increase in oxygen consumption and heart rate. However, the use of D-thyroxine, administered at a daily dose of 6 mg, was discontinued from the Coronary Drug Project in 1971 because it was associated with an increased frequency of deaths despite reduction in serum cholesterol and triglycerides (Coronary Drug Project Research Group 1972). Hitherto, it had been assumed that D-thyroxine was not contaminated with L-thyroxine. However, it was shown later (Young et al. 1984) that commercial D-thyroxine tablets

("Choloxin") were contaminated with 0.5%–2.3% L-thyroxine. This small degree of contamination resulted in physiologically significant doses of L-thyroxine being delivered to the patients, thus explaining the paradoxical findings from Coronary Drug Project. During treatment with two different lots of 8 mg "Choloxin," serum L-thyroxine accounted for 33%–53% of the measurable total serum thyroxine (YOUNG et al. 1984). This finding also cast doubt on previous studies of the thyromimetic and hypolipidaemic activities of D-thyroxine since the purity of the preparations used was, in retrospect, unknown. More recent studies suggest a more favorable reevaluation of the therapeutic role of D-thyroxine. D-thyroxine has been shown to be a potent hypolipidaemic agent (PAGNAN et al. 1984) with a highly favorable influence on lipid profile and only mild cardiac effects. Other workers (HAMON et al. 1993) have shown that 10 days of treatment with 24 mg of highly purified preparations of D-thyroxine ("Dynethol," contaminated with only 0.05% of L-thyroxine, equivalent to 12 μg of L-thyroxine daily) was associated with only modest increase in free thyroxine levels and decreases in serum thyroid-stimulating hormone despite profound decreases in serum cholesterol levels.

Likewise, since large doses are required, a similar contamination of the therapeutic D-penicillamine with its highly toxic L-isomer could quite easily have concealed the stereospecific toxicity of this important drug.

For the finished product containing the drug substance, information is required on the composition, the development of the formulation to be marketed, the control of its quality and the stability studies justifying the proposed shelf-life.

I. Synthesis of the Active Substance

In addition to providing the usual information concerning the manufacturing procedure, the step at which the chiral center is formed must be described in detail, and the measures taken to maintain the desired configuration during subsequent stages of synthesis must be shown. The ability of the process to provide adequate stereochemical control must be validated. The synthetic product must be fully characterized with respect to identity, related substances and other impurities as for any other drug substance, but with the additional requirement of establishing stereochemical purity.

Several synthetic strategies are possible and demand that different types of information be provided. In cases where the starting material, whether a racemate or an enantiomer, already contains the required chiral center, full characterization of that substance is required, including stereochemical purity and validation of chiral analytical procedures. Key intermediates, those compounds in which the essential molecular characteristics necessary for the desired pharmacological activity are first introduced into the structure, will often be those where a chiral center of the correct stereochemistry is introduced, and as such they should be subjected to quantitative tests to limit the content of undesired isomers.

Where a racemate or other nonracemic enantiomeric mixture is intended, evidence should be provided that these are the result, unless obvious from the synthetic route employed. Where the preferred enantiomer is obtained by isolation, the resolution step is considered part of the overall manufacturing process and the usual details of the procedure should be given, together with the number of cycles used. If a nonracemic mixture of enantiomers is needed, then the manufacturing process must be validated to ensure consistent composition of the active ingredient.

When it is not possible to obtain the required enantiomer at manufacturing scale either by synthesis or by isolation, all the experimental results available should be described and the reason for the failure given. Likewise, if enantiomeric material could not be obtained for preclinical and clinical studies (see Sect. C.), this should also be discussed.

II. Chemical Development

Proof of structure and configuration are required as part of the information on chemical development. It should be established whether a drug produced as a racemate is a true racemate or a conglomerate by investigating physical parameters such as melting point, solubility and crystal properties. The physicochemical properties of the drug substance should be characterized, e.g., crystallinity, polymorphism and rate of dissolution, as these may have a bearing on the bioavailability of the drug from the dosage form when administered to the patient.

Stereoisomeric reference substances may be required for test procedures for chiral drugs. The stereochemical purity of reference materials must be stated by giving a value for their assay determination. Care should be taken in the characterization of such materials when they are required to support the stereochemical identification of the drug substance. It is all too easy to fall into a circular argument when trying to establish the absolute configuration of a compound based on mechanistic arguments and/or the chirality of starting materials. Single crystal X-ray diffraction studies of the final drug substance with methods appropriate for the determination of absolute configuration provide the greatest confidence.

III. Quality of the Active Substance and Finished Product

The quality of a drug substance and any medicinal products containing that substance are controlled by their specifications which set out the acceptable limits for a range of tests to be carried out before the former may be used in the manufacture of the finished product or the latter released onto the market. The original European and FDA guidance notes on chiral active substances refer to the need to include stereospecific tests for identity and purity. However, there is now an internationally harmonized guideline on specifications and tests for chemical substances as active ingredients and in drug products which makes particular reference to chiral compounds.

For a drug substance, an identity test should be capable of distinguishing between the enantiomers and the racemate for a drug substance developed as a single enantiomer. A chiral assay or enantiomeric impurity procedure may serve to provide a chiral identity test. When the active ingredient is a racemate, a stereospecific test is appropriate where there is a significant possibility that substitution of an enantiomer for a racemate may occur or when preferential crystallization may lead to unintentional production of a non-racemic mixture. Such a test is generally not needed in the finished product specification if there is insignificant racemization during manufacture of the dosage form or during storage, and a test is included in the drug substance specification. If the opposite enantiomer is formed during storage, then a chiral assay or enantiomeric impurity testing will serve to identify the stereoisomer as well. The ICH guideline allows that appropriate testing of a starting material or intermediate, with suitable justification from studies conducted during development, could give assurance of control. This approach may be necessary, for example, when there are multiple chiral centers present in the drug molecule.

Limits for the other enantiomer in the finished product are needed unless racemization during manufacture of the dosage form or during storage is insignificant. The applicant should show that the manufacturing process produces no unacceptable changes in the stereochemical purity of the active ingredient, and that such changes do not occur during storage for the proposed shelf-life of the product. Determination of the drug substance content is expected to be enantioselective, and this may be achieved by including a chiral assay in the specification or an achiral assay together with appropriate methods of controlling the enantiomeric impurity. For a drug product where racemization does not occur during manufacture or storage, an achiral assay may suffice. If racemization does happen, then a chiral assay should be used or an achiral method combined with a validated procedure to control the presence of the other enantiomer.

Methods that may be used for the control of chiral drug substances range from the simpler ones such as optical rotation, melting point and chiral HPLC to the more sophisticated techniques including optical rotatory dispersion, circular dichroism or NMR with chiral shift reagents. This list is not expected to preclude the adoption of other methods or those that may be introduced in the future. It is the responsibility of applicants to decide on the techniques that are appropriate for the satisfactory control of each drug substance and to ensure that they are fully validated. Guidance on analytical validation has also been internationally harmonized.

IV. Status of Distomer as an Impurity

Stereoisomers may arise during synthesis of a drug substance or they may arise as degradation products during storage. The guidance on chiral active substances states that when a chiral drug substance is presented as a single enantiomer, the unwanted enantiomer is considered to be an impurity. The internationally harmonized guideline on impurities applies in principle to substances containing enantiomeric or diastereoisomeric impurities as it would to active ingredients containing any other organic impurities. However, the limits normally expected do not apply to chiral impurities. The limits for the control of enantiomers in drug substances are usually relaxed compared to tests using achiral methods because it is recognized that chiral analytical techniques may not be able to achieve the same sensitivity.

The use to which development batches have been put must also be detailed by the applicant so that each can be linked to a particular safety or clinical study. This information assists in the *qualification* of impurities, which is the process by which the biological safety of an individual impurity, or an impurity profile, is established at a specified level. If the new drug substance containing a particular level of impurity has been adequately tested in safety and/or clinical studies, then that level is considered to be qualified for use in the medicinal product. Impurities that are also significant human metabolites do not need further qualification as exposure to them would be automatic on administration of the drug in clinical trials. Together with the batch analysis data, the qualification studies should be used to justify the specification limits for individual known, unknown and total impurities. The applicant should demonstrate that unacceptable changes in stereochemical purity or enantiomeric ratio do not occur during storage of the active ingredient.

C. Preclinical and Clinical Requirements

In general, the preclinical and clinical sets of studies required are analogous to each other. The two sets of programs go hand in hand, and what is observed preclinically must be shown to be valid clinically by a corresponding set of clinical investigations.

The FDA policy statement focuses essentially on the development of racemates and of single enantiomers after the racemate is studied. In contrast, the EU guideline covers a number of scenarios.

Both the FDA and the EU have emphasized the need to develop chiral assays early in the drug development process for in vivo use. This enables an assessment of stereoselectivity in pharmacology, extent of chiral interconversion and comparison of preclinical and early clinical pharmacology data early in the process of drug development. These results should enable an informed decision to be made as to whether an individual enantiomer or the racemate should be developed. If the drug product is to contain a racemate and the pharmacokinetic profiles of the individual isomers are different, appropriate studies should be conducted to evaluate pharmacokinetic and pharmacodynamic characteristics such as the dose linearity, the effects of altered metabolism and excretion and drug–drug interactions for the individual enantiomers. An achiral assay or monitoring of only one enantiomer is acceptable if the pharmacokinetics of the optical isomers is the same or in a fixed ratio in the target population. Assessment of the data from toxicological studies is facilitated by in vivo measurement of individual enantiomers, but if this measurement is not possible then human pharmacokinetic studies would be sufficient.

The pharmacological activities of the isomers should be compared in vitro and in vivo in both animals and humans. Separate toxicological evaluation of the enantiomers is not usually required when the profile of the racemate is relatively benign in terms of safety. However, if unexpected effects are observed, especially if unusual or occurring near the effective doses in animals or near the potential human exposure, further studies with the individual enantiomers will be required.

I. Development of a Single Enantiomer as a New Active Substance

In some situations, development of a single enantiomer is particularly desirable (e.g., where only one of the enantiomers has a toxic or undesirable pharmacologic effect).

Two potentially good examples are baclofen and bupivacaine. (*R*)-baclofen is far the more potent as a skeletal muscle relaxant, while the (*S*)-isomer is not only toxic but also antagonizes the therapeutic effect of (*R*)-baclofen (Fromm and Terrence 1987). The cardiotoxicity of bupivacaine seems to reside essentially in its (+)-(*R*)-isomer. In healthy volunteer studies, (−)-(*S*)-bupivacaine has been shown to be clearly less cardiotoxic (Gristwood et al. 1994; McClellan and Spencer 1998). (+)-(*R*)-bupivacaine is much more toxic in terms of either the sodium channel blockade (by a factor of 1.6) (Valenzuela et al. 1995a) or potassium channel blockade (by a factor of 7) (Valenzuela et al. 1995b).

The case for developing a single enantiomer is less compelling when one is pharmacologically active while the other is inert. The development of a single enantiomer as a new active substance requires the same full preclinical documentation as any other new active substance.

All preclinical studies should be carried out with the proposed single enantiomer. The enantiomeric purity of the active substance used in preclinical studies should be defined. However, where development studies have commenced with a racemate, the studies conducted up to the decision to develop the enantiomer may be taken into account to determine the necessity for further studies.

The possibility of the formation of the other enantiomer in vivo should be considered at an early state in order to justify the need for any enantiospecific bioanalysis. If the other enantiomer is formed in vivo, it should be treated and evaluated as any other metabolite would be. In the case of endogenous human chiral compounds, enantiospecific methods may not be necessary to characterize their pharmacokinetic disposition.

A signal that should trigger further investigation of the properties of the individual enantiomers and their active metabolites is the occurrence at clinical doses of toxicity with the racemate that is not clearly expected from the pharmacology of the drug or the occurrence of any other unexpected pharmacologic effect with the racemate. These signals might be explored in animals but human testing may be essential. The FDA guideline warns that toxicity or unusual pharmacologic properties might reside not in the parent isomer, but in an isomer-specific metabolite.

The principles and requirements for clinical studies are analogous to those required for preclinical studies described above, that is, a full package of clinical development. The FDA considers that, in general, it is more important to evaluate both enantiomers clinically and consider developing only one when both enantiomers are pharmacologically active but differ significantly in potency, specificity or maximum effect, rather than when one isomer is essentially inert.

II. Development of a Racemate as a New Active Substance

The development of a new racemic drug as a new active substance requires the same full preclinical documentation as any other new active substance. The choice of the racemate instead of a single enantiomer should be explained. Where the difference observed in the pharmacodynamic activity and disposition of the enantiomers is slight, racemates may be developed. Many currently available drugs fit into this category. As stated earlier, the development of a racemate may also be appropriate when one of the enantiomers can be shown to be pharmacologically inert.

In addition, one of the following two situations which may occur should also be considered:

1. Rapid interconversion in vivo (e.g., the diuretic chlorthalidone).
   If the interconversion rate in vivo is appreciably higher than the apparent distribution and elimination rates of the enantiomers, only the racemate should be studied as the active substance.
2. Absent or slow interconversion in vivo (e.g., certain NSAIDs).
   In cases where such chiral interconversion is absent or slow, then differences in the pharmacodynamics and/or pharmacokinetics of the enantiomers may become apparent. However, the development of the racemate may still be justified if any toxicity is associated with the pharmacological action of the drug and the therapeutic index is the same for both isomers.

Consequently, regulatory approval of the racemate would require that (a) both enantiomers are almost equipotent in terms of their primary and secondary pharmacological activities, (b) one of the two enantiomers has a different but therapeutically desirable pharmacological activity which may be exploited to optimize efficacy, (c) there is substantial chiral interconversion in vivo, (d) the two enantiomers produce the same achiral active metabolite or (e) the pharmacokinetics of the drug is nonstereoselective. The possibility of an interenantiomeric interaction whereby one enantiomer may ameliorate the toxicity of the other (for example, by inhibiting the formation of a toxic metabolite) may constitute another reason for developing a racemate. Stereoselective assay of all the four diastereoisomers following administration of racemic labetalol has shown that the AUCs for (*S,S*)-, (*S,R*)-, (*R,S*)- and (*R,R*)-labetalol, as a percentage of the total AUC over 12h of study, are 41%, 25%, 25% and 9%, respectively (DAKERS et al. 1997). Although there are no data directly comparing the pharmacokinetics of (*R,R*)-labetalol (dilevalol) with and without the presence of other diastereoisomers of labetalol (FUJIMURA et al. 1989; TENORO et al. 1989), it is worth speculating whether an interaction between the diastereoisomers may explain the relative lack of hepatotoxicity of racemic labetalol in comparison to dilevalol (DAKERS et al. 1997).

Often, the pharmacological profile of the racemate may be superior to that of either enantiomers. The (–)-enantiomer of dobutamine, which possesses mainly α1-adrenoceptor agonist activity, produces marked increases in cardiac output, stroke volume, total peripheral resistance and mean arterial pressure, but does not significantly increase heart rate. In contrast, (+)-dobutamine, which possesses predominantly β1- and β2-adrenoceptor agonist activity (RUFFOLO et al. 1981), elicits only a modest increase in cardiac output, which is due entirely to increase in heart rate since stroke volume does not increase. This isomer has also been shown to have potent α-blocking activity. Total peripheral vascular resistance and mean arterial blood pressure are both reduced by (+)-dobutamine, characteristic of a β-adrenoceptor agonist. The net result following administration of racemic dobutamine is increase in stroke volume and cardiac output with little effect on systemic resistance and mild tachycardia (RUFFOLO and MESSICK 1985).

Therefore, in most situations, the following preclinical pharmacodynamic, pharmacokinetic and toxicology studies will be needed for evaluation.

The profile of the pharmacodynamic effects related to the therapeutic use should be provided for the racemate and each enantiomer. The racemate data related to the general pharmacodynamic properties should be extended with studies on individual enantiomers if necessary from the point of view of safety.

The effective exposure to the enantiomers after administration of the racemate in pivotal preclinical studies should be measured by enantiospecific analytical methods to allow extrapolation of the data to human exposure.

It is ordinarily sufficient to carry out toxicology studies on the racemate. However, relevant toxicity studies should be repeated with the individual enantiomers (when possible) if pharmacologically unpredictable toxicity occurs at relatively low multiples of the potential human exposure. Interactions between the enantiomers at either the pharmacokinetic level or at pharmacological targets may be the confounding factors which should also be considered.

The principles and requirements for clinical studies correspond to those for preclinical studies. In most situations the following clinical studies will be needed, in principle, for evaluation.

In terms of human pharmacodynamics and tolerance, the main effect should be studied with racemate. The results should be compared with those obtained in animal studies with racemate and enantiomers. Studies with separate enantiomers should also be undertaken if considered necessary from a safety point of view.

In terms of human pharmacokinetics, studies on healthy volunteers, patients and special at-risk groups should be carried out with enantiospecific methods, unless it has been shown that there is no qualitative and quantitative difference in the fate of both enantiomers.

The clinical studies necessary to demonstrate safety and efficacy should be carried out with the racemate. It may be useful to monitor the pharmacokinetics in these studies with enantiospecific methods to further evaluate any stereoselective differences in pharmacology.

III. Development of a Single Enantiomer from an Approved Racemate

"Chiral switching" represents one of the most common scenarios today. Examples of such chiral switches, either potential or accomplished, include the development of (*S*)-bupivacaine, (–)-cetirizine, (*S*)-citalopram, (*S*)-doxazosin, (+)-fenfluramine (now withdrawn because of cardiac valvulopathy), (*R*)-fluoxetine (for depression) (now terminated because of potential risk of QT interval prolongation), (*S*)-fluoxetine (for migraine), (*R*,*R*)-formoterol, (*S*)-ibuprofen, (*S*)-ketoprofen, (*S*)-ketamine, (*R*,*R*)-labetalol (now withdrawn because of hepatotoxicity), (*R*)-lansoprazole, (*S*)-ofloxacin, (*S*)-omeprazole, (*S*)-oxybutynin, (*R*)-salbutamol and (*R*)-sibutramine, among many others. In principle this is equivalent to the development of a new active substance requiring a complete new application. The decision to develop the enantiomer selected should be explained.

Although the commercial reasons for chiral switches are numerous and varied, regulatory approval of a single enantiomer of previously approved racemate can be best justified scientifically if (a) one of the enantiomers is highly potent and there is hardly any chiral interconversion; this allows reduction of daily dose and consequently the "metabolic load," (b) the secondary pharmacology of the drug, responsible for toxic effects, resides predominantly in one enantiomer, (c) the pharmacokinetics are enantioselective; this allows the choice of the isomer with long half-life in order to decrease dosing frequency or of the isomer with shorter half-life in order to diminish the risk of accumulation in susceptible population, e.g., the elderly, or if the overdose with the drug is a distinct clinical possibility, (d) there is great interindividual variability which could be eliminated by developing only the enantiomer which is not subject to polymorphic metabolism, (e) therapeutically "inactive" enantiomer alters in a detrimental manner the pharmacokinetics (or pharmacodynamics) of the therapeutically active enantiomer, (f) notwithstanding the slight loss in potency, one of the enantiomer has the required "site-specificity" and (g) the inhibition of drug-metabolizing CYP isoforms by the two enantiomers is enantioselective; this could be exploited to improve the drug–drug interaction profile of the racemic mixture if this is likely to prove to be a major clinical problem.

Consequently, as far as is applicable to the enantiomer, data on the corresponding racemate may be used in order to reduce the amount of new studies. It is assumed that the applicant can provide the full dossier of the racemate to the authorities, or an equivalent to the scientific content of the full dossier. Suitable bridging studies, determined on a case-by-case basis, should be carried out to link the complete racemate data to the incomplete data on the selected enantiomer. The way the racemate results are used should be explained. When deciding on the extent of bridging studies, it is important to consider the following:

The pharmacodynamic and the pharmacokinetic profiles of the selected enantiomer should be compared with that of the racemate. Evaluation should include the extent of chiral interconversion. In addition, it may be appropriate to study the other enantiomer to correct for interactions.

With respect to toxicology, a suitable program of bridging studies may consist of (a) an acute toxicity study of the selected enantiomer using the racemate as positive control, (b) a repeated dose study (up to 3 months) in a single most sensitive species and (c) a study for effects on pre- and postnatal development (including maternal function), with the modification of starting treatment at conception, not at implantation, in a single most appropriate species, with the selected enantiomer and with the racemate as a positive control using at least one effective dose level.

Results should be compared with corresponding previous racemate studies. If unexpected results are found, further studies on a case-by-case basis will be necessary. In the worst case, the enantiomer will need to be fully investigated.

The principles and requirements for clinical studies are identical to those for preclinical studies. The choice to develop an enantiomer should be explained (and justified in clinical context).

The extent of bridging studies should be defined on a case-by-case basis. The pharmacokinetic and the pharmacodynamic profiles of the selected enantiomer should be compared with those of the racemate, and with the results of animal studies. It may be appropriate to study the other enantiomer to correct for possible racemate interactions.

Clinically, in principle, the usual pharmacotherapeutic (efficacy) studies should all be carried out. However, with detailed consideration of the preclinical data on the racemate and the selected enantiomer, as well as the therapeutic trials with the racemate, it may be possible to extrapolate to the single enantiomer being developed. This may reduce the number of additional clinical studies needed. The clinical expert report (required in EU submissions) should reflect this situation and provide justification for the inference and give additional arguments to support extrapolation from one indication study with the single enantiomer to other indications.

IV. Development of a Racemate from an Approved Single Enantiomer

Although development of a racemate from an approved single enantiomer remains a possibility, the authors cannot think of any example to date where this has actually happened.

The choice of racemate to be developed should be explained. In this case a completely new application, with a full program of preclinical and clinical studies, will have to be submitted, as in case of development of a racemate as a new active substance described in Sect. C.II. Useful data on the approved enantiomer may be added to support the new application.

V. Development of a Nonracemic Mixture from an Approved Racemate or Single Enantiomer

In principle, a tailored (nonracemic or nonequimolar) mixture of enantiomers can be viewed as an approach towards the optimization of a pharmacotherapeutic profile.

Where both enantiomers are fortuitously found to carry desirable but different properties, development of a non-50:50 mixture of the two as a fixed combination might be reasonable.

In this case, the application concerns a fixed combination product of which one or both components may be unknown. Therefore, all the necessary data on the unknown enantiomer(s) should be provided, as well as those necessary for the justification of this fixed combination. Most regulatory authorities have a separate guidance note on the development of fixed combination products. These will apply equally to the development of nonracemic mixtures. Essentially, the regulatory requirements in such cases originate from the necessity to show that both components of the mixture are compatible with each other pharmacokinetically, particularly the duration of activity, and pharmacodynamically with the clinical benefits claimed.

The principles and requirements for clinical studies are analogous to those for preclinical studies. These studies will need to show the pharmacokinetics and pharmacodynamics of the individual components and of the proposed mixture. Preclinical and clinical studies should be planned so as to confirm the claimed benefit.

A potential example which immediately comes to mind is sotalol; an ideal mixture could be developed to complement the class III antiarrhythmic effect of (+)-(*S*)-sotalol with the β-blocking activity of (–)-(*R*)-sotalol. Another example could include carvedilol. β-blocking activity of carvedilol resides in the (–)-(*S*)-enantiomer, while the α_1-receptor antagonist activity is provided by both the stereoisomers. The metabolic pattern of the two enantiomers of carredilol suggests that an ideally profiled mixture may be superior to the racemate. This approach was also tried with the development of indacrinone, but an ideal mixture could not be profiled because of the influence of natriuresis, which depresses urinary pH and therefore causes a reduction in urinary solubility of uric acid.

VI. Generic Applications of Chiral Medicinal Products

Generic formulations must be both pharmaceutically and therapeutically equivalent to the innovator product.

The issue of chirality in bioequivalence was first addressed by the Swedish authority in 1991. Japan has not addressed the issue of stereochemistry in the context of generic products. The EU discusses such circumstances in broad terms. The US approach is based on a case-by-case basis.

In general, there are no regulatory requirements for enantioselective assays in uncomplicated bioequivalence studies. There are, however, some instances when enantioselective assays become important in determining bioequivalence.

Bioequivalence studies supporting generic applications of chiral medicinal products should be based upon enantiospecific bioanalytical methods, unless:

1. Both products contain the same, stable single enantiomer as the active substance or
2. Both products contain the racemate and both enantiomers show linear pharmacokinetics

The issues relevant to requirements for enantiospecific analytical methods for racemic drugs are (a) the difference in the rates of input of the two enantiomers into systemic circulation (enantioselectivity in absorption and first pass metabolism), (b) the rate of chiral inversion, (c) genetic polymorphisms of enzymes metabolizing the drug and (d) the release characteristics of the product. Enantiospecific analytical methods are of course not important even in these circumstances if the pharmacodynamics of the two enantiomers are the same or alternatively, if pharmacokinetic (based on racemic assay) and clinical equivalence can be shown.

With nonlinear active absorption, the enantiomeric ratio changes with the change in the rate of input.

In a case where the drug undergoes polymorphic first pass metabolism, the measurement of the racemic drug is unlikely to be adequate if the clearance of the individual enantiomers varies in phenotypically distinct subgroups during both treatments in a cross-over study.

A change in systemic enantiomeric ratio is more likely with modified release products. Thus, enantioselective assays are more likely to be required when establishing bioequivalence of modified release products with their conventional release counterparts (provided the two enantiomers differ normally in their pharmacokinetics or pharmacodynamics). Enantioselective assays may still be required if there is a likelihood of in vivo chiral inversion (which is synonymous with generating a metabolite) when establishing bioequivalence of two modified release products containing a single enantiomer, or when establishing bioequivalence of two conventional release products containing racemic drugs.

D. Status of Approved Racemic Drugs

The safety and efficacy of marketed medicinal products, the active substance of which is a racemate, is generally considered to be well established. Unless new evidence emerges indicating a relationship between one enantiomer and a safety or efficacy issue, the guidelines do not stipulate any requirements for further data on such products.

If new claims related to the chiral nature of the active substance are put forward, supporting studies with the separate enantiomers will be required.

E. The Effect of Regulatory Guidelines

The change in the attitudes of the regulatory authorities over the last 10 years with respect to chiral drugs has affected the number of submissions for single enantiomers.

An informal analysis of submissions to the UK Medicines Control Agency between July 1996 and June 2000 confirms the trend reported by other surveys (Shindo and Caldwell 1991, 1995; Ariëns et al. 1988; Millership and Fitzpatrick 1993). The proportion of synthetic chiral drugs developed as single enantiomers appears to have genuinely risen between 1982 and 2000, even though the figures have been obtained from different surveys. This increase reflects both the regulatory requirements introduced in the early 1990s and the availability of the necessary scientific techniques to synthesize and control the enantiopurity of chiral drugs. The proportion of synthetic drugs which are in an achiral form is somewhat variable over the period of time represented by the figures in Table 1. In the surveys conducted between 1982 and 1985, achiral drugs accounted for about 60%; between 1986 and 1991 it was about 50% whereas during the period 1992 to 1999 it appeared to increase from 30% to 40%, thus making any conclusions about trends rather difficult.

The area of "chiral switches," where a single enantiomer is developed subsequent to a corresponding racemate which is already on the market, has attracted much interest (Hutt and Tan 1996; Triggle 1997; Tucker 1999). A description of the preclinical and clinical development of dexketoprofen provides a detailed example of one of these chiral switches (Mauleon et al. 1996). The regulations in Europe and the US both allow for the development of a single enantiomer from a racemate by the use of bridging studies between the

Table 1. An analysis of the trends in the development of chiral (racemates and single enantiomers) and achiral drugs

	Drugs in use in 1982[a]	NCEs approved 1983–1985[a]	NCEs approved in Japan 1986–1989[b]	Drugs in use in 1991[c]	NCEs approved in Japan 1992–1993[d]	NCEs assessed by MCA 1996–2000
Total natural or semi-synthetic	475	39	53	147		19
Racemates	8	0	5	8		1
Single enantiomers	461	39	47	119		16
Achiral	6	0	1	2		2
Total synthetic	1200	91	47	521	47	76
Racemates	422	36	29	140	22	15
Single enantiomers	58 (12%)	2 (5%)	7 (19%)	110 (44%)	11 (33%)	30 (67%)
Achiral	720 (60%)	53 (58%)	11 (23%)	269 (52%)	14 (30%)	31 (41%)
Total, all natural, semi-synthetic and synthetic drugs	1675	130	100	668		95

[a] ARIENS et al. 1988.
[b] SHINDO and CALDWELL 1991.
[c] MILLERSHIP and FITZPATRICK 1993.
[d] SHINDO and CALDWELL 1995.

old and new applications. One problem at the forefront is how a company which was not responsible for the original development can provide equivalent data.

Apart from any intrinsically beneficial effects to patients from the administration of pure enantiomers, it has been speculated that such switches may provide a mechanism for extending the patent period of a racemic drug about to come off patent. While this may have been attempted, the theoretical advantages of single enantiomers in many cases are more hypothetical than of real clinical benefit. For other drugs, the source of toxicity has been traced to one enantiomer, and an overwhelming case can be made for developing the other enantiomer. In the case of dilevalol, however, the development of a single enantiomer (of the racemic labetalol) proved to be a disaster. These aspects are discussed in Chap. 17 of this volume.

It is obvious that there is a need for studies in future to ascertain whether switches from racemates to single enantiomers, and vice versa, have had any positive effect on the risk/benefit of the drugs concerned.

F. Conclusions

Regulation of chiral drugs is now well-established and has had the effect of increasing the ratio of single enantiomer to racemic compounds for new synthetic drugs on the market. New analytical and preparative techniques will make it easier in the future to develop single enantiomers. There is no real evidence that the number of achiral drugs is increasing to avoid the problems associated with enantiopurity. Therefore, it seems safe to assume that the technical methods for controlling chiral drugs have developed to a stage where many of the challenges presented to the industry can be solved.

The clinical aspects of developing chiral drugs with a view to improving risk/benefit are discussed in Chap. 17 of this volume.

The decision to develop a racemate or a single enantiomer (either de novo or as a switch) must be based on sound pharmacological principles. As far as switches from previously marketed racemates are concerned, a word of caution may be appropriate. It is a matter of fact that once the racemic drug goes out of patent, cheaper generic products may appear in competition. Since an abbreviated development plan for a single enantiomer requires access to previous full development data on the racemate, it is difficult to foresee how generic competition for the switched single enantiomer can emerge immediately on expiration of the exclusivity period of the racemate.

Intuitively, a switched single enantiomer is likely to be more expensive than the older racemic product. Therefore, in the interests of good clinical practice and health economics, the debate on generic substitution may well find its way to include a discussion on racemic versus single enantiomers. Despite their short-term advantages, decisions to develop single enantiomers of previous racemates, based on essentially commercial grounds, are likely to be counterproductive in the long-term, especially if the expensive single enantiomer is soon found to offer no tangible clinical advantage over the well established racemic product.

References

Ariëns EJ, Wuis EW, Verings EJ (1988) Stereoselectivity of bioactive xenobiotics. Biochem Pharmacol 37:9–15
Coronary Drug Project Research Group (1972) The Coronary Drug Project: Findings leading to further modifications of its protocol with respect to dextrothyroxine. JAMA 220:996–1008
Cushny AR (1926) Biological relations of optically isomeric substances. Balliere, Tindall and Cox, London
Dakers JM, Boulton DW, Fawcett JP (1997) Sensitive chiral high-performance liquid chromatographic assay for labetalol in biological fluids. J Chromatogr B Biomed Sci Appl 704:215–220
Eudralex (1998) The rules governing medicinal products in the European Union, Vols 1–9, Office for Official Publications of the European Communities, Luxembourg, 1998
Fromm GH, Terrence CF (1987) Comparison of L-baclofen and racemic baclofen in trigeminal neuralgia. Neurology 37:1725–1728
Fujimura A, Ohashi K, Tsuru M, Ebihara A, Kondo K (1989) Clinical pharmacology of dilevalol (I). Comparison of the pharmacokinetic and pharmacodynamic properties of dilevalol and labetalol after a single oral administration in healthy subjects. J Clin Pharmacol 29:635–642
Gristwood R, Bardsley H, Baker H, Dickens J (1994) Reduced cardiotoxicity of levobupivacaine compared to racemic bupivacaine (Marcaine): new clinical evidence. Exp Opin Invest Drugs 3:1209–1212
Hamon P, Dingeon B, Jiang *N*-S, Orgiazzi J (1993) Purified D-thyroxine in athyreotic patients. Lancet 341:1477
Hutt AJ, Tan SC (1996) Drug chirality and its clinical significance. Drugs 52 [Suppl 5]:1–12
Jenner P, Testa B (1973) Influence of stereochemical factors on drug disposition. Drug Metab Rev 2:117–184
Mauleon D, Artigas R, Garcia ML, Carganico G (1996) Pre-clinical and clinical development of dexketoprofen. Drugs 52 [Suppl 5]:24–46
McClellan KJ, Spencer CM (1998) Levobupivacaine. Drugs 56:355–362
Millership JS, Fitzpatrick A (1993) Commonly used chiral drugs: a survey. Chirality 5:573–576
Pagnan A, Busnardo B, Zanetti G, Simioni N, Braggion M, Ziron L, Girelli ME (1984) Effects of dextro-thyroxine, a preparation almost free of levo-thyroxine, on thyroid function, serum lipids and apolipoprotein A-1 in "double pre-beta lipoproteinemia". Atherosclerosis 50:191–202
Rauws AG, Groen K (1994) Current regulatory (draft) guidance on chiral medicinal products: Canada, EEC, Japan, United States. Chirality 6:72–75
Ruffolo RR Jr, Spradlin TA, Pollock GD, Waddell JE, Murphy PJ (1981) Alpha and beta adrenergic effects of the stereoisomers of dobutamine. J Pharmacol Exp Ther 219:447–452
Ruffolo RR Jr, Messick K (1985) Systemic hemodynamic effects of dopamine, (+/–)-dobutamine and the (+)-and (–)-enantiomers of dobutamine in anesthetized normotensive rats. Eur J Pharmacol 109:173–181
Shah RR, Midgley JM, Branch SK (1998) Stereochemical origin of some clinically significant drug safety concerns: lessons for future drug development. Adverse Drug React Toxicol Rev 17:145–190

Shindo H, Caldwell J (1991) Regulatory aspects of the development of chiral drugs in Japan: a status report. Chirality 3:91–93

Shindo H, Caldwell J (1995) Development of chiral drugs in Japan: an update on regulatory and industrial opinion. Chirality 7:349–352

Tenero DM, Bottorff MB, Given BD, Kramer WG, Affrime MB, Patrick JE, Lalonde RL (1989) Pharmacokinetics and pharmacodynamics of dilevalol. Clin Pharmacol Ther 46:648–656

Triggle DJ (1997) Drug Discovery Today 2:138–147

Tucker GT (1999) Chiral switches. Lancet 355:1085–1087

Valenzuela C, Snyders DJ, Bennet PB, Tamargo J, Hondeghem LM (1995a) Stereoselective block of cardiac sodium channels by bupivacaine in guinea-pig ventricular myocytes. Circulation 92:3014–3024

Valenzuela C, Delpon E, Tamkun MM, Tamargo J, Snyders DJ (1995b) Stereoselective block of a human cardiac potassium channel (Kv1.5) by bupivacaine enantiomers. Biophys J 69:418–427

Young WF Jr, Gorman CA, Jiang NS, Machacek D, Hay ID (1984) L-thyroxine contamination of pharmaceutical D-thyroxine: probable cause of therapeutic effect. Clin Pharmacol Ther 36:781–787

CHAPTER 17

Improving Clinical Risk/Benefit Through Stereochemistry

R.R. SHAH

The views expressed in this chapter are those of the author and do not necessarily represent the views or the opinions of Medicines Control Agency, other regulatory authorities or any of their advisory committees.

A. Introduction

This chapter will focus on a few selected drugs which best illustrate the principles relevant to stereochemical aspects of drug development with a view to improving clinical safety, efficacy and risk/benefit. With regard to stereochemical designation of the drugs discussed, attention is drawn to the fact that the current system of nomenclature of the enantiomers has evolved over time. Depending on direction of rotation of the plane polarized light, the enantiomers of older drugs have usually been designated only as (+) and (–) or (D) and (L) without any information on the corresponding designations based on the modern Cahn, Ingold and Prelog convention, which assigns the absolute configuration at an asymmetric center, indicated by the symbols (*R*) and (*S*). For newer drugs, only the absolute configurations of the enantiomers have been cited, (*R*) or (*S*) based on this convention. Since the absolute configuration is independent of optical rotation, it follows, for example, that (*S*)-enantiomer may rotate the light clockwise or anticlockwise. Consequently, in this chapter, enantiomers of a drug will be distinguished by those designations which have been used in the corresponding literature.

I. Stereochemistry and Pharmacogenetics

Stereoselectivity in metabolism appears to be responsible for much of the pharmacokinetic differences observed between the enantiomers of a drug. Receptors, channels and enzymes which mediate drug activity also have three dimensional asymmetry and therefore their interactions with the chiral drugs/metabolites are also stereosensitive.

An additional feature of stereoselective pharmacokinetics is the fact that a number of cytochrome P450 isozymes which mediate stereoselective drug metabolism display genetic polymorphism (e.g., CYP 2D6, CYP 2C9 and CYP 2C19). Often, therefore, the consequences of stereoselective pharmacokinetics are linked intricately with those of polymorphic drug metabolism. This is best illustrated by CYP 2C9 isozyme, which mediates 7-hydroxylation of (*S*)-warfarin. Of the two enantiomers, (*S*)-warfarin is pharmacologically the more potent. Its slow elimination in individuals deficient in CYP 2C9 activity predisposes these individuals to greater risk of serious hemorrhage when prescribed the normal doses (AITHAL et al. 1999).

The clinical consequences of this interaction between chirality and pharmacogenetics are obvious. Since the preferential metabolism of one or the other enantiomer depends on the relative affinities of the two enantiomers for the drug-metabolizing enzyme, this polymorphic distribution of drug-metabolizing capacity may either exaggerate or attenuate the differences in the pharmacokinetics of individual enantiomers. The net effect is that the clinical evidence for stereogenic basis of drug response (toxic or therapeutic) either becomes manifest most often in one phenotype or becomes difficult to discern.

II. Stereochemistry and Metabolites

Stereoselectivity in metabolism may refer to either the substrate (parent drug) or the product of its metabolism (metabolites). In case of substrate stereoselectivity, the two enantiomers are metabolized either by different routes or at different rates. In contrast, product selectivity implies the differential formation of stereoisomeric metabolites from a single substrate containing one or more prochiral centers. Parahydroxylation of the prochiral anticonvulsant drug phenytoin by CYP 2C9 or by CYP 2C19 appears to favor the formation of (*S*)-*p*-hydroxyphenytoin or of (*R*)-*p*-hydroxyphenytoin respectively. Clearly, the selectivity in this situation is very high and if the pathway involved is deficient – genetically or through inhibition – the pattern of resulting metabolite(s) is different. In the extensive metabolizers (EMs) of CYP 2D6, 99% of 4-hydroxydebrisoquine is of (*S*)-configuration. In contrast, the poor metabolizers (PMs) in whom this pathway is genetically deficient, 5%–36% of what little 4-hydroxydebrisoquine is formed is of (*R*)-configuration. This interaction between pharmacogenetics and stereoselectivity becomes important when the active metabolite is also secondarily metabolized stereoselectively by one of these polymorphic CYP isoforms. For example, (*S*)-norfluoxetine is metabolized by CYP 2D6 (half-lives in EMs and PMs of CYP 2D6 are 5.5 and 17.4 days respectively), but (*R*)-norfluoxetine is not (half-lives in EMs and PMs are 5.5 and 6.9 days, respectively) (FJORDSIDE et al. 1999). Consequently, (*S*)-norfluoxetine has a far greater potential for accumulation in the PMs, with all its attendant consequences (see Sect. D.VII.). Citalopram, discussed in Sect. E.V., provides another example of this complex interaction between chirality, pharmacogenetics and metabolite-related toxicity.

Similarly, all proton pump inhibitors (PPI) exist as two enantiomers because of the presence of an asymmetric sulphinyl group in their principal core structure. Parent PPI are inactive prodrugs. Following their accumulation in the acidic space of secretory canaliculi in the gastric lumen, these parent compounds are converted to achiral thiophilic sulphenamide which inactivates the H^+/K^+-ATPase pump, resulting in reduced gastric acid secretion. Since both enantiomers are transformed to the sulphenamide at the same rate, both parent enantiomers of PPIs are "equipotent." Therefore, the efficacy of these drugs correlate with their area under plasma concentration versus time curve (AUC). Their metabolism, however, is stereoselective for one of the enantiomers and this is principally mediated by the polymorphic CYP 2C19. Thus, when a racemate is administered, CYP 2C19-mediated metabolism introduces wide interindividual variability in the AUCs of the two enantiomers at a given dose. For example, in case of omeprazole, the ratios of AUC for PMs/EMs of CYP 2C19 are 7.5 for (+)-(*R*)-omeprazole and 3.1 for (–)-(*S*)-omeprazole (TYBRING et al. 1997). This variability can be reduced, with a more predictable efficacy, by administering only the enantiomer which is less subject to polymorphic metabolism. (–)-(*S*)-omeprazole, recently developed as a new chemical entity, is now approved and marketed as esomeprazole ("Nexium").

Similarly, investigation on the stereoselective pharmacokinetics of pantoprazole in EMs and PMs of CYP 2C19 has shown that the metabolism of (+)-pantoprazole is impaired to a much greater extent than (–)-pantoprazole in PMs. Thus, interindividual variability in exposure to this PPI arising from CYP 2C19 polymorphism, and therefore in its efficacy, can also be improved by development of (–)-pantoprazole (TANAKA et al. 2001).

Although the efficacy of (–)-(*S*)-omeprazole is not in doubt (SPENCER and FAULDS 2000), there are issues related to the relative dosing schemes of the racemic drug and its enantiomer. This may have possible safety implications. Compared to the racemic drug, (–)-(*S*)-omeprazole provides greater and more prolonged hypochlorhydria. A recent report of gastric carcinoid tumor associated with omeprazole may be a coincidence (DAWSON and MANSON 2000). If not, greater vigilance may be required during the clinical use of (–)-(*S*)-omeprazole. Pooled data have shown that although there is no increase in cell dysplasia following the administration of (–)-(*S*)-omeprazole, there is an increase in entero-chromaffin-like cells (DANIELS 2001).

B. Stereochemistry and Regulatory Control of Drugs

Before discussing other drugs, old or new, it is ironic that any discussion on the risks arising from clinical use of racemic drugs should begin with thalidomide. The thalidomide tragedy of 1959–1961 contrasts sharply with the scientifically guided introduction in 1961 of ethambutol to improve its risk/benefit. In stereochemical terms, ethambutol was well ahead of its time.

I. Thalidomide

Thalidomide probably represents a particularly dramatic example of the alleged role of stereoselectivity in drug toxicity. The consequences that fol-

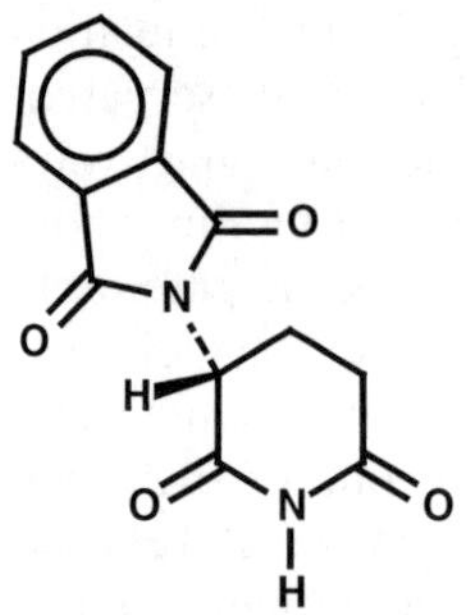

Fig. 1. (+)-(*R*)-thalidomide

lowed the clinical use of thalidomide as a novel sedative were to provide the impetus to the introduction, for the first time in most countries (including those in Western Europe) of regulatory control of drugs to be marketed for clinical use. The resulting tragedy (an "epidemic" of phocomelia) stimulated world-wide, not only an ever-increasing set of regulatory requirements on preclinical testing of drugs before administration to humans, but also the laws on liability and compensation.

Thalidomide has a chiral center (Fig. 1) and was marketed as a racemic drug. Since the first reports of its teratogenic effect, the stereochemical aspects of thalidomide-induced teratogenesis have been investigated extensively. Although earlier studies (FABRO et al. 1967) with New Zealand white rabbit showed that both enantiomers were teratogenic, later studies in rodents appeared to suggest that teratogenicity of thalidomide resided essentially in one of the two enantiomers, (–)-(*S*)-thalidomide. Since it is the (+)-(*R*)-enantiomer which has the therapeutically desirable sedative effect, it has been frequently argued that thalidomide-induced teratogenesis might have been averted or greatly minimized had only the (+)-(*R*)-enantiomer been marketed.

Since all the studies have been performed in animals, the relevance of the observed stereoselective toxicity in animals to the disaster in humans is difficult to evaluate. However, it is discussed here for no other reason than to emphasize the need to consider stereochemical factors during drug development. More recent studies (ERIKSSON et al. 1995) have shown that, after in vivo administration of either of the two enantiomers to the humans, there is a rapid interconversion between the two forms, thus making administration of a single enantiomer pointless.

The major route of thalidomide breakdown in humans and animals is through spontaneous hydrolysis with subsequent elimination in the urine. It is now thought that phocomelia results from inhibition of angiogenesis and vasculogenesis (which are essential for the formation of limb buds) by a metabolite of thalidomide (thought to be an arene oxide) (D'AMATO et al. 1994). This metabolite, which is formed both in humans and in rabbits but not in rodents,

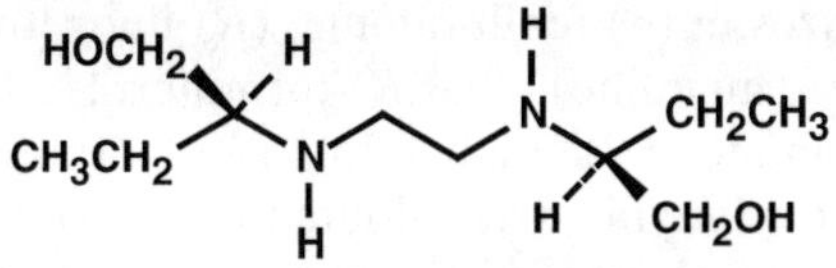

Fig. 2. (*S*,*S*)-ethambutol

racemizes rather rapidly in vivo, further questioning the role of stereochemical factors in the human disaster. For a more detailed discussion of this aspect, the reader is referred to Chap. 4 (this volume). Interestingly, the search for the mechanism of thalidomide-induced teratogenicity has uncovered more interesting pharmacological properties of the drug. Recent resurgence of clinical interest in thalidomide has focused on its efficacy in leprosy and other conditions requiring immunosuppression.

II. Ethambutol

Ethambutol (Blessington, 1997), a powerful and highly selective antituberculous drug, resulted from synthesis of a large number of *N*, *N'*-hydroxyalkylethylendiamines. Ethambutol introduced for clinical use is (+)-ethambutol with an (*S*,*S*)-stereochemical configuration (Fig. 2). Compared to the (+)-isomer, the (−)-isomer had about 1/500th of the required antibacterial activity while the optically inactive *meso*-isomer was only 1/12th as active. In contrast, all the three isomers were almost equipotent in terms of their potential to produce the major side effect of the drug, optic neuritis. This side effect was related to the dose and duration of treatment with the drug. Not surprisingly, the risk/benefit ratio of treatment with ethambutol was greatly enhanced by marketing only the (+)-isomer.

Other older drugs which have been used in chirally pure form are L-dopa and D-penicillamine. Although stereoselectivity in their toxicity was recognized, both were originally used as racemates until it was possible to produce pure enantiomers on a commercial scale. L-dopa is absorbed selectively by active processes and in addition, it is D-dopa which is associated with granulocytopenia (Cotzias et al. 1969). Likewise, D-penicillamine is far less prone to serious nephrotoxicity than L-penicillamine (Williams 1990). Other familiar but less well appreciated examples of stereochemically pure drugs in clinical use are (*S*)-timolol, (*S*)-naproxen, (+)-methorphan, (which is an over-the-counter antitussive while (−)-methorphan is a controlled narcotic) and (+)-α-propoxyphene, ("dextropropoxyphene" which is a potent analgesic while (−)-α-propoxyphene is an active antitussive).

C. Clinical Aspects of Stereochemistry

Drugs already developed or under development as single enantiomers from previously marketed racemates include (*S*)-bupivacaine, (−)-cetirizine, (*S*)-

citalopram, (*S*)-doxazosin, (+)-fenfluramine, (*R*)-fluoxetine (for depression), (*S*)-fluoxetine (for migraine), (*R*,*R*)-formoterol, (*S*)-ibuprofen, (*S*)-ketoprofen, (*S*)-ketamine, (*R*,*R*)-labetalol, (*R*)-lansoprazole, (*S*)-ofloxacin, (*S*)-omeprazole, (*S*)-oxybutynin, (*R*)-salbutamol and (*R*)-sibutramine.

Development of a single enantiomer by the innovator of a previously marketed racemic drug is often driven by commercial reasons – this may confer commercial advantages in terms of extended patent and market exclusivity. Leaving aside any commercial considerations, there are a number of sound scientific reasons for developing either a racemic mixture or a single enantiomer. However, the reasons for selecting a racemate or an enantiomer of a drug for development are often not immediately apparent. At times, the choice ought to be guided not only by the pharmacology of the enantiomers but also by the target population. This is well illustrated, for example, by propafenone and its β-blocking activity; while useful for its antiarrhythmic efficacy, this property is clearly undesirable in those with obstructive airways disease or cardiac failure.

Examination of the cases of racemic drugs where enantiomers have been evaluated separately for their pharmacological activities reveals three possibilities:

1. Both enantiomers may have similar activities.
2. One enantiomer is pharmacologically active while the other is inactive.
3. Both enantiomers have quite different (occasionally therapeutically complementary or antagonistic) activities or have different potency for a given property (different concentration-response relationship).

The first two possibilities are sufficiently well known and exemplified by a number of drugs. They have little impact on the safety of the drug concerned. It is the third possibility that has attracted considerable regulatory and academic interest. It offers exciting opportunities for harnessing differences in pharmacological or toxicological properties of the two enantiomers to improve the clinical risk/benefit of the drugs concerned. Failure to appreciate the significance of stereochemical factors or to address them satisfactorily has resulted in less than optimal development and clinical use of the drugs in terms of their safety, efficacy and risk/benefit. In some instances, there is sufficient evidence that stereoselective toxicity was responsible, either directly or indirectly, for suspension of further development of the drug or its withdrawal from the market. While pharmacological differences in the enantiomers of some older drugs withdrawn from the market due to safety concerns are not adequately investigated or documented, there are a few recent withdrawals (prenylamine, dilevalol and terodiline) where the stereogenic origin of safety concern is not much in doubt.

The scientific decision to develop a racemate or an enantiomer can be a complex one, depending on the relationship between stereoselectivity in drug metabolism, renal clearance, pharmacogenetics and the activities of parent drugs and metabolites. This is illustrated by two class I antiarrhythmic drugs

– flecainide and encainide – which act primarily by inhibiting sodium channels.

I. Flecainide and Encainide

Flecainide and encainide were shown in the Cardiac Arrhythmias Suppression Trial to increase mortality relative to placebo in the treatment of post-myocardial infarction ventricular arrhythmias. Both are optically active and metabolized by CYP 2D6 with the difference that for encainide, it is the metabolites which are primarily responsible for its therapeutic effects.

Flecainide is eliminated by metabolic and renal clearances. Metabolic clearance shows modest stereoselectivity for (–)-(*R*)-flecainide (Gross et al 1989). However, changes in metabolic clearance of (–)-(*R*)-flecainide are offset by inverse changes in renal clearance (Birgersdotter et al 1992) and therefore, the overall disposition of flecainide does not show any evidence of stereoselectivity. The differences in the pharmacodynamics of flecainide enantiomers are equivocal. Overall, it appears that both enantiomers produce a concentration- and frequency-dependent decrease in action potential amplitude and an increase in conduction time with no significant differences between the enantiomers (Banitt et al 1986; Vanhoutte et al 1991). Flecainide also diminishes the rapid component of delayed rectifier potassium current but again, the differences between the two enantiomers are of no clinical significance (Smallwood et al 1989).

Encainide metabolism is stereoselective for (–)-encainide (urinary +/– ratio 1.20 in the EMs). In PMs of CYP 2D6, the metabolism is nonstereoselective (urinary +/– ratio close to unity). In vitro, encainide enantiomers are pharmacologically almost equipotent, depressing the maximum rate of depolarization with similar frequency and concentration dependence (Turgeon et al 1991). The possibility of the metabolites exerting stereoselective electrophysiological effects has been considered and is thought to be remote. Therefore, this genetically determined stereoselective disposition is unlikely to play a major role in mediating the clinical actions of encainide.

Thus, despite the observed clinical safety concerns and evidence of stereoselective pharmacokinetics, there is no data to support the development of either drug as a single enantiomer.

D. Improving Risk/Benefit Through Stereochemistry

The most pragmatic reason for developing a racemic mixture is the evidence that one of the enantiomers is pharmacologically inactive, both in terms of therapeutic effect and toxicity. Other reasons for developing a racemic mixture could include the facts that (a) both enantiomers are almost equipotent in terms of their primary and secondary pharmacological activities, (b) one of the two enantiomers has a different but therapeutically desirable pharmacological activity which may be harnessed to optimize efficacy, (c) there is substantial chiral interconversion in vivo, (d) the two enantiomers produce the same achiral active metabolite or (e) the pharmacokinetics of the drug is nonstereoselective.

Even when the pharmacodynamic activities of two enantiomers are therapeutically complementary, it is perhaps too optimistic to expect that the 50:50 mixture of the enantiomers (which constitutes the racemate) represents the optimal ratio of the two enantiomers; it is more likely that the dose-concentration and/or the concentration-response curves of the two enantiomers will be different. When the two enantiomers are pharmacologically similar but have different potencies or different elimination profiles, it may be possible to develop each individual enantiomer separately for use in different subsets of populations, each with its own dose schedule.

I. Levofloxacin

At times, there is a genuine clinical reason for developing a single enantiomer of a previously marketed racemic drug. A typical example is the development of levofloxacin (Fig. 3a), the (–)-(*S*)-enantiomer of the previously marketed racemic antibacterial agent, ofloxacin (Fig. 3b). The (–)-(*S*)-enantiomer of ofloxacin had not only 8–125-fold greater antibacterial potency (HAYAKAWA et al. 1986) and higher water solubility than the (+)-(*R*)-isomer but also a markedly different metabolic fate (OKAZAKI et al. 1991). While ofloxacin is usually administered at a dose of 200–800 mg daily, levofloxacin is usually administered at a dose of 250–500 mg daily. There can be little doubt that efficacy is preserved but it may be premature to exclude the possibility that the pure enantiomer may be associated with some unexpected safety concern – there is now emerging a preliminary evidence of a proarrhythmic hazard.

However, a number of quinolone antibiotics such as grepafloxacin, sparfloxacin, gatifloxacin and moxifloxacin have been associated with cardiac repolarization abnormalities. Clinically relevant inhibition of HERG (human

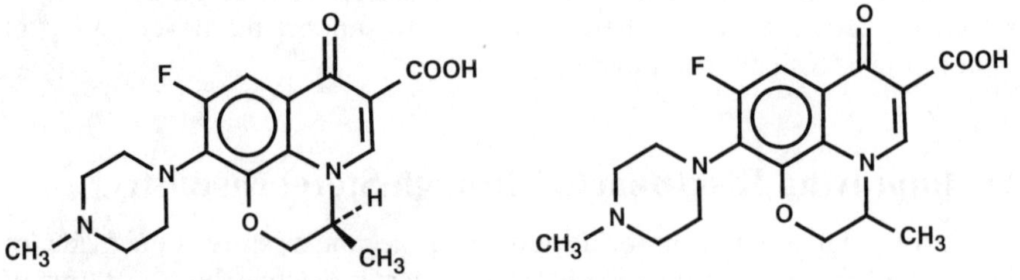

(a) (b)

Fig. 3. **a** Levofloxacin [(–)-(*S*)-ofloxacin] and **b** ofloxacin

Fig. 4. Temafloxacin

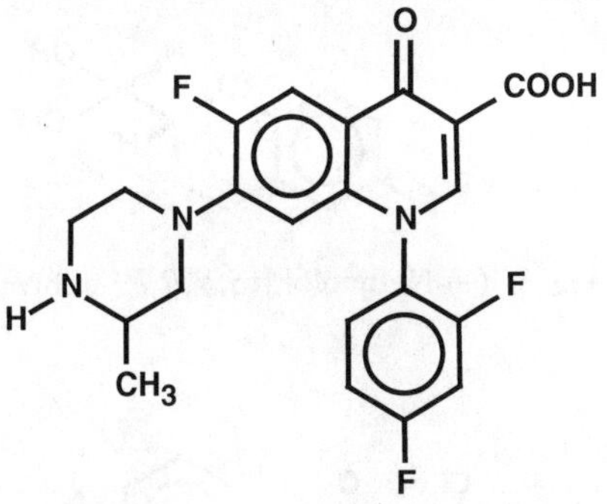

ether-a-go-go gene, responsible for encoding for delayed rectifier potassium channel) current is not a class effect of the fluoroquinolone antibacterials but is highly dependent upon specific substitutions within this series of compounds and therefore, HERG channel affinity should be an important criterion for the development of newer fluoroquinolones (KANG et al. 2001). Although in vitro electrophysiological data on the cardiac safety of levofloxacin relative to ofloxacin have been reassuring (ADAMANTIDIS et al. 1998), the significance of recently appearing isolated reports of QT interval prolongation and torsade de pointes associated with clinical use of levofloxacin needs to be evaluated, especially since no such events have been documented with ofloxacin.

In contrast, temafloxacin (Fig. 4) is a racemic quinolone with a chiral center at C-3′ position of the piperazinyl ring, and the racemate as well as its two enantiomers possess almost identical antibacterial activity in vitro (CHU et al. 1991). Temafloxacin was introduced in the market in 1992 and was withdrawn within a few months because of its toxic effects such as anaphylaxis, hemolytic anemia and renal failure, which were the most striking adverse events reported following its use. In addition, hypoglycemia and hepatic failure were also reported. Whether or not this toxicity profile has any stereogenic basis has not been studied.

II. Nebivolol

An innovative approach to combining various enantiomers to achieve a desired effect is represented by the *β*-blocker nebivolol, which has four asymmetric carbons. Nebivolol is the most *β*1-selective adrenoceptor antagonist currently available for clinical use in hypertension. The drug substance marketed is a racemic mixture of (+)-nebivolol and (−)-nebivolol. The (+)-isomer, which has the (*S*,*R*,*R*,*R*)-configuration (Fig. 5), is a highly selective *β*-blocker and has a long half-life (VAN PEER et al. 1991). The (−)-isomer (with its (*R*,*S*,*S*,*S*)-configuration) has no *β*-blocking activity but is a more potent endothelium-dependent vasodilator than the (+)-isomer (GAO et al. 1991). Inhibition of nitric oxide synthase abolishes this vasodilator activity, implicating the endothelial L-arginine/NO mechanism. The (−)-isomer with its

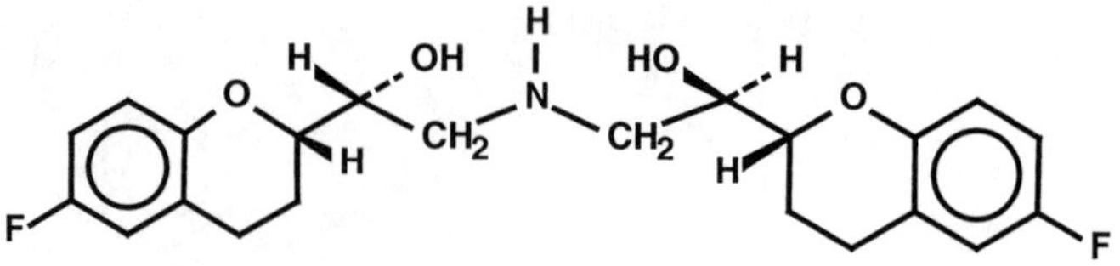

Fig. 5. (+)-Nebivolol [(*S*,*R*,*R*,*R*)-nebivolol]

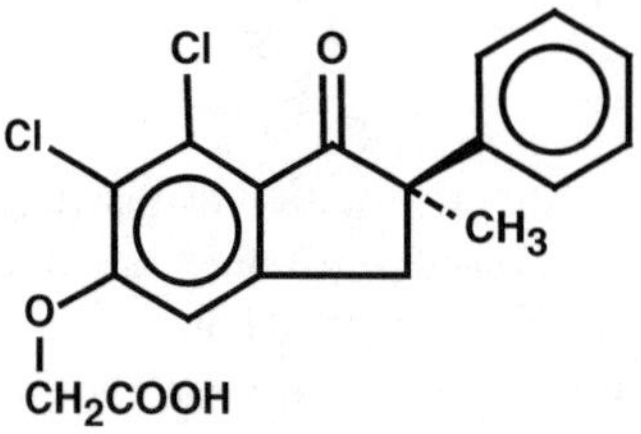

Fig. 6. (–)-(*R*)-indacrinone

vasodilatory properties is reported to moderate the negative inotropic effects of (+)-nebivolol (GILL et al. 1990; VAN DE WATER et al. 1988). Thus the mixture provides a good example of complementary therapeutic properties of the two stereoisomers in an equimolar mixture.

III. Indacrinone

The two enantiomers of indacrinone, a phenoxyacetic acid diuretic, have very different activities that may be therapeutically desirable. It was intended for the treatment of mild to moderate hypertension and congestive cardiac failure but never reached the market.

Many hypertensive patients have associated hyperuricemia as a comorbidity. Thiazide diuretics and frusemide have the propensity to raise serum uric acid, thus limiting their use in these patients. The (–)-(*R*)-enantiomer (Fig. 6) of indacrinone and its *p*-hydroxy-metabolite are potent natriuretic diuretics, while both enantiomers have potent uricosuric activity (IRVIN et al. 1980; VLASSES et al. 1981). One risk with potent uricosuric agents is precipitation of uric acid in the renal tubules, leading to renal side effects. Indeed, this was one of the reasons (the main reason being its hepatotoxicity, reported in 57 patients in the USA) why the clinically popular *achiral* uricosuric diuretic, ticrynafen (also a phenoxyacetic acid derivative and known as tienilic acid) was withdrawn from the market in 1980 (ANON 1980).

High doses of (+)-(*S*)-indacrinone to human volunteers resulted in a decrease of about 50% in serum uric acid (VLASSES et al. 1981). However, this property of racemic indacrinone is transient and clinically inadequate. Attempts to profile an ideal nonracemic mixture of (+)-(*S*)-indacrinone and

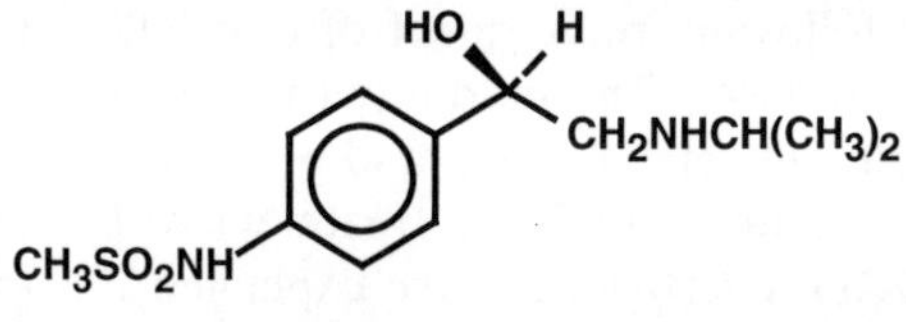

Fig. 7. (–)-(*R*)-sotalol

(–)-(*R*)-enantiomer proved to be insuperable on account of the influence of natriuresis, which depressed urinary pH and therefore caused a reduction in urinary solubility of uric acid (VLASSES et al. 1981). The development of indacrinone was therefore discontinued in the mid-1980s due to imbalance between the natriuretic and uricosuric activities, resulting in indacrinone-induced hyperuricemia.

IV. Sotalol

Sotalol, another β-blocker, has long been available for clinical use in the treatment of ventricular and supraventricular arrhythmias. Like all β-blockers of the arylethanolamine series, the carbon atom bearing the hydroxyl group is asymmetric and, therefore, sotalol has two enantiomers. The absolute configuration at this asymmetric carbon atom has to be (*R*) (Fig. 7) for β-blocking activity because, even at high concentrations, there is little β-blocking activity associated with the (*S*)-isomer (PATIL 1968). This is in contrast to β-blockers of the aryloxypropanolamine series (most of the currently widely used agents) in which the β-blocking activity resides in (*S*)-enantiomers (HOWE and SHANKS 1966).

However, the aryl group of sotalol bears the methanesulphonamide function, which is a potent class III pharmacophore which endows the molecule with QT interval prolonging properties, resulting from blockade of potassium channels. Consequently, the class III activity resides with both enantiomers of sotalol (KATO et al. 1986). This activity carries dose-related proarrhythmic risks with an incidence of 3%–5% at doses greater than 320mg daily.

In one randomized study in 1456 patients (JULIAN et al. 1982), enrolled within 5 days of an acute myocardial infarction, mortality was 18% lower in the patients treated with racemic sotalol than in those treated with placebo. Although this decrease in mortality was not statistically significant, there was also a 41% reduction in reinfarction rate in patients receiving racemic sotalol.

In contrast, the SWORD (Survival With ORal D-sotalol) study with (+)-(*S*)-sotalol(*S*), aimed at the prevention of sudden deaths and arrhythmic deaths in patients with a recent or remote myocardial infarction and a left ventricular ejection fraction of <40%, had to be discontinued prematurely. The choice of (+)-sotalol with pure class III but no β-blocking activity was prompted by the failure of class I antiarrhythmic drugs to provide any benefit to postmyocardial infarction patients. The SWORD study had to be terminated prematurely following recruitment of only 3121 of the planned 6400 patients (WALDO et al. 1996). The mortality (presumed to be due to arrhythmias) was 5% in the (+)-sotalol group and 3.1% in the placebo group – an increase of 65% in mortality following treatment with (+)-sotalol relative to placebo. The SWORD investigators have explained this excess on the basis that the effect of the drug on action potential is lost when there is an increase in sympathetic activity or during periods of myocardial ischemia which may trigger proarrhythmias.

However, β-blockers have been known to provide some protection from all-cause mortality and sudden cardiac deaths after myocardial infarction. The lack of protective effect of (+)-sotalol observed in SWORD is almost certainly due to the absence of the antiarrhythmic β-blocking activity (present in the racemic drug). Since there is little doubt that both class II and class III activities provide antiarrhythmic mechanisms and that class II and class III activities complement each other in a beneficial manner, sotalol raises interesting possibilities for investigating the antiarrhythmic efficacy of either the pure (–)-(*R*)-sotalol (providing a better balance of the β-blocking and class III activities) or an ideally profiled nonracemic mixture of the two enantiomers of sotalol.

V. Carvedilol

The complexity involved in evaluation of a racemic or nonracemic mixture is best illustrated by carvedilol, an aryloxypropanolamine-derived β-blocking agent.

Carvedilol brings together the issues arising from pharmacogenetics, chirality, multiple pharmacodynamic activities of the parent drug, the questions on contributions by the activity of metabolites and influence of various disease states on its pharmacokinetics.

Carvedilol is described as a nonselective β-blocking agent with vasodilating properties – both properties are desirable for therapeutic effect in hypertension and angina. These two activities, responsible for its approval for use in mild to moderate, initially and later also in severe, form of chronic heart failure, are both evident in the same dose range that is clinically recommended. Vasodilatation is predominantly mediated through α_1-receptor antagonism. Carvedilol also suppresses the renin-angiotensin system through its β-blocking activity.

The β-blocking activity resides in the (–)-(*S*)-enantiomer while the α_1-receptor antagonist activity is provided by both the stereoisomers (BARTSCH et al. 1990). The drug undergoes stereoselective metabolism with a preference for the β-blocking (–)-(*S*)-enantiomer (Fig. 8). The (+)-(*R*)-enantiomer with α_1-receptor antagonist activity is subject to polymorphic metabolism by CYP 2D6 (ZHOU et al. 1995) and therefore attains higher (about 2–3-fold) C_{max} and AUC in poor metabolizers of debrisoquine/sparteine. In view of the more sustained duration of the activity of (+)-(*R*)-enantiomer with α_1-receptor antag-

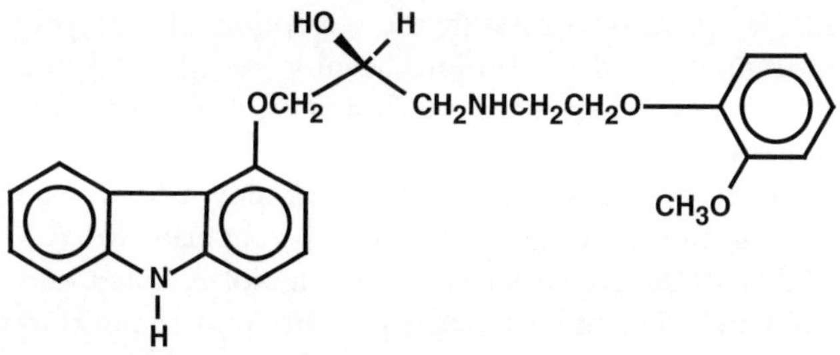

Fig. 8. (−)-(*S*)-carvedilol

onist activity in comparison to the short-lived duration of the activity of β-blocking (−)-(*S*)-enantiomer, racemic carvedilol is therefore better described as predominantly an α_1-receptor antagonist vasodilator with some β-blocking activity rather than as predominantly a β-blocker with vasodilating activity. Nevertheless, the point to be emphasized is that the two enantiomers complement each other's activity in producing antihypertensive effect. The β-blocking (−)-(*S*)-enantiomer may also protect against any tendency to produce reflex tachycardia by the other vasodilating enantiomer.

Following chronic dosing, the above metabolic profile results in preferential elimination of β-blocking (−)-(*S*)-enantiomer in both extensive and poor metabolizer phenotypes while the exposure to the vasodilatory (+)-(*R*)-enantiomer with α_1-receptor antagonist activity becomes relatively more dominant, particularly in those which are poor metabolizers. These individuals may be at risk of severe hypotension due to higher plasma concentrations of (+)-(*R*)-isomer and the resultant greater α-blockade (Zhou et al. 1995; Meadowcroft et al. 1997).

VI. Timolol

Timolol is particularly interesting since the drug on market is pure (*S*)-timolol. The futility of developing a pure enantiomer of β-adrenoceptor antagonists is illustrated by the fact that the safety profile of (*S*)-timolol in clinical use is similar to that of other β-adrenoceptor antagonists marketed as racemates. One of its indications is glaucoma because of its blockade of ocular β_2-receptors. Inadvertent administration of timolol eye drops to susceptible individuals (e.g., asthmatics) often results in potentially fatal adverse cardiac and pulmonary effects (Fraunfelder and Barker 1984).

The (*R*)-stereoisomers of aryloxypropanolamine β-adrenergic antagonists are usually much less active than their corresponding (*S*)-forms. In the eye, however, some studies have shown that these (*R*)-stereoisomers are unexpectedly potent in altering intraocular pressure. In one study of six stereoisomeric pairs of β-adrenergic antagonists, all six pairs of antagonists demonstrated the expected increased potency of (*S*)-forms in the heart.

However, all (*R*)-enantiomers (with the exception of metoprolol) demonstrated a substantially higher absolute affinity for ciliary process receptors (known to be almost exclusively of the β_2 subtype) than for cardiac receptors (NATHANSON 1988).

This difference in potency at various sites may at first suggest that the risk/benefit of (*R*)-timolol in glaucoma may be superior to (*R,S*)-timolol or (*S*)-timolol. In man, the (*R*)-enantiomer of timolol is four times less potent than (*S*)-timolol in reducing intraocular pressure and 13 times less potent on the airways of *normal* subjects. If the same were to be true in susceptible patients, (*R*)-timolol would have potential benefits over timolol in the treatment of glaucoma, *particularly in those patients with bronchial hyperreactivity.* However, this has not proved to be the case.

In subjects with *mild asthma* who bronchoconstricted to timolol eye drops (0.25% or 0.5%), both (*R*)-timolol and (*S*)-timolol caused a dose-dependent fall in specific airways conductance and FEV1 with (*R*)-timolol was approximately four times less potent than (*S*)-timolol (RICHARDS and TATTERSFIELD 1987). Since the difference in potency of respiratory effects is similar to the reported difference in potency of the two drugs on intraocular pressure, it was concluded that (*R*)-timolol would not have a greater safety margin than timolol. These findings do not support a better risk/benefit ratio for (*R*)-timolol compared to (*S*)-timolol in the treatment of glaucoma. Findings such as these illustrate the risks of extrapolating data from animals to man and from normal individuals to at risk groups.

VII. Fluoxetine

This widely used antidepressant is marketed as a racemic mixture. One potential difficulty may be the long elimination half-life of the drug. Stereosensitive pharmacokinetic studies have shown that the clearance of (*R*)-fluoxetine is about fourfold greater and the duration of its activity threefold shorter than its (*S*)-antipode (ROBERTSON et al. 1988). The clinical availability of (*R*)-fluoxetine might represent a great advantage for clinical use of the drug in individuals in whom a greater dosing flexibility is required, e.g., in the elderly. This isomer has also been claimed to show significantly enhanced activity at the $5HT_{2A}$ and $5HT_{2C}$ receptors while maintaining the same potency as a serotonin reuptake inhibitor – properties which could offer enhanced efficacy.

Furthermore, both (*R*)- and (*S*)-fluoxetine are competitive inhibitors of CYP 2D6. In vitro data suggest that (*R*)-norfluoxetine is less potent than (*S*)-norfluoxetine in this respect (STEVENS and WRIGHTON 1993). Thus, (*R*)-fluoxetine may be less prone to inhibit CYP 2D6 compared to (*S*)-fluoxetine; this may prove to be an advantage for the elderly who are likely to be in receipt of other CYP 2D6 substrates.

A case for developing (*R*)-fluoxetine on the basis of relative clearances would have to be very carefully made. (*R*)-fluoxetine and (*S*)-fluoxetine have half-lives of 2.6 days and 1.1 days, respectively, in EMs and 9.5 days and 6.1 days, respectively, in PMs of CYP 2D6 (Fjordside et al. 1999). The eudismic ratio for the fluoxetine enantiomers is near unity. However, both are metabolized to their corresponding metabolite, norfluoxetine, which has longer half-lives. In contrast to the parent enantiomers, (*S*)-norfluoxetine is the active *N*-demethylated metabolite responsible for the persistently potent and selective inhibition of serotonin uptake in vivo. Additionally, results from early clinical trials suggest that the risk/benefit ratio of (*R*)-fluoxetine may warrant careful reevaluation since its use in about 2000 patients has raised concerns over its potential to prolong the QT interval at the highest dose administered. Consequently, the sponsors of (*R*)-fluoxetine are reported to have terminated its further development.

E. Stereogenic Origin of Clinical Safety Concerns

Very often, the two enantiomers differ sufficiently in their pharmacology that the toxicity of the racemic mixture is associated with one of the enantiomers (which may not necessarily be the therapeutically active isomer). When one of the two enantiomers has a different and often undesirable pharmacological activity, it may be possible to greatly reduce certain risks. While the perceived therapeutic benefits of a proposed racemic mixture are emphasized all too often, the risks of administering racemic mixtures appear to be poorly appreciated or evaluated. Equally, when a pure enantiomer of a previously marketed racemic drug is developed, claims for the superiority of the enantiomer are often based on poorly documented evidence.

Reasons for developing a single enantiomer could include the facts that (a) one of the enantiomers is highly potent and there is hardly any chiral interconversion; this allows reduction of daily dose and consequently the "metabolic load," (b) the secondary pharmacology of the drug, responsible for toxic effects, resides predominantly in one enantiomer, (c) the pharmacokinetics are enantioselective; this allows the choice of the isomer with long half-life in order to decrease dosing frequency or of the isomer with shorter half-life in order to diminish the risk of accumulation in susceptible population, e.g., the elderly, or if the overdose with the drug is a distinct clinical possibility, (d) there is great interindividual variability which could be eliminated by developing only the enantiomer which is not subject to polymorphic metabolism, (e) therapeutically "inactive" enantiomer alters in a detrimental manner the pharmacokinetics (or pharmacodynamics) of the therapeutically active enantiomer (enantiomer–enantiomer interaction), (f) notwithstanding the slight loss in potency, one of the enantiomers has the required "site-specificity" and (g) the inhibition of drug-metabolizing CYP isoforms by the two enantiomers is enantioselective; this could be exploited to improve the drug–drug interaction profile of the racemic mixture if this is likely to prove to be a major clinical problem.

I. Disopyramide

Disopyramide, a class Ia antiarrhythmic drug, is used for the treatment of a variety of arrhythmias, including those following acute myocardial infarction. The principal serious side effects of the drug are anticholinergic and negative inotropic effects. These may precipitate glaucoma or urinary retention and cardiac failure in susceptible patients. Of particular interest are the reports of ventricular tachycardia or fibrillation or torsade de pointes in patients receiving disopyramide. Usually, but not always, these proarrhythmias have been associated with high doses of the drug which cause significant widening of QRS complex or QT interval prolongation. It is recommended that if the QT interval is prolonged by more than 25%, treatment with disopyramide should be discontinued. The adverse effect of disopyramide on cardiac repolarization may not have attracted much clinical attention since prolongation of the QT interval itself, when not excessive, can be therapeutically beneficial and the enantiomer responsible for this effect is generally eliminated more rapidly than its antipode. Also, the concerns may have focused on its anticholinergic and negative inotropic effects.

Disopyramide, a 2-phenylbutyramide derivative, is optically active (Fig. 9). Both enantiomers possess the antiarrhythmic sodium channel blocking class I activity. The negative inotropic effect of the (+)-(*S*)-enantiomer is approximately 20%–25% of that of the (−)-(*R*)-antipode. This suggests a good case for developing only the (+)-(*S*)-enantiomer. However, (+)-(*S*)-disopyramide has class III (KIDWELL et al. 1987; MIRRO et al. 1981; POLLICK et al. 1982) and potent anticholinergic activities (GIACOMINI et al. 1980) in addition to sharing class I activity with its (−)-(*R*)-antipode. The anticholinergic effect may, to a certain extent, mitigate the class III activity. The case for developing a single enantiomer becomes less compelling in view of this complex mixture of pharmacology.

II. Propafenone

Propafenone is a class Ic antiarrhythmic drug for the treatment and prophylaxis of ventricular arrhythmias and paroxysmal supraventricular tachyarrhythmias.

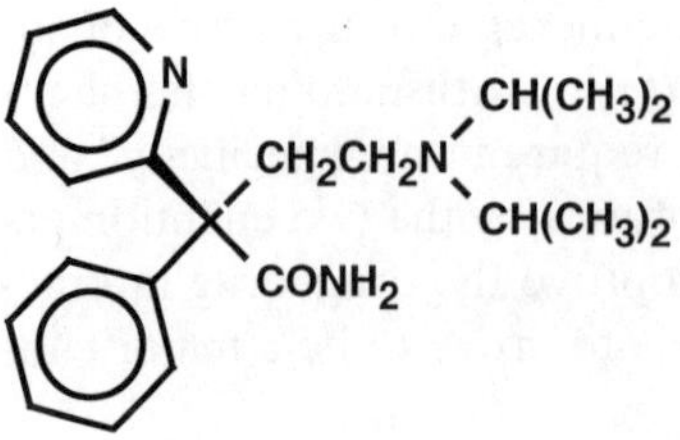

Fig. 9. (+)-(*S*)-disopyramide

Fig. 10. (–)-(*S*)-propafenone

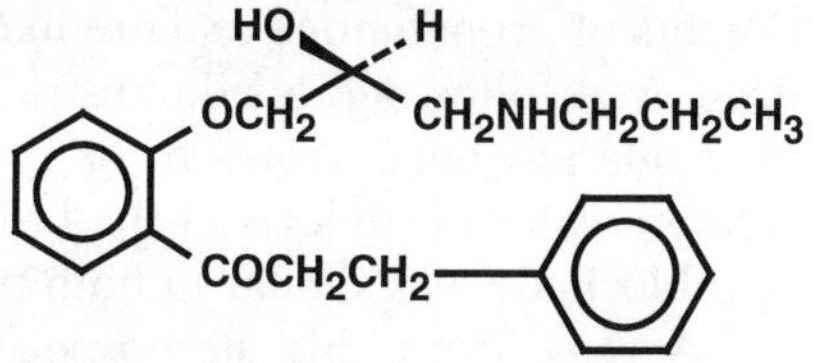

Pharmacologically, it has a membrane-stabilizing effect and reduces the fast inward sodium current. In addition, it has mild to moderate β-adrenoceptor blocking activity (Ledda et al. 1981) which probably contributes to its antiarrhythmic effects. However, this activity may become clinically significant and prove harmful in a subset of particularly susceptible patients (e.g., those with impaired LV function or chronic obstructive airways disease).

Propafenone is extensively metabolized, predominantly by hydroxylation to 5-hydroxypropafenone (by CYP 2D6). The resultant metabolite is as active as, if not more so than, the parent drug in its sodium channel blocking activity but is much less potent as a β-adrenoceptor blocking agent than the parent drug in animal studies. Since CYP 2D6 displays genetic polymorphism in the population, the PMs (at a given dose) attain much higher plasma levels of propafenone than do the EMs (Siddoway et al. 1987).

Chemically, propafenone is an aryloxypropanolamine derivative with one chiral center and both of the enantiomers have similar activity on sodium channels (Kroemer et al. 1989). However, its β-adrenoceptor blocking activity is due almost exclusively to the (*S*)-enantiomer (Fig. 10) (Groschner et al. 1991). In terms of β-blockade, (*R,S*)-propafenone was as active as (*S*)-propafenone at half the dose while (*R*)-propafenone was inactive in volunteers (Stoschitzky et al. 1990). Using bronchial provocation tests in patients with mild asthma, propafenone has been shown to predispose these individuals to bronchial hyperreactivity (Hill et al. 1986). Potentially relevant increase in airway reactivity occurred in 7 of the 12 volunteers following administration of propafenone 300 mg 8 hourly and in 1 of the 12 volunteers following 150 mg 8 hourly for at least 2 days.

The metabolism of propafenone by CYP 2D6 is enantioselective, favoring the elimination of (*R*)-propafenone (Kroemer et al. 1989) with the plasma levels of (*S*)-enantiomer about twofold greater than those of the (*R*)-enantiomer at any given time. Plasma concentrations of (*S*)-propafenone associated with β-blockade are in the range of 800–1000 ng/ml. However, since (*R*)-propafenone inhibits the metabolism of its (*S*)-antipode, this enantiomer–enantiomer interaction is beneficial for the antiarrhythmic effect of the drug but harmful to those with LV dysfunction or with obstructive airways disease. It is also obvious that, due to a markedly diminished clearance of the drug in PMs, these individuals will attain much higher levels of both the enantiomers of propafenone and are likely to display the β-blocking effects of the drug even following its administration at low doses. Even at lower doses, β-blockade has been shown to be present in both phenotypes but was significantly greater in subjects of the PM phenotype. At higher doses, the intensity of β-blockade was similar in both phenotypes (Lee et al. 1990).

Arising from this stereospecific β-adrenoceptor blocking effect, propafenone is contraindicated in patients with uncontrolled congestive heart failure and in those with severe bradycardia, as well as those with severe obstructive pulmonary disease; caution is advised when using it in patients with asthma.

III. Salbutamol

Salbutamol, a β_2-agonist, is a particularly interesting example of a drug in which the two enantiomers display antagonist activity. While (–)-(*R*)-salbutamol prevents bronchoconstriction, (+)-(*S*)-salbutamol enhances bronchoconstriction caused by carbachol or leukotriene C4. In addition, the clearance of the therapeutic (*R*)-isomer is faster than that of the harmful (*S*)-enantiomer; thus, on chronic administration, the accumulation of toxic isomer may be expected to occur. This may explain paradoxic bronchoconstriction is some individuals. Although the data are equivocal, (+)-(*S*)-salbutamol may increase the bioavailability of (–)-(*R*)-salbutamol by competitively inhibiting its metabolism.

Endogenous adrenaline is produced exclusively as the single isomer, (*R*)-adrenaline, although all selective β_2-agonists are marketed as racemic drugs. The (*R*)-isomers of these drugs, essentially all congeners of (*R*)-adrenaline, produce the observed bronchodilatation and clinical benefit of the racemate. The (*S*)-isomer of adrenaline is inert and those of the racemic β_2-agonists are devoid of clinical benefit, are assumed to be benign and have not been studied until recently. Data from carbachol-induced contraction studies on isolated tracheal strips from guinea-pig have shown that both racemic formoterol and (*R*,*R*)-formoterol produced an immediate relaxation, followed by a slow recovery of tone. In contrast, (*S*,*S*)-formoterol had no effect on smooth muscle tone. Similar results were obtained with the enantiomers of terbutaline. The eutomers, (*R*,*R*)-formoterol and (*R*)-terbutaline, inhibited cholinergic-induced contractions in a concentration-dependent manner. The distomers, (*S*,*S*)-formoterol and (*S*)-terbutaline, showed qualitatively the same effects but were about 1000 times less potent than the corresponding eutomer.

In contradistinction to its assumed benign status, extensive studies with (+)-(*S*)-salbutamol have shown that it opposes the bronchodilatory effects of (–)-(*R*)-salbutamol. In one small study (Boulton and Fawcett 1997), a single oral dose of (–)-(*R*)-salbutamol was shown to produce larger pharmacodynamic effects in the absence of (+)-(*S*)-salbutamol. In another study of 10 healthy subjects, using inhaled racemic salbutamol with a metered-dose inhaler (MDI) alone and with a MDI and holding chamber, the results strongly suggested that there was a preferential retention of (+)-(*S*)-salbutamol in the lung. This could lead to accumulation of the (*S*)-enantiomer after long-term use of racemic salbutamol. It has been observed that (+)-(*S*)-salbutamol may cause mobilization of intracellular Ca2+, apparently by means of a cholinergic mechanism.

The antagonist interaction between the two isomers is elegantly demonstrated in a study of 362 asthmatics (Nelson et al. 1998). Improvement in FEV1 was similar on 0.63 mg (–)-(*R*)-salbutamol and 2.5 mg racemic salbutamol, and greatest on 1.25 mg (–)-(*R*)-salbutamol, especially in subjects with severe asthma. Racemic salbutamol 1.25 mg demonstrated the weakest bronchodilatory effect, particularly after chronic dosing. All active treatments were well tolerated. β-adrenergic side effects after administration of 0.63 mg of (–)-(*R*)-salbutamol were reduced relative to 1.25 mg (–)-(*R*)-salbutamol or 2.5 mg racemic salbutamol. If (+)-(*S*)-salbutamol was inactive, 1.25 mg of racemic drug would have been equivalent to 0.63 mg of (–)-(*R*)-salbutamol.

Given that the distomer (+)-(*S*)-salbutamol has an adverse effect on the pharmacodynamics and a possible beneficial effect on the pharmacokinetics of the eutomer, (–)-(*R*)-salbutamol, a definite clinical benefit of this chiral switch remains to be shown. Not surprisingly, no claims of superiority over the racemate have been allowed in the product information of (–)-(*R*)-salbutamol.

IV. Oxybutynin

Oxybutynin is widely used for urinary incontinence. Typical doses are 15–20 mg daily. The therapeutic benefit is mainly due to anticholinergic *and* antispasmodic activities. Its clinical use is associated with a high frequency of dose-dependent anticholinergic side effects such as blurred vision, dry mouth and constipation. In one study, it was determined that the ratio of anticholinergic to antispasmodic activities was 50 for (*R*)-oxybutynin and only 2 for (*S*)-isomer. Furthermore, (*R*)-oxybutynin is metabolized to (*R*)-deoxybutynin, which too has anticholinergic activity (Noronha-Blob and Kachur 1991; Smith et al. 1998).

These data would suggest that it may be possible to achieve the same degree of efficacy as racemic drug by administration of (*S*)-oxybutynin (at doses much higher than those of (*R*,*S*)-oxybutynin) but without the accompanying anticholinergic side effects. Indeed, in one study, (*S*)-oxybutynin (at doses of 320–480 mg three times daily) significantly improved urinary frequency in a dose-dependent manner. Only 14%–16% of the patients receiving this high dose of (*S*)-oxybutynin complained of anticholinergic side effects. This contrasts with up to 50% of patients experiencing these effects following the routine clinical use of (*R*,*S*)-oxybutynin.

V. Citalopram

Citalopram is a potent and selective serotonin reuptake inhibitor used for depression. Following a suicidal overdose, it has been reported to cause sudden death (ÖSTRÖM et al. 1996) and QT interval prolongation. The pharmacological activity of citalopram resides in the (+)-(*S*)-enantiomer with the eudismic ratios of 167 and 6.6 for citalopram and its metabolite *N*-demethylcitalopram (DCT), respectively. The pharmacological profiles of the eutomers of citalopram and *N*-demethylcitalopram very much resemble the profile of the respective racemates (HYTTEL et al. 1992). Citalopram is metabolized principally by CYP 2C19. A modest, but statistically significant, stereoselectivity has been observed in the disposition of citalopram and its two main metabolites, DCT and didemethylcitalopram (DDCT). Serum levels of the (+)-(*S*)-enantiomers of citalopram, DCT and DDCT throughout the steady-state dosing interval investigated were 37%, 42% and 32%, respectively, of their total racemic serum concentrations. The (+)-(*S*)-enantiomers of citalopram, DCT and DDCT were eliminated faster than their antipodes. For (+)-(*S*)-citalopram and (–)-(*R*)-citalopram, the mean serum half-lives averaged 35 h and 47 h respectively (SIDHU et al. 1997). Although DDCT is normally a minor metabolite in man, it may accumulate following an overdose. Although in vitro studies show that citalopram does not prolong action potential duration (PACHER et al. 2000), DDCT is cardiotoxic and prolongs QT interval (ÖSTRÖM et al. 1996). Recent studies have shown that the racemic metabolite DDCT caused a dose-dependent QT interval prolongation and this was almost exclusively attributable to (–)-(*R*)-DDCT. These pharmacological differences between the enantiomers raise the possibility of developing (+)-(*S*)-citalopram with a potentially better risk/benefit ratio. In one 4-week clinical study of 468 patients, (+)-(*S*)-citalopram 10 mg/daily was shown to have a superior therapeutic effect from week 1 onwards when compared with racemic citalopram 20 mg/daily, which was no different at week 1 from placebo (MONTGOMERY et al. 2001). The superior efficacy of (+)-(*S*)-citalopram over racemic citalopram suggests a likelihood of an antagonistic effect of (–)-(*R*)-citalopram at a pharmacodynamic level. There would appear to be a good case for developing the more rapidly eliminated (+)-(*S*)-citalopram if its cardiac safety can be confirmed. Rapid elimination is also an advantage in case of drugs likely to be prescribed to a population prone to suicidal overdose attempts. Indeed, (+)-(*S*)-citalopram has recently been approved ("Cipralex") but without any claims of superiority.

VI. Halofantrine

Halofantrine is an effective antimalarial drug of the phenanthrenemethanol group. Its clinical use is associated with a relatively high frequency of dose-dependent prolongation of QT interval and potentially fatal ventricular tachyarrhythmias, including torsade de pointes. Despite its efficacy in chloroquine-resistant malaria, regulatory authorities were prompted to issue warnings urging caution in its use (Anon 1994). As of June 1995, the Food and Drug Administration (FDA) in the USA had received 17 reports of QT interval prolongation, of which 11 had developed ventricular tachyarrhythmias and 2 had syncope. In addition, there were 13 reports in which QT information was not available, but included 10 sudden cardiac deaths and 2 cardiac arrests. Halofantrine is optically active and undergoes CYP 3A4-mediated stereoselective metabolism to *N*-desbutylhalofantrine, which is also optically active. Following multiple dosing, the concentrations and AUC of (+)-(*R*)-halofantrine are significantly higher (Gimenez et al. 1994). Following 500 mg of racemic halofantrine once daily in the fasted state for 42 days, the mean steady-state concentrations were 97.6 ng/ml for (+)-(*R*)-halofantrine and 48.5 ng/ml for (–)-(*S*)-halofantrine (Abernethy et al. 2001). With regard to therapeutic activity, the metabolite has been shown in in vitro studies to be as active as the corresponding enantiomer of the parent drug (Basco et al. 1994). A recent in vitro study investigating stereoselective aspects of cardiotoxicity of halofantrine has shown that this potentially fatal toxic activity resides predominantly in (+)-(*R*)-halofantrine, whereas *N*-desbutyl metabolite was virtually inactive in this respect (WESCHE et al. 2000). Interestingly, the cardiotoxic potential of the racemic mixture was much greater than either enantiomer.

This finding is consistent with previous observations that the QTc interval following halofantrine administration is significantly correlated with the plasma halofantrine level but not with the plasma *N*-desbutylhalofantrine level (Touze et al. 1996) and that in 14 of 16 subjects who received active drug had QTc interval prolongation that was positively correlated with both (+)-(*R*)- and (–)-(*S*)-halofantrine concentrations (Abernethy et al. 2001), the correlation being stronger with (+)-(*R*)-halofantrine. These data would favor the development of (–)-(*S*)-halofantrine or preferably of *N*-desbutylhalofantrine as potentially safer antimalarial alternatives to racemic halofantrine.

F. Stereoselectivity and Drug Withdrawals

There are many drugs which have been withdrawn from the market due to significant clinical safety concerns. For some drugs, these concerns have resulted from toxicity attributable to one of the enantiomers in the racemate. Not surprisingly, there is not much information available concerning the drugs whose clinical development has been suspended following early findings of enantioselective toxicity.

I. Bufenadrine

Bufenadrine was never marketed but it is discussed briefly as an introduction to dilevalol, which is discussed next. Both these drugs have shown enantiospecific hepatotoxicity.

Bufenadrine (Fig. 11a) is a structural analogue of diphenhydramine (Fig. 11b), an antihistamine used for the treatment of motion sickness. Substi-

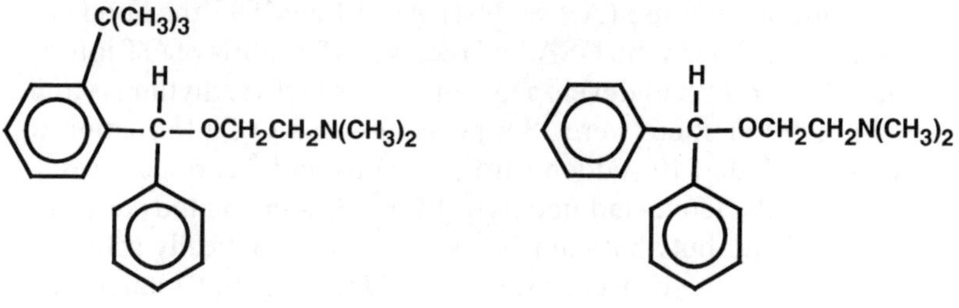

(a) **(b)**

Fig. 11. **a** Bufenadrine and **b** diphenhydramine

tution of a *tert*-butyl side chain into one of the phenyl rings of diphenhydramine yields bufenadrine and introduces a chiral center, giving rise to a pair of enantiomers. Chronic toxicity studies in rats showed that administration of bufenadrine was associated with hepatotoxicity, considered to have resulted from the accumulation of the drug in liver. Studies with the individual enantiomers showed that hepatotoxicity was associated with the (–)-isomer, which was not only much less potent as an antihistamine but also had more potent anticholinergic activity than its antipode (HESPE et al. 1972). The data suggested that the liver was unable to metabolize (–)-bufenadrine effectively, leading to its accumulation and resulting toxicity. Because of this stereoselective hepatotoxicity, the development of bufenadrine was terminated (POWELL et al. 1988).

II. Dilevalol

Dilevalol is the (*R*,*R*)-diastereoisomer of labetalol (Fig. 12). Labetalol has been in clinical use since 1977 for the treatment of hypertension. Compared to the racemate, this isomer has fourfold β_1-adrenoceptor blocking and sevenfold vasodilatory activity due to its agonist activity at β_2-adrenoreceptors (SYBERTZ et al. 1981). Clinically, introduction of dilevalol generated great excitement as it represented a new generation of β-blockers with vasodilating properties. However, it was known that after 1 year of therapy in clinical trials, serum transaminase levels were elevated in about 0.8%–1.2% of patients receiving dilevalol; this was a major concern to the advisory committee of the FDA (ANON 1990a). It was recommended that a postmarketing study be carried out to better characterize the risk of hepatotoxicity during the long-term use of the drug.

Soon after its introduction onto the market in December 1989, reports of hepatotoxicity began to appear and by August 1990, there had been 17 reports of hepatotoxicity (one fatal) associated with its clinical use (ANON 1990b). The

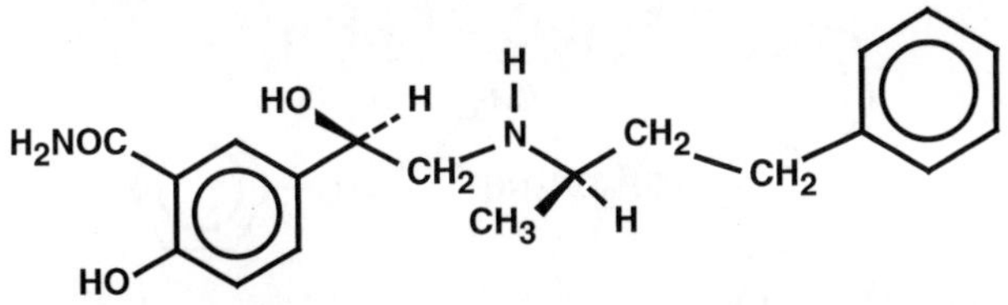

Fig. 12. Dilevalol [(–)-(*R*,*R*)-labetalol]

incidence rate of dilevalol-induced hepatotoxicity was calculated to be 1 in 1015 of those patients exposed to the drug for a period longer than 3 months (ANON 1990c). Although hepatotoxicity is also known to occur with labetalol, the incidence rate appears to be much lower than that associated with dilevalol (CLARK et al. 1990). Dilevalol was withdrawn from the market in August 1990.

III. Prenylamine

Prenylamine, an antianginal agent, was introduced in the market in the early 1960s. Reports linking prenylamine with prolongation of the QT interval, ventricular tachycardia, ventricular fibrillation and torsade de pointes began to appear from 1971 onwards (PICARD et al. 1971). Despite warnings and advice to increase the dose more gradually, reports of these ventricular arrhythmias continued to appear. By 1988, 158 cases of polymorphous ventricular tachycardia were reported in association with prenylamine. Some of these events had a fatal outcome and the drug was withdrawn from the market world-wide in 1988 (ANON 1988).

Prenylamine is a potent inhibitor of calmodulin-dependent enzymes, relaxes smooth muscle and reduces slow inward current. It has negative inotropic effects in addition to class I and class III electrophysiological activities (HASHIMOTO et al. 1978; BAYER et al. 1988). The two enantiomers of prenylamine have been shown to possess enantiospecific pharmacodynamic effects (RODENKIRCHEN et al. 1980). In cat papillary muscle preparations, (+)-prenylamine (Fig. 13) has positive inotropic action and prolongs action potential duration while the (–)-isomer had a negative inotropic effect and shortened the action potential duration but only to a minor extent (BAYER et al. 1988). (+)-prenylamine also caused dysrrhythmias in 4 of the 12 isolated papillary muscle preparations.

Prenylamine is extensively metabolized primarily by ring hydroxylation in man and less than 0.1% of the dose is excreted unchanged. The metabolism displays wide interindividual variation and is enantioselective, favoring the elimination of the (+)-(*S*)-enantiomer (PAAR et al. 1990; GEITL et al. 1990). In two of the eight volunteers, plasma half-lives for the (+)-(*S*)-enantiomer following the single dose were extremely long (82h and 83h in contrast to the mean values of 24h) (GEITL et al.1990). None of the eight subjects were phe-

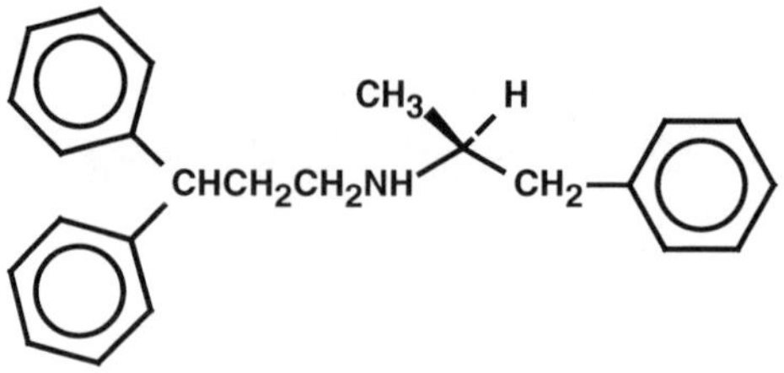

Fig. 13. (+)-(*S*)-prenylamine

notyped, but these extreme values raise the possibility that prenylamine hydroxylation may be genetically controlled (probably by CYP 2D6). Together with its structural similarity to terodiline, prenylamine meets all other criteria for being a CYP 2D6 substrate. While the (–)-(*R*)- prenylamine may shorten the duration of action potential, its (+)-(*S*)-enantiomer has positive inotropic effect and prolongs the action potential duration. Therefore, overall, the data suggest that the proarrhythmic effect of prenylamine may have been mediated by (+)-(*S*)-prenylamine which, at low concentrations (like quinidine), prolongs action potential plateau and total action potential duration.

IV. Terodiline

Terodiline represents perhaps the best documented example of a drug which had to be withdrawn from the market as a result of proven stereospecific toxicity. Structurally, and in many respects pharmacologically, it bears a striking resemblance to prenylamine. It was first marketed as an antianginal agent in Scandinavia but urinary retention proved to be a frequent and troublesome side effect. Therefore, the drug was redeveloped and marketed in 1986 for clinical use in urinary incontinence. Beginning 1988, reports of QT interval prolongation and torsade de pointes following the clinical use of terodiline began to appear. Following 3 more reports over the next 2 years, the number of reports of serious cardiotoxicity had increased by October 1991 to 69 (including 24 reports of torsade de pointes). Of these, 14 had a fatal outcome. In September 1991, the drug was withdrawn from the market world-wide (ANON 1991).

Terodiline has potent calcium antagonist and antimuscarinic activities. Since terodiline is related structurally to prenylamine, it is not surprising that its pharmacological effects are also enantioselective. Calcium antagonist activity resides predominantly in (–)-(*S*)-terodiline, while the anticholinergic activity is predominantly found in (+)-(*R*)-terodiline (Fig. 14) (LARSSON-BACKSTRÖM et al. 1985; ANDERSSON et al. 1988). Both activities probably contribute to the therapeutic effect to a variable extent. Recently, it has been shown that the prolongation of the QT interval associated with racemic terodiline is caused exclusively by the (+)-(*R*)-enantiomer. Therefore, this antipode is responsible

Fig. 14. (+)-(*R*)-terodiline

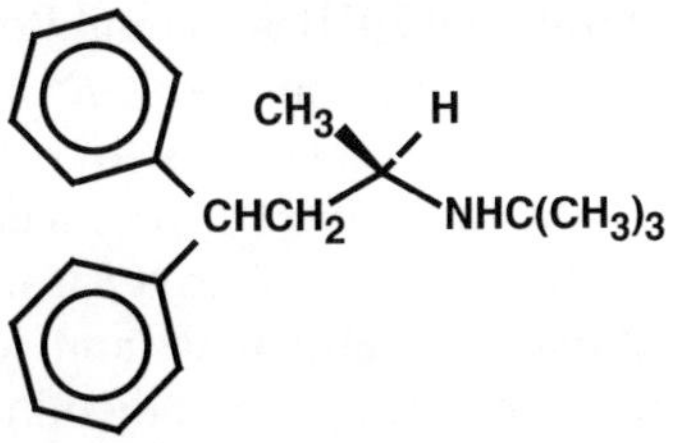

for the serious ventricular tachyarrhythmias observed with the clinical use of this drug (HARTIGAN-GO et al. 1996). It has also been shown to block delayed rectifier potassium channels (JONES et al. 1998). Unfortunately, however, these workers did not investigate the two enantiomers separately.

The half-life of the parent drug is much longer in the elderly (130h versus 60h) (HALLÉN et al. 1989). The metabolism of terodiline is shown to be stereoselective for the (+)-(*R*)-enantiomer (NORÉN et al. 1989; HALLÉN et al. 1995). One study reported that the formation of *p*-hydroxyterodiline from (+)-(*R*)-terodiline was impaired in the only poor metabolizer of debrisoquine in the study (HALLÉN et al. 1993). This observation suggests that the metabolism of terodiline is most likely mediated principally by CYP 2D6 (probably with some contribution by CYP 3A4) and therefore, would display genetically controlled polymorphism of the debrisoquine/sparteine type. The consequence of this stereosensitive polymorphic metabolism is that (–)-(*S*)-terodiline (which is a calcium antagonist) would accumulate in all patients over time but, in addition, there would be an accumulation of the cardiotoxic (+)-(*R*)-terodiline in the poor and intermediate metabolizers of debrisoquine or sparteine. An analysis of predisposing factors in the 69 reports of cardiotoxicity due to terodiline had shown that in 12 cases (18%), there were no clinically identifiable risk factors at all. Torsade de pointes on terodiline have been reported in patients without any risk factors and in whom plasma terodiline levels were markedly elevated (CONNOLLY et al. 1991; ANDREWS and BEVAN 1991). Thus, in a number of patients, it seems reasonable to conclude that the cardiotoxicity of terodiline originated from impaired metabolism of the (+)-(*R*)-enantiomer.

The marketing authorization holder of terodiline has recently introduced the (*R*)-enantiomer of tolterodine, a newly synthesized structural analogue of terodiline. This enantiomer has antimuscarinic properties, is primarily metabolized by CYP 2D6 and is marketed for the treatment of urinary incontinence.

V. Levacetylmethadol

Stereoselective interactions of drugs at human voltage-gated potassium channels are being reported with an increasing number of other drugs. For example, chromanol 293B is an inhibitor of the slowly activating, delayed rectifier potassium channel but (–)-(3*R*, 4*S*)-chromanol 293B is far more potent in this respect than is (+)-(3*S*, 4*R*)-chromanol 293B (YANG et al. 2000).

In this context, the recent withdrawal of levacetylmethadol is of great interest. Marketed as "Orlaam," it is a synthetic opioid analgesic, structurally similar to methadone, and was approved for the substitution maintenance treatment of opiate addiction in adults previously treated with methadone, as part of a comprehensive treatment plan including medical, social and psychological care. Its only advantage was its long half-life facilitating its administration 3 days a week instead of daily administration of methadone. There is no information on its antipode but levacetylmethadol, which is (–)-(3*S*, 6*S*)-isomer of acetylmethadol, has been reported to cause QT interval prolongation and induce potentially fatal torsade de pointes. There were 10 cases of this tachyarrhythmia after exposure of about 33000 patients to the drug world-wide (HAEHL 2001). This reporting rate represents a very high torsadogenic risk. (–)-(3*S*, 6*S*)-acetylmethadol was approved in the USA in 1994 and in the EU in 1997. Following these reports of proarrhythmias (a vast majority of these being from the USA), the Committee for Proprietary Medicinal Products (CPMP) decided to suspend the drug in the EU in April 2001 while the FDA was content with strengthening the labeling in the USA (ANON 2001a), thus illustrating how the perception of risk/benefit of a drug, notwithstanding the lack of adequate information on any enantioselective toxicity and/or benefit, can vary between different regulatory authorities.

G. Conclusions

Advances in medicinal chemistry, pharmacology and pharmaceutical technology are proceeding at a very rapid pace; innovative chemical structures which target novel receptors or have novel pharmacological actions are being developed. Unless the full physiological functions of these targets and the consequences of their pharmacological modulation are fully appreciated, there is a real risk that many of these new drugs may either not reach the market or, should an unexpected hazard emerge, may end up with a very limited market life. It could be argued that, in parallel with developing drugs with innovative chemical structures which target novel receptors or have novel pharmacological actions, a more pragmatic approach may be to reevaluate the potential of the existing drugs and, when appropriate, improve their risk/benefit ratio through stereochemistry.

The decision to develop a racemate or a single enantiomer (either de novo or as a switch) should be based on sound pharmacological principles. Unfortunately, however, the decisions for chiral switches are often driven commercially by the attraction of exclusivity in marketing the pure enantiomer. Often, there are poorly substantiated assumptions of a superior pharmacological profile of the single enantiomer of interest over the corresponding racemate without adequately addressing the potential hazards that may be associated with its use.

Once the racemic drug is off patent, cheaper generic products may appear in competition. Chiral switching in this setting subdues this competition. Nexium was launched in September 2000. Sales data from the UK and Germany have shown that as of February 2001, 47% of 500 sample prescriptions were for patients new to PPIs, while 9.4% were switches from omeprazole and an additional 10.6% were switches from other PPIs (Anon 2001b).

Despite their short-term advantages, commercially driven decisions for chiral switches are likely to be counterproductive in the long-term, especially if the (possibly more expensive) single enantiomer is soon found to offer no tangible clinical advantage over the well-established racemic product.

References

Abernethy DR, Wesche DL, Barbey JT, Ohrt C, Mohanty S, Pezzullo JC, Schuster BG (2001) Stereoselective halofantrine disposition and effect: concentration-related QTc prolongation. Br J Clin Pharmacol 51:231–237

Adamantidis MM, Dumotier BM, Caron JF, Bordet R (1998) Sparfloxacin but not levofloxacin or ofloxacin prolongs cardiac repolarization in rabbit Purkinje fibers. Fundam Clin Pharmacol 12:70–76

Aithal GP, Day CP, Kesteven PJ, Daly AK (1999) Association of polymorphisms in the cytochrome P450 CYP2C9 with warfarin dose requirement and risk of bleeding complications. Lancet 353:717–719

Andersson K-E, Ekström B, Mattiasson A (1988) Actions of terodiline, its isomers and main metabolite on isolated detrusor muscle from rabbit and man. Pharmacol & Toxicol 63:390–395

Andrews NP, Bevan J (1991) Torsade de pointes and terodiline Lancet, 338:633

Anon (1980) Ticrynafen Recalled. FDA Drug Bulletin 10(1):3–4

Anon (1988) Prenylamine withdrawn in UK. Scrip (England) 1300:26

Anon (1990a) Schering's UNICARD recommended for once-daily use in hypertension by FDA. F-D-C Reports. The Pink Sheet. 52(5):18

Anon (1990b) Dilevalol cost Schering $100million. Scrip (England) 1543:19

Anon (1990c) Report on the analysis of cases of drug induced liver injury associated to dilevalol. Schering-Plough, Kenilworth, NJ

Anon (1991) Withdrawal of terodiline. Current Problems (Issue 32) Committee on Safety of Medicines, London 1–2

Anon (1994) Cardiac arrhythmias with halofantrine (Halfan). Current Problems, Committee on Safety of Medicines, London, 20:3

Anon (2001a) EMEA Press Release announcing suspension of "Orlaam" http://www.emea.eu.int/pdfs/human/press/pus/877601en.pdf

Anon (2001b) AstraZeneca Nexium data shows 47% of scripts are first-time PPI users F-D-C Reports. The Pink Sheet. 63(7):20

Banitt EH, Schmid JR, Newmark RA (1986) Resolution of flecainide acetate, *N*-(2-piperidylmethyl)-2,5-bis(2,2,2-trifluoroethoxy)benzam ide acetate, and antiarrhythmic properties of the enantiomers. J Med Chem 29:299–302

Bartsch W, Sponer G, Strein K, Muller-Beckmann B, Kling L, Bohm E, Martin U, Borbe HO (1990) Pharmacological characteristics of the stereoisomers of carvedilol. Eur J Clin Pharmacol 38 [Suppl 2]:S104–S107

Basco LK, Peytavin G, Gimenez F, Genissel B, Farinotti R, Le Bras J (1994) In vitro activity of the enantiomers of *N*-desbutyl derivative of halofantrine. Trop Med Parasitol 45:45–46

Bayer R, Schwarzmaler J, Pernice R (1988) Basic mechanism underlying prenylamine-induced torsade de pointes: differences between prenylamine and fendiline due to basic actions of the isomers. Curr Med Res Opin 11:254–272

Birgersdotter UM, Wong W, Turgeon J, Roden DM (1992) Stereoselective genetically-determined interaction between chronic flecainide and quinidine in patients with arrhythmias. Br J Clin Pharmacol 33:275–280

Blessington B (1997) Ethambutol and tuberculosis, a neglected and confused chiral puzzle. In: Aboul-Enein HY, Wainer IW (eds) The impact of stereochemistry on drug development and use. John Wiley & Sons Inc, New York, pp 235–261

Boulton DW, Fawcett JP (1997) Pharmacokinetics and pharmacodynamics of single oral doses of albuterol and its enantiomers in humans. Clin Pharmacol Ther 62:138–144

Chu DTW, Nordeen CW, Hardy DJ, Swanson RN, Giardina WJ, Pernet AG, Plattner JJ (1991) Synthesis, antimicrobial activities and pharmacological properties of enantiomers of temafloxacin hydrochloride. J Med Chem 34:168–174

Clark JA, Zimmerman HJ, Tanner LA (1990) Labetalol hepatotoxicity. Ann Intern Med 113:210–213

Connolly MJ, Astridge PS, White EG, Morley CA, Campbell-Cowan J (1991) Torsade de pointes, ventricular tachycardia and terodiline Lancet, 338:344–345

Cotzias GC, Papavasiliou PS, Gellene R (1969) Modification of Parkinsonism – chronic treatment with L-dopa. N Engl J Med 280:337–345

D'Amato RJ, Loughnan MS, Flynn E, Folkman J (1994) Thalidomide is an inhibitor of angiogenesis. Proc Nat Acad Sci 91:4082–4085

Daniels IR (2001) Omeprazole, its isomers and the carcinoid question. Lancet 357:1290–1291

Dawson R, Manson JMcK (2000) Omeprazole in oesophageal reflux disease. Lancet 356:1770–1771

Eriksson T, Bjorkman S, Roth B, Fyge A, Hoglund P (1995) Stereospecific determination, chiral inversion in vitro and pharmacokinetics in humans of the enantiomers of thalidomide. Chirality 7:44–52

Fabro S, Smith RL, Williams RT (1967) Toxicity and teratogenicity of optical isomers of thalidomide. Nature (London) 215:296

Fjordside L, Jeppesen U, Eap CB, Powell K, Baumann P, Brosen K (1999) The stereoselective metabolism of fluoxetine in poor and extensive metabolizers of sparteine. Pharmacogenetics 9:55–60

Fraunfelder FT, Barker AF (1984) Respiratory effects of timolol. N Engl J Med 311:1441

Giacomini KM, Cox BM, Blaschke TF (1980) Comparative anticholinergic potencies of *R*- and *S*-disopyramide in longitudinal muscle strips from guinea pig ileum. Life Sci 27:1191–1197

Gao YS, Nagao T, Bond RA, Janssens WJ, Vanhoutte PM (1991) Nebivolol induces endothelium-dependent relaxations of canine coronary arteries. J Cardiovasc Pharmacol 17:964–969

Geitl Y, Spahn H, Knauf H, Mutschler E (1990) Single and multiple dose pharmacokinetics of *R*-(−)- and *S*-(+)-prenylamine in man. Eur J Clin Pharmacol 38:587–593

Gill A, Addo M, Lewis J (1990) Therapeutic interaction of L- and D-nebivolol in SHR. FASEB Journal 4: A747

Gimenez F, Gillotin C, Basco LK, Bouchaud O, Aubry AF, Wainer IW, Le Bras J, Farinotti R (1994) Plasma concentrations of the enantiomers of halofantrine and its main metabolite in malaria patients. Eur J Clin Pharmacol 46:561–562

Groschner K, Lindner W, Schnedl H, Kukovetz WR (1991) The effects of the stereoisomers of propafenone and diprafenone in guinea-pig heart. Br J Pharmacol 102:669–674

Gross AS, Mikus G, Fischer C, Hertrampf R, Gundert-Remy U, Eichelbaum M (1989) Stereoselective disposition of flecainide in relation to the sparteine/debrisoquine metaboliser phenotype. Br J Clin Pharmacol 28:555–566

Haehl M, Roxane Laboratories, Inc "Dear Healthcare Professional" letter dated 11 April 2001, http://www.fda.gov/medwatch/safety/2001/orlaam_deardoc.pdf

Hallén B, Bogentoft S, Sandquist S, Strömberg S, Setterberg G, Ryd-Kjellén E (1989) Tolerability and steady-state pharmacokinetics of terodiline and its main metabolites in elderly patients with urinary incontinence. Eur J Clin Pharmacol 36:487–493

Hallén B, Gabrielsson J, Palmér L, Ekström B (1993) Pharmacokinetics of *R*(+)-terodiline given intravenously and orally to healthy volunteers. Pharmacol & Toxicol 73:153–158

Hallén B, Gabrielsson J, Nyambati S, Johansson A, Larsson E, Guilbaud O (1995) Concomitant single-dose and multiple-dose pharmacokinetics of terodiline in man, with a note on its enantiomers and major metabolites. Pharmacol & Toxicol 76:171–177

Hartigan-Go K, Bateman ND, Daly AK, Thomas SHL (1996) Stereoselective cardiotoxic effects of terodiline. Clin Pharmacol Ther 60:89–98

Hashimoto K, Nakagawa Y, Nabata H, Imai S (1978) In vitro analysis of Ca-antagonistic effects of prenylamine as mechanisms for its cardiac actions. Arch Int Pharmacodyn Ther 273:212–221

Hayakawa I, Atarashi S, Yokohama S, Imamura M, Sakano K, Furukawa M (1986) Synthesis and antibacterial activities of optically active ofloxacin. Antimicrob Agents Chemother 29:163–164

Hespe W, Mulder D, van Eeken CJ (1972) Differences in metabolic behaviour and liver toxicity between the optical isomers of bufenadrine hydrochloride, a substituted diphenhydramine, in the rat. Acta Pharmacol Toxicol 31:369–379

Hill MR, Gotz VP, Harman E, McLeod I, Hendeles L (1986) Evaluation of the asthmogenicity of propafenone, a new antiarrhythmic drug: comparison of spirometry with metacholine challenge. Chest 90:698–702

Howe R, Shanks RG (1966) Optical isomers of propranalol. Nature 210:1336–1338

Hyttel J, Bogeso KP, Perregaard J, Sanchez C (1992) The pharmacological effect of citalopram resides in the (*S*)-(+)-enantiomer. J Neural Transm Gen Sect 88:157–160

Irvin JD, Vlasses PH, Huber PB, Feinberg JA, Ferguson RK, Scrogie JJ, Davies RO (1980) Different pharmacodynamic effects of the (+) and (–) enantiomers of indacrinone in man. Clin Pharmacol Ther 27:260 (abst)

Jones SE, Ogura T, Shuba LM, McDonald TF (1998) Inhibition of the rapid component of the delayed-rectifier K+ current by therapeutic concentrations of the antispasmodic agent terodiline. Br J Pharmacol 125:1138–1143

Julian DG, Prescott RJ, Jackson FS, Szekely P (1982) Controlled trial of sotalol for one year after myocardial infarction Lancet 1:1142–1147

Kato R, Ikeda N, Yabek S, Kannan R, Singh BN (1986) Electrophysiologic effects of the levo- and dextrorotatory isomers of sotalol in isolated cardiac muscle and their in vivo pharmacokinetics. J Am Coll Cardiol 7:116–125

Kang J, Wang L, Chen XL, Triggle DJ, Rampe D (2001) Interactions of a series of fluoroquinolone antibacterial drugs with the human cardiac K+ channel HERG. Mol Pharmacol 59:122–126

Kidwell GA, Schaal SF, Muir WW (1987) Stereospecific effects of disopyramide enantiomers following pretreatment of canine cardiac Purkinje fibers with verapamil and nisoldipine. J Cardiovasc Pharmacol 9:276–284

Kroemer HK, Funck-Brentano C, Silberstein DJ, Wood AJ, Eichelbaum M, Woosley RL, Roden DM (1989) Stereoselective disposition and pharmacologic activity of propafenone enantiomers. Circulation 79:1068–1076

Larsson-Backström C, Arrhenius E, Sagge K (1985) Comparison of the calcium antagonistic effects of terodiline, nifedipine and verapamil. Acta Pharmacol et Toxicol 57:8–17

Ledda F, Mantelli L, Manzini S, Amerini S, Mugell A (1981) Electrophysiological and antiarrhythmic properties of propafenone in isolated cardiac preparations. J Cardiovasc Pharmacol 3:1162–1173

Lee JT, Kroemer HK, Silberstein DJ, Funck-Brentano C, Lineberry MD, Wood AJ, Roden DM, Woosley RL (1990) The role of genetically determined polymorphic drug metabolism in the beta-blockade produced by propafenone. N Eng J Med 322:1764–1769

Meadowcroft AM, Williamson KM, Patterson JH, Pieper JA (1997) Pharmacogenetics and heart failure: A convergence with carvedilol. Pharmacotherapy 17:637–639

Mirro MJ, Watanabe AM, Bailey JC (1981) Electrophysiological effects of the optical isomers of disopyramide and quinidine in the dog. Circ Res 48:867–874

Montgomery SA, Loft H, Sánchez C, Reines EH, Papp M (2001) Escitalopram (*S*-enantiomer of citalopram): Clinical efficacy and onset of action predicted from a rat model. Pharmacol & Toxicol 88:282–286

Nathanson JA (1988) Stereospecificity of beta adrenergic antagonists: *R*-enantiomers show increased selectivity for beta-2 receptors in ciliary process. J Pharmacol Exp Ther 245:94–101

Nelson HS, Bensch G, Pleskow WW, DiSantostefano R, DeGraw S, Reasner DS, Rollins TE, Rubin PD (1998) Improved bronchodilation with levalbuterol compared with racemic albuterol in patients with asthma. J Allergy Clin Immunol 102 (Pt 1):943–952

Norén B, Strömberg S, Ericsson O, Lindeke B (1989) Biotransformation of terodiline V. Stereoselectivity in hydroxylation by human liver microsomes. Chem Biol Interactions 71:325–337

Noronha-Blob L, Kachur JF (1991) Enantiomers of oxybutynin: in vitro pharmacological characterization at M1, M2 and M3 muscarinic receptors and in vivo effects on urinary bladder contraction, mydriasis and salivary secretion in guinea pigs. J Pharmacol Exp Ther 256:562–567

Okazaki O, Kojima C, Hakusui H, Nakashima M (1991) Enantioselective disposition of ofloxacin in humans. Antimicrob Agents Chemother 35:2106–2109

Öström M, Eriksson A, Thorson J, Spigset O (1996) Fatal overdose with citalopram. Lancet 348:339–340

Paar WD, Brockmeier D, Hirzebruch M, Schmidt EK, von Unruh GE, Dengler HJ (1990) Pharmacokinetics of prenylamine racemate and enantiomers in man. Arzneim Forsch 40:657–661

Pacher P, Bagi Z, Lako-Futo Z, Ungvari Z, Nanasi PP, Kecskemeti V (2000) Cardiac electrophysiological effects of citalopram in guinea pig papillary muscle comparison with clomipramine. Gen Pharmacol 34:17–23

Patil PN (1968) Steric aspects of adrenergic drugs. VIII. Optical isomers of beta adrenergic receptor antagonists. J Pharmacol Exp Ther 160:308–314

Picard R, Auzepy P, Chauvin JP (1971) Syncopes a repetition au cours d'un traitement prolonge par la prenylamine (Segontine 60). Presse Med 79:145

Pollick C, Giacomini KM, Blaschke TF, Nelson WL, Turner-Tamiyasu K, Briskin V, Popp RL (1982) The cardiac effects of *d*- and *l*-disopyramide in normal subjects: a noninvasive study. Circulation 66:447–453

Powell JR, Ambre JJ, Ruo TI (1988) The efficacy and toxicity of drug stereoisomers. In: Wainer IW, Drayer DE (eds) Drug Stereochemistry: Analytical Methods and Pharmacology. Marcel Dekker Inc, New York, pp 245–270

Richards R, Tattersfield AE (1987) Comparison of the airway response to eye drops of timolol and its isomer L-714465 in asthmatic subjects. Br J Clin Pharmacol 24:485–491

Robertson DW, Krushinski JH, Fuller RW, Leander JD (1988) Absolute configurations and pharmacological activities of the optical isomers of fluoxetine, a selective serotonin-uptake inhibitor. J Med Chem 31:1412–1417

Rodenkirchen R, Bayer R, Mannhold R (1980) On the stereospecific negative inotropic action of prenylamine. Naunyn-Schmiedeberg's Arch Pharmacol 313 [Suppl]: R41

Siddoway LA, Thompson KA, McAllister B, Wang T, Wilkinson GR, Roden DM, Woosley RL (1987) Polymorphism of propafenone metabolism and disposition in man: clinical and pharmacokinetic consequences. Circulation 75:785–791

Sidhu J, Priskorn M, Poulsen M, Segonzac A, Grollier G, Larsen F (1997) Steady-state pharmacokinetics of the enantiomers of citalopram and its metabolites in humans. Chirality 9:686–692

Smallwood JK, Robertson DW, Steinberg MI (1989) Electrophysiological effects of flecainide enantiomers in canine Purkinje fibres. Naunyn Schmiedebergs Arch Pharmacol 339:625–629

Smith ER, Wright SE, Aberg G, Fang Y, McCullough JR (1998) Comparison of the antimuscarinic and antispasmodic actions of racemic oxybutynin and desethyloxybutynin and their enantiomers with those of racemic terodiline. Arzneimittelforschung 48:1012–1018

Spencer CM, Faulds D (2000) Esomeprazole. Drugs 60:321–329

Stevens JC, Wrighton SA (1993) Interaction of the enantiomers of fluoxetine and norfluoxetine with human liver cytochromes P450. J Pharmacol Exp Ther 266:964–971

Stoschitzky K, Klein W, Stark G, Stark U, Zernig G, Graziadei I, Lindner W (1990) Different stereoselective effects of (*R*)- and (*S*)-propafenone; clinical, pharmacologic, electrophysiologic and radioligand binding studies. Clin Pharmacol Ther 47: 740–746

Sybertz EJ, Sabin CS, Pula KK, Vliet GV, Glennon J, Gold EH, Baum T (1981) Alpha and beta adrenoceptor blocking properties of labetalol and its *R*,*R*-isomer, SCH 19927. J Pharmacol Exp Ther 218:435–443

Tanaka M, Ohkubo T, Otani K, Suzuki A, Kaneko S, Sugawara K, Ryokawa Y, Ishizaki T (2001) Stereoselective pharmacokinetics of pantoprazole, a proton pump inhibitor, in extensive and poor metabolizers of *S*-mephenytoin. Clin Pharmacol Ther 69:108–113

Touze JE, Bernard J, Keundjian A, Imbert P, Viguier A, Chaudet H, Doury JC (1996) Electrocardiographic changes and halofantrine plasma level during acute falciparum malaria. Am J Trop Med Hyg 54:225–228

Turgeon J, Funck-Brentano C, Gray HT, Pavlou HN, Prakash C, Blair IA, Roden DM (1991) Genetically determined stereoselective excretion of encainide in humans and electrophysiologic effects of its enantiomers in canine cardiac Purkinje fibers. Clin Pharmacol Ther 49:488–496

Tybring G, Bottiger Y, Widen J, Bertilsson L (1997) Enantioselective hydroxylation of omeprazole catalyzed by CYP2C19 in Swedish white subjects. Clin Pharmacol Ther 62:129–137

Van de Water A, Xhonneux R, Reneman RS, Janssen PAJ (1988) Cardiovascular effects of dl-nebivolol and its enantiomers, a comparison with atenolol. Eur J Pharmacol 156:95–103

Vanhoutte F, Vereecke J, Carmeliet E, Verbeke N (1991) Effects of the enantiomers of flecainide on action potential characteristics in the guinea-pig papillary muscle. Arch Int Pharmacodyn Ther 310:102–115

Van Peer A, Snoeck E, Woestenborghs R, van de Velde V, Mannens G, Meuldermans W, Heykants J (1991) Clinical pharmacokinetics of nebivolol: A review. Drug Invest 3 [Suppl 1]:25–30

Vlasses PH, Irvin JD, Huber PB, Lee RB, Ferguson RK, Schrogie JJ, Zacchei AG, Davies RO, Abrams WB (1981) Pharmacology of enantiomers and (–) *p*-OH metabolite of indacrinone. Clin Pharmacol Ther 29:798–807

Waldo AL, Camm AJ, deRuyter H, Friedman PL, MacNeil DJ, Pauls KF, Pitt B, Pratt CM, Schwartz PJ, Veltri EP (1996) for the SWORD Investigators. Effect of *d*-sotalol on mortality in patients with left ventricular dysfunction after recent and remote myocardial infarction. Lancet 348:7–12

Wesche DL, Schuster BG, Wang W-X, Woosley RL (2000) Mechanism of cardiotoxicity of halofantrine. Clin Pharmacol Ther 67:521–529
Williams KM (1990) Enantiomers in arthritic disorders. Pharmacol Ther 47:273–295
Yang IC-H, Scherz MW, Bahinski A, Bennet PB, Murray KT (2000) Stereoselective interactions of the enantiomers of chromanol 293B with human voltage-gated potassium channels. J Pharmacol Exp Ther 294:955–962
Zhou HH, Wood AJJ (1995) Stereoselective disposition of carvedilol is determined by CYP2D6. Clin Pharmacol Ther 57:518–524

Subject Index

Printing (Computer to Plate): Saladruck Berlin
Binding: Stürtz AG, Würzburg